Introduction to Mathematical Logic

THIRD EDITION

THE WADSWORTH & BROOKS/COLE MATHEMATICS SERIES

M. Adams, and V. Guillemin, *Measure Theory and Probability*
W. Beckner, A. Calderón, R. Fefferman, and P. Jones, *Conference on Harmonic Analysis in Honor of Antoni Zygmund*
G. Chartrand, and L. Lesniak, *Graphs & Digraphs, Second Edition*
J. Cochran, *Applied Mathematics: Principles, Techniques, and Applications*
W. Derrick, *Complex Analysis and Applications, Second Edition*
J. Dieudonné, *History of Algebraic Geometry*
R. Durrett, *Brownian Motion and Martingales in Analysis*
S. Fisher, *Complex Variables*
A. Garsia, *Topics in Almost Everywhere Convergence*
R. McKenzie, G. McNulty, and W. Taylor, *Algebras, Lattices, Varieties, Volume I*
E. Mendelson, *Introduction to Mathematical Logic, Third Edition*
R. Salem, *Algebraic Numbers and Fourier Analysis,* and L. Carleson, *Selected Problems on Exceptional Sets*
R. Stanley, *Enumerative Combinatorics, Volume I*
K. Stromberg, *An Introduction to Classical Real Analysis*

Introduction to Mathematical Logic

THIRD EDITION

Elliott Mendelson
QUEENS COLLEGE OF THE CITY UNIVERSITY OF NEW YORK

WADSWORTH & BROOKS/COLE ADVANCED BOOKS & SOFTWARE
MONTEREY, CALIFORNIA

To Arlene

Wadsworth & Brooks/Cole Advanced Books & Software
A Division of Wadsworth, Inc.

Printed in the United States of America

10 9 8 7 6 5 4 3 2

Library of Congress Cataloging-in-Publication Data

Mendelson, Elliott.
Introduction to mathematical logic.

(The Wadsworth & Brooks/Cole mathematics series)
Bibliography: p.
Includes index.
1. Logic, Symbolic and mathematical. I. Title.
II. Series.
QA9.M4 1987 511'.3 86-11084
ISBN 0-534-06624-0

Sponsoring Editor: *John Kimmel*
Editorial Assistant: *Maria Rosillo Alsadi*
Production Editor: *Phyllis Larimore*
Manuscript Editor: *Carol Reitz*
Interior and Cover Design: *Vernon T. Boes*
Art Coordinator: *Lisa Torri*
Interior Illustration: *Lori Heckelman*
Typesetting: *ASCO Trade Typesetting Ltd., Hong Kong*
Printing and Binding: *Maple-Vail Book Mfg. Group, York, Pennsylvania*

Preface

This is a compact introduction to some of the principal topics of mathematical logic. In the belief that beginners should be exposed to the most natural and easiest proofs, I have used free-swinging set-theoretic methods. The significance of a demand for constructive proofs can be evaluated only after a certain amount of experience with mathematical logic has been obtained. If we are to be expelled from "Cantor's paradise" (as nonconstructive set theory was called by Hilbert), at least we should know what we are missing.

The major changes in this new edition are the following. (1) In Chapter 5, Effective Computability, Turing-computability is now the central notion, and diagrams (flow-charts) are used to construct Turing machines. There are also treatments of Markov algorithms, Herbrand-Gödel-computability, register machines, and random access machines. Recursion theory is gone into a little more deeply, including the *s-m-n* theorem, the recursion theorem, and Rice's Theorem. (2) The proofs of the Incompleteness Theorems are now based upon the Diagonalization Lemma. Löb's Theorem and its connection with Gödel's Second Theorem are also studied. (3) In Chapter 2, Quantification Theory, Henkin's proof of the completeness theorem has been postponed until the reader has gained more experience in proof techniques. The exposition of the proof itself has been improved by breaking it down into smaller pieces and using the notion of a scapegoat theory. There is also an entirely new section on semantic trees. (4) Chapter 4, Axiomatic Set Theory, has been enlarged to include a section on theories other than NBG (ZF, Morse-Kelley, type theory, Quine's NF and ML). (5) The notation for connectives and quantifiers has been changed to conform to current usage. In addition, the exposition has been improved in many places, there are more exercises, and the Bibliography has been updated.

The five chapters of the book can be covered in two semesters, but, for a one-semester course Chapters 1–3 are quite adequate (omitting, if hurried, Sections 5 and 6 of Chapter 1 and Sections 10–15 of Chapter 2). I adopted the convention of prefixing a D to any section or exercise that will probably be difficult for a beginner, and an A to any section or exercise that presupposes familiarity with a topic that has not been carefully explained in the text. Bibliographical references are given to the best source of information, which is not always the earliest paper; hence, these references give no indication as to priority.

I believe this book can be read with ease by anyone with some experience in abstract mathematical thinking. There is, however, no specific prerequisite.

This book owes an obvious debt to the standard works of Hilbert and Bernays (1934, 1939), Kleene (1952), Rosser (1953), and Church (1956). I am grateful to many people for their help, and would especially like to thank John Corcoran for a great deal of useful advice and criticism, and Frank Cannonito and Robert Cowen for valuable suggestions. I would also like to express my thanks to John Kimmel and Phyllis Larimore at Wadsworth & Brooks/Cole for their assistance in the writing and editing of this new edition. Also, I appreciate the advice of the following reviewers: Richard Duke, Georgia Institute of Technology; Robert Frankel, Massachusetts Institute of Technology; and Martin G. Kalin, DePaul University.

Elliott Mendelson

Contents

Introduction

One of the most popular definitions of logic is that it is the analysis of methods of reasoning. In studying these methods, logic is interested in the form rather than the content of the argument. For example, consider the two arguments:

1. All men are mortal. Socrates is a man. Hence Socrates is mortal.
2. All rabbits like carrots. Sebastian is a rabbit. Hence Sebastian likes carrots.

Both have the same form: All A are B. S is an A. Hence S is a B. The truth or falsity of the particular premisses and conclusions is of no concern to logicians. They want to know only whether the premisses imply the conclusion. The systematic formalization and cataloguing of valid methods of reasoning are a main task of logicians. If the work uses mathematical techniques and if it is primarily devoted to the study of mathematical reasoning, then it may be called *mathematical logic*. We can narrow the domain of mathematical logic if we define its principal aim to be a precise and adequate understanding of the notion of *mathematical proof*.

Impeccable definitions have little value at the beginning of the study of a subject. The best way to find out what mathematical logic is about is to start doing it, and students are advised to begin reading the book even though (or especially if) they have qualms about the meaning and purpose of the subject.

Although logic is basic to all other studies, its fundamental and apparently self-evident character discouraged any deep logical investigations until the late 19th century. Then, under the impetus of the discovery of non-Euclidean geometry and the desire to provide a rigorous foundation for calculus and higher analysis, interest in logic revived. This new interest, however, was still rather unenthusiastic until, around the turn of the century, the mathematical world was shocked by the discovery of the paradoxes—that is, arguments that lead to contradictions. The most important paradoxes are described here.

LOGICAL PARADOXES

1. (Russell, 1902) By a set, we mean any collection of objects—for example, the set of all even integers or the set of all saxophone players in Brooklyn. The objects that make up a set are called its members. Sets may themselves be members of sets; for example, the set of all sets of integers has sets as its members. Most sets

are not members of themselves; the set of cats, for example, is not a member of itself because the set of cats is not a cat. However, there may be sets that do belong to themselves—for example, the set of all sets. Now, consider the set A of all those sets X such that X is not a member of X. Clearly, by definition, A is a member of A if and only if A is not a member of A. So, if A is a member of A, then A is also not a member of A; and if A is not a member of A, then A is a member of A. In any case, A is a member of A and A is not a member of A.

2. (Cantor, 1899) This paradox involves the theory of cardinal numbers and may be skipped by those having no previous acquaintance with that theory. The cardinal number $\overline{\overline{Y}}$ of a set Y is defined to be the set of all sets X that are equinumerous with Y (i.e., for which there is a one–one correspondence between Y and X, see page 7). We define $\overline{\overline{Y}} \leqslant \overline{\overline{Z}}$ to mean that Y is equinumerous with a subset of Z; by $\overline{\overline{Y}} < \overline{\overline{Z}}$ we mean $\overline{\overline{Y}} \leqslant \overline{\overline{Z}}$ and $\overline{\overline{Y}} \neq \overline{\overline{Z}}$. Cantor proved that, if $\mathscr{P}(Y)$ is the set of all subsets of Y, then $\overline{\overline{Y}} < \overline{\overline{\mathscr{P}(Y)}}$ (see page 199). Let C be the universal set—that is, the set of all sets. Now, $\mathscr{P}(C)$ is a subset of C, so it follows easily that $\overline{\overline{\mathscr{P}(C)}} \leqslant \overline{\overline{C}}$. On the other hand, by Cantor's Theorem, $\overline{\overline{C}} < \overline{\overline{\mathscr{P}(C)}}$. The Schröder-Bernstein Theorem (see page 198) asserts that if $\overline{\overline{Y}} \leqslant \overline{\overline{Z}}$ and $\overline{\overline{Z}} \leqslant \overline{\overline{Y}}$, then $\overline{\overline{Y}} = \overline{\overline{Z}}$. Hence, $\overline{\overline{C}} = \overline{\overline{\mathscr{P}(C)}}$, contradicting $\overline{\overline{C}} < \overline{\overline{\mathscr{P}(C)}}$.

3. (Burali-Forti, 1897) This paradox is the analogue in the theory of ordinal numbers of Cantor's Paradox and will make sense only to those already familiar with ordinal number theory. Given any ordinal number, there is a still larger ordinal number. But the ordinal number determined by the set of all ordinal numbers is the largest ordinal number.

SEMANTIC PARADOXES

4. *The Liar Paradox*. A man says, "I am lying." If he is lying, then what he says is true, and so he is not lying. If he is not lying, then what he says is true, and so he is lying. In any case, he is lying and he is not lying.*

5. (Richard, 1905) Some phrases of the English language denote real numbers; for example, "the ratio between the circumference and diameter of a circle" denotes the number π. All phrases of the English language can be enumerated in a standard way: order all phrases that have k letters lexicographically (as in a dictionary), and then place all phrases with k letters before all phrases with a larger number of letters. Hence, all phrases of the English language that denote real numbers can be enumerated merely by omitting all other phrases in the given standard enumeration. Call the nth real number in this enumeration the nth Richard number. Consider the phrase: "the real number whose nth decimal place is 1 if the nth decimal place of the nth Richard number is not 1, and whose nth decimal place is 2 if the nth decimal

*The Cretan "paradox", known in antiquity, is similar to the Liar Paradox. The Cretan philosopher Epimenides said, "All Cretans are liars." If what he said is true, then, since Epimenides is a Cretan, it must be false. Hence, what he said is false. Thus, there must be some Cretan who is not a liar. This is not logically impossible, so we do not have a genuine paradox. However, the fact that the utterance by Epimenides of that false sentence could imply the existence of some Cretan who is not a liar is rather unsettling.

place of the nth Richard number is 1". This phrase defines a Richard number—say, the kth Richard number; but, by its definition, it differs from the kth Richard number in the kth decimal place.

6. (Berry, 1906) There are only a finite number of symbols (letters, punctuation signs, etc.) in the English language. Hence, there are only a finite number of English expressions that contain fewer than 200 occurrences of symbols (allowing repetitions). There are, therefore, only a finite number of positive integers that are denoted by an English expression containing fewer than 200 occurrences of symbols. Let k be *the least positive integer that is not denoted by an English expression containing fewer than 200 occurrences of symbols.* The italicized English phrase contains fewer than 200 occurrences of symbols and denotes the integer k.

7. (Grelling, 1908) An adjective is called *autological* if the property denoted by the adjective holds for the adjective itself. An adjective is called *heterological* if the property denoted by the adjective does not apply to the adjective itself. For example, "polysyllabic" and "English" are autological, whereas "monosyllabic", "French", and "blue" are heterological. Consider the adjective "heterological". If "heterological" is heterological, then it is not heterological. If "heterological" is not heterological, then it is heterological. In any case, "heterological" is both heterological and not heterological.

8. (Löb, 1955) Let A be any sentence. Let B be the sentence: "If this sentence is true, then A". So, B asserts: "If B is true, then A". Now consider the following argument: Assume B is true. Then, by B, since B is true, A is true. This argument shows that, if B is true, then A. But this is exactly what B asserts. Hence, B is true. Therefore, by B, since B is true, A is true. Thus, every sentence is true.

All of these paradoxes are genuine in the sense that they contain no obvious logical flaws. The logical paradoxes involve only notions from the theory of sets, whereas the semantic paradoxes also make use of concepts like "denote", "true", and "adjective", which need not occur within our standard mathematical language. For this reason, the logical paradoxes are a much greater threat to a mathematician's peace of mind than the semantic paradoxes.

Analysis of the paradoxes has led to various proposals for avoiding them. All of these proposals are restrictive in one way or another of the "naive" concepts that enter into the derivation of the paradoxes. Russell noted the self-reference present in all the paradoxes and suggested that every object must have a definite non-negative integer as its "type". Then an expression "x is a member of the set y" is *meaningful* if and only if the type of y is one greater than the type of x.

This approach, known as the theory of types and systematized and developed by Whitehead-Russell (1910–1913), is successful in eliminating the known paradoxes,* but it is clumsy in practice and has certain other drawbacks as well. A different criticism of the logical paradoxes is aimed at their assumption that, for every property $P(x)$, there exists a corresponding set of all objects x that satisfy $P(x)$.

*Russell's Paradox, for example, depends on the existence of the set A of all sets that are not members of themselves. Because, according to the theory of types, it is meaningless to say that a set belongs to itself, there can be no such set A.

If we reject this assumption, then the logical paradoxes are no longer derivable.* It is necessary, however, to provide new postulates that will enable us to prove the existence of those sets that are a daily necessity to the practicing mathematician. The first such axiomatic set theory was invented by Zermelo (1908). In Chapter 4 we shall present an axiomatic theory of sets that is a descendant of Zermelo's system (with some new twists given to it by von Neumann, R. Robinson, Bernays, and Gödel). There are also various hybrid theories combining some aspects of type theory and axiomatic set theory—for example, Quine's system NF (see Rosser, 1953).

A more radical interpretation of the paradoxes has been advocated by Brouwer and his intuitionist school (see Heyting, 1956). They refuse to accept the universality of certain basic logical laws, such as the law of excluded middle: P or not-P. Such a law, they claim, is true for finite sets, but it is invalid to extend it on a wholesale basis to all sets. Likewise, they say it is invalid to conclude that "there exists an object x such that not-$P(x)$" follows from "not-(for all $x, P(x)$)"; we are justified in asserting the existence of an object having a certain property only if we know an effective method for constructing (or finding) such an object. The paradoxes are, of course, not derivable (or even meaningful) if we obey the intuitionist strictures, but, alas, so are many beloved theorems of everyday mathematics, and, for this reason, intuitionism has found few converts among mathematicians.

Whatever approach one takes to the paradoxes, it is necessary first to examine the language of logic and mathematics to see what symbols may be used, to determine the ways in which these symbols are put together to form terms, formulas, sentences, and proofs, and to find out what can and cannot be proved if certain axioms and rules of inference are assumed. This is one of the tasks of mathematical logic, and, until it is done, there is no basis for comparing rival foundations of logic and mathematics. The deep and devastating results of Gödel, Tarski, Church, Rosser, Kleene, and many others have been ample reward for the labor invested and have earned for mathematical logic its status as an independent branch of mathematics.

For the absolute novice a summary will be given here of some of the basic ideas and results used in the text. The reader is urged to skip these explanations now and, if necessary, to refer to them later on.

A *set* is a collection of objects.† The objects in the collection are called *elements* or *members* of the set, and we shall write "$x \in y$" for the statement that x is a member of y. (Synonymous expressions are "x belongs to y" and "y contains x".) The negation of "$x \in y$" will be written "$x \notin y$".

* Russell's Paradox then proves that there is no set A of all sets that do not belong to themselves; the paradoxes of Cantor and Burali-Forti show that there is no universal set and no set that contains all ordinal numbers. The semantic paradoxes cannot even be formulated, since they involve notions not expressible within the system.

† Which collections of objects form sets will not be specified here. Care will be exercised to avoid using any ideas or procedures that may lead to the paradoxes; all the results can be formalized in the axiomatic set theory of Chapter 4. The term "class" is sometimes used as a synonym for "set", but it will be avoided here because it has a different meaning in Chapter 4. If the property $P(x)$ does determine a set, this set is often denoted $\{x|P(x)\}$.

By "$x \subseteq y$" we mean that every member of x is also a member of y, or, in other words, that x is a *subset* of y (or, synonymously, that x is *included* in y). We shall write "$t = s$" to mean that "t" and "s" denote the same object. As usual, "$t \neq s$" is the negation of "$t = s$". For sets x and y, we assume that $x = y$ if and only if $x \subseteq y$ and $y \subseteq x$—that is, if and only if x and y have the same members. A set x is called a *proper* subset of a set y, written "$x \subset y$", if $x \subseteq y$ but $x \neq y$.*

The *union* $x \cup y$ of sets x and y is defined to be the set of all elements that are members of x or y or both. Hence, $x \cup x = x$, $x \cup y = y \cup x$, and $(x \cup y) \cup z = x \cup (y \cup z)$. The *intersection* $x \cap y$ is the set of elements that x and y have in common. It is easy to verify that $x \cap x = x$, $x \cap y = y \cap x$, $x \cap (y \cap z) = (x \cap y) \cap z$, $x \cap (y \cup z) = (x \cap y) \cup (x \cap z)$, and $x \cup (y \cap z) = (x \cup y) \cap (x \cup z)$. The *relative complement* $x - y$ is the set of members of x that are not members of y. We also postulate the existence of the *empty set* (or *null set*) 0—that is, a set that has no members at all. Then, $x \cap 0 = 0$, $x \cup 0 = x$, $x - 0 = x$, $0 - x = 0$, and $x - x = 0$. Two sets x and y are called *disjoint* if $x \cap y = 0$.

Given any objects $b_1, \ldots, b_k$, the set that contains $b_1, \ldots, b_k$ as its only members is denoted $\{b_1, \ldots, b_k\}$. In particular, $\{x, y\}$ is a set having x and y as its only members and, if $x \neq y$, is called the *unordered pair* of x and y. The set $\{x, x\}$ is identical with $\{x\}$ and is called the *unit set* of x. Notice that $\{x, y\} = \{y, x\}$. On the other hand, by $\langle b_1, \ldots, b_k \rangle$ we mean the *ordered k-tuple* of $b_1, \ldots, b_k$. The basic property of ordered k-tuples is that $\langle b_1, \ldots, b_k \rangle = \langle c_1, \ldots, c_k \rangle$ if and only if $b_1 = c_1, b_2 = c_2, \ldots, b_k = c_k$. Thus, $\langle b_1, b_2 \rangle = \langle b_2, b_1 \rangle$ if and only if $b_1 = b_2$. Ordered 2-tuples are called *ordered pairs*. If X is a set and k is a positive integer greater than 1, we denote by X^k the set of all ordered k-tuples $\langle b_1, \ldots, b_k \rangle$ of elements $b_1, \ldots, b_k$ of X. We also make the convention that X^1 stands for X. X^k is called the *Cartesian product* of X with itself k times. If Y and Z are sets, then by $Y \times Z$ we denote the set of all ordered pairs $\langle y, z \rangle$ such that $y \in Y$ and $z \in Z$. $Y \times Z$ is called the Cartesian product of Y and Z.

An *n-place relation* (or a *relation with n arguments*) on a set X is a subset of X^n—that is, a set of ordered n-tuples of elements of X. For example, the 3-place relation of betweenness for points on a line is the set of all 3-tuples $\langle x, y, z \rangle$ such that the point x lies between the points y and z. A 2-place relation is called a *binary* relation; for example, the binary relation of fatherhood on the set of human beings is the set of all ordered pairs $\langle x, y \rangle$ such that x and y are human beings and x is the father of y. A 1-place relation on X is a subset of X and is called a *property* on X.

Given a binary relation R on a set X, the *domain* of R is defined to be the set of all y such that $\langle y, z \rangle \in R$ for some z; the *range* of R is the set of all z such that $\langle y, z \rangle \in R$ for some y; and the *field* of R is the union of the domain and range of R. The *inverse* relation R^{-1} of R is the set of all ordered pairs $\langle y, z \rangle$ such that $\langle z, y \rangle \in R$. For example, the domain of the relation $<$ on the set ω of nonnegative integers is ω,† its range is $\omega - \{0\}$, and the inverse of $<$ is $>$. Notation: Very often xRy is written instead of $\langle x, y \rangle \in R$. Thus, in the example just given, we usually write $x < y$ instead of $\langle x, y \rangle \in <$.

*The notation $x \subsetneq y$ is often used instead of $x \subset y$.

†ω will also be referred to as the set of *natural numbers*.

A binary relation R is said to be *reflexive* if xRx for all x in the field of R. R is *symmetric* if xRy implies yRx, and R is *transitive* if xRy and yRz imply xRz. Examples: The relation $\leqslant$ on the set of integers is reflexive and transitive but not symmetric. The relation "having at least one parent in common" on the set of human beings is reflexive and symmetric but not transitive.

A binary relation that is reflexive, symmetric, and transitive is called an *equivalence relation*. Examples of equivalence relations are (1) the *identity relation* I_X on a set X, consisting of all pairs $\langle y, y\rangle$, where $y \in X$; (2) the relation of parallelism between lines in a plane; (3) given a fixed positive integer n, the relation $x \equiv y \pmod n$ holds when x and y are integers and $x - y$ is divisible by n; (4) the relation between directed line segments in three-dimensional space that holds when and only when they have the same length and the same direction; (5) the congruence relation on the set of triangles in a plane; and (6) the similarity relation on the set of triangles in a plane. Given an equivalence relation R on a set X, and given any $y \in X$, define $[y]$ as the set of all z in X such that yRz. Then $[y]$ is called the *R-equivalence class* of y. It is easy to check that $[y] = [z]$ if and only if yRz and that, if $[y] \neq [z]$, then $[y] \cap [z] = 0$; that is, different R-equivalence classes have no elements in common. Hence, the set X is completely partitioned into the R-equivalence classes. For some of the preceding examples, (1) the equivalence classes are just the unit sets $\{y\}$, where $y \in X$; (2) the equivalence classes can be considered to be the directions in the plane; (3) there are n equivalence classes, the kth equivalence class ($k = 0, 1, \ldots, n-1$) being the set of all numbers that leave the remainder k upon division by n; and (4) the equivalence classes are the three-dimensional vectors.

A *function* f is a binary relation such that $\langle x, y\rangle \in f$ and $\langle x, z\rangle \in f$ imply $y = z$. Thus, for any element x of the domain of a function f, there is a unique y such that $\langle x, y\rangle \in f$; this unique element y is denoted $f(x)$. If x is in the domain of f, then $f(x)$ is said to be defined. A function f with domain X and range Y is said to be a *function from X onto Y*. If f is a function from X onto Y and $Y \subseteq Z$, then f is called a *function from X into Z*. For example, if $f(x) = 2x$ for every integer x, f is a function from the set of integers onto the set of even integers, and f is a function from the set of integers into the set of integers. A function the domain of which consists of n-tuples is said to be a *function of n arguments*. A *total function of n arguments on a set X* is a function f whose domain is X^n. We usually write $f(x_1, \ldots, x_n)$ instead of $f(\langle x_1, \ldots, x_n\rangle)$, and we refer to $f(x_1, \ldots, x_n)$ as the *value* of f for the *arguments* $x_1, \ldots, x_n$. A *partial* function of n arguments on a set X is a function whose domain is a subset of X^n; for example, ordinary division is a partial, but not total, function of two arguments on the set of integers (since division by zero is not defined). If f is a function with domain X and range Y, then the *restriction* f_Z of f to a set Z is the function $f \cap (Z \times Y)$. Clearly, $f_Z(u) = v$ if and only if $u \in Z$ and $f(u) = v$. The *image* of the set Z under the function f is the range of f_Z. The *inverse image* of a set W under the function f is the set of all elements u of the domain of f such that $f(u) \in W$. We say that f *maps* X onto (into) Y if X is a subset of the domain of f and the image of X under f is (a subset of) Y. By an *n-place operation* (or *operation with n arguments*) *on a set* X we mean a function from X^n into X. For example, ordinary addition is a binary (i.e., 2-place) operation on the set of natural numbers $\{0, 1, 2, \ldots\}$.

But ordinary subtraction is not a binary operation on the set of natural numbers, though it is a binary operation on the set of integers.

Given two functions f and g, the *composition* $f \circ g$ (also sometimes denoted fg) is the function such that $(f \circ g)(x) = f(g(x))$; $(f \circ g)(x)$ is defined if and only if $g(x)$ is defined and $f(g(x))$ is defined. For example, if $g(x) = x^2$ and $f(x) = x + 1$ for every integer x, then $(f \circ g)(x) = x^2 + 1$ and $(g \circ f)(x) = (x + 1)^2$. Also, if $h(x) = -x$ for every real number x and $f(x) = \sqrt{x}$ for every nonnegative real number x, then $(f \circ h)(x)$ is defined only for $x \leqslant 0$, and, for such x, $(f \circ h)(x) = \sqrt{-x}$. A function f such that $f(x) = f(y)$ implies $x = y$ is called a 1–1 (*one–one*) function. For example, (1) the identity relation I_X on a set X is a 1–1 function, since $I_X(y) = y$ for any $y \in X$; (2) the function $g(x) = 2x$, for every integer x, is a 1–1 function; and (3) the function $h(x) = x^2$, for every integer x, is not 1–1, since $h(-1) = h(1)$. Notice that a function f is 1–1 if and only if its inverse relation f^{-1} is a function. If the domain and range of a 1–1 function f are X and Y, respectively, then f is said to be a *1–1 (one–one) correspondence between X and Y*; then f^{-1} is a 1–1 correspondence between Y and X, and $(f^{-1} \circ f) = I_X$ and $(f \circ f^{-1}) = I_Y$. If f is a 1–1 correspondence between X and Y and g is a 1–1 correspondence between Y and Z, then $g \circ f$ is a 1–1 correspondence between X and Z. Sets X and Y are said to be *equinumerous* (written $X \cong Y$) if and only if there is a 1–1 correspondence between X and Y. Clearly, $X \cong X$, $X \cong Y$ implies $Y \cong X$, and $X \cong Y$ and $Y \cong Z$ imply $X \cong Z$. One can prove (see Schröder-Bernstein Theorem, page 198) that if $X \cong Y_1 \subseteq Y$ and $Y \cong X_1 \subseteq X$, then $X \cong Y$. If $X \cong Y$, one sometimes says that X and Y *have the same cardinal number*, and if X is equinumerous with a subset of Y but Y is not equinumerous with a subset of X, one says that the cardinal number of X *is smaller than* the cardinal number of Y.*

A set X is *denumerable* if it is equinumerous with the set of positive integers. A denumerable set is said to have cardinal number $\aleph_0$, and any set equinumerous with the set of all subsets of a denumerable set is said to have the cardinal number $2^{\aleph_0}$ (or to have the *power of the continuum*). A set X is *finite* if it is empty or if it is equinumerous with the set of all positive integers $\{1, 2, \ldots, n\}$ that are less than or equal to some positive integer n. A set that is not finite is said to be *infinite*. A set is *countable* if it is either finite or denumerable. Clearly, any subset of a denumerable set is countable. A *denumerable sequence* is a function s whose domain is the set of positive integers; one usually writes s_n instead of $s(n)$. A *finite sequence* is a function whose domain is the empty set or $\{1, 2, \ldots, n\}$, for some positive integer n.

Let $P(x, y_1, \ldots, y_k)$ be some relation on the set of nonnegative integers. In particular, P may involve only the variable x and thus be a property. If $P(0, y_1, \ldots, y_k)$ holds, and, if, for any n, $P(n, y_1, \ldots, y_k)$ implies $P(n + 1, y_1, \ldots, y_k)$, then $P(x, y_1, \ldots, y_k)$ is true for all nonnegative integers x (*Principle of Mathematical Induction*). In applying this principle, one usually proves that, for any n, $P(n, y_1, \ldots, y_k)$ implies $P(n + 1, y_1, \ldots, y_k)$ by assuming $P(n, y_1, \ldots, y_k)$ and then

*One can attempt to define the cardinal number of a set X as the collection $[X]$ of all sets equinumerous with X. However, in certain systems of set theory, $[X]$ does not exist, whereas in others (see page 200), $[X]$ exists but is not a set. For cardinal numbers $[X]$ and $[Y]$, one can define $[X] \leqslant [Y]$ to mean that X is equinumerous with a subset of Y.

deducing $P(n+1, y_1, \ldots, y_k)$; in the course of this deduction, $P(n, y_1, \ldots, y_k)$ is called the *inductive hypothesis*. If the relation P actually involves variables $y_1, \ldots, y_k$ other than x, then the proof of "for all x, $P(x)$" is said to proceed by *induction on* x. A similar induction principle holds for the set of integers greater than some fixed integer j. An example is: to prove by mathematical induction that the sum of the first n odd integers $1 + 3 + 5 + \cdots + (2n - 1)$ is n^2, first show that $1 = 1^2$ [i.e., $P(1)$], and then, that if $1 + 3 + 5 + \cdots + (2n - 1) = n^2$, then $1 + 3 + 5 + \cdots + (2n - 1) + (2n + 1) = (n + 1)^2$ [i.e., if $P(n)$, then $P(n + 1)$]. From the Principle of Mathematical Induction one can prove the *Principle of Complete Induction*: if, for every nonnegative integer x the assumption that $P(u, y_1, \ldots, y_k)$ is true for all $u < x$ implies that $P(x, y_1, \ldots, y_k)$ holds, then, for all nonnegative integers x, $P(x, y_1, \ldots, y_k)$ is true. (Exercise: Show, by complete induction, that every integer greater than 1 is divisible by a prime number.)

A *partial order* is a binary relation R such that R is transitive and, for every x in the field of R, xRx is false. If R is a partial order, then the relation R' that is the union of R and the set of all ordered pairs $\langle x, x \rangle$, where x is in the field of R, we shall call a *reflexive partial order*; in the literature, "partial order" is used for either partial order or reflexive partial order. Notice that (xRy and yRx) is impossible if R is a partial order, whereas (xRy and yRx) implies $x = y$ if R is a reflexive partial order. A (reflexive) *total order* is a (reflexive) partial order R such that, for any x and y in the field of R, either $x = y$ or xRy or yRx. Examples are (1) the relation $<$ on the set of integers is a total order, whereas $\leqslant$ is a reflexive total order and (2) the relation $\subset$ on the set of all subsets of the set of positive integers is a partial order but not a total order, whereas the relation $\subseteq$ is a reflexive partial order but not a reflexive total order. If C is the field of a relation R, and if B is a subset of C, then an element y of B is called an *R-least element* of B if yRz for every element z of B different from y. A *well-order* (or *well-ordering relation*) is a total order R such that every nonempty subset of the field of R has an R-least element. Examples are (1) the relation $<$ on the set of nonnegative integers is a well-order; (2) the relation $<$ on the set of nonnegative rational numbers is a total order but not a well-order; and (3) the relation $<$ on the set of integers is a total order but not a well-order. Associated with every well-order R having field X there is a corresponding *Complete Induction Principle*: if P is a property such that, for any u in X, whenever all z in X such that zRu have the property P, then u has the property P, then it follows that all members of X have the property P. If the set X is infinite, a proof using this principle is called a proof by *transfinite induction*. One says that a *set X can be well-ordered* if there exists a well-order whose field includes X. An assumption that is useful in modern mathematics but about the validity of which there has been considerable controversy is the *Well-Ordering Principle*: every set can be well-ordered. The Well-Ordering Principle is equivalent (given the usual axioms of set theory) to the *Axiom of Choice* (*Multiplicative Axiom*): given any set X of nonempty pairwise disjoint sets, there is a set Y (called a *choice set*) that contains exactly one element in common with each set in X.

Let B be a nonempty set, f a function from B into B, and g a function from B^2 into B. Let us write x' for $f(x)$, and $x \cap y$ for $g(x, y)$. Then $\langle B, f, g \rangle$ is called a *Boolean algebra* if and only if B contains at least two elements and the following condition are satisfied:

1. $x \cap y = y \cap x$ for all x, y in B
2. $(x \cap y) \cap z = x \cap (y \cap z)$ for all x, y, z in B
3. $x \cap y' = z \cap z'$ if and only if $x \cap y = x$ for any x, y, z in B

We let $x \cup y$ stand for $(x' \cap y')'$, and we write $x \leqslant y$ for $x \cap y = x$. It is easily proved that $z \cap z' = w \cap w'$ for any w, z in B; we denote the value of $z \cap z'$ by 0. (The symbols $\cap$, $\cup$, 0 should not be confused with the corresponding symbols used in set theory.) We let 1 stand for $0'$. Then $z \cup z' = 1$ for all z in B; $\leqslant$ is a reflexive partial order on B; and $\langle B, f, \cup \rangle$ is a Boolean algebra. An *ideal* in $\langle B, f, g \rangle$ is a nonempty subset J of B such that (1) if $x \in J$ and $y \in J$, then $x \cup y \in J$, and (2) if $x \in J$ and $y \in B$, then $x \cap y \in J$. Clearly, $\{0\}$ and B are ideals. An ideal different from B is called a *proper ideal.* A *maximal ideal* is a proper ideal that is included in no other proper ideal. It can be shown that a proper ideal J is maximal if and only if, for any u in B, $u \in J$ or $u' \in J$. From the Well-Ordering Principle (or the Axiom of Choice) it can be proved that every Boolean algebra contains a maximal ideal or, equivalently, that every proper ideal is included in some maximal ideal. For example, let B be the set of all subsets of a set X; for $Y \in B$, let $Y' = X - Y$, and for Y, Z in B, let $Y \cap Z$ be the ordinary set-theoretic intersection of Y and Z. Then $\langle B, ', \cap \rangle$ is a Boolean algebra. The 0 of B is the empty set 0, and 1 is X. Given an element u in X, let J_u be the set of all subsets of X that do not contain u. Then J_u is a maximal ideal. For a detailed study of Boolean algebras, see Sikorski (1960), Halmos (1963), and Mendelson (1970).

CHAPTER ONE

The Propositional Calculus

1. PROPOSITIONAL CONNECTIVES. TRUTH TABLES

Sentences may be combined in various ways to form more complicated sentences. Let us consider only *truth-functional* combinations, in which the truth or falsity of the new sentence is determined by the truth or falsity of its component sentences.

Negation is one of the simplest operations on sentences. Although a sentence in a natural language may be negated in many ways, we shall adopt a uniform procedure: placing a sign for negation, the symbol $\neg$, in front of the entire sentence. Thus, if A is a sentence, then $\neg A$ denotes the negation of A.

The truth-functional character of negation is made apparent in the following *truth table*:

A	$\neg A$
T	F
F	T

When A is true, $\neg A$ is false; when A is false, $\neg A$ is true. We use T and F to denote the *truth values* truth and falsity.

Another common truth-functional operation is *conjunction*: "and". The conjunction of sentences A and B will be designated by $A \wedge B$ and has the following truth table:

A	B	$A \wedge B$
T	T	T
F	T	F
T	F	F
F	F	F

$A \wedge B$ is true when and only when both A and B are true. A and B are called the *conjuncts* of $A \wedge B$. Note that there are four rows in the table, corresponding to the number of possible assignments of truth values to A and B.

In natural languages, there are two distinct uses of "or": the inclusive and the exclusive. According to the inclusive usage, "A or B" means "A or B or both", whereas according to the exclusive usage, the meaning is "A or B, but not both".

We shall introduce a special sign, $\vee$, for the inclusive connective. Its truth table is as follows:

A	B	$A \vee B$
T	T	T
F	T	T
T	F	T
F	F	F

Thus $A \vee B$ is false when and only when both A and B are false. "$A \vee B$" is called a *disjunction*, with the *disjuncts* A and B.

Another important truth-functional operation is the *conditional*: "if A, then B." Ordinary usage is unclear here. Surely, "if A, then B" is false when the *antecedent* A is true and the *consequent* B is false. However, in other cases, there is no well-defined truth value. For example, the following sentences would be considered neither true nor false:

1. If $1 + 1 = 2$, then Paris is the capital of France.
2. If $1 + 1 \neq 2$, then Paris is the capital of France.
3. If $1 + 1 \neq 2$, then Rome is the capital of France.

Their meaning is unclear, since we are accustomed to the assertion of some sort of relationship (usually causal) between the antecedent and the consequent. We shall make the convention that "if A, then B" is false when and only when A is true and B false. Thus, sentences 1–3 are assumed to be true. Let us denote "if A, then B" by "$A \Rightarrow B$". An expression "$A \Rightarrow B$" is called a *conditional*. Then $\Rightarrow$ has the following truth table:

A	B	$A \Rightarrow B$
T	T	T
F	T	T
T	F	F
F	F	T

This sharpening of the meaning of "if A, then B" involves no conflict with ordinary usage but rather only an extension of that usage.*

A justification of the truth table for $\Rightarrow$ is the fact that we wish "if A and B, then B" to be true in all cases. Thus, the case in which A and B are true justifies the first line of our truth table for $\Rightarrow$, since (A and B) and B are both true. If A is false and

*There seems to be a common non-truth-functional interpretation of "if A, then B" connected with causal laws. The sentence "if this piece of iron is placed in water at time t, then the iron will dissolve" is regarded as false even in the case that the piece of iron is not placed in water at time t—that is, even when the antecedent is false. Another non-truth-functional usage of "if..., then ____" occurs in so-called counterfactual conditionals, such as "if Sir Walter Scott had not written any novels, then there would have been no War Between the States." (This was Mark Twain's contention in *Life on the Mississippi*: "Sir Walter had so large a hand in making Southern character, as it existed before the war, that he is in great measure responsible for the war".) This sentence might be asserted to be false even though the antecedent is admittedly false. Fortunately, causal laws and counterfactual conditionals are not needed in mathematics and logic. For a clear treatment of conditionals and other connectives, see Quine (1951). (The quotation from *Life on the Mississippi* was brought to my attention by Professor J. C. Owings, Jr.)

B true, then (A and B) is false while B is true. This corresponds to the second line of the truth table. Finally, if A is false and B is false, (A and B) is false and B is false. This gives the fourth line of the truth table. Still more support for our definition comes from the meaning of statements such as "for every x, if x is an odd positive integer, then x^2 is an odd positive integer." This asserts that, for every x, the statement "if x is an odd positive integer, then x^2 is an odd positive integer" is true. Now we certainly do not want to consider cases in which x is not an odd positive integer as counterexamples to our general assertion. This provides us with the second and fourth lines of our truth table. In addition, any case in which x is an odd positive integer and x^2 is an odd positive integer confirms our general assertion. This corresponds to the first line of the truth table.

Let us denote "A if and only if B" by "$A \Leftrightarrow B$". Such an expression is called a *biconditional*. Clearly, $A \Leftrightarrow B$ is true when and only when A and B have the same truth value. Its truth table, therefore, is

A	B	$A \Leftrightarrow B$
T	T	T
F	T	F
T	F	F
F	F	T

The symbols $\neg$, $\wedge$, $\vee$, $\Rightarrow$, and $\Leftrightarrow$ will be called *propositional connectives*.* Any sentence built up by application of these connectives has a truth value that depends on the truth values of the constituent sentences. In order to make this dependence apparent, let us apply the name *statement form* to an expression built up from the *statement letters*, A, B, C, and so on by appropriate applications of the propositional connectives.

1. All statement letters (capital italic letters) and such letters with numerical subscripts† are statement forms.
2. If $\mathscr{A}$ and $\mathscr{B}$ are statement forms, then so are $(\neg\mathscr{A})$, $(\mathscr{A} \wedge \mathscr{B})$, $(\mathscr{A} \vee \mathscr{B})$, $(\mathscr{A} \Rightarrow \mathscr{B})$, and $(\mathscr{A} \Leftrightarrow \mathscr{B})$.
3. Only those expressions are statement forms that are determined to be so by means of conditions 1 and 2.‡

Some examples of statement forms are B, $(\neg C_2)$, $(D_3 \wedge (\neg B))$, $(((\neg B_1) \vee B_2) \Rightarrow (A_1 \wedge C_2))$, and $(((\neg A) \Leftrightarrow A) \Leftrightarrow (C \Rightarrow (B \vee C)))$.

For every assignment of truth values T or F to the statement letters that occur in a statement form, there corresponds, by virtue of the truth tables for the propositional connectives, a truth value for the statement form. Thus, each statement form

*We shall avoid the use of quotation marks to form names whenever this is not likely to cause confusion. Strictly speaking, the given sentence should have quotation marks around each of the connectives. See Quine (1951, pp. 23–27).

† For example, $A_1, A_2, A_{17}, B_{31}, C_2, \ldots$.

‡ This can be rephrased as follows: $\mathscr{C}$ is a statement form if and only if there is a finite sequence $\mathscr{A}_1, \ldots, \mathscr{A}_n$ $(n \geqslant 1)$ such that $\mathscr{A}_n = \mathscr{C}$ and, if $1 \leqslant i \leqslant n$, $\mathscr{A}_i$ is either a statement letter or a negation, conjunction, disjunction, conditional, or biconditional constructed from previous expressions in the sequence. Notice that we use script letters $\mathscr{A}, \mathscr{B}, \mathscr{C}, \ldots$ to stand for arbitrary expressions, whereas italic letters are used as statement letters.

determines a *truth function*, which can be graphically represented by a truth table for the statement form. For example, the statement form $(((\neg A) \vee B) \Rightarrow C)$ has the following truth table:

A	B	C	$(\neg A)$	$((\neg A) \vee B)$	$(((\neg A) \vee B) \Rightarrow C)$
T	T	T	F	T	T
F	T	T	T	T	T
T	F	T	F	F	T
F	F	T	T	T	T
T	T	F	F	T	F
F	T	F	T	T	F
T	F	F	F	F	T
F	F	F	T	T	F

Each row represents an assignment of truth values to the letters A, B, and C and the corresponding truth values assumed by the statement forms that appear in the construction of $(((\neg A) \vee B) \Rightarrow C)$.

The truth table for $((A \Leftrightarrow B) \Rightarrow ((\neg A) \wedge B))$ is as follows:

A	B	$(A \Leftrightarrow B)$	$(\neg A)$	$((\neg A) \wedge B)$	$((A \Leftrightarrow B) \Rightarrow ((\neg A) \wedge B))$
T	T	T	F	F	F
F	T	F	T	T	T
T	F	F	F	F	T
F	F	T	T	F	F

If there are n distinct letters in a statement form, then there are 2^n possible assignments of truth values to the statement letters and, hence, 2^n rows in the truth table.

A truth table can be abbreviated by writing only the full statement form, putting the truth values of the statement letters underneath all occurrences of these letters, and writing, step by step, the truth values of each component statement form under the principal connective of the form.* As an example, for $((A \Leftrightarrow B) \Rightarrow ((\neg A) \wedge B))$, we obtain:

$((A$	$\Leftrightarrow$	$B)$	$\Rightarrow$	$((\neg$	$A)$	$\wedge$	$B))$
T	T	T	F	F	T	F	T
F	F	T	T	T	F	T	T
T	F	F	T	F	T	F	F
F	T	F	F	T	F	F	F

Exercises

1.1 Write the truth table for the exclusive usage of "or".

1.2 Construct truth tables for the statement forms $((A \Rightarrow B) \vee (\neg A))$ and $((A \Rightarrow (B \Rightarrow C)) \Rightarrow ((A \Rightarrow B) \Rightarrow (A \Rightarrow C)))$.

1.3 Write abbreviated truth tables for $((A \Rightarrow B) \wedge A)$ and $((A \vee (\neg C)) \Leftrightarrow B)$.

1.4 Write the following sentences as statement forms, using statement letters to stand for the *atomic sentences*—that is, those sentences that are not built up out of other sentences.

* The *principal connective* of a statement form is the one that is applied last in constructing the form.

(a) If Mr. Jones is happy, Mrs. Jones is not happy, and if Mr. Jones is not happy, Mrs. Jones is not happy.
(b) Either Sam will come to the party and Max will not, or Sam will not come to the party and Max will enjoy himself.
(c) A necessary and sufficient condition for the sheik to be happy is that he has wine, women, and song.
(d) Fiorello goes to the movies only if a comedy is playing.
(e) A sufficient condition for x to be odd is that x is prime.
(f) A necessary condition for a sequence s to converge is that s be bounded.
(g) The bribe will be paid if and only if the goods are delivered.
(h) Karpov will win the chess tournament unless Kasparov wins today.
(i) If x is positive, x^2 is positive.

2. TAUTOLOGIES

A *truth function* of n arguments is defined to be a function of n arguments, the arguments and values of which are the truth values T or F. As we have seen, any statement form determines a corresponding truth function.*

A statement form that is always true, no matter what the truth values of its statement letters may be, is called a *tautology*. A statement form is a tautology if and only if its corresponding truth function takes only the value T, or, equivalently, if, in its truth table, the column under the statement form contains only T's. An example of a tautology is $(A \vee (\neg A))$, the so-called Law of the Excluded Middle. Other simple examples are $(\neg(A \wedge (\neg A)))$, $(A \Leftrightarrow (\neg(\neg A)))$, $((A \wedge B) \Rightarrow A)$, and $(A \Rightarrow (A \vee B))$.

If $(\mathscr{A} \Rightarrow \mathscr{B})$ is a tautology, $\mathscr{A}$ is said to *logically imply* $\mathscr{B}$, or, alternatively, $\mathscr{B}$ is said to be a *logical consequence* of $\mathscr{A}$. For example, $(A \wedge B)$ logically implies A, A logically implies $(A \vee B)$, and $(A \wedge (A \Rightarrow B))$ logically implies B.

If $(\mathscr{A} \Leftrightarrow \mathscr{B})$ is a tautology, $\mathscr{A}$ and $\mathscr{B}$ are said to be *logically equivalent*. For example, B and $(\neg(\neg B))$ are logically equivalent, as are $(A \Rightarrow B)$ and $((\neg A) \vee B)$.

By means of truth tables, we have effective procedures for determining whether

*To be precise, enumerate all statement letters as follows: $A, B, \ldots, Z; A_1, B_1, \ldots, Z_1; A_2, \ldots$. If a statement form contains the i_1th, ..., i_nth statement letters in this enumeration, where $i_1 < \cdots < i_n$, then the corresponding truth function is to have $x_{i_1}, \ldots, x_{i_n}$, in that order, as its arguments, where x_{i_j} corresponds to the i_jth statement letter. For example, $A \Rightarrow B$ generates the truth function

x_1	x_2	$f(x_1, x_2)$
T	T	T
F	T	T
T	F	F
F	F	T

whereas $B \Rightarrow A$ generates the truth function

x_1	x_2	$g(x_1, x_2)$
T	T	T
F	T	F
T	F	T
F	F	T

a statement form is a tautology and for determining whether a statement form logically implies or is logically equivalent to another statement form.

Examples To see whether a statement form is a tautology, there is another method that is often shorter than the construction of the truth table.

1. Determine whether $((A \Leftrightarrow ((\neg B) \vee C)) \Rightarrow ((\neg A) \Rightarrow B))$ is a tautology. Assume that the statement form sometimes is F (line 1). Then $(A \Leftrightarrow ((\neg B) \vee C))$ is T and $((\neg A) \Rightarrow B)$ is F (line 2). Since $((\neg A) \Rightarrow B)$ is F, $(\neg A)$ is T, and B is F (line 3). Since $(\neg A)$ is T, A is F (line 4). Since A is F and $(A \Leftrightarrow ((\neg B) \vee C))$ is T, $((\neg B) \vee C)$ is F (line 5). Since $((\neg B) \vee C)$ is F, $(\neg B)$ and C are F (line 6). Since $(\neg B)$ is F, B is T (line 7). But B is both T and F (lines 7 and 3). Hence, it is impossible for the form to be false.

$((A \Leftrightarrow ((\neg B) \vee C)) \Rightarrow ((\neg A) \Rightarrow B))$

A	$\Leftrightarrow$	$\neg$	B	$\vee$	C	$\Rightarrow$	$\neg$	A	$\Rightarrow$	B	
						F					1
	T								F		2
							T			F	3
F								F			4
				F							5
		F			F						6
			T								7

2. Determine whether $((A \Rightarrow (B \vee C)) \vee (A \Rightarrow B))$ is a tautology. Assume the form is F (line 1). Then, $(A \Rightarrow (B \vee C))$ and $(A \Rightarrow B)$ are F (line 2). Since $(A \Rightarrow B)$ is F, A is T and B is F (line 3). Since $(A \Rightarrow (B \vee C))$ is F, A is T and $(B \vee C)$ is F (line 4). Since $(B \vee C)$ is F, B and C are F (line 5). Thus, when A is T, B is F, and C is F, the form is F. Therefore, it is not a tautology.

$((A \Rightarrow (B \vee C)) \vee (A \Rightarrow B))$

A	$\Rightarrow$	B	$\vee$	C	$\vee$	A	$\Rightarrow$	B	
					F				1
	F						F		2
						T		F	3
T			F						4
		F		F					5

Exercises

1.5 Determine whether the following are tautologies.

(a) $(((A \Rightarrow B) \Rightarrow B) \Rightarrow B)$
(b) $((A \Leftrightarrow B) \Leftrightarrow (A \Leftrightarrow (B \Leftrightarrow A)))$
(c) $((A \vee (\neg(B \wedge C))) \Rightarrow ((A \Leftrightarrow C) \vee B))$
(d) $(((B \Rightarrow C) \Rightarrow (A \Rightarrow B)) \Rightarrow (A \Rightarrow B))$
(e) $(A \Rightarrow (B \Rightarrow (B \Rightarrow A)))$
(f) $((A \wedge B) \Rightarrow (A \vee C))$
(g) $(((A \Rightarrow B) \Rightarrow A) \Rightarrow A)$
(h) $((A \Rightarrow B) \vee (B \Rightarrow A))$

1.6 Verify or disprove:

(a) $(A \Leftrightarrow B)$ is logically equivalent to $((A \Rightarrow B) \wedge (B \Rightarrow A))$.
(b) $((\neg A) \vee B)$ is logically equivalent to $((\neg B) \vee A)$.
(c) $(\neg(A \Leftrightarrow B))$ is logically equivalent to $(A \Leftrightarrow (\neg B))$.

1.7 Prove that $\mathscr{A}$ logically implies $\mathscr{B}$ if and only if every truth assignment to the variables of $\mathscr{A}$ and $\mathscr{B}$ that makes $\mathscr{A}$ true also makes $\mathscr{B}$ true. (Hence, to show that $\mathscr{A}$ does not logically imply $\mathscr{B}$, it suffices to find a truth assignment under which $\mathscr{A}$ is true and $\mathscr{B}$ is false.)

1.8 Prove that $\mathscr{A}$ and $\mathscr{B}$ are logically equivalent if and only if, under every truth assignment to the variables of $\mathscr{A}$ and $\mathscr{B}$, $\mathscr{A}$ and $\mathscr{B}$ receive the same truth value.

1.9 Prove that $\mathscr{A}$ is logically equivalent to $\mathscr{B}$ if and only if $\mathscr{A}$ logically implies $\mathscr{B}$ and $\mathscr{B}$ logically implies $\mathscr{A}$.

1.10 Show that $\mathscr{A}$ and $\mathscr{B}$ are logically equivalent if and only if, in their truth tables, the columns under $\mathscr{A}$ and $\mathscr{B}$ are the same.

1.11 Which of the following statement forms are logically implied by $(A \wedge B)$?

(a) A	(d) $((\neg A) \vee B)$	(g) $(A \Rightarrow B)$
(b) B	(e) $((\neg B) \Rightarrow A)$	(h) $((\neg B) \Rightarrow (\neg A))$
(c) $(A \vee B)$	(f) $(A \Leftrightarrow B)$	(i) $(A \wedge (\neg B))$

1.12 Repeat Exercise 1.11, with $(A \wedge B)$ replaced by $(A \Rightarrow B)$.
1.13 Repeat Exercise 1.11, with $(A \wedge B)$ replaced by $(A \vee B)$.
1.14 Repeat Exercise 1.11, with $(A \wedge B)$ replaced by $(A \Leftrightarrow B)$.

A statement form that is false for all possible truth values of its statement letters is called a *contradiction*. Its truth table has only F's in the column under the statement form.

Example $(A \Leftrightarrow (\neg A))$

A	$(\neg A)$	$(A \Leftrightarrow (\neg A))$
T	F	F
F	T	F

Another example of a contradiction is $(A \wedge (\neg A))$.

Notice that a statement form $\mathscr{A}$ is a tautology if and only if $(\neg \mathscr{A})$ is a contradiction, and vice versa.

A sentence (in some natural language like English, or in a formal theory*) that arises from a tautology by the substitution of sentences for all the statement letters, with occurrences of the same letter being replaced by the same sentence, is said to be *logically true* (according to the propositional calculus). Such a sentence may be said to be true by virtue of its truth-functional structure alone. An example is the English sentence, "If it is raining or it is snowing, and it is not snowing, then it is raining", which arises by substitution from the tautology $(((A \vee B) \wedge (\neg B)) \Rightarrow A)$. A sentence that comes from a contradiction by means of substitution is said to be *logically false* (according to the propositional calculus).

Now let us prove a few general facts about tautologies.

PROPOSITION 1.1 If $\mathscr{A}$ and $(\mathscr{A} \Rightarrow \mathscr{B})$ are tautologies, then so is $\mathscr{B}$.

Proof Assume that $\mathscr{A}$ and $(\mathscr{A} \Rightarrow \mathscr{B})$ are tautologies. If $\mathscr{B}$ took the value F for some assignment of truth values to the statement letters of $\mathscr{A}$ and $\mathscr{B}$, then, since $\mathscr{A}$ is a tautology, $\mathscr{A}$ would take the value T and, therefore, $(\mathscr{A} \Rightarrow \mathscr{B})$ would have the value F for that assignment. This contradicts the assumption that $(\mathscr{A} \Rightarrow \mathscr{B})$ is a tautology. Hence, $\mathscr{B}$ never takes the value F.

PROPOSITION 1.2 If $\mathscr{A}$ is a tautology containing as statement letters $A_1, A_2, \ldots, A_n$, and $\mathscr{B}$ arises from $\mathscr{A}$ by substituting statement forms $\mathscr{A}_1, \mathscr{A}_2, \ldots, \mathscr{A}_n$ for $A_1, A_2, \ldots, A_n$, respectively, then $\mathscr{B}$ is a tautology; that is, substitution in a tautology yields a tautology.

* By a formal theory we mean an artificial language in which the notions of *meaningful expressions*, *axioms*, and *rules of inference* are precisely described (see page 28).

Example Let $\mathscr{A}$ be $((A_1 \wedge A_2) \Rightarrow A_1)$, let $\mathscr{A}_1$ be $(B \vee C)$, and let $\mathscr{A}_2$ be $(C \wedge D)$. Then $\mathscr{B}$ is $(((B \vee C) \wedge (C \wedge D)) \Rightarrow (B \vee C))$.

Proof Assume that $\mathscr{A}$ is a tautology. For any assignment of truth values to the statement letters in $\mathscr{B}$, the forms $\mathscr{A}_1, \ldots, \mathscr{A}_n$ have truth values $x_1, \ldots, x_n$ (where each x_i is T or F). If we assign the values $x_1, \ldots, x_n$ to $A_1, \ldots, A_n$, respectively, then the resulting truth value of $\mathscr{A}$ is the truth value of $\mathscr{B}$ for the given assignment of truth values. Since $\mathscr{A}$ is a tautology, this truth value must be T. Thus, $\mathscr{B}$ always takes the value T.

PROPOSITION 1.3 If $\mathscr{B}_1$ arises from $\mathscr{A}_1$ by substitution of $\mathscr{B}$ for one or more occurrences of $\mathscr{A}$, then $((\mathscr{A} \Leftrightarrow \mathscr{B}) \Rightarrow (\mathscr{A}_1 \Leftrightarrow \mathscr{B}_1))$ is a tautology. Hence, if $\mathscr{A}$ and $\mathscr{B}$ are logically equivalent, then so are $\mathscr{A}_1$ and $\mathscr{B}_1$.

Example Let $\mathscr{A}_1$ be $(C \vee D)$, let $\mathscr{A}$ be C, and let $\mathscr{B}$ be $(\neg(\neg C))$. Then $\mathscr{B}_1$ is $((\neg(\neg C)) \vee D)$. Since C and $(\neg(\neg C))$ are logically equivalent, $(C \vee D)$ and $((\neg(\neg C)) \vee D)$ are also logically equivalent.

Proof Consider any assignment of truth values to the statement letters. If $\mathscr{A}$ and $\mathscr{B}$ have opposite truth values under this assignment, then $(\mathscr{A} \Leftrightarrow \mathscr{B})$ takes the value F, and so $((\mathscr{A} \Leftrightarrow \mathscr{B}) \Rightarrow (\mathscr{A}_1 \Leftrightarrow \mathscr{B}_1))$ is T. If $\mathscr{A}$ and $\mathscr{B}$ take the same truth values, then so do $\mathscr{A}_1$ and $\mathscr{B}_1$, since $\mathscr{B}_1$ differs from $\mathscr{A}_1$ only in containing $\mathscr{B}$ in some places where $\mathscr{A}_1$ contains $\mathscr{A}$. Hence, in this case, $(\mathscr{A} \Leftrightarrow \mathscr{B})$ is T, $(\mathscr{A}_1 \Leftrightarrow \mathscr{B}_1)$ is T, and, therefore, $((\mathscr{A} \Leftrightarrow \mathscr{B}) \Rightarrow (\mathscr{A}_1 \Leftrightarrow \mathscr{B}_1))$ is T.

PARENTHESES

It is profitable at this point to agree on some conventions to avoid the use of so many parentheses in writing formulas. This will make the reading of complicated expressions easier.

First, we may omit the outer pair of parentheses of a statement form. (In the case of statement letters, there is no outer pair of parentheses.)

Second, the connectives are ordered as follows: $\neg$, $\wedge$, $\vee$, $\Rightarrow$, and $\Leftrightarrow$, and parentheses are eliminated according to the rule that, first, $\neg$ applies to the smallest statement form following it, then $\wedge$ is to connect the smallest statement forms surrounding it, then $\vee$ connects the smallest forms surrounding it, and similarly for $\Rightarrow$ and $\Leftrightarrow$.

Example Parentheses are restored to $A \Leftrightarrow \neg B \vee C \Rightarrow A$ in the following steps:

$$A \Leftrightarrow (\neg B) \vee C \Rightarrow A$$

$$A \Leftrightarrow ((\neg B) \vee C) \Rightarrow A$$

$$A \Leftrightarrow (((\neg B) \vee C) \Rightarrow A)$$

$$(A \Leftrightarrow (((\neg B) \vee C) \Rightarrow A))$$

In applying this rule to occurrences of the same connective, we proceed from left to right. For example,

$$A \Rightarrow \neg B \Rightarrow C$$
$$A \Rightarrow (\neg B) \Rightarrow C$$
$$(A \Rightarrow (\neg B)) \Rightarrow C$$
$$((A \Rightarrow (\neg B)) \Rightarrow C)$$

However, consecutive occurrences of $\neg$ are processed from right to left:

$$B \Rightarrow \neg \neg A$$
$$B \Rightarrow \neg(\neg A)$$
$$B \Rightarrow (\neg(\neg A))$$
$$(B \Rightarrow (\neg(\neg A)))$$

Not every form can be represented without the use of parentheses. For example, parentheses cannot be further eliminated from $A \Rightarrow (B \Rightarrow C)$, since $A \Rightarrow B \Rightarrow C$ stands for $((A \Rightarrow B) \Rightarrow C)$. Likewise, the remaining parentheses cannot be removed from $\neg(A \vee B)$ or from $A \wedge (B \Rightarrow C)$.

Exercises

1.15 Eliminate as many parentheses as possible from the following forms.

(a) $((B \Rightarrow (\neg A)) \wedge C)$
(b) $(A \vee (B \vee C))$
(c) $(((A \wedge (\neg B)) \wedge C) \vee D)$
(d) $((B \vee (\neg C)) \vee (A \wedge B))$
(e) $((A \Leftrightarrow B) \Leftrightarrow (\neg(C \vee D)))$
(f) $((\neg(\neg(\neg(B \vee C)))) \Leftrightarrow (B \Leftrightarrow C))$
(g) $(\neg((\neg(\neg(B \vee C))) \Leftrightarrow (B \Leftrightarrow C)))$
(h) $((((A \Rightarrow B) \Rightarrow (C \Rightarrow D)) \wedge (\neg A)) \vee C)$

1.16 Restore parentheses to the following forms.

(a) $C \vee \neg A \wedge B$
(b) $B \Rightarrow \neg \neg \neg A \vee C$
(c) $C \Rightarrow \neg(A \vee C) \wedge A \Leftrightarrow B$
(d) $C \Rightarrow A \Rightarrow A \Leftrightarrow \neg A \vee B$

1.17 Determine whether the following expressions are abbreviations of statement forms, and, if so, restore all parentheses.

(a) $\neg \neg A \Leftrightarrow A \Leftrightarrow B \vee C$
(b) $\neg(\neg A \Leftrightarrow A) \Leftrightarrow B \vee C$
(c) $\neg(A \Rightarrow B) \vee C \vee D \Rightarrow B$
(d) $A \Leftrightarrow (\neg A \vee B) \Rightarrow (A \wedge (B \vee C)))$
(e) $\neg A \vee B \vee C \wedge D \Leftrightarrow A \wedge \neg A$

1.18 If we write $\neg \mathscr{A}$ instead of $(\neg \mathscr{A})$, $\Rightarrow \mathscr{A}\mathscr{B}$ instead of $(\mathscr{A} \Rightarrow \mathscr{B})$, $\wedge \mathscr{A}\mathscr{B}$ instead of $(\mathscr{A} \wedge \mathscr{B})$, $\vee \mathscr{A}\mathscr{B}$ instead of $(\mathscr{A} \vee \mathscr{B})$, and $\Leftrightarrow \mathscr{A}\mathscr{B}$ instead of $(\mathscr{A} \Leftrightarrow \mathscr{B})$, then there is no need for parentheses. For example, $((\neg A) \Rightarrow (B \vee (\neg D)))$ becomes $\Rightarrow \neg A \vee B \neg D$. This way of writing forms is called *Polish notation.*

(a) Write $((C \Rightarrow (\neg A)) \vee B)$ and $(C \vee ((B \wedge (\neg D)) \Rightarrow C))$ in this notation.
(b) If we count $\Rightarrow$, $\wedge$, $\vee$, and $\Leftrightarrow$ each as $+1$, each statement letter as -1, and $\neg$ as 0, prove that an expression $\mathscr{A}$ in this parenthesis-free notation is a statement form if and only if (i) the sum of the symbols of $\mathscr{A}$ is -1 and (ii) the sum of the symbols in any proper initial segment of $\mathscr{A}$ is nonnegative. (If an expression $\mathscr{A}$ can be written in the form $\mathscr{B}\mathscr{C}$, where $\mathscr{B} \neq \mathscr{A}$, then $\mathscr{B}$ is called a *proper initial segment* of $\mathscr{A}$.)
(c) Write the statement forms of Exercise 1.15 in Polish notation.
(d) Determine whether the following expressions are statement forms in Polish notation. If so, write the statement forms in the standard way.

(i) $\neg \Rightarrow ABC \vee AB \neg C$ (iii) $\vee \wedge \vee \neg A \neg BC \wedge \vee AC \vee \neg C \neg A$

(ii) $\Rightarrow \Rightarrow AB \Rightarrow \Rightarrow BC \Rightarrow \neg AC$

1.19 Determine whether each of the following is a tautology, a contradiction, or neither.

(a) $B \Leftrightarrow (B \vee B)$
(b) $((A \Rightarrow B) \wedge B) \Rightarrow A$
(c) $(\neg A) \Rightarrow (A \wedge B)$
(d) $(A \Rightarrow B) \Rightarrow ((B \Rightarrow C) \Rightarrow (A \Rightarrow C))$
(e) $A \wedge (\neg(A \vee B))$
(f) $(A \Rightarrow B) \Leftrightarrow ((\neg A) \vee B)$
(g) $(A \Rightarrow B) \Leftrightarrow \neg(A \wedge (\neg B))$
(h) $(B \Leftrightarrow (B \Rightarrow A)) \Rightarrow A$

1.20 If A and B are true and C is false, what are the truth values of the following statement forms?

(a) $A \vee C$
(b) $A \wedge C$
(c) $\neg A \wedge \neg C$
(d) $A \Leftrightarrow \neg B \vee C$
(e) $B \vee \neg C \Rightarrow A$
(f) $(B \vee A) \Rightarrow (B \Rightarrow \neg C)$
(g) $(B \Leftrightarrow \neg A) \Leftrightarrow (A \Leftrightarrow C)$
(h) $(B \Rightarrow A) \Rightarrow ((A \Rightarrow \neg C) \Rightarrow (\neg C \Rightarrow B))$

1.21 If $A \Rightarrow B$ is T, what can be deduced about the truth values of:

(a) $A \vee C \Rightarrow B \vee C$ (b) $A \wedge C \Rightarrow B \wedge C$ (c) $\neg A \wedge B \Leftrightarrow A \vee B$

1.22 What further truth values can be deduced from those shown?

(a) $\neg A \vee (A \Rightarrow B)$
F (under $\Rightarrow$)

(b) $\neg(A \wedge B) \Leftrightarrow \neg A \Rightarrow \neg B$
T (under $\wedge$)

(c) $(\neg A \vee B) \Rightarrow (A \Rightarrow \neg C)$
F (under the second $\Rightarrow$)

1.23 If $A \Leftrightarrow B$ is F, what can be deduced about the truth values of:

(a) $A \wedge B$ (b) $A \vee B$ (c) $A \Rightarrow B$ (d) $A \wedge C \Leftrightarrow B \wedge C$

1.24 Repeat Exercise 1.23, except assume $A \Leftrightarrow B$ is T.

1.25 What further truth values can be deduced from those given?

(a) $(A \wedge B) \Leftrightarrow (A \vee B)$
F F (under $\Leftrightarrow$ and $\vee$)

(b) $(A \Rightarrow \neg B) \Rightarrow (C \Rightarrow B)$
F (under the second $\Rightarrow$)

1.26 (a) Apply Proposition 1.2 when $\mathscr{A}$ is $A_1 \Rightarrow A_1 \vee A_2$, $\mathscr{A}_1$ is $B \wedge D$, and $\mathscr{A}_2$ is $\neg B$.

(b) Apply Proposition 1.3 when $\mathscr{A}_1$ is $(B \Rightarrow C) \vee D$, $\mathscr{A}$ is $B \Rightarrow C$, and $\mathscr{B}$ is $\neg B \vee C$.

1.27 Show that each statement form in column I is logically equivalent to the form next to it in column II.

	I	II		I	II
(a)	$A \Rightarrow (B \Rightarrow C)$	$(A \wedge B) \Rightarrow C$	(j)	$\neg(A \vee B)$	$(\neg A) \wedge (\neg B)$
(b)	$A \wedge (B \vee C)$	$(A \wedge B) \vee (A \wedge C)$	(k)	$\neg(A \wedge B)$	$(\neg A) \vee (\neg B)$
(c)	$A \vee (B \wedge C)$	$(A \vee B) \wedge (A \vee C)$	(l)	$A \vee (A \wedge B)$	A
(d)	$(A \wedge B) \vee \neg B$	$A \vee \neg B$	(m)	$A \wedge (A \vee B)$	A
(e)	$(A \vee B) \wedge \neg B$	$A \wedge \neg B$	(n)	$A \wedge B$	$B \wedge A$
(f)	$A \Rightarrow B$	$\neg B \Rightarrow \neg A$	(o)	$A \vee B$	$B \vee A$
(g)	$A \Leftrightarrow B$	$B \Leftrightarrow A$	(p)	$(A \wedge B) \wedge C$	$A \wedge (B \wedge C)$
(h)	$(A \Leftrightarrow B) \Leftrightarrow C$	$A \Leftrightarrow (B \Leftrightarrow C)$	(q)	$(A \vee B) \vee C$	$A \vee (B \vee C)$
(i)	$A \Leftrightarrow B$	$(A \wedge B) \vee (\neg A \wedge \neg B)$	(r)	$\neg(A \Leftrightarrow B)$	$A \Leftrightarrow \neg B$

1.28 Show the logical equivalence of the following pairs.

(a) $\mathscr{T} \wedge \mathscr{A}$ and $\mathscr{A}$, where $\mathscr{T}$ is a tautology.
(b) $\mathscr{T} \vee \mathscr{A}$ and $\mathscr{T}$, where $\mathscr{T}$ is a tautology.
(c) $\mathscr{F} \wedge \mathscr{A}$ and $\mathscr{F}$, where $\mathscr{F}$ is a contradiction.
(d) $\mathscr{F} \vee \mathscr{A}$ and $\mathscr{A}$, where $\mathscr{F}$ is a contradiction.

1.29 (a) Show the logical equivalence of $\neg(A \Rightarrow B)$ and $A \wedge \neg B$.

(b) Show the logical equivalence of $\neg(A \Leftrightarrow B)$ and $(A \wedge \neg B) \vee (\neg A \wedge B)$.

(c) For each of the following statement forms, find a statement form that is logically equivalent to its negation and in which negation signs apply only to statement letters.

(i) $A \Rightarrow (B \Leftrightarrow \neg C)$ (ii) $\neg A \vee (B \Rightarrow C)$ (iii) $A \wedge (B \vee \neg C)$

1.30 (*Duality*)

(a) If $\mathscr{A}$ is a statement form involving only $\neg$, $\wedge$, and $\vee$, and $\mathscr{A}'$ arises from $\mathscr{A}$ by replacing each $\wedge$ by $\vee$ and each $\vee$ by $\wedge$, show that $\mathscr{A}$ is a tautology if and only if $\neg\mathscr{A}'$ is a tautology. Then prove that, if $\mathscr{A} \Rightarrow \mathscr{B}$ is a tautology, so is $\mathscr{B}' \Rightarrow \mathscr{A}'$, and if $\mathscr{A} \Leftrightarrow \mathscr{B}$ is a tautology, so is $\mathscr{A}' \Leftrightarrow \mathscr{B}'$.

(b) Among the logical equivalences in Exercise 1.27, derive (c) from (b), (e) from (d), (k) from (j), (m) from (l), (o) from (n), and (q) from (p).

(c) If $\mathscr{A}$ is a statement form involving only $\neg$, $\wedge$, and $\vee$, and $\mathscr{A}^*$ results from $\mathscr{A}$ by interchanging $\wedge$ and $\vee$ and replacing every statement letter by its negation, show that $\mathscr{A}^*$ is logically equivalent to $\neg\mathscr{A}$. Find a statement form that is logically equivalent to the negation of $(A \vee B \vee C) \wedge (\neg A \vee \neg B \vee D)$, in which $\neg$ applies only to statement letters.

1.31 (a) Prove that a statement form that contains $\Leftrightarrow$ as its only connective is a tautology if and only if each statement letter occurs an even number of times.

(b) Prove that a statement form that contains $\neg$ and $\Leftrightarrow$ as its only connectives is a tautology if and only if $\neg$ and each statement letter occur an even number of times.

1.32 (Shannon, 1938) An electric circuit containing only on–off switches (when a switch is on, it passes current; otherwise, it does not) can be represented by a diagram in which, next to each switch, we put a letter representing a necessary and sufficient condition for the switch to be on (see Figure 1.1). The condition that a current flows through this network can be given by a statement form: $(A \wedge B) \vee (C \wedge \neg A)$. A statement form representing the circuit shown in Figure 1.2 is $(A \wedge B) \vee ((C \vee A) \wedge \neg B)$, which is logically equivalent to each of the following forms by virtue of the indicated logical equivalence of Exercise 1.27.

$((A \wedge B) \vee (C \vee A)) \wedge ((A \wedge B) \vee \neg B)$	(c)
$((A \wedge B) \vee (C \vee A)) \wedge (A \vee \neg B)$	(d)
$((A \wedge B) \vee (A \vee C)) \wedge (A \vee \neg B)$	(o)
$(((A \wedge B) \vee A) \vee C) \wedge (A \vee \neg B)$	(q)
$(A \vee C) \wedge (A \vee \neg B)$	(o), (l)
$A \vee (C \wedge \neg B)$	(c)

Hence, the given circuit is equivalent to the simpler circuit shown in Figure 1.3. (Two circuits are said to be *equivalent* if current flows through one if and only if it flows through the other, and one circuit is simpler if it contains fewer switches.)

(a) Find simpler equivalent circuits for those shown in Figures 1.4, 1.5, and 1.6.

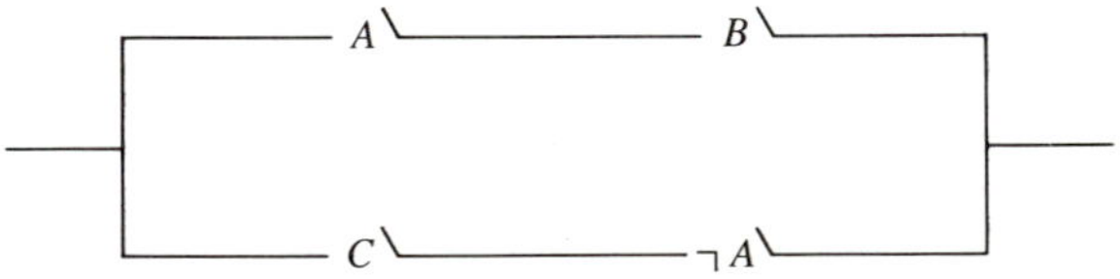

Figure 1.1

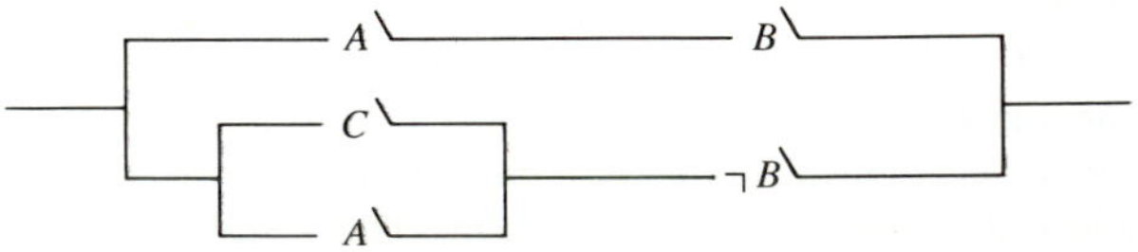

Figure 1.2

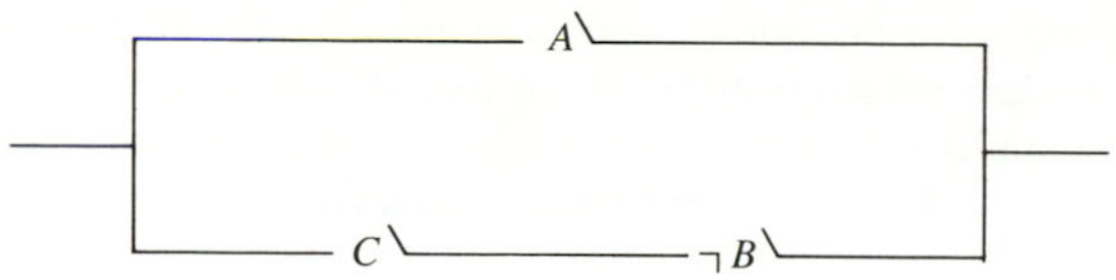

Figure 1.3

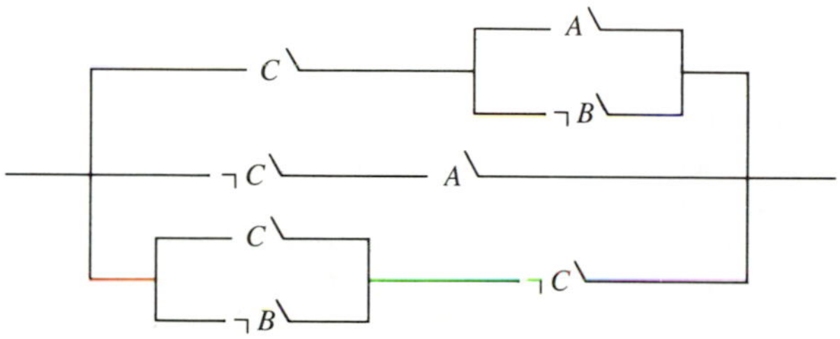

Figure 1.4

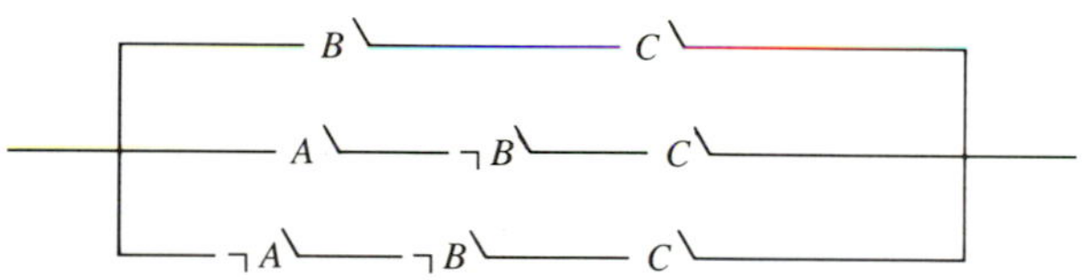

Figure 1.5

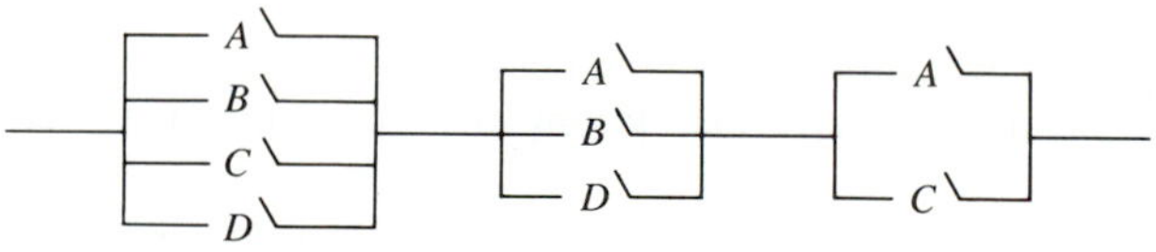

Figure 1.6

(b) Assume that each of the three members of a committee votes *yes* on a proposal by pressing a button. Devise as simple a circuit as you can that will allow current to pass when and only when at least two of the members vote in the affirmative.

(c) We wish a light to be controlled by three different switches in a room in such a way that flicking any one of these switches will turn the light on if it is off and turn it off if it is on. Construct a simple circuit to do the required job.

1.33 Determine whether the following arguments are logically correct by representing each sentence as a statement form and checking whether the conclusion is logically implied by the conjunction of the assumptions. (To do this, assign T to each assumption and F to the conclusion, and determine whether a contradiction results.)

(a) If Jones is a Communist, Jones is an atheist. Jones is an atheist. Hence, Jones is a Communist.

(b) If the temperature and air pressure remained constant, there was no rain. The temperature did remain constant. Therefore, if there was rain, then the air pressure did not remain constant.

(c) If Forbush wins the election, then taxes will increase if the deficit will remain high. If Forbush wins the election, the deficit will remain high. Therefore, if Forbush wins the election, taxes will increase.

(d) If the number x ends in 0, it is divisible by 5. x does not end in 0. Hence, x is not divisible by 5.

(e) If the number x ends in 0, it is divisible by 5. x is not divisible by 5. Hence, x does not end in 0.

(f) If $a = 0$ or $b = 0$, then $ab = 0$. But $ab \neq 0$. Hence, $a \neq 0$ and $b \neq 0$.

(g) If x is positive, then y is negative. If z is negative, then y is negative. Therefore, if x is positive or z is negative, then y is negative.

(h) If fallout shelters are built, other countries will feel endangered and our people will get a false sense of security. If other countries will feel endangered, they may start a preventive war. If our people will get a false sense of security, they will put less effort into preserving peace. If fallout shelters are not built, we run the risk of tremendous losses in the event of war. Hence, either other countries may start a preventive war and our people will put less effort into preserving peace, or we run the risk of tremendous losses in the event of war.

(i) If capital investment remains constant, then government spending will increase or unemployment will result. If government spending will not increase, taxes can be reduced. If taxes can be reduced and capital investment remains constant, then unemployment will not result. Hence, government spending will increase.

1.34 Which of the following sets of statement forms are consistent, in the sense that there is an assignment of truth values to the statement letters that makes all the forms in the set true?

(a)	(b)	(c)
$A \Rightarrow B$	$\neg(\neg B \vee A)$	$D \Rightarrow B$
$B \Leftrightarrow C$	$A \vee \neg C$	$A \vee \neg B$
$C \vee D \Leftrightarrow \neg B$	$B \Rightarrow \neg C$	$\neg(D \wedge A)$
		D

1.35 Check each of the following sets of statements for consistency by representing the sentences as statement forms and then testing their conjunction to see whether it is a contradiction.

(a) Either the witness was not intimidated, or, if Doherty committed suicide, a note was found. If the witness was intimidated, then Doherty did not commit suicide. If a note was found, then Doherty committed suicide.

(b) The contract is satisfied if and only if the building is completed by November 30. The building is completed by November 30 if and only if the electrical subcontractor completes his work by November 10. The bank loses money if and only if the contract is not satisfied. Yet the electrical subcontractor completes his work by November 10 if and only if the bank loses money.

3. ADEQUATE SETS OF CONNECTIVES

Every statement form containing n statement letters generates a corresponding truth function of n arguments. The arguments and values of the function are T or F. Logically equivalent forms generate the same truth function. The natural question is whether all truth functions are so generated.

PROPOSITION 1.4 Every truth function is generated by a statement form involving the connectives $\neg$, $\wedge$, and $\vee$.

Proof (Refer to Examples (1) and (2) below for clarification.) Let $f(x_1, \ldots, x_n)$ be a truth function. Clearly f can be represented by a truth table of 2^n rows, where each row represents some assignment of truth values to the variables $x_1, \ldots, x_n$, followed by the corresponding value of $f(x_1, \ldots, x_n)$. If $1 \leqslant i \leqslant 2^n$, let C_i be the conjunction $U_1^i \wedge U_2^i \wedge \cdots \wedge U_n^i$, where U_j^i is A_j if, in the ith row of the truth table, x_j takes the value T, and U_j^i is $\neg A_j$ if x_j takes the value F. Let D be the disjunction of all those C_i's such that f has the value T for the ith row of the truth table. (If there are no such rows, then f always takes the value F, and we let D be $A_1 \wedge \neg A_1$, which satisfies the theorem.) As its corresponding truth function, D has f. Let there be given an assignment of truth values to the statement letters $A_1, \ldots, A_n$, and assume that the corresponding assignment to the variables $x_1, \ldots, x_n$ is row k of the truth table for f. Then C_k has the value T for this assignment, whereas every other C_i has the value F. If f has the value T for row k, then C_k is a disjunct of D. Hence, D would also have the value T for this assignment. If f has the value F for row k, then C_k is not a disjunct of D and all the disjuncts of D take the value F for this assignment. Therefore, D would also have the value F. Thus, D generates the truth function f.

Examples

1.

x_1	x_2	$f(x_1, x_2)$
T	T	F
F	T	T
T	F	T
F	F	T

D is $(\neg A_1 \wedge A_2) \vee (A_1 \wedge \neg A_2) \vee (\neg A_1 \wedge \neg A_2)$.

2.

x_1	x_2	x_3	$g(x_1, x_2, x_3)$
T	T	T	T
F	T	T	F
T	F	T	T
F	F	T	T
T	T	F	F
F	T	F	F
T	F	F	F
F	F	F	T

D is $(A_1 \wedge A_2 \wedge A_3) \vee (A_1 \wedge \neg A_2 \wedge A_3) \vee (\neg A_1 \wedge \neg A_2 \wedge A_3)$
$\vee (\neg A_1 \wedge \neg A_2 \wedge \neg A_3)$.

Exercise

1.36 Find statement forms in the connectives $\neg$, $\wedge$, and $\vee$ that have the following truth functions.

x_1	x_2	x_3	$f(x_1, x_2, x_3)$	$g(x_1, x_2, x_3)$	$h(x_1, x_2, x_3)$
T	T	T	T	T	F
F	T	T	T	T	T
T	F	T	T	T	F
F	F	T	F	F	F
T	T	F	F	T	T
F	T	F	F	F	T
T	F	F	F	T	F
F	F	F	T	F	T

COROLLARY 1.5 Every truth function corresponds to a statement form containing as connectives only $\wedge$ and $\neg$, or only $\vee$ and $\neg$, or only $\Rightarrow$ and $\neg$.

Proof Notice that $A \vee B$ is logically equivalent to $\neg(\neg A \wedge \neg B)$. Hence, by Proposition 1.3 (second part), any statement form in $\wedge$, $\vee$, and $\neg$ is logically equivalent to a statement form in only $\wedge$ and $\neg$ [obtained by replacing all expressions $\mathscr{A} \vee \mathscr{B}$ by $\neg(\neg\mathscr{A} \wedge \neg\mathscr{B})$]. The other parts of the corollary are similar consequences of the following tautologies:

$$A \wedge B \Leftrightarrow \neg(\neg A \vee \neg B)$$

$$A \vee B \Leftrightarrow (\neg A) \Rightarrow B$$

$$A \wedge B \Leftrightarrow \neg(A \Rightarrow \neg B)$$

We have just seen that there are certain pairs of connectives—for example, $\neg$ and $\wedge$—in terms of which all other truth functions are definable (in the sense of Corollary 1.5). It turns out that there is a single connective, $\downarrow$ (joint denial), that will do the same job. Its truth table is

A	B	$A \downarrow B$
T	T	F
F	T	F
T	F	F
F	F	T

$A \downarrow B$ is true when and only when neither A nor B is true. Clearly, $\neg A \Leftrightarrow (A \downarrow A)$ and $(A \wedge B) \Leftrightarrow ((A \downarrow A) \downarrow (B \downarrow B))$ are tautologies. Hence, the adequacy of $\downarrow$ for the construction of all truth functions follows from Corollary 1.5.

Another connective, $|$ (alternative denial), is also adequate for this purpose. Its truth table is

A	B	$A \mid B$
T	T	F
F	T	T
T	F	T
F	F	T

$A|B$ is true when and only when not both A and B are true. The adequacy of $|$ follows from the tautologies $\neg A \Leftrightarrow (A|A)$ and $(A \vee B) \Leftrightarrow ((A|A)|(B|B))$.

PROPOSITION 1.6 The only binary connectives that alone are adequate for the construction of all truth functions are $\downarrow$ and $|$.

Proof Assume that $h(A, B)$ is an adequate connective. Now, if $h(T, T)$ were T, then any statement form built up using only h would take the value T when all its statement letters take the value T. Hence, $\neg A$ would not be definable in terms of h. So, $h(T, T) = F$. Likewise, $h(F, F) = T$. Thus, we have the partial truth table

A	B	$h(A, B)$
T	T	F
F	T	
T	F	
F	F	T

If the second and third entries in the last column are F, F or T, T, then h is $\downarrow$ or $|$. If they are F, T, then $h(A, B) \Leftrightarrow \neg B$ would be a tautology; and if they are T, F, then $h(A, B) \Leftrightarrow \neg A$ is a tautology. In both cases, h would be definable in terms of $\neg$. But $\neg$ is not adequate by itself because the only truth functions of one variable definable from it are the identity function and negation itself, whereas the truth function that is always T would not be definable.

Exercises

1.37 Prove that each of the pairs $\Rightarrow$, $\vee$ and $\neg$, $\Leftrightarrow$ is not alone adequate to express all truth functions.

1.38 (a) Prove that $A \vee B$ can be expressed in terms of $\Rightarrow$ alone.

(b) Prove that $A \wedge B$ cannot be expressed in terms of $\Rightarrow$ alone.

(c) Prove that $A \Leftrightarrow B$ cannot be expressed in terms of $\Rightarrow$ alone.

1.39 With one variable A, there are four truth functions:

A	$\neg A$	$A \vee \neg A$	$A \wedge \neg A$
T	F	T	F
F	T	T	F

(a) With two variables A and B, how many truth functions are there?

(b) How many truth functions of n variables are there?

1.40 Show that the truth function h determined by $(A \vee B) \Rightarrow \neg C$ generates all truth functions.

1.41 By a *literal* we mean a statement letter or a negation of a statement letter. A statement form is said to be in *disjunctive normal form* (dnf) if it is a disjunction consisting of one or more disjuncts, each of which is a conjunction of one or more literals—for example, $(A \wedge B) \vee (\neg A \wedge C)$, $(A \wedge B \wedge \neg A) \vee (C \wedge \neg B) \vee (A \wedge \neg C)$, A, $A \wedge B$, and $A \vee (B \wedge C)$. A form is in *conjunctive normal form* (cnf) if it is a conjunction of one or more conjuncts, each of which is a disjunction of one or more literals—for example, $(B \vee \neg C) \wedge (A \vee D)$, A, $A \wedge B$, $A \vee \neg B$, and $A \wedge (B \vee A) \wedge (\neg B \vee A)$. Note that literals are considered (degenerate) conjunctions and disjunctions.

(a) The proof of Proposition 1.4 shows that every statement form $\mathscr{A}$ is logically equivalent to one in disjunctive normal form. By applying this result to $\neg \mathscr{A}$, prove that $\mathscr{A}$ is also logically equivalent to a form in conjunctive normal form.

(b) Find logically equivalent dnfs and cnfs for $\neg(A \Rightarrow B) \vee (\neg A \wedge C)$ and $A \Leftrightarrow ((B \wedge \neg A) \vee C)$. [*Hint*: Instead of relying on Proposition 1.4, it is usually easier to use Exercise 1.27(b) and (c).]

(c) A dnf (cnf) is called *full* if no disjunct (conjunct) contains two occurrences of literals with the same letter and if a letter that occurs in one disjunct (conjunct) also occurs in all the others. For example, $(A \wedge \neg A \wedge B) \vee (A \wedge B)$, $(B \wedge B \wedge C) \vee (B \wedge C)$, and $(B \wedge C) \vee B$ are not full, whereas $(A \wedge B \wedge \neg C) \vee (A \wedge B \wedge C) \vee (A \wedge \neg B \wedge \neg C)$ and $(A \wedge \neg B) \vee (B \wedge A)$ are full dnfs.

(i) Find full dnfs and cnfs for $(A \wedge B) \vee \neg A$ and $\neg(A \Rightarrow B) \vee (\neg A \wedge C)$.

(ii) Prove that every noncontradictory (nontautologous) statement form $\mathscr{A}$ is logically equivalent to a full dnf (cnf) $\mathscr{B}$, and, if $\mathscr{B}$ contains exactly n letters, then $\mathscr{A}$ is a tautology (contradiction) if and only if $\mathscr{B}$ has 2^n disjuncts (conjuncts).

(d) For each of the following, find a logically equivalent dnf (cnf), and then find a logically equivalent full dnf (cnf).

(i) $(A \vee B) \wedge (\neg B \vee C)$ (iii) $(A \wedge \neg B) \vee (A \wedge C)$

(ii) $\neg A \vee (B \Rightarrow \neg C)$ (iv) $(A \vee B) \Leftrightarrow \neg C$

(e) Construct statement forms in $\neg$ and $\wedge$ (respectively, in $\neg$ and $\vee$ or in $\neg$ and $\Rightarrow$) logically equivalent to the statement forms in part (d).

1.42 A statement form is said to be *satisfiable* if it is true for some assignment of truth values to its statement letters. The problem of determining the satisfiability of an arbitrary cnf plays an important role in the theory of computational complexity; it is an example of a so-called *$\mathscr{N}\mathscr{P}$-complete* problem (see Garey & Johnson, 1978).

(a) Show that $\mathscr{A}$ is satisfiable if and only if $\neg\mathscr{A}$ is not logically valid.

(b) Determine whether the following are satisfiable:

(i) $(A \vee B) \wedge (\neg A \vee B \vee C) \wedge (\neg A \vee \neg B \vee \neg C)$

(ii) $((A \Rightarrow B) \vee C) \Leftrightarrow (\neg B \wedge (A \vee C))$

(c) Given a disjunction $\mathscr{D}$ of four or more literals: $L_1 \vee L_2 \vee \cdots \vee L_n$, let $C_1, \ldots, C_{n-2}$ be statement letters that do not occur in $\mathscr{D}$, and construct the cnf $\mathscr{E}$:

$$(L_1 \vee L_2 \vee C_1) \wedge (\neg C_1 \vee L_3 \vee C_2) \wedge (\neg C_2 \vee L_4 \vee C_3) \wedge \cdots$$
$$\wedge (\neg C_{n-3} \vee L_{n-1} \vee C_{n-2}) \wedge (\neg C_{n-2} \vee L_n \vee \neg C_1)$$

Show that any truth assignment satisfying $\mathscr{D}$ can be extended to a truth assignment satisfying $\mathscr{E}$ and, conversely, any truth assignment satisfying $\mathscr{E}$ is an extension of a truth assignment satisfying $\mathscr{D}$. (This permits the reduction of the problem of satisfying cnfs to the corresponding problem for cnfs with each conjunct containing at most three literals.)

(d) For a disjunction $\mathscr{D}$ of three literals $L_1 \vee L_2 \vee L_3$, show that a form that has the properties of $\mathscr{E}$ in part (c) cannot be constructed, with $\mathscr{E}$ a cnf in which each conjunct contains at most two literals (R. Cowen).

1.43 *Resolution.* Let $\mathscr{A}$ be a cnf and let C be a statement letter. If C is a disjunct of a disjunction $\mathscr{D}_1$ in $\mathscr{A}$ and $\neg C$ is a disjunct of another disjunction $\mathscr{D}_2$ in $\mathscr{A}$, then a nonempty disjunction obtained by eliminating C from $\mathscr{D}_1$ and $\neg C$ from $\mathscr{D}_2$ and forming the disjunction of the remaining literals (dropping repetitions) is said to be obtained from $\mathscr{A}$ by *resolution on C*. For example, If $\mathscr{A}$ is $(A \vee \neg C \vee \neg B) \wedge (\neg A \vee D \vee \neg B) \wedge (C \vee D \vee A)$, then the first and third conjuncts yield $A \vee \neg B \vee D$ by resolution on C. In addition, the first and second conjuncts yield $\neg C \vee \neg B \vee D$ by resolution on A, and the second and third conjuncts yield $D \vee \neg B \vee C$ by resolution on A. If we conjoin to $\mathscr{A}$ any new disjuncts obtained by resolution on all variables, and if we apply the same procedure to the new cnf and keep on iterating this operation, the process must eventually stop, and the final result is denoted $\mathscr{R}es(\mathscr{A})$. In the example, $\mathscr{R}es(\mathscr{A})$ is:

$$(A \vee \neg C \vee \neg B) \wedge (\neg A \vee D \vee \neg B) \wedge (C \vee D \vee A) \wedge (\neg C \vee \neg B \vee D)$$
$$\wedge (D \vee \neg B \vee C) \wedge (A \vee \neg B \vee D) \wedge (D \vee \neg B)$$

(Notice that we have not been careful about specifying the order in which conjuncts or disjuncts are written, since any two arrangements will be logically equivalent.)

(a) Find $\mathscr{R}es(\mathscr{A})$ when $\mathscr{A}$ is each of the following:
(i) $(A \vee \neg B) \wedge B$ (iii) $(A \vee C) \wedge (\neg A \vee B) \wedge (A \vee \neg C) \wedge (\neg A \vee \neg B)$
(ii) $(A \vee B \vee C) \wedge (A \vee \neg B \vee C)$

(b) Show that $\mathscr{A}$ logically implies $\mathscr{R}es(\mathscr{A})$.

(c) If $\mathscr{A}$ is a cnf, let $\mathscr{A}_C$ be the cnf obtained from $\mathscr{A}$ by deleting those conjuncts that contain C or $\neg C$. Let $r_C(\mathscr{A})$ be the cnf that is the conjunction of $\mathscr{A}_C$ and all those disjunctions obtained from $\mathscr{A}$ by resolution on C. For example, if $\mathscr{A}$ is the cnf in the example above, then $r_C(\mathscr{A})$ is $(\neg A \vee D \vee \neg B) \wedge (A \vee \neg B \vee D)$. Prove that if $r_C(\mathscr{A})$ is satisfiable, then so is $\mathscr{A}$ (R. Cowen).

(d) A cnf $\mathscr{A}$ is said to be a *blatant contradiction* if it contains some letter C and its negation $\neg C$ as conjuncts. An example of a blatant contradiction is $(A \vee B) \wedge B \wedge (C \vee D) \wedge \neg B$. Prove that if $\mathscr{A}$ is unsatisfiable, then $\mathscr{R}es(\mathscr{A})$ is a blatant contradiction. [*Hint*: Use induction on the number n of letters that occur in $\mathscr{A}$. In the induction step, use part (c).]

(e) Prove that $\mathscr{A}$ is unsatisfiable if and only if $\mathscr{R}es(\mathscr{A})$ is a blatant contradiction.

1.44 (a) A certain country is inhabited only by people who either always tell the truth or always tell lies, and who will respond only to *yes or no* questions. A tourist comes to a fork in a road where one branch leads to the capital and the other does not. There is no sign indicating which branch to take, but there is a native standing at the fork. What yes or no question should the tourist ask in order to determine which branch to take? [*Hint*: Let A stand for "You always tell the truth" and let B stand for "The left-hand branch leads to the capital". Construct, by means of a suitable truth table, a statement form involving A and B such that the native's answer to the question as to whether this statement form is true will be *yes* when and only when B is true.]

(b) In a certain country, there are three kinds of people: workers (who always tell the truth), capitalists (who never tell the truth), and students (who sometimes tell the truth and sometimes lie). At a fork in a road, one branch leads to the capital. A worker, a capitalist, and a student are standing at the side of the road but are not identifiable in any obvious way. By asking two yes or no questions, find out which fork leads to the capital. (Each question may be addressed to any one of the three.)

More puzzles of this kind may be found in Smullyan (1978, chap. 3; 1985, chaps. 2, 4–8).

4. AN AXIOM SYSTEM FOR THE PROPOSITIONAL CALCULUS

Truth tables enable us to answer most of the significant questions concerning the truth-functional connectives, such as whether a given statement form is a tautology, contradiction, or neither, and whether it logically implies or is logically equivalent to some other given statement form. The more complex parts of logic we shall treat later cannot be handled by truth tables or by any other similar effective procedure. Consequently, another approach, by means of formal theories, will have to be tried.

Although, as we have seen, the propositional calculus surrenders completely to the truth table method, it will be instructive to illustrate the axiomatic method in this simple branch of logic.

A formal theory $\mathscr{S}$ is defined when the following conditions are satisfied:

1. A countable set of symbols is given as the symbols of $\mathscr{S}$.* A finite sequence of symbols of $\mathscr{S}$ is called an *expression* of $\mathscr{S}$.
2. There is a subset of the expressions of $\mathscr{S}$ called the set of *well-formed formulas* (abbreviated "wfs") of $\mathscr{S}$. (There is usually an effective procedure to determine whether a given expression is a wf.)
3. A set of wfs is set aside and called the set of *axioms* of $\mathscr{S}$. (Most often, one can effectively decide whether a given wf is an axiom, and, in such a case, $\mathscr{S}$ is called an *axiomatic* theory.)
4. There is a finite set $R_1, \ldots, R_n$ of relations among wfs, called *rules of inference*. For each R_i, there is a unique positive integer j such that, for every set of j wfs and each wf $\mathscr{A}$, one can effectively decide whether the given j wfs are in the relation R_i to $\mathscr{A}$, and, if so, $\mathscr{A}$ is called a *direct consequence* of the given wfs by virtue of R_i.

A *proof* in $\mathscr{S}$ is a sequence $\mathscr{A}_1, \ldots, \mathscr{A}_n$ of wfs such that, for each i, either $\mathscr{A}_i$ is an axiom of $\mathscr{S}$ or $\mathscr{A}_i$ is a direct consequence of some of the preceding wfs by virtue of one of the rules of inference.

A *theorem* of $\mathscr{S}$ is a wf $\mathscr{A}$ of $\mathscr{S}$ such that there is a proof, the last wf of which is $\mathscr{A}$. Such a proof is called a *proof of* $\mathscr{A}$.

Even if $\mathscr{S}$ is axiomatic—that is, if there is an effective procedure for checking any given wf to see whether it is an axiom—the notion of "theorem" is not necessarily effective, since, in general, there is no mechanical method (effective procedure) for determining, given any wf $\mathscr{A}$, whether there is a proof of $\mathscr{A}$. A theory for which there is such a mechanical method is said to be *decidable*; otherwise, it is called *undecidable*. A decidable theory is, roughly speaking, one for which a machine can be devised to test wfs for theoremhood, whereas, in an undecidable theory, ingenuity is required to determine whether wfs are theorems.

A wf $\mathscr{A}$ is said to be a *consequence* in $\mathscr{S}$ of a set Γ of wfs if and only if there is a sequence $\mathscr{A}_1, \ldots, \mathscr{A}_n$ of wfs such that $\mathscr{A} = \mathscr{A}_n$ and, for each i, either $\mathscr{A}_i$ is an axiom or $\mathscr{A}_i$ is in Γ, or $\mathscr{A}_i$ is a direct consequence by some rule of inference of some of the preceding wfs in the sequence. Such a sequence is called a *proof* (or *deduction*) *of* $\mathscr{A}$ *from* Γ. The members of Γ are called the *hypotheses* or *premisses* of the proof. We use $\Gamma \vdash \mathscr{A}$ as an abbreviation for "$\mathscr{A}$ is a consequence of Γ." In order to avoid confusion when dealing with more than one theory, we write $\Gamma \vdash_{\mathscr{S}} \mathscr{A}$, adding the subscript $\mathscr{S}$ to indicate the theory in question.

If Γ is a finite set $\{\mathscr{B}_1, \ldots, \mathscr{B}_n\}$, we write $\mathscr{B}_1, \ldots, \mathscr{B}_n \vdash \mathscr{A}$ instead of $\{\mathscr{B}_1, \ldots, \mathscr{B}_n\} \vdash \mathscr{A}$. If Γ is the empty set 0, then $0 \vdash \mathscr{A}$ if and only if $\mathscr{A}$ is a theorem. It is customary to omit the sign "0" and simply write $\vdash \mathscr{A}$. Thus, $\vdash \mathscr{A}$ is another way of asserting that $\mathscr{A}$ is a theorem.

*These "symbols" may be thought of as arbitrary objects rather than just linguistic objects. This will become absolutely necessary when we deal with theories with uncountably many symbols in Chapter 2, Section 12.

The following are simple properties of the notion of consequence:

1. If $\Gamma \subseteq \Delta$ and $\Gamma \vdash \mathscr{A}$, then $\Delta \vdash \mathscr{A}$.
2. $\Gamma \vdash \mathscr{A}$ if and only if there is a finite subset Δ of Γ such that $\Delta \vdash \mathscr{A}$.
3. If $\Delta \vdash \mathscr{A}$, and, for each $\mathscr{B}$ in Δ, $\Gamma \vdash \mathscr{B}$, then $\Gamma \vdash \mathscr{A}$.

Assertion 1 represents the fact that if $\mathscr{A}$ is provable from a set Γ of premisses, then, if we add still more premisses, $\mathscr{A}$ is still provable. Half of 2 follows from 1. The other half is obvious when we notice that any proof of $\mathscr{A}$ from Γ uses only a finite number of premisses from Γ. Proposition 3 is also quite simple: if $\mathscr{A}$ is provable from premisses in Δ, and each premiss in Δ is provable from the premisses in Γ, then $\mathscr{A}$ is provable from premisses in Γ.

We now introduce a formal axiomatic theory L for the propositional calculus.

1. The symbols of L are $\neg$, $\Rightarrow$, (,), and the letters A_i with positive integers i as subscripts: $A_1, A_2, A_3, \ldots$. The symbols $\neg$ and $\Rightarrow$ are called *primitive connectives*, and the letters A_i are called *statement letters*.
2. (a) All statement letters are wfs.
   (b) If $\mathscr{A}$ and $\mathscr{B}$ are wfs, so are $(\neg\mathscr{A})$ and $(\mathscr{A} \Rightarrow \mathscr{B})$.* (Thus, a wf of L is just a statement form built up from the statement letters A_i by means of the connectives $\neg$ and $\Rightarrow$.)
3. If $\mathscr{A}$, $\mathscr{B}$, and $\mathscr{C}$ are any wfs of L, then the following are axioms of L:
   (A1) $(\mathscr{A} \Rightarrow (\mathscr{B} \Rightarrow \mathscr{A}))$
   (A2) $((\mathscr{A} \Rightarrow (\mathscr{B} \Rightarrow \mathscr{C})) \Rightarrow ((\mathscr{A} \Rightarrow \mathscr{B}) \Rightarrow (\mathscr{A} \Rightarrow \mathscr{C})))$
   (A3) $(((\neg\mathscr{B}) \Rightarrow (\neg\mathscr{A})) \Rightarrow (((\neg\mathscr{B}) \Rightarrow \mathscr{A}) \Rightarrow \mathscr{B}))$
4. The only rule of inference of L is *modus ponens*: $\mathscr{B}$ is a direct consequence of $\mathscr{A}$ and $(\mathscr{A} \Rightarrow \mathscr{B})$. We shall abbreviate application of this rule by MP.

We shall use our conventions for eliminating parentheses.

Notice that the infinite set of axioms of L is given by means of three axiom schemas (A1)–(A3), with each schema standing for an infinite number of axioms. One can easily check for any given wf whether or not it is an axiom; therefore, L is axiomatic. It is our intention, in setting up the system L, to obtain as theorems precisely the class of all tautologies.

We introduce other connectives by definition:

(D1) $(\mathscr{A} \wedge \mathscr{B})$ for $\neg(\mathscr{A} \Rightarrow \neg\mathscr{B})$
(D2) $(\mathscr{A} \vee \mathscr{B})$ for $(\neg\mathscr{A}) \Rightarrow \mathscr{B}$
(D3) $(\mathscr{A} \Leftrightarrow \mathscr{B})$ for $(\mathscr{A} \Rightarrow \mathscr{B}) \wedge (\mathscr{B} \Rightarrow \mathscr{A})$

The meaning of (D1), for example, is that, for any wfs $\mathscr{A}$ and $\mathscr{B}$, "$(\mathscr{A} \wedge \mathscr{B})$" is an abbreviation for "$\neg(\mathscr{A} \Rightarrow \neg\mathscr{B})$".†

LEMMA 1.7 $\vdash_L \mathscr{A} \Rightarrow \mathscr{A}$ for all wfs $\mathscr{A}$.

*To be precise, we should add the so-called extremal clause: (c) An expression is a wf only if it can be shown to be a wf on the basis of clauses (a) and (b). This definition can be made rigorous using as a model the definition in the last footnote on page 12.

† When we say that "$(\mathscr{A} \wedge \mathscr{B})$" is an abbreviation for "$\neg(\mathscr{A} \Rightarrow \neg\mathscr{B})$", we mean that "$(\mathscr{A} \wedge \mathscr{B})$" is to be taken as another name in the English language (or in whatever language $\mathscr{L}$ we happen to be using to talk about the theory L) for the expression "$\neg(\mathscr{A} \Rightarrow \neg\mathscr{B})$".

Proof* We shall construct a proof in L of $\mathscr{A} \Rightarrow \mathscr{A}$.

1. $(\mathscr{A} \Rightarrow ((\mathscr{A} \Rightarrow \mathscr{A}) \Rightarrow \mathscr{A})) \Rightarrow ((\mathscr{A} \Rightarrow (\mathscr{A} \Rightarrow \mathscr{A})) \Rightarrow (\mathscr{A} \Rightarrow \mathscr{A}))$	Instance of axiom schema (A2)
2. $\mathscr{A} \Rightarrow ((\mathscr{A} \Rightarrow \mathscr{A}) \Rightarrow \mathscr{A})$	Axiom schema (A1)
3. $(\mathscr{A} \Rightarrow (\mathscr{A} \Rightarrow \mathscr{A})) \Rightarrow (\mathscr{A} \Rightarrow \mathscr{A})$	From 1 and 2 by MP
4. $\mathscr{A} \Rightarrow (\mathscr{A} \Rightarrow \mathscr{A})$	Axiom schema (A1)
5. $\mathscr{A} \Rightarrow \mathscr{A}$	From 3 and 4 by MP

Exercise

1.45 Prove:

(a) $\vdash_L (\neg\mathscr{A} \Rightarrow \mathscr{A}) \Rightarrow \mathscr{A}$
(b) $\mathscr{A} \Rightarrow \mathscr{B}, \mathscr{B} \Rightarrow \mathscr{C} \vdash_L \mathscr{A} \Rightarrow \mathscr{C}$
(c) $\mathscr{A} \Rightarrow (\mathscr{B} \Rightarrow \mathscr{C}) \vdash_L \mathscr{B} \Rightarrow (\mathscr{A} \Rightarrow \mathscr{C})$
(d) $\vdash_L (\neg\mathscr{B} \Rightarrow \neg\mathscr{A}) \Rightarrow (\mathscr{A} \Rightarrow \mathscr{B})$

In mathematical arguments, one often proves a statement $\mathscr{B}$ on the assumption of some other statement $\mathscr{A}$ and then concludes that "if $\mathscr{A}$, then $\mathscr{B}$" is true. This procedure is justified for the system L by the following theorem.

PROPOSITION 1.8 (DEDUCTION THEOREM)[†] If Γ is a set of wfs, and $\mathscr{A}$ and $\mathscr{B}$ are wfs, and $\Gamma, \mathscr{A} \vdash \mathscr{B}$, then $\Gamma \vdash \mathscr{A} \Rightarrow \mathscr{B}$. In particular, if $\mathscr{A} \vdash \mathscr{B}$, then $\vdash \mathscr{A} \Rightarrow \mathscr{B}$ (Herbrand, 1930).

Proof Let $\mathscr{B}_1, \ldots, \mathscr{B}_n$ be a proof of $\mathscr{B}$ from $\Gamma \cup \{\mathscr{A}\}$, where $\mathscr{B}_n = \mathscr{B}$. Let us prove, by induction on i, that $\Gamma \vdash \mathscr{A} \Rightarrow \mathscr{B}_i$ for $1 \leqslant i \leqslant n$. First of all, $\mathscr{B}_1$ must be either in Γ or an axiom of L or $\mathscr{A}$ itself. By axiom schema (A1), $\mathscr{B}_1 \Rightarrow (\mathscr{A} \Rightarrow \mathscr{B}_1)$ is an axiom. Hence, in the first two cases, by MP, $\Gamma \vdash \mathscr{A} \Rightarrow \mathscr{B}_1$. For the third case,

* The word "proof" is used in two distinct senses. First, it has a precise meaning defined above as a certain kind of finite sequence of wfs of L. However, in another sense, it also designates certain sequences of sentences of the English language (supplemented by various technical terms) that are supposed to serve as an argument justifying some assertion about the language L (or other formal theories). In general, the language we are studying (in this case L) is called the *object language*, while the language in which we formulate and prove results about the object language is called the *metalanguage*. The metalanguage might also be formalized and made the subject of study, which we would carry out in a metameta-language, and so on. However, we shall use the English language as our (unformalized) metalanguage, although, for a substantial part of this book, we use only a mathematically weak portion of the English language. The contrast between object language and metalanguage is also present in the study of a foreign language; for example, in a German class, German is the object language, while the metalanguage, the language we use, is English. The distinction between *proof* and *metaproof* (i.e., a proof in the meta-language) leads to a distinction between theorems of the object language and *metatheorems* of the metalanguage. To avoid confusion, we generally use "proposition" instead of "metatheorem". The word "metamathematics" refers to the study of logical and mathematical object languages; sometimes the word is restricted to those investigations that use what appear to the metamathematician to be constructive (or so-called finitary) methods.

† For the remainder of the chapter, unless something is said to the contrary, we shall omit the subscript L in $\vdash_L$. In addition, we shall use $\Gamma, \mathscr{A} \vdash \mathscr{B}$ to stand for $\Gamma \cup \{\mathscr{A}\} \vdash \mathscr{B}$. In general, we let $\Gamma, \mathscr{A}_1, \ldots, \mathscr{A}_n \vdash \mathscr{B}$ stand for $\Gamma \cup (\mathscr{A}_1, \ldots, \mathscr{A}_n) \vdash \mathscr{B}$.

when $\mathscr{B}_1$ is $\mathscr{A}$, we have $\vdash \mathscr{A} \Rightarrow \mathscr{B}_1$ by Lemma 1.7, and, therefore, $\Gamma \vdash \mathscr{A} \Rightarrow \mathscr{B}_1$. This takes care of the case $i = 1$. Assume now that $\Gamma \vdash \mathscr{A} \Rightarrow \mathscr{B}_k$ for all $k < i$. Either $\mathscr{B}_i$ is an axiom or $\mathscr{B}_i$ is in Γ, or $\mathscr{B}_i$ is $\mathscr{A}$, or $\mathscr{B}_i$ follows by modus ponens from some $\mathscr{B}_j$ and $\mathscr{B}_m$, where $j < i$, $m < i$, and $\mathscr{B}_m$ has the form $\mathscr{B}_j \Rightarrow \mathscr{B}_i$. In the first three cases, $\Gamma \vdash \mathscr{A} \Rightarrow \mathscr{B}_i$ as in the case $i = 1$ above. In the last case, we have, by inductive hypothesis, $\Gamma \vdash \mathscr{A} \Rightarrow \mathscr{B}_j$ and $\Gamma \vdash \mathscr{A} \Rightarrow (\mathscr{B}_j \Rightarrow \mathscr{B}_i)$. But, by axiom schema (A2), $\vdash (\mathscr{A} \Rightarrow (\mathscr{B}_j \Rightarrow \mathscr{B}_i)) \Rightarrow ((\mathscr{A} \Rightarrow \mathscr{B}_j) \Rightarrow (\mathscr{A} \Rightarrow \mathscr{B}_i))$. Hence, by MP, $\Gamma \vdash (\mathscr{A} \Rightarrow \mathscr{B}_j) \Rightarrow (\mathscr{A} \Rightarrow \mathscr{B}_i)$, and, again by MP, $\Gamma \vdash \mathscr{A} \Rightarrow \mathscr{B}_i$. Thus, the inductive proof is complete. The case $i = n$ is the desired result. [Notice that, given a deduction of $\mathscr{B}$ from Γ and $\mathscr{A}$, the proof just given enables us to construct a deduction of $\mathscr{A} \Rightarrow \mathscr{B}$ from Γ. Also note that only axiom schemas (A1) and (A2) are used in proving the Deduction Theorem.]

COROLLARY 1.9

(a) $\mathscr{A} \Rightarrow \mathscr{B}, \mathscr{B} \Rightarrow \mathscr{C} \vdash \mathscr{A} \Rightarrow \mathscr{C}$
(b) $\mathscr{A} \Rightarrow (\mathscr{B} \Rightarrow \mathscr{C}), \mathscr{B} \vdash \mathscr{A} \Rightarrow \mathscr{C}$

Proof For part (a):

1. $\mathscr{A} \Rightarrow \mathscr{B}$	Hyp (abbreviation for "hypothesis")
2. $\mathscr{B} \Rightarrow \mathscr{C}$	Hyp
3. $\mathscr{A}$	Hyp
4. $\mathscr{B}$	1, 3, MP
5. $\mathscr{C}$	2, 4, MP

Thus, $\mathscr{A} \Rightarrow \mathscr{B}, \mathscr{B} \Rightarrow \mathscr{C}, \mathscr{A} \vdash \mathscr{C}$. So, by the Deduction Theorem, $\mathscr{A} \Rightarrow \mathscr{B}, \mathscr{B} \Rightarrow \mathscr{C} \vdash \mathscr{A} \Rightarrow \mathscr{C}$.

To prove (b) use the Deduction Theorem.

LEMMA 1.10 For any wfs $\mathscr{A}$ and $\mathscr{B}$, the following are theorems of L:

(a) $\neg\neg\mathscr{B} \Rightarrow \mathscr{B}$
(b) $\mathscr{B} \Rightarrow \neg\neg\mathscr{B}$
(c) $\neg\mathscr{A} \Rightarrow (\mathscr{A} \Rightarrow \mathscr{B})$
(d) $(\neg\mathscr{B} \Rightarrow \neg\mathscr{A}) \Rightarrow (\mathscr{A} \Rightarrow \mathscr{B})$
(e) $(\mathscr{A} \Rightarrow \mathscr{B}) \Rightarrow (\neg\mathscr{B} \Rightarrow \neg\mathscr{A})$
(f) $\mathscr{A} \Rightarrow (\neg\mathscr{B} \Rightarrow \neg(\mathscr{A} \Rightarrow \mathscr{B}))$
(g) $(\mathscr{A} \Rightarrow \mathscr{B}) \Rightarrow ((\neg\mathscr{A} \Rightarrow \mathscr{B}) \Rightarrow \mathscr{B})$

Proof

(a) $\vdash \neg\neg\mathscr{B} \Rightarrow \mathscr{B}$

1. $(\neg\mathscr{B} \Rightarrow \neg\neg\mathscr{B}) \Rightarrow ((\neg\mathscr{B} \Rightarrow \neg\mathscr{B}) \Rightarrow \mathscr{B})$	Axiom schema (A3)
2. $\neg\mathscr{B} \Rightarrow \neg\mathscr{B}$	Lemma 1.7*
3. $(\neg\mathscr{B} \Rightarrow \neg\neg\mathscr{B}) \Rightarrow \mathscr{B}$	1, 2, Corollary 1.9(b)
4. $\neg\neg\mathscr{B} \Rightarrow (\neg\mathscr{B} \Rightarrow \neg\neg\mathscr{B})$	Axiom (A1)
5. $\neg\neg\mathscr{B} \Rightarrow \mathscr{B}$	3, 4, Corollary 1.9(a)

*Instead of writing the complete proof of $\neg\mathscr{B} \Rightarrow \neg\mathscr{B}$, we simply cite Lemma 1.7. In this way, we indicate how the proof of $\neg\neg\mathscr{B} \Rightarrow \mathscr{B}$ could be written if we wished to take the time and space to do so. This, of course, is nothing more than the ordinary application of previously proved theorems.

(b) $\vdash \mathscr{B} \Rightarrow \neg\neg\mathscr{B}$

1. $(\neg\neg\neg\mathscr{B} \Rightarrow \neg\mathscr{B}) \Rightarrow ((\neg\neg\neg\mathscr{B} \Rightarrow \mathscr{B}) \Rightarrow \neg\neg\mathscr{B})$	Axiom schema (A3)
2. $\neg\neg\neg\mathscr{B} \Rightarrow \neg\mathscr{B}$	Part (a)
3. $(\neg\neg\neg\mathscr{B} \Rightarrow \mathscr{B}) \Rightarrow \neg\neg\mathscr{B}$	1, 2, MP
4. $\mathscr{B} \Rightarrow (\neg\neg\neg\mathscr{B} \Rightarrow \mathscr{B})$	Axiom schema (A1)
5. $\mathscr{B} \Rightarrow \neg\neg\mathscr{B}$	3, 4, Corollary 1.9(a)

(c) $\vdash \neg\mathscr{A} \Rightarrow (\mathscr{A} \Rightarrow \mathscr{B})$

1. $\neg\mathscr{A}$	Hyp
2. $\mathscr{A}$	Hyp
3. $\mathscr{A} \Rightarrow (\neg\mathscr{B} \Rightarrow \mathscr{A})$	Axiom schema (A1)
4. $\neg\mathscr{A} \Rightarrow (\neg\mathscr{B} \Rightarrow \neg\mathscr{A})$	Axiom schema (A1)
5. $\neg\mathscr{B} \Rightarrow \mathscr{A}$	2, 3, MP
6. $\neg\mathscr{B} \Rightarrow \neg\mathscr{A}$	1, 4, MP
7. $(\neg\mathscr{B} \Rightarrow \neg\mathscr{A}) \Rightarrow ((\neg\mathscr{B} \Rightarrow \mathscr{A}) \Rightarrow \mathscr{B})$	Axiom schema (A3)
8. $(\neg\mathscr{B} \Rightarrow \mathscr{A}) \Rightarrow \mathscr{B}$	6, 7, MP
9. $\mathscr{B}$	5, 8, MP
10. $\neg\mathscr{A}, \mathscr{A} \vdash \mathscr{B}$	1–9
11. $\neg\mathscr{A} \vdash \mathscr{A} \Rightarrow \mathscr{B}$	10, Deduction Theorem
12. $\vdash \neg\mathscr{A} \Rightarrow (\mathscr{A} \Rightarrow \mathscr{B})$	11, Deduction Theorem

(d) $\vdash (\neg\mathscr{B} \Rightarrow \neg\mathscr{A}) \Rightarrow (\mathscr{A} \Rightarrow \mathscr{B})$

1. $\neg\mathscr{B} \Rightarrow \neg\mathscr{A}$	Hyp
2. $(\neg\mathscr{B} \Rightarrow \neg\mathscr{A}) \Rightarrow ((\neg\mathscr{B} \Rightarrow \mathscr{A}) \Rightarrow \mathscr{B})$	Axiom schema (A3)
3. $\mathscr{A} \Rightarrow (\neg\mathscr{B} \Rightarrow \mathscr{A})$	Axiom schema (A1)
4. $(\neg\mathscr{B} \Rightarrow \mathscr{A}) \Rightarrow \mathscr{B}$	1, 2, MP
5. $\mathscr{A} \Rightarrow \mathscr{B}$	3, 4, Corollary 1.9(a)
6. $\neg\mathscr{B} \Rightarrow \neg\mathscr{A} \vdash \mathscr{A} \Rightarrow \mathscr{B}$	1–5
7. $\vdash (\neg\mathscr{B} \Rightarrow \neg\mathscr{A}) \Rightarrow (\mathscr{A} \Rightarrow \mathscr{B})$	6, Deduction Theorem

(e) $\vdash (\mathscr{A} \Rightarrow \mathscr{B}) \Rightarrow (\neg\mathscr{B} \Rightarrow \neg\mathscr{A})$

1. $\mathscr{A} \Rightarrow \mathscr{B}$	Hyp
2. $\neg\neg\mathscr{A} \Rightarrow \mathscr{A}$	Part (a)
3. $\neg\neg\mathscr{A} \Rightarrow \mathscr{B}$	1, 2, Corollary 1.9(a)
4. $\mathscr{B} \Rightarrow \neg\neg\mathscr{B}$	Part (b)
5. $\neg\neg\mathscr{A} \Rightarrow \neg\neg\mathscr{B}$	3, 4, Corollary 1.9(a)
6. $(\neg\neg\mathscr{A} \Rightarrow \neg\neg\mathscr{B}) \Rightarrow (\neg\mathscr{B} \Rightarrow \neg\mathscr{A})$	Part (d)
7. $(\neg\mathscr{B} \Rightarrow \neg\mathscr{A})$	5, 6, MP
8. $\mathscr{A} \Rightarrow \mathscr{B} \vdash \neg\mathscr{B} \Rightarrow \neg\mathscr{A}$	1–7
9. $\vdash (\mathscr{A} \Rightarrow \mathscr{B}) \Rightarrow (\neg\mathscr{B} \Rightarrow \neg\mathscr{A})$	8, Deduction Theorem

(f) $\vdash \mathscr{A} \Rightarrow (\neg\mathscr{B} \Rightarrow \neg(\mathscr{A} \Rightarrow \mathscr{B}))$.

Clearly, $\mathscr{A}, \mathscr{A} \Rightarrow \mathscr{B} \vdash \mathscr{B}$ by MP. Hence, $\vdash \mathscr{A} \Rightarrow ((\mathscr{A} \Rightarrow \mathscr{B}) \Rightarrow \mathscr{B})$ by two uses of the Deduction Theorem. Now, $\vdash ((\mathscr{A} \Rightarrow \mathscr{B}) \Rightarrow \mathscr{B}) \Rightarrow (\neg\mathscr{B} \Rightarrow \neg(\mathscr{A} \Rightarrow \mathscr{B}))$ by part (e). Hence, by Corollary 1.9(a), $\vdash \mathscr{A} \Rightarrow (\neg\mathscr{B} \Rightarrow \neg(\mathscr{A} \Rightarrow \mathscr{B}))$.

(g) $\vdash (\mathscr{A} \Rightarrow \mathscr{B}) \Rightarrow ((\neg\mathscr{A} \Rightarrow \mathscr{B}) \Rightarrow \mathscr{B})$

1. $\mathscr{A} \Rightarrow \mathscr{B}$	Hyp
2. $\neg\mathscr{A} \Rightarrow \mathscr{B}$	Hyp
3. $(\mathscr{A} \Rightarrow \mathscr{B}) \Rightarrow (\neg\mathscr{B} \Rightarrow \neg\mathscr{A})$	Part (e)

4. $\neg\mathscr{B} \Rightarrow \neg\mathscr{A}$	1, 3, MP
5. $(\neg\mathscr{A} \Rightarrow \mathscr{B}) \Rightarrow (\neg\mathscr{B} \Rightarrow \neg\neg\mathscr{A})$	Part (e)
6. $\neg\mathscr{B} \Rightarrow \neg\neg\mathscr{A}$	2, 5, MP
7. $(\neg\mathscr{B} \Rightarrow \neg\neg\mathscr{A}) \Rightarrow ((\neg\mathscr{B} \Rightarrow \neg\mathscr{A}) \Rightarrow \mathscr{B})$	Axiom schema (A3)
8. $(\neg\mathscr{B} \Rightarrow \neg\mathscr{A}) \Rightarrow \mathscr{B}$	6, 7, MP
9. $\mathscr{B}$	4, 8, MP

Thus, $\mathscr{A} \Rightarrow \mathscr{B}, \neg\mathscr{A} \Rightarrow \mathscr{B} \vdash \mathscr{B}$. Two applications of the Deduction Theorem yield part (g).

Exercises

1.46 Show that the following are theorems of L.

(a) $\mathscr{A} \Rightarrow (\mathscr{A} \vee \mathscr{B})$

(b) $\mathscr{A} \Rightarrow (\mathscr{B} \vee \mathscr{A})$

(c) $\mathscr{B} \vee \mathscr{A} \Rightarrow \mathscr{A} \vee \mathscr{B}$

(d) $\mathscr{A} \wedge \mathscr{B} \Rightarrow \mathscr{A}$

(e) $\mathscr{A} \wedge \mathscr{B} \Rightarrow \mathscr{B}$

(f) $(\mathscr{A} \Rightarrow \mathscr{C}) \Rightarrow ((\mathscr{B} \Rightarrow \mathscr{C}) \Rightarrow (\mathscr{A} \vee \mathscr{B} \Rightarrow \mathscr{C}))$

(g) $((\mathscr{A} \Rightarrow \mathscr{B}) \Rightarrow \mathscr{A}) \Rightarrow \mathscr{A}$

(h) $\mathscr{A} \Rightarrow (\mathscr{B} \Rightarrow (\mathscr{A} \wedge \mathscr{B}))$

1.47 Exhibit a complete proof in L of Lemma 1.10(c). [*Hint*: Apply the procedure used in the proof of the Deduction Theorem to the demonstration given earlier of Lemma 1.10(c).] Greater fondness for the Deduction Theorem will result if the reader tries to prove Lemma 1.10 without using the Deduction Theorem.

It is our purpose to show that a wf of L is a theorem of L if and only if it is a tautology. Half of this is very easy.

PROPOSITION 1.11 Every theorem of L is a tautology.

Proof As an exercise, verify that all the axioms of L are tautologies. By Proposition 1.1, modus ponens leads from tautologies to other tautologies. Hence, every theorem of L is a tautology.

The following lemma is to be used in the proof that every tautology is a theorem of L.

LEMMA 1.12 Let $\mathscr{A}$ be a wf, and let $B_1, \ldots, B_k$ be the statement letters that occur in $\mathscr{A}$. For a given assignment of truth values to $B_1, \ldots, B_k$, let B_i' be B_i if B_i takes the value T; and let B_i' be $\neg B_i$ if B_i takes the value F. Let $\mathscr{A}'$ be $\mathscr{A}$ if $\mathscr{A}$ takes the value T under the assignment, and let $\mathscr{A}'$ be $\neg\mathscr{A}$ if $\mathscr{A}$ takes the value F. Then $B_1', \ldots, B_k' \vdash \mathscr{A}'$.

For example, let $\mathscr{A}$ be $\neg(\neg A_2 \Rightarrow A_5)$. Then for each row of the truth table

A_2	A_5	$\neg(\neg A_2 \Rightarrow A_5)$
T	T	F
F	T	F
T	F	F
F	F	T

Lemma 1.12 asserts a corresponding deducibility relation. For instance, corresponding to the third row there is $A_2, \neg A_5 \vdash \neg\neg(\neg A_2 \Rightarrow A_5)$, and to the fourth row, $\neg A_2, \neg A_5 \vdash \neg(\neg A_2 \Rightarrow A_5)$.

Proof The proof is by induction on the number n of occurrences of $\neg$ and $\Rightarrow$ in $\mathscr{A}$. (We assume $\mathscr{A}$ written without abbreviations.) If $n = 0$, $\mathscr{A}$ is just a statement letter B_1, and then the lemma reduces to $B_1 \vdash B_1$ and $\neg B_1 \vdash \neg B_1$. Assume now that the lemma holds for all $j < n$.

Case 1. $\mathscr{A}$ is $\neg\mathscr{B}$. Then $\mathscr{B}$ has fewer than n occurrences of $\neg$ and $\Rightarrow$.

Subcase 1a. Let $\mathscr{B}$ take the value T under the given truth value assignment. Then $\mathscr{A}$ takes the value F. So, $\mathscr{B}'$ is $\mathscr{B}$ and $\mathscr{A}'$ is $\neg\mathscr{A}$. By the inductive hypothesis applied to $\mathscr{B}$, $B'_1, \ldots, B'_k \vdash \mathscr{B}$. Then, by Lemma 1.10(b) and MP, $B'_1, \ldots, B'_k \vdash \neg\neg\mathscr{B}$. But, $\neg\neg\mathscr{B}$ is $\mathscr{A}'$.

Subcase 1b. Let $\mathscr{B}$ take the value F. Then $\mathscr{B}'$ is $\neg\mathscr{B}$ and $\mathscr{A}'$ is $\mathscr{A}$. By inductive hypothesis, $B'_1, \ldots, B'_k \vdash \neg\mathscr{B}$. But, $\neg\mathscr{B}$ is $\mathscr{A}'$.

Case 2. $\mathscr{A}$ is $\mathscr{B} \Rightarrow \mathscr{C}$. Then $\mathscr{B}$ and $\mathscr{C}$ have fewer occurrences of $\neg$ and $\Rightarrow$ than $\mathscr{A}$. So, by inductive hypothesis, $B'_1, \ldots, B'_k \vdash \mathscr{B}'$ and $B'_1, \ldots, B'_k \vdash \mathscr{C}'$.

Case 2a. $\mathscr{B}$ takes the value F. Hence, $\mathscr{A}$ takes the value T. Then $\mathscr{B}'$ is $\neg\mathscr{B}$ and $\mathscr{A}'$ is $\mathscr{A}$. So, $B'_1, \ldots, B'_k \vdash \neg\mathscr{B}$. By Lemma 1.10(c), $B'_1, \ldots, B'_k \vdash \mathscr{B} \Rightarrow \mathscr{C}$. But, $\mathscr{B} \Rightarrow \mathscr{C}$ is $\mathscr{A}'$.

Case 2b. $\mathscr{C}$ takes the value T. Hence $\mathscr{A}$ takes the value T. Then $\mathscr{C}'$ is $\mathscr{C}$ and $\mathscr{A}'$ is $\mathscr{A}$. Now, $B'_1, \ldots, B'_k \vdash \mathscr{C}$. Then, by axiom schema (A1), $B'_1, \ldots, B'_k \vdash \mathscr{B} \Rightarrow \mathscr{C}$. But, $\mathscr{B} \Rightarrow \mathscr{C}$ is $\mathscr{A}'$.

Case 2c. $\mathscr{B}$ takes the value T and $\mathscr{C}$ takes the value F. Then $\mathscr{A}$ takes the value F, $\mathscr{B}'$ is $\mathscr{B}$, $\mathscr{C}'$ is $\neg\mathscr{C}$, and $\mathscr{A}'$ is $\neg\mathscr{A}$. Now, $B'_1, \ldots, B'_k \vdash \mathscr{B}$ and $B'_1, \ldots, B'_k \vdash \neg\mathscr{C}$. So, by Lemma 1.10(f), $B'_1, \ldots, B'_k \vdash \neg(\mathscr{B} \Rightarrow \mathscr{C})$. But, $\neg(\mathscr{B} \Rightarrow \mathscr{C})$ is $\mathscr{A}'$.

PROPOSITION 1.13 (COMPLETENESS THEOREM) If a wf $\mathscr{A}$ of L is a tautology, then it is a theorem of L.

Proof (Kalmar, 1935) Assume $\mathscr{A}$ is a tautology, and let $B_1, \ldots, B_k$ be the statement letters in $\mathscr{A}$. For any truth value assignment to $B_1, \ldots, B_k$, we have, by Lemma 1.12, $B'_1, \ldots, B'_k \vdash \mathscr{A}$. ($\mathscr{A}'$ is $\mathscr{A}$ because $\mathscr{A}$ always takes the value T.) Hence, if B_k is given the value T, $B'_1, \ldots, B'_{k-1}, B_k \vdash \mathscr{A}$, and, if B_k is given the value F, $B'_1, \ldots, B'_{k-1}, \neg B_k \vdash \mathscr{A}$. So, by the Deduction Theorem, $B'_1, \ldots, B'_{k-1} \vdash B_k \Rightarrow \mathscr{A}$ and $B'_1, \ldots, B'_{k-1} \vdash \neg B_k \Rightarrow \mathscr{A}$. Then, by Lemma 1.10(g) and MP, $B'_1, \ldots, B'_{k-1} \vdash \mathscr{A}$. Similarly, B_{k-1} may be chosen to be T or F, and, again applying the Deduction Theorem, Lemma 1.10(g), and MP, we can eliminate B'_{k-1} just as we eliminated B'_k. After k such steps, we finally obtain $\vdash \mathscr{A}$.

Exercise

1.48 $B_1 \wedge B_2 \Rightarrow B_1$ is a tautology. By the method of the proof of Proposition 1.13, show that $\vdash B_1 \wedge B_2 \Rightarrow B_1$.

COROLLARY 1.14 If $\mathscr{B}$ is an expression involving the signs $\neg$, $\Rightarrow$, $\wedge$, $\vee$, and $\Leftrightarrow$ that is an abbreviation for a wf $\mathscr{A}$ of L, then $\mathscr{B}$ is a tautology if and only if $\mathscr{A}$ is a theorem of L.

Proof In definitions (D1)–(D3), the abbreviating formulas replace wfs to which they are logically equivalent. Hence, by Proposition 1.3, $\mathscr{A}$ and $\mathscr{B}$ are logically equivalent, and $\mathscr{B}$ is a tautology if and only if $\mathscr{A}$ is. The corollary now follows from Propositions 1.11 and 1.13.

COROLLARY 1.15 The system L is consistent; that is, there is no wf $\mathscr{A}$ such that both $\mathscr{A}$ and $\neg\mathscr{A}$ are theorems of L.

Proof By Proposition 1.11, every theorem of L is a tautology. The negation of a tautology cannot be a tautology, and, therefore, it is impossible for both $\mathscr{A}$ and $\neg\mathscr{A}$ to be theorems of L.

Notice that L is consistent if and only if not all wfs of L are theorems. Clearly, if L is consistent, then there are wfs that are not theorems (e.g., the negations of theorems). On the other hand, by Lemma 1.10(c), $\vdash_L \neg\mathscr{A} \Rightarrow (\mathscr{A} \Rightarrow \mathscr{B})$, and so, if L were inconsistent—that is, if some wf $\mathscr{A}$ and its negation $\neg\mathscr{A}$ were provable—then, by MP, any wf $\mathscr{B}$ would be provable. [This equivalence holds for any theory that has modus ponens as a rule of inference and in which Lemma 1.10(c) is provable.] A theory in which not all wfs are theorems is said to be *absolutely consistent*, and this definition is applicable even to theories that do not contain a negation sign.

Exercise

1.49 Let $\mathscr{A}$ be a statement form that is not a tautology. Let L^+ be the formal theory obtained from L by adding as new axioms all wfs obtainable from $\mathscr{A}$ by substituting arbitrary statement forms for the statement letters in $\mathscr{A}$, with the same forms being substituted for all occurrences of a statement letter. Show that L^+ is inconsistent.

5. INDEPENDENCE. MANY-VALUED LOGICS

Given a theory, a subset Y of the axioms is said to be *independent* if some wf in Y cannot be proved by means of the rules of inference from the set of those axioms not in Y.

PROPOSITION 1.16 Each of axiom schemas (A1)–(A3) is independent.

Proof To prove the independence of axiom schema (A1), consider the following tables:

A	$\neg A$
0	1
1	1
2	0

A	B	$A \Rightarrow B$
0	0	0
1	0	2
2	0	0
0	1	2
1	1	2
2	1	0
0	2	2
1	2	0
2	2	0

Given an assignment of the values 0, 1, and 2 to the statement letters of a wf $\mathscr{A}$, these tables determine a corresponding value of $\mathscr{A}$. If $\mathscr{A}$ always takes the value 0, $\mathscr{A}$ is called *select*. Now, modus ponens preserves selectness, since it is obvious that, if $\mathscr{A}$ and $\mathscr{A} \Rightarrow \mathscr{B}$ are select, so is $\mathscr{B}$. Verify also that all instances of axioms (A2) and (A3) are select. Hence, any wf derivable from axioms (A2) and (A3) by modus ponens is select. However, $A_1 \Rightarrow (A_2 \Rightarrow A_1)$, which is an instance of axiom (A1), is not select, since it takes the value 2 when A_1 is 1 and A_2 is 2.

To prove the independence of axiom schema (A2), consider the following tables:

A	$\neg A$
0	1
1	0
2	1

A	B	$A \Rightarrow B$
0	0	0
1	0	0
2	0	0
0	1	2
1	1	2
2	1	0
0	2	1
1	2	0
2	2	0

Let us call a wf that always takes the value 0 according to these tables *grotesque*. Modus ponens preserves grotesqueness, and it is easy to verify that all instances of (A1) and (A3) are grotesque. However, the instance $(A_1 \Rightarrow (A_2 \Rightarrow A_3)) \Rightarrow ((A_1 \Rightarrow A_2) \Rightarrow (A_1 \Rightarrow A_3))$ of (A2) takes the value 2 when A_1 is 0, A_2 is 0, and A_3 is 1 and, therefore, is not grotesque.

The following proves the independence of (A3): If $\mathscr{A}$ is any wf, let $h(\mathscr{A})$ be the wf obtained by erasing all negation signs in $\mathscr{A}$. For each instance $\mathscr{A}$ of axiom schemas (A1) and (A2), $h(\mathscr{A})$ is a tautology. Also, modus ponens preserves the property of a wf $\mathscr{A}$ that $h(\mathscr{A})$ is a tautology; for, if $h(\mathscr{A} \Rightarrow \mathscr{B})$ and $h(\mathscr{A})$ are tautologies, then $h(\mathscr{B})$ is a tautology. [Just note that $h(\mathscr{A} \Rightarrow \mathscr{B})$ is $h(\mathscr{A}) \Rightarrow h(\mathscr{B})$.] Hence, every wf $\mathscr{A}$ derivable from (A1) and (A2) by modus ponens has the property that $h(\mathscr{A})$ is a tautology. But, $h((\neg A_1 \Rightarrow \neg A_1) \Rightarrow ((\neg A_1 \Rightarrow A_1) \Rightarrow A_1))$ is $(A_1 \Rightarrow A_1) \Rightarrow ((A_1 \Rightarrow A_1) \Rightarrow A_1)$, which is not a tautology. Hence, $(\neg A_1 \Rightarrow \neg A_1) \Rightarrow ((\neg A_1 \Rightarrow A_1) \Rightarrow A_1)$, an instance of axiom schema (A3), is not derivable from (A1) and (A2) by modus ponens.

The idea used in the proof of independence of axiom schemas (A1) and (A2) may be generalized to the notion of a many-valued logic. Call the numbers 0, 1, 2, $\ldots$, n *truth values*, and let $0 \leqslant m < n$. The numbers $0, 1, \ldots, m$ are called *designated values*. Take a finite number of "truth tables" representing functions from sets of the form $\{0, 1, \ldots, n\}^k$ into $\{0, 1, \ldots, n\}$. For each truth table, introduce a sign, called the corresponding *connective*. Using these connectives and statement letters, we may construct "statement forms," and every such statement form containing j distinct letters determines a "truth function" from $\{0, 1, \ldots, n\}^j$ into $\{0, 1, \ldots, n\}$. A statement form whose corresponding truth function takes only designated values is said to be *exceptional*. The numbers m and n and the basic truth tables are said to define a (finite) *many-valued logic* M. An axiomatic theory involving statement letters and the connectives of M is said to be *suitable* for M if and only if the theorems of the theory coincide with the exceptional statement forms of M. All these notions obviously can be generalized to the case of an infinite number of truth values. If $n = 1$ and $m = 0$ and the truth tables are those given for $\neg$ and $\Rightarrow$ in Section 1, the corresponding two-valued logic is that studied in this chapter. The exceptional wfs in this case were called tautologies. The system L is suitable for this logic, as proved in Propositions 1.11 and 1.13. In the proofs of the independence of axiom schemas (A1) and (A2), two three-valued logics were used.

Exercises

1.50 Prove the independence of axiom schema (A3) by constructing appropriate "truth tables" for $\neg$ and $\Rightarrow$.

1.51 (McKinsey & Tarski, 1948) Consider the axiom system P in which there is exactly one binary connective $*$, the only rule of inference is modus ponens (that is, $\mathscr{B}$ follows from $\mathscr{A}$ and $\mathscr{A} * \mathscr{B}$), and the axioms are all wfs of the form $\mathscr{A} * \mathscr{A}$. Show that P is not suitable for any (finite) many-valued logic.

1.52 For any (finite) many-valued logic M, prove that there is an axiomatic theory suitable for M.

Further information about many-valued logics can be gained from the monograph of Rosser and Turquette (1952) and from Rescher (1969).

6. OTHER AXIOMATIZATIONS

Although the axiom system L is quite simple, there are many other systems that would do as well. We can use, instead of $\neg$ and $\Rightarrow$, any collection of primitive connectives as long as these are adequate for the definition of all other truth-functional connectives.

Examples

L_1: $\vee$ and $\neg$ are the primitive connectives. We use $\mathscr{A} \Rightarrow \mathscr{B}$ as an abbreviation for $\neg\mathscr{A} \vee \mathscr{B}$. We have four axiom schemas: (1) $\mathscr{A} \vee \mathscr{A} \Rightarrow \mathscr{A}$, (2) $\mathscr{A} \Rightarrow \mathscr{A} \vee \mathscr{B}$, (3) $\mathscr{A} \vee \mathscr{B} \Rightarrow \mathscr{B} \vee \mathscr{A}$, and (4) $(\mathscr{B} \Rightarrow \mathscr{C}) \Rightarrow (\mathscr{A} \vee \mathscr{B} \Rightarrow \mathscr{A} \vee \mathscr{C})$. The only rule of inference is modus ponens. This system is developed in Hilbert and Ackermann (1950).

L_2: $\wedge$ and $\neg$ are the primitive connectives. $\mathscr{A} \Rightarrow \mathscr{B}$ is an abbreviation for $\neg(\mathscr{A} \wedge \neg\mathscr{B})$. There are three axiom schemas: (1) $\mathscr{A} \Rightarrow (\mathscr{A} \wedge \mathscr{A})$, (2) $\mathscr{A} \wedge \mathscr{B} \Rightarrow \mathscr{A}$, and (3) $(\mathscr{A} \Rightarrow \mathscr{B}) \Rightarrow (\neg(\mathscr{B} \wedge \mathscr{C}) \Rightarrow \neg(\mathscr{C} \wedge \mathscr{A}))$. Modus ponens is the only rule. Consult Rosser (1953) for a detailed study.

L_3: This is just like our original system L except that, instead of the axiom schemas (A1)–(A3), we have three specific axioms: (1) $A_1 \Rightarrow (A_2 \Rightarrow A_1)$, (2) $(A_1 \Rightarrow (A_2 \Rightarrow A_3)) \Rightarrow ((A_1 \Rightarrow A_2) \Rightarrow (A_1 \Rightarrow A_3))$, and (3) $(\neg A_2 \Rightarrow \neg A_1) \Rightarrow ((\neg A_2 \Rightarrow A_1) \Rightarrow A_2)$. In addition to modus ponens, we have a substitution rule: we may substitute any wf for all occurrences of a statement letter in a given wf.

L_4: The primitive connectives are $\Rightarrow$, $\wedge$, $\vee$, and $\neg$. Modus ponens is the only rule, and we have ten axiom schemas:

(1) $\mathscr{A} \Rightarrow (\mathscr{B} \Rightarrow \mathscr{A})$

(2) $(\mathscr{A} \Rightarrow (\mathscr{B} \Rightarrow \mathscr{C})) \Rightarrow ((\mathscr{A} \Rightarrow \mathscr{B}) \Rightarrow (\mathscr{A} \Rightarrow \mathscr{C}))$

(3) $\mathscr{A} \wedge \mathscr{B} \Rightarrow \mathscr{A}$

(4) $\mathscr{A} \wedge \mathscr{B} \Rightarrow \mathscr{B}$
(5) $\mathscr{A} \Rightarrow (\mathscr{B} \Rightarrow (\mathscr{A} \wedge \mathscr{B}))$
(6) $\mathscr{A} \Rightarrow (\mathscr{A} \vee \mathscr{B})$
(7) $\mathscr{B} \Rightarrow (\mathscr{A} \vee \mathscr{B})$
(8) $(\mathscr{A} \Rightarrow \mathscr{C}) \Rightarrow ((\mathscr{B} \Rightarrow \mathscr{C}) \Rightarrow (\mathscr{A} \vee \mathscr{B} \Rightarrow \mathscr{C}))$
(9) $(\mathscr{A} \Rightarrow \mathscr{B}) \Rightarrow ((\mathscr{A} \Rightarrow \neg \mathscr{B}) \Rightarrow \neg \mathscr{A})$
(10) $\neg \neg \mathscr{A} \Rightarrow \mathscr{A}$

We define, as usual, $\mathscr{A} \Leftrightarrow \mathscr{B}$ to be $(\mathscr{A} \Rightarrow \mathscr{B}) \wedge (\mathscr{B} \Rightarrow \mathscr{A})$. This system is discussed in Kleene (1952).

Axiomatizations can be found for the propositional calculus that contain only one axiom schema. For example, if $\neg$ and $\Rightarrow$ are the primitive connectives and modus ponens the only rule of inference, the axiom schema

$$[(((\mathscr{A} \Rightarrow \mathscr{B}) \Rightarrow (\neg \mathscr{C} \Rightarrow \neg \mathscr{D})) \Rightarrow \mathscr{C}) \Rightarrow \mathscr{E}] \Rightarrow [(\mathscr{E} \Rightarrow \mathscr{A}) \Rightarrow (\mathscr{D} \Rightarrow \mathscr{A})]$$

is sufficient (Meredith, 1953). Another single-axiom formulation, due to Nicod (1917), uses only alternative denial $|$. Its rule of inference is: $\mathscr{C}$ follows from $\mathscr{A}|(\mathscr{B}|\mathscr{C})$ and $\mathscr{A}$, and its axiom schema is

$$(\mathscr{A}|(\mathscr{B}|\mathscr{C}))|\{[\mathscr{D}|(\mathscr{D}|\mathscr{D})]|[(\mathscr{E}|\mathscr{B})|((\mathscr{A}|\mathscr{E})|(\mathscr{A}|\mathscr{E}))]\}$$

Further information, including historical background, may be found in Church (1956) and in a paper by Lukasiewicz and Tarski in Tarski (1956, IV).

Exercises

1.53 (Hilbert & Ackermann, 1950) Prove the following results about the theory L_1:

(a) $\mathscr{A} \Rightarrow \mathscr{B} \vdash_{L_1} \mathscr{C} \vee \mathscr{A} \Rightarrow \mathscr{C} \vee \mathscr{B}$
(b) $\vdash_{L_1} (\mathscr{A} \Rightarrow \mathscr{B}) \Rightarrow ((\mathscr{C} \Rightarrow \mathscr{A}) \Rightarrow (\mathscr{C} \Rightarrow \mathscr{B}))$
(c) $\mathscr{C} \Rightarrow \mathscr{A}, \mathscr{A} \Rightarrow \mathscr{B} \vdash_{L_1} \mathscr{C} \Rightarrow \mathscr{B}$
(d) $\vdash_{L_1} \mathscr{A} \Rightarrow \mathscr{A}$ (i.e., $\vdash_{L_1} \neg \mathscr{A} \vee \mathscr{A}$)
(e) $\vdash_{L_1} \mathscr{A} \vee \neg \mathscr{A}$
(f) $\vdash_{L_1} \mathscr{A} \Rightarrow \neg \neg \mathscr{A}$
(g) $\vdash_{L_1} \neg \mathscr{B} \Rightarrow (\mathscr{B} \Rightarrow \mathscr{C})$
(h) $\vdash_{L_1} \mathscr{A} \vee (\mathscr{B} \vee \mathscr{C}) \Rightarrow ((\mathscr{B} \vee (\mathscr{A} \vee \mathscr{C})) \vee \mathscr{A})$
(i) $\vdash_{L_1} (\mathscr{B} \vee (\mathscr{A} \vee \mathscr{C})) \vee \mathscr{A} \Rightarrow \mathscr{B} \vee (\mathscr{A} \vee \mathscr{C})$
(j) $\vdash_{L_1} \mathscr{A} \vee (\mathscr{B} \vee \mathscr{C}) \Rightarrow \mathscr{B} \vee (\mathscr{A} \vee \mathscr{C})$
(k) $\vdash_{L_1} (\mathscr{A} \Rightarrow (\mathscr{B} \Rightarrow \mathscr{C})) \Rightarrow (\mathscr{B} \Rightarrow (\mathscr{A} \Rightarrow \mathscr{C}))$
(l) $\vdash_{L_1} (\mathscr{C} \Rightarrow \mathscr{A}) \Rightarrow ((\mathscr{A} \Rightarrow \mathscr{B}) \Rightarrow (\mathscr{C} \Rightarrow \mathscr{B}))$
(m) $\mathscr{A} \Rightarrow (\mathscr{B} \Rightarrow \mathscr{C}), \mathscr{A} \Rightarrow \mathscr{B} \vdash_{L_1} \mathscr{A} \Rightarrow (\mathscr{A} \Rightarrow \mathscr{C})$
(n) $\mathscr{A} \Rightarrow (\mathscr{B} \Rightarrow \mathscr{C}), \mathscr{A} \Rightarrow \mathscr{B} \vdash_{L_1} \mathscr{A} \Rightarrow \mathscr{C}$
(o) If $\Gamma, \mathscr{A} \vdash_{L_1} \mathscr{B}$, then $\Gamma \vdash_{L_1} \mathscr{A} \Rightarrow \mathscr{B}$ (Deduction Theorem)
(p) $\mathscr{B} \Rightarrow \mathscr{A}, \neg \mathscr{B} \Rightarrow \mathscr{A} \vdash_{L_1} \mathscr{A}$
(q) $\vdash_{L_1} \mathscr{A}$ if and only if $\mathscr{A}$ is a tautology

1.54 (Rosser, 1953) Prove the following facts about the theory L_2:

(a) $\mathscr{A} \Rightarrow \mathscr{B}, \mathscr{B} \Rightarrow \mathscr{C} \vdash_{L_2} \neg(\neg \mathscr{C} \wedge \mathscr{A})$
(b) $\vdash_{L_2} \neg(\neg \mathscr{A} \wedge \mathscr{A})$
(c) $\vdash_{L_2} \neg \neg \mathscr{A} \Rightarrow \mathscr{A}$
(d) $\vdash_{L_2} \neg(\mathscr{A} \wedge \mathscr{B}) \Rightarrow (\mathscr{B} \Rightarrow \neg \mathscr{A})$
(e) $\vdash_{L_2} \mathscr{A} \Rightarrow \neg \neg \mathscr{A}$
(f) $\vdash_{L_2} (\mathscr{A} \Rightarrow \mathscr{B}) \Rightarrow (\neg \mathscr{B} \Rightarrow \neg \mathscr{A})$

(g) $\neg\mathscr{A} \Rightarrow \neg\mathscr{B} \vdash_{L_2} \mathscr{B} \Rightarrow \mathscr{A}$
(h) $\mathscr{A} \Rightarrow \mathscr{B} \vdash_{L_2} \mathscr{C} \wedge \mathscr{A} \Rightarrow \mathscr{B} \wedge \mathscr{C}$
(i) $\mathscr{A} \Rightarrow \mathscr{B}, \mathscr{B} \Rightarrow \mathscr{C}, \mathscr{C} \Rightarrow \mathscr{D} \vdash_{L_2} \mathscr{A} \Rightarrow \mathscr{D}$
(j) $\vdash_{L_2} \mathscr{A} \Rightarrow \mathscr{A}$
(k) $\vdash_{L_2} \mathscr{A} \wedge \mathscr{B} \Rightarrow \mathscr{B} \wedge \mathscr{A}$
(l) $\mathscr{A} \Rightarrow \mathscr{B}, \mathscr{B} \Rightarrow \mathscr{C} \vdash_{L_2} \mathscr{A} \Rightarrow \mathscr{C}$
(m) $\mathscr{A} \Rightarrow \mathscr{B}, \mathscr{C} \Rightarrow \mathscr{D} \vdash_{L_2} \mathscr{A} \wedge \mathscr{C} \Rightarrow \mathscr{B} \wedge \mathscr{D}$
(n) $\mathscr{B} \Rightarrow \mathscr{C} \vdash_{L_2} \mathscr{A} \wedge \mathscr{B} \Rightarrow \mathscr{A} \wedge \mathscr{C}$
(o) $\vdash_{L_2} (\mathscr{A} \Rightarrow (\mathscr{B} \Rightarrow \mathscr{C})) \Rightarrow ((\mathscr{A} \wedge \mathscr{B}) \Rightarrow \mathscr{C})$
(p) $\vdash_{L_2} ((\mathscr{A} \wedge \mathscr{B}) \Rightarrow \mathscr{C}) \Rightarrow (\mathscr{A} \Rightarrow (\mathscr{B} \Rightarrow \mathscr{C}))$
(q) $\mathscr{A} \Rightarrow \mathscr{B}, \mathscr{A} \Rightarrow (\mathscr{B} \Rightarrow \mathscr{C}) \vdash_{L_2} \mathscr{A} \Rightarrow \mathscr{C}$
(r) $\vdash_{L_2} \mathscr{A} \Rightarrow (\mathscr{B} \Rightarrow \mathscr{A} \wedge \mathscr{B})$
(s) $\vdash_{L_2} \mathscr{A} \Rightarrow (\mathscr{B} \Rightarrow \mathscr{A})$
(t) If $\Gamma, \mathscr{A} \vdash_{L_2} \mathscr{B}$, then $\Gamma \vdash_{L_2} \mathscr{A} \Rightarrow \mathscr{B}$ (Deduction Theorem)
(u) $\vdash_{L_2} (\neg\mathscr{A} \Rightarrow \mathscr{A}) \Rightarrow \mathscr{A}$
(v) $\mathscr{A} \Rightarrow \mathscr{B}, \neg\mathscr{A} \Rightarrow \mathscr{B} \vdash_{L_2} \mathscr{B}$
(w) $\vdash_{L_2} \mathscr{A}$ if and only if $\mathscr{A}$ is a tautology

1.55 Show that the theory L_3 has the same theorems as the theory L.

1.56 (Kleene, 1952) Derive the following facts about the theory L_4:
(a) $\vdash_{L_4} \mathscr{A} \Rightarrow \mathscr{A}$
(b) If $\Gamma, \mathscr{A} \vdash_{L_4} \mathscr{B}$, then $\Gamma \vdash_{L_4} \mathscr{A} \Rightarrow \mathscr{B}$ (Deduction Theorem)
(c) $\mathscr{A} \Rightarrow \mathscr{B}, \mathscr{B} \Rightarrow \mathscr{C} \vdash_{L_4} \mathscr{A} \Rightarrow \mathscr{C}$
(d) $\vdash_{L_4} (\mathscr{A} \Rightarrow \mathscr{B}) \Rightarrow (\neg\mathscr{B} \Rightarrow \neg\mathscr{A})$
(e) $\mathscr{B}, \neg\mathscr{B} \vdash_{L_4} \mathscr{C}$
(f) $\vdash_{L_4} \mathscr{B} \Rightarrow \neg\neg\mathscr{B}$
(g) $\vdash_{L_4} \neg\mathscr{B} \Rightarrow (\mathscr{B} \Rightarrow \mathscr{C})$
(h) $\vdash_{L_4} \mathscr{B} \Rightarrow (\neg\mathscr{C} \Rightarrow \neg(\mathscr{B} \Rightarrow \mathscr{C}))$
(i) $\vdash_{L_4} \neg\mathscr{B} \Rightarrow (\neg\mathscr{C} \Rightarrow \neg(\mathscr{B} \vee \mathscr{C}))$
(j) $\vdash_{L_4} (\neg\mathscr{B} \Rightarrow \mathscr{A}) \Rightarrow ((\mathscr{B} \Rightarrow \mathscr{A}) \Rightarrow \mathscr{A})$
(k) $\vdash_{L_4} \mathscr{A}$ if and only if $\mathscr{A}$ is a tautology

1.57[D] Consider the following axiomatization of the propositional calculus $\mathscr{L}$ (due to Lukasiewicz). $\mathscr{L}$ has the same wfs as our system L. Its only rule of inference is modus ponens. Its axiom schemas are:
(a) $(\neg\mathscr{A} \Rightarrow \mathscr{A}) \Rightarrow \mathscr{A}$,
(b) $\mathscr{A} \Rightarrow (\neg\mathscr{A} \Rightarrow \mathscr{B})$, and
(c) $(\mathscr{A} \Rightarrow \mathscr{B}) \Rightarrow ((\mathscr{B} \Rightarrow \mathscr{C}) \Rightarrow (\mathscr{A} \Rightarrow \mathscr{C}))$.

Prove that a wf $\mathscr{A}$ of $\mathscr{L}$ is provable in $\mathscr{L}$ if and only if $\mathscr{A}$ is a tautology. [*Hint:* Show that L and $\mathscr{L}$ have the same theorems. However, remember that none of the results proved about L (such as Propositions 1.7–1.12) automatically carries over to $\mathscr{L}$. In particular, the Deduction Theorem is not available until it is proved for $\mathscr{L}$.]

1.58 Show that axiom schema (A3) of L can be replaced by the schema $(\neg\mathscr{A} \Rightarrow \neg\mathscr{B}) \Rightarrow (\mathscr{B} \Rightarrow \mathscr{A})$ without altering the class of theorems.

1.59 If, in L_4, axiom schema (10) is replaced by the schema (10)′: $\neg\mathscr{A} \Rightarrow (\mathscr{A} \Rightarrow \mathscr{B})$, then the new system L_I is called the *intuitionistic* propositional calculus.* Prove the following results about L_I.

* The principal origin of intuitionistic logic was L. E. J. Brouwer's belief that classical logic is wrong. According to Brouwer, $\mathscr{A} \vee \mathscr{B}$ is proved only when a proof of $\mathscr{A}$ or a proof of $\mathscr{B}$ has been found. As a consequence, various tautologies, such as $\mathscr{A} \vee \neg\mathscr{A}$, are not generally acceptable. For further information, consult Brouwer (1976), Heyting (1956), Kleene (1952), Troelstra (1969), and Dummett (1977). Jaśkowski (1936) showed that L_I is suitable for a many-valued logic with denumerably many values.

(a) Consider an $(n + 1)$-valued logic with these connectives: $\neg\mathscr{A}$ is 0 when $\mathscr{A}$ is n, and otherwise it is n; $\mathscr{A} \wedge \mathscr{B}$ has the maximum of the values of $\mathscr{A}$ and $\mathscr{B}$, whereas $\mathscr{A} \vee \mathscr{B}$ has the minimum of these values; and $\mathscr{A} \Rightarrow \mathscr{B}$ is 0 if $\mathscr{A}$ has a value not less than that of $\mathscr{B}$, and otherwise it has the same value as $\mathscr{B}$. If we take 0 as the only designated value, all theorems of L_I are exceptional.

(b) $A_1 \vee \neg A_1$ and $\neg\neg A_1 \Rightarrow A_1$ are not theorems of L_I.

(c) For any m, the wf

$$(A_1 \Leftrightarrow A_2) \vee \cdots \vee (A_1 \Leftrightarrow A_m) \vee (A_2 \Leftrightarrow A_3) \vee \cdots \vee (A_2 \Leftrightarrow A_m) \vee \cdots \vee (A_{m-1} \Leftrightarrow A_m)$$

is not a theorem of L_I.

(d) (Gödel, 1933) L_I is not suitable for any finite many-valued logic.

(e) (i) If $\Gamma, \mathscr{A} \vdash_{L_I} \mathscr{B}$, then $\Gamma \vdash_{L_I} \mathscr{A} \Rightarrow \mathscr{B}$ (Deduction Theorem)

(ii) $\mathscr{A} \Rightarrow \mathscr{B}, \mathscr{B} \Rightarrow \mathscr{C} \vdash_{L_I} \mathscr{A} \Rightarrow \mathscr{C}$

(iii) $\vdash_{L_I} \mathscr{A} \Rightarrow \neg\neg\mathscr{A}$

(iv) $\vdash_{L_I} (\mathscr{A} \Rightarrow \mathscr{B}) \Rightarrow (\neg\mathscr{B} \Rightarrow \neg\mathscr{A})$

(v) $\vdash_{L_I} \mathscr{A} \Rightarrow (\neg\mathscr{A} \Rightarrow \mathscr{B})$

(vi) $\vdash_{L_I} \neg\neg(\neg\neg\mathscr{A} \Rightarrow \mathscr{A})$

(vii) $\neg\neg(\mathscr{A} \Rightarrow \mathscr{B}), \neg\neg\mathscr{A} \vdash_{L_I} \neg\neg\mathscr{B}$

(viii) $\vdash_{L_I} \neg\neg\neg\mathscr{A} \Rightarrow \neg\mathscr{A}$

(f)[D] $\vdash_{L_I} \neg\neg\mathscr{A}$ if and only if $\mathscr{A}$ is a tautology

(g) $\vdash_{L_I} \neg\mathscr{A}$ if and only if $\neg\mathscr{A}$ is a tautology

(h)[D] If $\mathscr{A}$ has $\wedge$ and $\neg$ as its only connectives, $\vdash_{L_I} \mathscr{A}$ if and only if $\mathscr{A}$ is a tautology.

1.60[A] Let $\mathscr{A}$ and $\mathscr{B}$ be in the relation R if and only if $\vdash_L \mathscr{A} \Leftrightarrow \mathscr{B}$. Show that R is an equivalence relation. Given equivalence classes $[\mathscr{A}]$ and $[\mathscr{B}]$, let $[\mathscr{A}] \cup [\mathscr{B}] = [\mathscr{A} \vee \mathscr{B}]$, $[\mathscr{A}] \cap [\mathscr{B}] = [\mathscr{A} \wedge \mathscr{B}]$, and $\overline{[\mathscr{A}]} = [\neg\mathscr{A}]$. Show that the equivalence classes under R form a Boolean algebra with respect to $\cup$, $\cap$, and $\overline{}$, called the *Lindenbaum algebra* L^* determined by L. The element 0 of L^* is the equivalence class consisting of all contradictions (i.e., negations of tautologies). The unit element 1 of L^* is the equivalence class consisting of all tautologies. Notice that $\vdash_L \mathscr{A} \Rightarrow \mathscr{B}$ if and only if $[\mathscr{A}] \leqslant [\mathscr{B}]$ in L^*, and that $\vdash_L \mathscr{A} \Leftrightarrow \mathscr{B}$ if and only if $[\mathscr{A}] = [\mathscr{B}]$. Show that a Boolean function f (built up from variables, 0, and 1, using $\cup$, $\cap$, and $\overline{}$) is equal to the constant function 1 in all Boolean algebras if and only if $\vdash_L f^{\#}$, where $f^{\#}$ is obtained from f by changing $\cup$, $\cap$, $\overline{}$, 0, and 1 to $\vee$, $\wedge$, $\neg$, $A_1 \wedge \neg A_1$, and $A_1 \vee \neg A_1$, respectively.

CHAPTER TWO

Quantification Theory

1. QUANTIFIERS

There are various kinds of logical inference that obviously cannot be justified on the basis of the propositional calculus; for example:

1. Any friend of Martin is a friend of John.
   Peter is not John's friend.
   Hence, Peter is not Martin's friend.
2. All human beings are rational.
   Some animals are human beings.
   Hence, some animals are rational.
3. The successor of an even integer is odd.
   2 is an even integer.
   Hence, the successor of 2 is odd.

The correctness of these inferences rests not only upon the meanings of the truth-functional connectives but also upon the meaning of such expressions as "any", "all", and "some".

In order to make the structure of complex sentences more transparent, it is convenient to introduce special notation to represent frequently occurring expressions. If $P(x)$ asserts that x has the property P, then $(\forall x)P(x)$ means that property P holds for all x or, in other words, that everything has the property P. On the other hand, $(\exists x)P(x)$ means that some x has the property P—that is, that there is at least one object having the property P. In $(\forall x)P(x)$, "$(\forall x)$" is called a *universal quantifier*; in $(\exists x)P(x)$, "$(\exists x)$" is called an *existential quantifier*. The study of quantifiers and related concepts is the principal subject of this chapter—hence, the title "Quantification Theory".

Examples

1′. Inference 1 above can be represented symbolically:

$$\begin{array}{l} (\forall x)(F(x,m) \Rightarrow F(x,j)) \\ \neg F(p,j) \\ \hline \neg F(p,m) \end{array}$$

Here, $F(x, y)$ means that x is a friend of y, while m, j, and p denote Martin, John, and Peter, respectively.

2′. Inference 2 becomes:

$$\frac{\begin{array}{l}(\forall x)(H(x) \Rightarrow R(x))\\ (\exists x)(A(x) \wedge H(x))\end{array}}{(\exists x)(A(x) \wedge R(x))}$$

Here, H, R, and A designate the properties of being human, rational, and an animal, respectively.

3′. Inference 3 can be symbolized as follows:

$$\frac{\begin{array}{l}(\forall x)(I(x) \wedge E(x) \Rightarrow D(s(x)))\\ I(b) \wedge E(b)\end{array}}{D(s(b))}$$

Here, I, E, and D designate, respectively, the properties of being an integer, even, and odd; $s(x)$ denotes the successor of x; and b denotes the integer 2.

Notice that the validity of these inferences does not depend upon the particular meanings of F, m, j, p, H, R, A, I, E, D, s, and b.

Just as statement forms were used to indicate logical structure dependent upon the propositional connectives, so also the form of inferences involving quantifiers, such as inferences 1–3, can be represented abstractly, as in 1′–3′. For this purpose, we shall use commas, parentheses, the symbols $\neg$ and $\Rightarrow$ of the propositional calculus, the universal quantifier symbol $\forall$, and the following groups of symbols:

Individual variables: $x_1, x_2, \ldots, x_n, \ldots$
Individual constants: $a_1, a_2, \ldots, a_n, \ldots$
Predicate letters: A_k^n (n and k are any positive integers)
Function letters: f_k^n (n and k are any positive integers)

The positive integer n that is a superscript of a predicate letter A_k^n or of a function letter f_k^n indicates the number of arguments, whereas the subscript k is just an indexing number to distinguish different predicate or function letters with the same number of arguments.

In the preceding examples, x plays the role of an individual variable; m, j, p, and b play the role of individual constants; F is a binary predicate letter (i.e., a letter with two arguments); H, R, A, I, E, and D are monadic predicate letters (i.e., letters with one argument); and s is a function letter with one argument.

The function letters applied to the variables and individual constants generate the *terms*:

1. Variables and individual constants are terms.
2. If f_k^n is a function letter and $t_1, \ldots, t_n$ are terms, then $f_k^n(t_1, \ldots, t_n)$ is a term.
3. An expression is a term only if it can be shown to be a term on the basis of conditions 1 and 2.

Terms correspond to what in ordinary languages are nouns and noun phrases—for example, "two", "two plus three", and "two plus x".

The predicate letters applied to terms yield the *atomic formulas*; that is, if A_k^n is a predicate letter and $t_1, \ldots, t_n$ are terms, then $A_k^n(t_1, \ldots, t_n)$ is an atomic formula.

The *well-formed formulas (wfs)* of quantification theory are defined as follows:

1. Every atomic formula is a wf.
2. If $\mathscr{A}$ and $\mathscr{B}$ are wfs and y is a variable, then $(\neg \mathscr{A})$, $(\mathscr{A} \Rightarrow \mathscr{B})$, and $((\forall y)\mathscr{A})$ are wfs.
3. An expression is a wf only if it can be shown to be a wf on the basis of conditions 1 and 2.

In $((\forall y)\mathscr{A})$, "$\mathscr{A}$" is called the *scope* of the quantifier "$(\forall y)$". Notice that $\mathscr{A}$ need not contain the variable y. In that case, we ordinarily understand $((\forall y)\mathscr{A})$ to mean the same thing as $\mathscr{A}$.

The expressions $\mathscr{A} \wedge \mathscr{B}$, $\mathscr{A} \vee \mathscr{B}$, and $\mathscr{A} \Leftrightarrow \mathscr{B}$ are defined as in system L (see page 29). It was unnecessary for us to use the symbol $\exists$ as a primitive symbol because we can define existential quantification as follows:

$((\exists x)\mathscr{A})$ stands for $(\neg((\forall x)(\neg \mathscr{A})))$

This definition is obviously faithful to the meaning of the quantifiers: $\mathscr{A}(x)$ is true for some x if and only if it is not the case that $\mathscr{A}(x)$ is false for all x.

PARENTHESES

The same conventions made in Chapter 1 (pages 17–18) about the omission of parentheses are made here, with the additional convention that quantifiers $(\forall y)$ and $(\exists y)$ rank in strength between $\neg$, $\wedge$, $\vee$, and $\Rightarrow$, $\Leftrightarrow$.

Examples

$(\forall x_1)A_1^1(x_1) \Rightarrow A_1^2(x_1, x_2)$ stands for $(((\forall x_1)A_1^1(x_1)) \Rightarrow A_1^2(x_1, x_2))$
$(\forall x_1)A_1^1(x_1) \vee A_1^2(x_1, x_2)$ stands for $((\forall x_1)(A_1^1(x_1) \vee A_1^2(x_1, x_2)))$

As an additional convention, we omit parentheses around quantified formulas when they are preceded by other quantifiers.

Example

$(\forall x_1)(\exists x_2)(\forall x_4)A_1^3(x_1, x_2, x_4)$ stands for $((\forall x_1)((\exists x_2)((\forall x_4)A_1^3(x_1, x_2, x_4))))$

Exercises

2.1 Restore parentheses to the following.
(a) $(\forall x_1)A_1^1(x_1) \wedge \neg A_1^1(x_2)$
(b) $(\forall x_2)A_1^1(x_2) \Leftrightarrow A_1^1(x_2)$
(c) $(\forall x_2)(\exists x_1)A_1^2(x_1, x_2)$
(d) $(\forall x_1)(\forall x_3)(\forall x_4)A_1^1(x_1) \Rightarrow A_1^1(x_2) \wedge \neg A_1^1(x_1)$
(e) $(\exists x_1)(\forall x_2)(\exists x_3)A_1^1(x_1) \vee (\exists x_2)\neg(\forall x_3)A_1^2(x_3, x_2)$
(f) $(\forall x_2)\neg A_1^1(x_1) \Rightarrow A_1^3(x_1, x_1, x_2) \vee (\forall x_1)A_1^1(x_1)$
(g) $\neg(\forall x_1)A_1^1(x_1) \Rightarrow (\exists x_2)A_1^1(x_2) \Rightarrow A_1^2(x_1, x_2) \wedge A_1^1(x_2)$

2.2 Eliminate parentheses from the following wfs as far as is possible.
(a) $(((\forall x_1)(A_1^1(x_1) \Rightarrow A_1^1(x_1))) \vee ((\exists x_1)A_1^1(x_1)))$
(b) $((\neg((\exists x_2)(A_1^1(x_2) \vee A_1^1(a_1)))) \Leftrightarrow A_1^1(x_2))$
(c) $(((\forall x_1)(\neg(\neg A_1^1(a_3)))) \Rightarrow (A_1^1(x_1) \Rightarrow A_1^1(x_2)))$

An occurrence of a variable x is said to be *bound* in a wf $\mathscr{A}$ if either it is the occurrence of x in a quantifier "$(\forall x)$" in $\mathscr{A}$ or it lies within the scope of a quantifier "$(\forall x)$" in $\mathscr{A}$. Otherwise, the occurrence is said to be *free* in $\mathscr{A}$.

Examples

1. $A_1^2(x_1, x_2)$
2. $A_1^2(x_1, x_2) \Rightarrow (\forall x_1)A_1^1(x_1)$
3. $(\forall x_1)(A_1^2(x_1, x_2) \Rightarrow (\forall x_1)A_1^1(x_1))$
4. $(\exists x_1)A_1^2(x_1, x_2)$

In Example 1 the single occurrence of x_1 is free. In Example 2 the first occurrence of x_1 is free, but the second and third occurrences are bound. In Example 3 all occurrences of x_1 are bound. And in Example 4 both occurrences of x_1 are bound. [Remember that $(\exists x_1)A_1^2(x_1, x_2)$ is an abbreviation of $\neg(\forall x_1)\neg A_1^2(x_1, x_2)$.] In all four wfs, every occurrence of x_2 is free. Notice that, as in Example 2, a variable may have both free and bound occurrences in the same wf. Also observe that an occurrence of a variable may be bound in some wf $\mathscr{A}$ but free in a subformula of $\mathscr{A}$. For example, the first occurrence of x_1 is free in Example 2 but bound in the larger wf of Example 3.

A variable is said to be *free* (*bound*) in a wf $\mathscr{A}$ if it has a free (bound) occurrence in $\mathscr{A}$. Thus, a variable may be both free and bound in the same wf; for example, x_1 is free and bound in Example 2.

Exercises

2.3 Pick out the free and bound occurrences of variables in the following wfs.
(a) $(\forall x_3)(((\forall x_1)A_1^2(x_1, x_2)) \Rightarrow A_1^2(x_3, a_1))$
(b) $(\forall x_2)A_1^2(x_3, x_2) \Rightarrow (\forall x_3)A_1^2(x_3, x_2)$
(c) $((\forall x_2)(\exists x_1)A_1^3(x_1, x_2, f_1^2(x_1, x_2))) \vee \neg(\forall x_1)A_1^2(x_2, f_1^1(x_1))$

2.4 Indicate the free and bound occurrences of all variables in the wfs of Exercises 2.1 and 2.2.

2.5 Indicate the free variables and the bound variables in the wfs of Exercises 2.1–2.3.

We shall often indicate that a wf $\mathscr{A}$ has some of the free variables $x_{i_1}, \ldots, x_{i_k}$ by writing it as $\mathscr{A}(x_{i_1}, \ldots, x_{i_k})$. This does not mean that $\mathscr{A}$ contains these variables as free variables, nor does it mean that $\mathscr{A}$ does not contain other free variables. This notation is convenient because we can then agree to write as $\mathscr{A}(t_1, \ldots, t_k)$ the result of substituting in $\mathscr{A}$ the terms $t_1, \ldots, t_k$ for all free occurrences (if any) of $x_{i_1}, \ldots, x_{i_k}$, respectively.

If $\mathscr{A}$ is a wf and t is a term, then t is said to be *free for* x_i *in* $\mathscr{A}$ if no free occurrence of x_i in $\mathscr{A}$ lies within the scope of any quantifier $(\forall x_j)$, where x_j is a variable in t. This concept of t being free for x_i in a wf $\mathscr{A}(x_i)$ will have certain technical applications later on. It means that, if t is substituted for all free occurrences of x_i in $\mathscr{A}(x_i)$, no occurrence of a variable in t becomes a bound occurrence in $\mathscr{A}(t)$.

Examples

1. The term x_2 is free for x_1 in $A_1^1(x_1)$, but x_2 is not free for x_1 in $(\forall x_2)A_1^1(x_1)$.
2. The term $f_1^2(x_1, x_3)$ is free for x_1 in $(\forall x_2)A_1^2(x_1, x_2) \Rightarrow A_1^1(x_1)$ but is not free for x_1 in $(\exists x_3)(\forall x_2)A_1^2(x_1, x_2) \Rightarrow A_1^1(x_1)$.

The following facts are obvious:

1. A term that contains no variables is free for any variable in any wf.
2. A term t is free for any variable in $\mathscr{A}$ if none of the variables of t is bound in $\mathscr{A}$.
3. x_i is free for x_i in any wf.
4. Any term is free for x_i in $\mathscr{A}$ if $\mathscr{A}$ contains no free occurrences of x_i.

Exercises

2.6 Is the term $f_1^2(x_1, x_2)$ free for x_1 in the following wfs?
(a) $A_1^2(x_1, x_2) \Rightarrow (\forall x_2)A_1^1(x_2)$
(b) $((\forall x_2)A_1^2(x_2, a_1)) \vee (\exists x_2)A_1^2(x_1, x_2)$
(c) $(\forall x_1)A_1^2(x_1, x_2)$
(d) $(\forall x_2)A_1^2(x_1, x_2)$
(e) $(\forall x_2)A_1^1(x_2) \Rightarrow A_1^2(x_1, x_2)$

2.7 Justify facts 1–4 above.

When English sentences are translated into formulas, two general guidelines will be useful:

1. A sentence of the form "All A's are B's" becomes $(\forall x)(A(x) \Rightarrow B(x))$.
2. A sentence of the form "Some A's are B's" becomes $(\exists x)(A(x) \wedge B(x))$.

Notice that, in informal discussions, to make formulas easier to read we use capital italic letters $A, B, C, \ldots$ instead of our official notation A_k^n for predicate letters.

Exercises

2.8 Translate the following sentences into wfs.
(a) Anyone who is persistent can learn logic.
(b) No politician is honest.
(c) Not all birds can fly.
(d) Some people are witty only if they are drunk.
(e) If anyone can do it, Jones can.
(f) Nobody in the statistics class is smarter than everyone in the logic class.
(g) John hates everyone who does not hate himself.
(h) Everyone loves somebody and no one loves everybody, or somebody loves everybody and someone loves nobody.
(i) You can fool some of the people all of the time, and you can fool all the people some of the time, but you can't fool all the people all the time.
(j) Any sets that have the same members are equal.
(k) Anyone who knows Julia loves her.
(l) There is no set containing precisely those sets that do not contain themselves.
(m) There is no barber who shaves precisely those men who do not shave themselves.

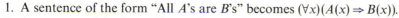

2.9 Translate the following into everyday English. Note that everyday English does not use variables.

(a) $(\forall x)(M(x) \wedge (\forall y)\neg W(x, y) \Rightarrow U(x))$, where $M(x)$ means x *is a man*, $W(x, y)$ means x *is married to* y, and $U(x)$ means x *is unhappy*.

(b) $(\forall x)(V(x) \wedge P(x) \Rightarrow A(x, b))$, where $V(x)$ means x *is an even integer*, $P(x)$ means x *is a prime integer*, $A(x, y)$ means $x = y$, and b denotes 2.

(c) $\neg(\exists y)(I(y) \wedge (\forall x)(I(x) \Rightarrow L(x, y)))$, where $I(y)$ means y *is an integer* and $L(x, y)$ means $x \leqslant y$.

(d) In the following wfs, $A_1^1(x)$ means x *is a person* and $A_1^2(x, y)$ means x *hates* y.

(i) $(\exists x)(A_1^1(x) \wedge (\forall y)(A_1^1(y) \Rightarrow A_1^2(x, y)))$

(ii) $(\forall x)(A_1^1(x) \Rightarrow (\forall y)(A_1^1(y) \Rightarrow A_1^2(x, y)))$

(iii) $(\exists x)(A_1^1(x) \wedge (\forall y)(A_1^1(y) \Rightarrow (A_1^2(x, y) \Leftrightarrow A_1^2(y, y))))$

(e) $(\forall x)(H(x) \Rightarrow (\exists y)(\exists z)(\neg A(y, z) \wedge (\forall u)(P(u, x) \Leftrightarrow (A(u, y) \vee A(u, z)))))$, where $H(x)$ means x *is a person*, $A(u, v)$ means $u = v$, and $P(u, x)$ means u *is a parent of* x.

2. INTERPRETATIONS. SATISFIABILITY AND TRUTH. MODELS

Well-formed formulas have meaning only when an interpretation is given for the symbols. An *interpretation* M consists of a nonempty set D, called the *domain* of the interpretation, and an assignment to each predicate letter A_j^n of an n-place relation $(A_j^n)^{\mathrm{M}}$ in D, to each function letter f_j^n of an n-place operation $(f_j^n)^{\mathrm{M}}$ in D (that is, a function from D^n into D), and to each individual constant a_i of some fixed element $(a_i)^{\mathrm{M}}$ of D. Given such an interpretation, variables are thought of as ranging over the set D, and $\neg$, $\Rightarrow$, and quantifiers are given their usual meaning. [Remember that an n-place relation in D can be thought of as a subset of D^n, the set of all n-tuples of elements of D. For example, if D is the set of human beings, then the relation "father of" can be identified with the set of all ordered pairs $\langle x, y\rangle$ such that x is the father of y.]

For a given interpretation, a wf without free variables (called a *closed wf* or a *sentence*) represents a proposition that is true or false, whereas a wf with free variables may be satisfied (i.e., true) for some values in the domain and not satisfied (i.e., false) for the others.

Examples Consider the following wfs:

1. $A_1^2(x_1, x_2)$
2. $(\forall x_2)A_1^2(x_1, x_2)$
3. $(\exists x_1)(\forall x_2)A_1^2(x_1, x_2)$

Let us take as domain the set of all positive integers and interpret $A_1^2(y, z)$ as $y \leqslant z$. Then wf 1 represents the expression "$x_1 \leqslant x_2$", which is satisfied by all the ordered pairs $\langle a, b\rangle$ of positive integers such that $a \leqslant b$. Wf 2 represents the expression "For all positive integers x_2, $x_1 \leqslant x_2$", which is satisfied only by the integer 1. Wf 3 is a true sentence asserting that there is a smallest positive integer. If we were to take as domain the set of all integers, then wf 3 would be false.

Exercises

2.10 For the following wfs and for the given interpretations, indicate for what values the wfs are satisfied (if they contain free variables) or whether they are true or false (if they are closed wfs).

(i) $A_1^2(f_1^2(x_1, x_2), a_1)$
(ii) $A_1^2(x_1, x_2) \Rightarrow A_1^2(x_2, x_1)$
(iii) $(\forall x_1)(\forall x_2)(\forall x_3)(A_1^2(x_1, x_2) \wedge A_1^2(x_2, x_3) \Rightarrow A_1^2(x_1, x_3))$

(a) The domain is the set of positive integers, $A_1^2(y, z)$ is $y \geqslant z$, $f_1^2(y, z)$ is $y \cdot z$, and a_1 is 2.
(b) The domain is the set of integers, $A_1^2(y, z)$ is $y = z$, $f_1^2(y, z)$ is $y + z$, and a_1 is 0.
(c) The domain is the set of all sets of integers, $A_1^2(y, z)$ is $y \subseteq z$, $f_1^2(y, z)$ is $y \cap z$, and a_1 is the empty set 0.

2.11 Describe in everyday English the assertions determined by the following wfs and interpretations.

(a) $(\forall x)(\forall y)(A_1^2(x, y) \Rightarrow (\exists z)(A_1^1(z) \wedge A_1^2(x, z) \wedge A_1^2(z, y)))$, where the domain D is the set of real numbers, $A_1^2(x, y)$ means $x < y$, and $A_1^1(z)$ means *z is a rational number*.
(b) $(\forall x)(A_1^1(x) \Rightarrow (\exists y)(A_2^1(y) \wedge A_1^2(y, x)))$, where D is the set of all days and people, $A_1^1(x)$ means *x is a day*, $A_2^1(y)$ means *y is a sucker*, and $A_1^2(y, x)$ means *y is born on day x*.
(c) $(\forall x)(\forall y)(A_1^1(x) \wedge A_1^1(y) \Rightarrow A_2^1(f_1^2(x, y)))$, where D is the set of integers, $A_1^1(x)$ means *x is odd*, $A_2^1(x)$ means *x is even*, and $f_1^2(x, y)$ denotes $x + y$.
(d) For the following wfs, D is the set of all people and $A_1^2(u, v)$ means *u loves v*.
(i) $(\exists x)(\forall y)A_1^2(x, y)$
(ii) $(\forall y)(\exists x)A_1^2(x, y)$
(iii) $(\exists x)(\forall y)((\forall z)A_1^2(y, z) \Rightarrow A_1^2(x, y))$
(iv) $(\exists x)(\forall y)\neg A_1^2(x, y)$

The concepts of satisfiability and truth are intuitively clear, but, following Tarski (1936), we also can provide a rigorous definition. Such a definition is necessary for carrying out precise proofs of many metamathematical results.

Satisfiability will be the fundamental notion, on the basis of which the notion of truth will be defined. Moreover, instead of talking about the n-tuples of objects that satisfy a wf that has n free variables, it is much more convenient from a technical standpoint to deal uniformly with denumerable sequences. What we have in mind is that a denumerable sequence $s = (s_1, s_2, s_3, \ldots)$ is to be thought of as satisfying a wf $\mathscr{A}$ that has $x_{j_1}, x_{j_2}, \ldots, x_{j_n}$ as free variables (where $j_1 < j_2 < \cdots < j_n$) if the n-tuple $\langle s_{j_1}, s_{j_2}, \ldots, s_{j_n} \rangle$ satisfies $\mathscr{A}$ in the usual sense. For example, a denumerable sequence $(s_1, s_2, s_3, \ldots)$ of objects in the domain of an interpretation M will turn out to satisfy the wf $A_1^2(x_2, x_5)$ if and only if the ordered pair $\langle s_2, s_5 \rangle$ is in the relation $(A_1^2)^{\mathrm{M}}$ assigned to the predicate letter A_1^2 by the interpretation M.

Let there be given an interpretation M with domain D. Let Σ be the set of all denumerable sequences of elements of D. We shall define what it means for a sequence $s = (s_1, s_2, \ldots)$ in Σ to satisfy a wf $\mathscr{A}$ in M. As a preliminary step, for a given s in Σ we shall define a function s^* that assigns to each term t an element $s^*(t)$ in D.

1. If t is a variable x_j, let $s^*(t)$ be s_j.
2. If t is an individual constant a_j, then $s^*(t)$ is the interpretation $(a_j)^{\mathrm{M}}$ of this constant.

3. If f_k^n is a function letter, $(f_k^n)^M$ is the corresponding operation in D, and $t_1, \ldots, t_n$ are terms, then

$$s^*(f_k^n(t_1, \ldots, t_n)) = (f_k^n)^M(s^*(t_1), \ldots, s^*(t_n))$$

Intuitively, $s^*(t)$ is the element of D obtained by substituting, for each j, a name of s_j for all occurrences of x_j in t and then performing the operations of the interpretation corresponding to the function letters of t. For instance, if t is $f_2^2(x_3, f_1^2(x_1, a_1))$, the interpretation has the set of integers as its domain, f_2^2 and f_1^2 are interpreted as ordinary multiplication and addition, respectively, and a_1 is interpreted as 2, then, for any sequence $s = (s_1, s_2, \ldots)$ of integers, $s^*(t)$ is the integer $s_3 \cdot (s_1 + 2)$.

Now we proceed to the definition of satisfaction, which will be an inductive definition.

1. If $\mathscr{A}$ is an atomic wf $A_k^n(t_1, \ldots, t_n)$ and $(A_k^n)^M$ is the corresponding n-place relation of the interpretation, then the sequence $s = (s_1, s_2, \ldots)$ satisfies $\mathscr{A}$ if and only if $(A_k^n)^M(s^*(t_1), \ldots, s^*(t_n))$—that is, if the n-tuple $\langle s^*(t_1), \ldots, s^*(t_n)\rangle$ is in the relation $(A_k^n)^M$.†
2. s satisfies $\neg\mathscr{B}$ if and only if s does not satisfy $\mathscr{B}$.
3. s satisfies $\mathscr{B} \Rightarrow \mathscr{C}$ if and only if either s does not satisfy $\mathscr{B}$ or s satisfies $\mathscr{C}$.
4. s satisfies $(\forall x_i)\mathscr{B}$ if and only if every sequence that differs from s in at most the ith component satisfies $\mathscr{B}$.‡

Intuitively, a sequence $s = (s_1, s_2, \ldots)$ satisfies a wf $\mathscr{A}$ if and only if, when, for each i, we replace all free occurrences of x_i in $\mathscr{A}$ by a symbol representing s_i, the resulting proposition is true under the given interpretation.

Now we can define the notions of truth and falsity of wfs for a given interpretation.

DEFINITIONS A wf $\mathscr{A}$ is *true for the interpretation* M (written $\models_M \mathscr{A}$) if and only if every sequence in Σ satisfies $\mathscr{A}$. $\mathscr{A}$ is said to be *false* for M if and only if no sequence in Σ satisfies $\mathscr{A}$. An interpretation M is said to be a *model* for a set Γ of wfs if and only if every wf in Γ is true for M.

The plausibility of our definition of truth will be strengthened by the fact that we can derive all of the following expected properties I–XI of the notions of truth, falsity, and satisfaction. Proofs that are not explicitly given are left to the reader (or

† For example, if the domain of the interpretation is the set of real numbers, the interpretation of A_1^2 is the relation $\leqslant$, and the interpretation of f_1^1 is the function e^x, then a sequence $s = (s_1, s_2, \ldots)$ of real numbers satisfies $A_1^2(f_1^1(x_2), x_5)$ if and only if $e^{s_2} \leqslant s_5$. If the domain is the set of points in a plane, the interpretation of $A_1^3(x, y, z)$ is *x and y are equidistant from z*, and the interpretation of $f_1^2(x, y)$ is *the midpoint of the line segment connecting x and y*, then a sequence $s = (s_1, s_2, \ldots)$ of points in the plane satisfies $A_1^3(f_1^2(x_1, x_2), f_1^2(x_3, x_1), x_4)$ if and only if the midpoint of the line segment between s_1 and s_2 is at the same distance from s_4 as the midpoint of the line segment between s_3 and s_1. If the domain is the set of integers, the interpretation of $A_1^4(x, y, u, v)$ is $x \cdot v = u \cdot y$, and the interpretation of a_1 is 3, then a sequence $s = (s_1, s_2, \ldots)$ of integers satisfies $A_1^4(x_3, a_1, x_1, x_3)$ if and only if $(x_3)^2 = 3x_1$.

‡ In other words, a sequence $s = (s_1, s_2, \ldots, s_i, \ldots)$ satisfies $(\forall x_i)\mathscr{B}$ if and only if, for every element c of the domain of the interpretation, the sequence $(s_1, s_2, \ldots, c, \ldots)$ satisfies $\mathscr{B}$. Here, $(s_1, s_2, \ldots, c, \ldots)$ denotes the sequence obtained from $(s_1, s_2, \ldots, s_i, \ldots)$ by replacing s_i by c.

may be found in the answer to Exercise 2.12; see page 310). Most of the results are also obvious if one wishes to use only the ordinary intuitive understanding of the notions of truth and satisfaction.

(I) (a) $\mathscr{A}$ is false for an interpretation M if and only if $\neg\mathscr{A}$ is true for M.

(b) $\mathscr{A}$ is true for M if and only if $\neg\mathscr{A}$ is false for M.

(II) It is not the case that both $\models_{\mathrm{M}}\mathscr{A}$ and $\models_{\mathrm{M}}\neg\mathscr{A}$; that is, no wf can be both true and false for M.

(III) If $\models_{\mathrm{M}}\mathscr{A}$ and $\models_{\mathrm{M}}\mathscr{A}\Rightarrow\mathscr{B}$, then $\models_{\mathrm{M}}\mathscr{B}$.

(IV) $\mathscr{A}\Rightarrow\mathscr{B}$ is false for M if and only if $\models_{\mathrm{M}}\mathscr{A}$ and $\models_{\mathrm{M}}\neg\mathscr{B}$.

(V)† (a) A sequence s satisfies $\mathscr{A}\wedge\mathscr{B}$ if and only if s satisfies $\mathscr{A}$ and s satisfies $\mathscr{B}$.

(b) s satisfies $\mathscr{A}\vee\mathscr{B}$ if and only if s satisfies $\mathscr{A}$ or s satisfies $\mathscr{B}$.

(c) s satisfies $\mathscr{A}\Leftrightarrow\mathscr{B}$ if and only if s satisfies both $\mathscr{A}$ and $\mathscr{B}$ or s satisfies neither $\mathscr{A}$ nor $\mathscr{B}$.

(d) s satisfies $(\exists x_i)\mathscr{B}$ if and only if there is a sequence s' that differs from s in at most the ith component such that s' satisfies $\mathscr{B}$. [In other words, $s=(s_1,s_2,\ldots,s_i,\ldots)$ satisfies $(\exists x_i)\mathscr{B}$ if and only if there is an element c in the domain D such that the sequence $(s_1,s_2,\ldots,c,\ldots)$ satisfies $\mathscr{B}$.]

(VI) $\models_{\mathrm{M}}\mathscr{A}$ if and only if $\models_{\mathrm{M}}(\forall x_i)\mathscr{A}$. We can extend this result in the following way. By the *closure* of $\mathscr{A}$ we mean the closed wf obtained from $\mathscr{A}$ by prefixing in universal quantifiers those variables, in order of descending subscripts, that are free in $\mathscr{A}$. If $\mathscr{A}$ has no free variables, the closure of $\mathscr{A}$ is defined to be $\mathscr{A}$ itself. For example, if $\mathscr{A}$ is $A_1^2(x_2,x_5)\Rightarrow\neg(\exists x_2)A_1^3(x_1,x_2,x_3)$, its closure is $(\forall x_5)(\forall x_3)(\forall x_2)(\forall x_1)\mathscr{A}$. It follows from (VI) that a wf $\mathscr{A}$ is true if and only if its closure is true.

(VII) Every instance of a tautology is true for any interpretation. [An *instance* of a statement form is a wf obtained from the statement form by substituting wfs for all statement letters, with all occurrences of the same statement letter being replaced by the same wf. Thus, an instance of $A_1\Rightarrow\neg A_2\vee A_1$ is $A_1^1(x_2)\Rightarrow\neg(\forall x_1)A_1^1(x_1)\vee A_1^1(x_2)$.] To prove (VII), show that all instances of the axioms of the system L are true and then use (III) and Proposition 1.13.

(VIII) If the free variables (if any) of a wf $\mathscr{A}$ occur in the list $x_{i_1},\ldots,x_{i_k}$, and if the sequences s and s' have the same components in the i_1th, $\ldots$, i_kth places, then s satisfies $\mathscr{A}$ if and only if s' satisfies $\mathscr{A}$. [*Hint:* Use induction on the number of connectives and quantifiers in $\mathscr{A}$. First prove this lemma: if the variables in a term t occur in the list $x_{i_1},\ldots,x_{i_k}$, and if s and s' have the same components in the i_1th, $\ldots$, i_kth places, then $s^*(t)=(s')^*(t)$. In particular, if t contains no variables at all, $s^*(t)=(s')^*(t)$ for any sequences s and s'.]

Although, by (VIII), a particular wf $\mathscr{A}$ with k free variables is essentially satisfied or not only by k-tuples, rather than by denumerable sequences, it is more convenient for a general treatment of satisfaction to deal with infinite rather than finite sequences. If we were to define satisfaction using finite sequences, conditions 3 and 4 of the definition of satisfaction would become much more complicated.

† Remember that $\mathscr{A}\wedge\mathscr{B}$, $\mathscr{A}\vee\mathscr{B}$, $\mathscr{A}\Leftrightarrow\mathscr{B}$, and $(\exists x_i)\mathscr{B}$ are abbreviations for $\neg(\mathscr{A}\Rightarrow\neg\mathscr{B})$, $\neg\mathscr{A}\Rightarrow\mathscr{B}$, $(\mathscr{A}\Rightarrow\mathscr{B})\wedge(\mathscr{B}\Rightarrow\mathscr{A})$, and $\neg(\forall x_i)\neg\mathscr{B}$, respectively.

Let $x_{i_1}, \ldots, x_{i_k}$ be k distinct variables in order of increasing subscripts. Let $\mathscr{A}(x_{i_1},\ldots,x_{i_k})$ be a wf that has $x_{i_1}, \ldots, x_{i_k}$ as its only free variables. The set of k-tuples $\langle b_1,\ldots,b_k\rangle$ of the domain D such that any sequence with $b_1, \ldots, b_k$ in its i_1th, ..., i_kth places, respectively, satisfies $\mathscr{A}(x_{i_1},\ldots,x_{i_k})$ is called the *relation* (or *property*) *of the interpretation defined by* $\mathscr{A}$. Extending our terminology, we shall say that every k-tuple $\langle b_1,\ldots,b_k\rangle$ in this relation *satisfies* $\mathscr{A}(x_{i_1},\ldots,x_{i_k})$ in the interpretation M; this will be written $\models_{\mathrm{M}} \mathscr{A}[b_1,\ldots,b_k]$. This extended notion of satisfaction corresponds to the original intuitive notion.

Examples

1. If the domain D of M is the set of human beings, $A_1^2(x, y)$ is interpreted as *x is a brother of y*, and $A_2^2(x, y)$ is interpreted as *x is a parent of y*, then the binary relation on D corresponding to the wf $\mathscr{A}(x_1, x_2)$: $(\exists x_3)(A_1^2(x_1, x_3) \wedge A_2^2(x_3, x_2))$ is the relation of unclehood. $\models_{\mathrm{M}} \mathscr{A}[b, c]$ when and only when b is an uncle of c.
2. If the domain is the set of positive integers, A_1^2 is interpreted as $=$, f_1^2 is interpreted as multiplication, and a_1 is interpreted as 1, then the wf $\mathscr{B}(x_1)$:

$$\neg A_1^2(x_1, a_1) \wedge (\forall x_2)((\exists x_3)A_1^2(x_1, f_1^2(x_2, x_3)) \Rightarrow A_1^2(x_2, x_1) \vee A_1^2(x_2, a_1))$$

determines the property of being a prime number. Thus, $\models_{\mathrm{M}} \mathscr{B}[k]$ if and only if k is a prime number.

(IX) If $\mathscr{A}$ is a closed wf, then, for any interpretation M, either $\models_{\mathrm{M}} \mathscr{A}$ or $\models_{\mathrm{M}} \neg\mathscr{A}$—that is, either $\mathscr{A}$ is true for M or $\mathscr{A}$ is false for M. [*Hint:* Use (VIII).] Of course, $\mathscr{A}$ may be true for some interpretations and false for others. [As an example, consider $A_1^1(a_1)$.]

If $\mathscr{A}$ is not closed—that is, if $\mathscr{A}$ contains free variables—$\mathscr{A}$ may be neither true nor false for some interpretations. For example, if $\mathscr{A}$ is $A_1^2(x_1, x_2)$ and we consider an interpretation in which the domain is the set of integers and $A_1^2(y, z)$ is interpreted as $y < z$, then $\mathscr{A}$ is satisfied by only those sequences $s = (s_1, s_2, \ldots)$ of integers in which $s_1 < s_2$. Hence, $\mathscr{A}$ is neither true nor false for this interpretation. On the other hand, there are wfs that are not closed but that nevertheless are true or false for every interpretation. A simple example is the wf $A_1^1(x_1) \vee \neg A_1^1(x_1)$, which is true for every interpretation.

(X) Assume t is free for x_i in $\mathscr{A}(x_i)$. Then

$$(\forall x_i)\mathscr{A}(x_i) \Rightarrow \mathscr{A}(t)$$

is true for all interpretations.

The proof of (X) is based upon the following lemmas.

LEMMA 1 If t and u are terms, s is a sequence in Σ, t' results from t by replacing all occurrences of x_i by u, and s' results from s by replacing the ith component of s by $s^*(u)$, then $s^*(t') = (s')^*(t)$. [*Hint:* Use induction on the length of t.†]

†The *length* of an expression is the number of occurrences of symbols in the expression.

LEMMA 2 Let t be free for x_i in $\mathscr{A}(x_i)$. Then:

(a) A sequence $s = (s_1, s_2, \ldots)$ satisfies $\mathscr{A}(t)$ if and only if the sequence s', obtained from s by substituting $s^*(t)$ for s_i in the ith place, satisfies $\mathscr{A}(x_i)$. [*Hint:* Use induction on the number of connectives and quantifiers in $\mathscr{A}(x_i)$, applying Lemma 1.]

(b) If $(\forall x_i)\mathscr{A}(x_i)$ is satisfied by the sequence s, then $\mathscr{A}(t)$ also is satisfied by s.

(XI) If $\mathscr{A}$ does not contain x_i free, then $(\forall x_i)(\mathscr{A} \Rightarrow \mathscr{B}) \Rightarrow (\mathscr{A} \Rightarrow (\forall x_i)\mathscr{B})$ is true for all interpretations.

Proof Assume (XI) is not correct. Then $(\forall x_i)(\mathscr{A} \Rightarrow \mathscr{B}) \Rightarrow (\mathscr{A} \Rightarrow (\forall x_i)\mathscr{B})$ is not true for some interpretation. By condition 3 of the definition of satisfaction, there is a sequence s such that s satisfies $(\forall x_i)(\mathscr{A} \Rightarrow \mathscr{B})$ and s does not satisfy $\mathscr{A} \Rightarrow (\forall x_i)\mathscr{B}$. From the latter and condition 3, s satisfies $\mathscr{A}$ and s does not satisfy $(\forall x_i)\mathscr{B}$. Hence, by condition 4, there is a sequence s', differing from s in at most the ith place, such that s' does not satisfy $\mathscr{B}$. Since x_i is free in neither $(\forall x_i)(\mathscr{A} \Rightarrow \mathscr{B})$ nor $\mathscr{A}$, and since s satisfies both of these wfs, it follows by (VIII) that s' also satisfies both $(\forall x_i)(\mathscr{A} \Rightarrow \mathscr{B})$ and $\mathscr{A}$. Since s' satisfies $(\forall x_i)(\mathscr{A} \Rightarrow \mathscr{B})$, it follows by condition 4 that s' satisfies $\mathscr{A} \Rightarrow \mathscr{B}$. Since s' satisfies $\mathscr{A} \Rightarrow \mathscr{B}$ and $\mathscr{A}$, condition 3 implies that s' satisfies $\mathscr{B}$, which contradicts the fact that s' does not satisfy $\mathscr{B}$. Hence, (XI) is established.

Exercises

2.12 Verify (I)–(X).

2.13 Prove that a closed wf $\mathscr{A}$ is true for M if and only if $\mathscr{A}$ is satisfied by some sequence s in Σ. (Remember that Σ is the set of denumerable sequences of elements of the domain of M.)

2.14 Find the properties or relations determined by the following wfs and interpretations.

(a) $[(\exists u)A_1^2(f_1^2(x,u),y)] \wedge [(\exists v)A_1^2(f_1^2(x,v),z)]$, where the domain D is the set of integers, A_1^2 is $=$, and f_1^2 is multiplication.

(b) Here, D is the set of nonnegative integers, A_1^2 is $=$, a_1 denotes 0, f_1^2 is addition, and f_2^2 is multiplication.

(i) $(\exists z)(\neg A_1^2(z,a_1) \wedge A_1^2(f_1^2(x,z),y))$

(ii) $(\exists y)(A_1^2(x, f_2^2(y,y)))$

(c) $(\exists x_3)A_1^2(f_1^2(x_1,x_3),x_2)$, where D is the set of positive integers, A_1^2 is $=$, and f_1^2 is multiplication.

(d) $A_1^1(x_1) \wedge (\forall x_2)\neg A_1^2(x_1,x_2)$, where D is the set of all living people, $A_1^1(x)$ means x *is a man*, and $A_1^2(x,y)$ means x *is married to* y.

(e) $(\exists x_1)(\exists x_2)(A_1^2(x_1,x_3) \wedge A_1^2(x_2,x_4) \wedge A_2^2(x_1,x_2))$, where D is the set of all people, $A_1^2(x,y)$ means x *is a parent of* y, and $A_2^2(x,y)$ means x *and* y *are siblings*.

(f) $(\forall x_3)((\exists x_4)A_1^2(f_1^2(x_4,x_3),x_1) \wedge (\exists x_4)(A_1^2(f_1^2(x_4,x_3),x_2)) \Rightarrow A_1^2(x_3,a_1))$, where D is the set of positive integers, A_1^2 is $=$, f_1^2 is multiplication, and a_1 denotes 1.

2.15 For each of the following wfs and interpretations, write a translation into ordinary English and determine its truth or falsity.

(a) The domain D is the set of nonnegative integers, A_1^2 is $=$, f_1^2 is addition, f_2^2 is multiplication, a_1 denotes 0, and a_2 denotes 1.

(i) $(\forall x)(\exists y)(A_1^2(x, f_1^2(y,y)) \vee A_1^2(x, f_1^2(f_1^2(y,y),a_2)))$

(ii) $(\forall x)(\forall y)(A_1^2(f_2^2(x,y),a_1) \Rightarrow A_1^2(x,a_1) \vee A_1^2(y,a_1))$

(iii) $(\exists y)A_1^2(f_1^2(y,y),a_2)$

(b) Here, D is the set of all integers, A_1^2 is $=$, and f_1^2 is addition.

(i) $(\forall x_1)(\forall x_2)A_1^2(f_1^2(x_1, x_2), f_1^2(x_2, x_1))$

(ii) $(\forall x_1)(\forall x_2)(\forall x_3)A_1^2(f_1^2(x_1, f_1^2(x_2, x_3)), f_1^2(f_1^2(x_1, x_2), x_3))$

(iii) $(\forall x_1)(\forall x_2)(\exists x_3)A_1^2(f_1^2(x_1, x_3), x_2)$

(c) The wfs are the same as in part (b), but the domain is the set of all positive integers, A_1^2 is $=$, and $f_1^2(x, y)$ is x^y.

(d) The domain is the set of all rational numbers, A_1^2 is $=$, A_2^2 is $<$, f_1^2 is multiplication, $f_1^1(x)$ is $x + 1$, and a_1 denotes 0.

(i) $(\exists x)(A_1^2(f_1^2(x, x), f_1^1(f_1^1(a_1))))$

(ii) $(\forall x)(\forall y)(A_2^2(x, y) \Rightarrow (\exists z)(A_2^2(x, z) \wedge A_2^2(z, y)))$

(iii) $(\forall x)(\neg A_1^2(x, a_1) \Rightarrow (\exists y)(A_1^2(f_1^2(x, y), f_1^1(a_1))))$

(e) The domain is the set of nonnegative integers, A_1^2 is $\leqslant$, and $A_1^3(u, v, w)$ means $u + v = w$.

(i) $(\forall x)(\forall y)(\forall z)(A_1^3(x, y, z) \Rightarrow A_1^3(y, x, z))$

(ii) $(\forall x)(\forall y)(A_1^3(x, x, y) \Rightarrow A_1^2(x, y))$

(iii) $(\forall x)(\forall y)(A_1^2(x, y) \Rightarrow A_1^3(x, x, y))$

(iv) $(\exists x)(\forall y)A_1^3(x, y, y)$

(v) $(\exists y)(\forall x)A_1^2(x, y)$

(vi) $(\forall x)(\forall y)(A_1^2(x, y) \Leftrightarrow (\exists z)A_1^3(x, z, y))$

DEFINITIONS A wf $\mathscr{A}$ is said to be *logically valid* if and only if $\mathscr{A}$ is true for every interpretation.*

$\mathscr{A}$ is said to be *satisfiable* if and only if there is an interpretation for which $\mathscr{A}$ is satisfied by at least one sequence.

It is obvious that $\mathscr{A}$ is logically valid if and only if $\neg\mathscr{A}$ is not satisfiable, and $\mathscr{A}$ is satisfiable if and only if $\neg\mathscr{A}$ is not logically valid.

If $\mathscr{A}$ is a closed wf, then we know that $\mathscr{A}$ is either true or false for any given interpretation; that is, $\mathscr{A}$ is satisfied by all sequences or by none. Therefore, if $\mathscr{A}$ is closed, then $\mathscr{A}$ is satisfiable if and only if $\mathscr{A}$ is true for some interpretation.

We say that $\mathscr{A}$ is *contradictory* if and only if $\mathscr{A}$ is false for every interpretation or, equivalently, if and only if $\neg\mathscr{A}$ is logically valid.

$\mathscr{A}$ is said to *logically imply* $\mathscr{B}$ if and only if, in every interpretation, every sequence that satisfies $\mathscr{A}$ also satisfies $\mathscr{B}$. More generally, $\mathscr{B}$ is said to be a *logical consequence* of a set Γ of wfs if and only if, in every interpretation, every sequence that satisfies every wf in Γ also satisfies $\mathscr{B}$.

$\mathscr{A}$ and $\mathscr{B}$ are said to be *logically equivalent* if and only if they logically imply each other.

The following assertions are easy consequences of these definitions.

1. $\mathscr{A}$ logically implies $\mathscr{B}$ if and only if $\mathscr{A} \Rightarrow \mathscr{B}$ is logically valid.
2. $\mathscr{A}$ and $\mathscr{B}$ are logically equivalent if and only if $\mathscr{A} \Leftrightarrow \mathscr{B}$ is logically valid.
3. If $\mathscr{A}$ logically implies $\mathscr{B}$ and $\mathscr{A}$ is true in a given interpretation, then so is $\mathscr{B}$.
4. If $\mathscr{B}$ is a logical consequence of a set Γ of wfs and all wfs in Γ are true in a given interpretation, then so is $\mathscr{B}$.

*The mathematician and philosopher G. W. Leibniz (1646–1716) gave a similar definition: $\mathscr{A}$ is logically valid if $\mathscr{A}$ is true in all "possible worlds".

Examples

1. Every instance of a tautology is logically valid. (VII)
2. If t is free for x in $\mathscr{A}(x)$, then $(\forall x)\mathscr{A}(x) \Rightarrow \mathscr{A}(t)$ is logically valid. (X)
3. If $\mathscr{A}$ does not contain x free, then $(\forall x)(\mathscr{A} \Rightarrow \mathscr{B}) \Rightarrow (\mathscr{A} \Rightarrow (\forall x)\mathscr{B})$ is logically valid. (XI)
4. $\mathscr{A}$ is logically valid if and only if $(\forall y_1) \dots (\forall y_n)\mathscr{A}$ is logically valid. (VI)
5. The wf $(\forall x_2)(\exists x_1)A_1^2(x_1, x_2) \Rightarrow (\exists x_1)(\forall x_2)A_1^2(x_1, x_2)$ is not logically valid. As a counterexample, let the domain D be the set of integers and let $A_1^2(y, z)$ mean $y < z$. Then, $(\forall x_2)(\exists x_1)A_1^2(x_1, x_2)$ is true but $(\exists x_1)(\forall x_2)A_1^2(x_1, x_2)$ is false.

Exercises

2.16 Show that the following wfs are not logically valid.

(a) $[(\forall x_1)A_1^1(x_1) \Rightarrow (\forall x_1)A_2^1(x_1)] \Rightarrow [(\forall x_1)(A_1^1(x_1) \Rightarrow A_2^1(x_1))]$

(b) $[(\forall x_1)(A_1^1(x_1) \vee A_2^1(x_1))] \Rightarrow [((\forall x_1)A_1^1(x_1)) \vee (\forall x_1)A_2^1(x_1)]$

2.17 Show that the following wfs are logically valid.

(a) $\mathscr{A}(t) \Rightarrow (\exists x_i)\mathscr{A}(x_i)$ if t is free for x_i in $\mathscr{A}(x_i)$

(b) $(\forall x_i)\mathscr{A} \Rightarrow (\exists x_i)\mathscr{A}$

(c) $(\forall x_i)(\forall x_j)\mathscr{A} \Leftrightarrow (\forall x_j)(\forall x_i)\mathscr{A}$

(d) $(\forall x_i)\mathscr{A} \Leftrightarrow \neg(\exists x_i)\neg\mathscr{A}$

(e) $(\forall x_i)(\mathscr{A} \Rightarrow \mathscr{B}) \Rightarrow ((\forall x_i)\mathscr{A} \Rightarrow (\forall x_i)\mathscr{B})$

(f) $((\forall x_i)\mathscr{A}) \wedge (\forall x_i)\mathscr{B} \Leftrightarrow (\forall x_i)(\mathscr{A} \wedge \mathscr{B})$

(g) $((\forall x_i)\mathscr{A}) \vee (\forall x_i)\mathscr{B} \Rightarrow (\forall x_i)(\mathscr{A} \vee \mathscr{B})$

(h) $(\exists x_i)(\exists x_j)\mathscr{A} \Leftrightarrow (\exists x_j)(\exists x_i)\mathscr{A}$

(i) $(\exists x_i)(\forall x_j)\mathscr{A} \Rightarrow (\forall x_j)(\exists x_i)\mathscr{A}$

2.18 (a) If $\mathscr{A}$ is a closed wf, show that $\mathscr{A}$ logically implies $\mathscr{B}$ if and only if $\mathscr{B}$ is true for every interpretation for which $\mathscr{A}$ is true.

(b) Although, by (VI), $(\forall x_1)A_1^1(x_1)$ is true whenever $A_1^1(x_1)$ is true, find an interpretation for which $A_1^1(x_1) \Rightarrow (\forall x_1)A_1^1(x_1)$ is not true. [Hence, the hypothesis that $\mathscr{A}$ is a closed wf is essential in part (a).]

2.19 Produce counterexamples to show that the following wfs are not logically valid.

(a) $$[(\forall x)(\forall y)(\forall z)(A_1^2(x, y) \wedge A_1^2(y, z) \Rightarrow A_1^2(x, z)) \wedge (\forall x)\neg A_1^2(x, x)] \Rightarrow (\exists x)(\forall y)\neg A_1^2(x, y)$$

(b) $(\forall x)(\exists y)A_1^2(x, y) \Rightarrow (\exists y)A_1^2(y, y)$

(c) $(\exists x)(\exists y)A_1^2(x, y) \Rightarrow (\exists y)A_1^2(y, y)$

(d) $[(\exists x)A_1^1(x) \Leftrightarrow (\exists x)A_2^1(x)] \Rightarrow (\forall x)(A_1^1(x) \Leftrightarrow A_2^1(x))$

(e) $(\exists x)(A_1^1(x) \Rightarrow A_2^1(x)) \Rightarrow ((\exists x)A_1^1(x) \Rightarrow (\exists x)A_2^1(x))$

(f) $$[(\forall x)(\forall y)(A_1^2(x, y) \Rightarrow A_1^2(y, x)) \wedge (\forall x)(\forall y)(\forall z)(A_1^2(x, y) \wedge A_1^2(y, z) \Rightarrow A_1^2(x, z))] \Rightarrow (\forall x)A_1^2(x, x)$$

(g)[D] $(\exists x)(\forall y)(A_1^2(x, y) \wedge \neg A_1^2(y, x) \Rightarrow [A_1^2(x, x) \Leftrightarrow A_1^2(y, y)])$

(h) $(\forall x)(\forall y)(\forall z)(A_1^2(x, x) \wedge (A_1^2(x, z) \Rightarrow A_1^2(x, y) \vee A_1^2(y, z))) \Rightarrow (\exists y)(\forall z)A_1^2(y, z)$

(i) $(\exists x)(\forall y)(\exists z)((A_1^2(y, z) \Rightarrow A_1^2(x, z)) \Rightarrow (A_1^2(x, x) \Rightarrow A_1^2(y, x)))$

2.20 Prove that, if the free variables of $\mathscr{A}$ are $y_1, \dots, y_n$, then $\mathscr{A}$ is satisfiable if and only if $(\exists y_1) \dots (\exists y_n)\mathscr{A}$ is satisfiable.

2.21 By introducing appropriate notation, write the sentences of the following arguments as wfs and determine whether the conclusion is logically implied by the conjunction of the premisses.

(a) All scientists are neurotic. No vegetarians are neurotic. Therefore, no vegetarians are scientists.
(b) All men are animals. Some animals are carnivorous. Therefore, some men are carnivorous.
(c) Some geniuses are celibate. Some students are not celibate. Therefore, some students are not geniuses.
(d) Any barber in Jonesville shaves exactly those men in Jonesville who do not shave themselves. Hence, there is no barber in Jonesville.
(e) For any numbers x, y, and z, if $x > y$ and $y > z$, then $x > z$. $x > x$ is false for all numbers x. Therefore, for any numbers x and y, if $x > y$, then it is not the case that $y > x$.
(f) No student in the statistics class is smarter than every student in the logic class. Hence, some student in the logic class is smarter than every student in the statistics class.
(g) Everyone who is sane can understand mathematics. None of Hegel's sons can understand mathematics. No madmen are fit to vote. Hence, none of Hegel's sons is fit to vote.
(h) For every set x, there is a set y such that the cardinality of y is greater than the cardinality of x. If x is included in y, the cardinality of x is not greater than the cardinality of y. Every set is included in V. Hence, V is not a set.

2.22 Determine whether the following sets of wfs or sentences are compatible—that is, whether their conjunction is satisfiable.

(a) $(\exists x)(\forall y)A_1^2(x, y)$
$(\forall x)(\forall y)(\exists z)(A_1^2(x, z) \wedge A_1^2(z, y))$
(b) $(\forall x)(\exists y)A_1^2(y, x)$
$(\forall x)(\forall y)(A_1^2(x, y) \Rightarrow \neg A_1^2(y, x))$
$(\forall x)(\forall y)(\forall z)(A_1^2(x, y) \wedge A_1^2(y, z) \Rightarrow A_1^2(x, z))$
(c) All unicorns are animals.
No unicorns are animals.

2.23 Determine whether the following wfs are logically valid.

(a) $\neg(\exists y)(\forall x)(A_1^2(x, y) \Leftrightarrow \neg A_1^2(x, x))$
(b) $[(\exists x)A_1^1(x) \Rightarrow (\exists x)A_2^1(x)] \Rightarrow (\exists x)(A_1^1(x) \Rightarrow A_2^1(x))$
(c) $(\exists x)(A_1^1(x) \Rightarrow (\forall y)A_1^1(y))$
(d) $(\forall x)(A_1^1(x) \vee A_2^1(x)) \Rightarrow (((\forall x)A_1^1(x)) \vee (\exists x)A_2^1(x))$

2.24 Exhibit a logically valid wf that is not an instance of a tautology. However, show that any logically valid *open* wf (that is, a wf without quantifiers) must be an instance of a tautology.

2.25 (a) Find a satisfiable closed wf that is not true in any interpretation whose domain has only one member.
(b) Find a satisfiable closed wf that is not true in any interpretation whose domain has fewer than three members.

3. FIRST-ORDER THEORIES

In the case of the propositional calculus, the method of truth tables provides an effective test as to whether any given statement form is a tautology. However, there does not seem to be any effective process to determine whether a given wf is logically valid, since, in general, one has to check the truth of a wf for interpretations with arbitrarily large finite or infinite domains. In fact, we shall see later that, according to a plausible definition of "effective", it may actually be proved that there is no effective way to test for logical validity. The axiomatic method, which was a luxury in the study of the propositional calculus, thus appears to be a necessity in the study of wfs involving quantifiers,* and we therefore turn now to the consideration of *first-order theories*.†

The symbols of a first-order theory K are essentially those introduced earlier in this chapter: the propositional connectives $\neg$ and $\Rightarrow$; the punctuation marks (,), and , (the comma is not strictly necessary but is convenient for ease in reading formulas); denumerably many individual variables $x_1, x_2, \ldots$; a finite or denumerable, possibly empty, set of function letters f_j^n $(n, j \geqslant 1)$; a finite or denumerable, possibly empty, set of individual constants a_i $(i \geqslant 1)$; and a nonempty set of predicate letters A_j^n. Thus, in the language of a theory K, some or all of the function letters and individual constants may be absent, and some (but not all) of the predicate letters may be absent. Different theories may differ in which of these symbols they possess.

The definitions given in Section 1 for *term*, *wf*, and the propositional connectives $\wedge$, $\vee$, and $\Leftrightarrow$ are adopted for any first-order theory. Of course, for a particular theory K, only those symbols that occur in K are used in the formation of terms and wfs. Thus, the terms and wfs of K are just those terms and wfs whose symbols are symbols of K. Notice that we require that a theory contain at least one predicate letter; otherwise, it would be impossible to construct any wfs.

The axioms of K are divided into two classes: the logical axioms and the proper (or nonlogical) axioms.

LOGICAL AXIOMS If $\mathscr{A}$, $\mathscr{B}$, and $\mathscr{C}$ are wfs of K, then the following are logical axioms of K:

(A1) $\mathscr{A} \Rightarrow (\mathscr{B} \Rightarrow \mathscr{A})$
(A2) $(\mathscr{A} \Rightarrow (\mathscr{B} \Rightarrow \mathscr{C})) \Rightarrow ((\mathscr{A} \Rightarrow \mathscr{B}) \Rightarrow (\mathscr{A} \Rightarrow \mathscr{C}))$
(A3) $(\neg \mathscr{B} \Rightarrow \neg \mathscr{A}) \Rightarrow ((\neg \mathscr{B} \Rightarrow \mathscr{A}) \Rightarrow \mathscr{B})$

*There is still another reason for a formal axiomatic approach. Concepts and propositions that involve the notion of interpretation and related ideas such as truth and model are often called *semantical* to distinguish them from *syntactical* concepts, which refer to simple relations among symbols and expressions of precise formal languages. Since semantical notions are set-theoretic in character, and since set theory, because of the paradoxes, is considered a rather shaky foundation for the study of mathematical logic, many logicians consider a syntactical approach, consisting of a study of formal axiomatic theories using only rather weak number-theoretic methods, to be much safer. For further discussions, see the pioneering study on semantics by Tarski (1936), Kleene (1952), Church (1956), and Hilbert and Bernays (1934).

†The adjective "first-order" is used to distinguish the theories we shall study from those in which there are predicates having other predicates or functions as arguments or in which predicate quantifiers or function quantifiers are permitted, or both. First-order theories suffice for the expression of known mathematical theories, and, in any case, most higher-order theories can be suitably "translated" into first-order theories. Examples of higher-order theories are discussed by Church (1940), Gödel (1931), Tarski (1933), Hasenjaeger and Scholz (1961), and Van Benthem and Doets (1983). Differences between first-order and higher-order theories are discussed in Corcoran (1980).

(A4) $(\forall x_i)\mathscr{A}(x_i) \Rightarrow \mathscr{A}(t)$ if $\mathscr{A}(x_i)$ is a wf of K and t is a term of K that is free for x_i in $\mathscr{A}(x_i)$. Note here that t may be identical with x_i so that all wfs $(\forall x)\mathscr{A} \Rightarrow \mathscr{A}$ are axioms by virtue of axiom (A4).

(A5) $(\forall x_i)(\mathscr{A} \Rightarrow \mathscr{B}) \Rightarrow (\mathscr{A} \Rightarrow (\forall x_i)\mathscr{B})$ if $\mathscr{A}$ is a wf of K that contains no free occurrences of x_i.

PROPER AXIOMS These cannot be specified, since they vary from theory to theory. A first-order theory in which there are no proper axioms is called a first-order *predicate calculus*.

RULES OF INFERENCE The rules of inference of any first-order theory are:

1. Modus ponens: $\mathscr{B}$ follows from $\mathscr{A}$ and $\mathscr{A} \Rightarrow \mathscr{B}$.
2. Generalization: $(\forall x_i)\mathscr{A}$ follows from $\mathscr{A}$.

We shall use the abbreviations MP and Gen, respectively, to indicate applications of these rules.

Every first-order theory is clearly a formal theory in the sense specified on page 28. Thus, all the terminology (proof, theorem, consequence, etc.) and notation $(\Gamma \vdash \mathscr{A}, \vdash \mathscr{A})$ applicable to formal theories can be used here with respect to first-order theories.

By a *model of a first-order theory K* we mean an interpretation for which all the axioms of K are true.* By (III) and (VI) on page 49, if the rules of modus ponens and generalization are applied to wfs that are true for a given interpretation, then the results of these applications are also true. Hence, *every theorem of K is true in any model of K.*

As we shall see, the logical axioms are so designed that the logical consequences (in the sense defined on page 52) of the closures of the axioms of K are precisely the theorems of K. In particular, if K is a first-order predicate calculus, it turns out that the theorems of K are just those wfs of K that are logically valid.

Some explanation is needed for the restrictions in axiom schemas (A4) and (A5). In the case of (A4), if t were not free for x_i in $\mathscr{A}(x_i)$, the following unpleasant result would arise: let $\mathscr{A}(x_1)$ be $\neg(\forall x_2)A_1^2(x_1, x_2)$ and let t be x_2. Notice that t is not free for x_1 in $\mathscr{A}(x_1)$. Consider the following pseudo-instance of axiom (A4):

$$(\nabla) \quad (\forall x_1)(\neg(\forall x_2)A_1^2(x_1, x_2)) \Rightarrow \neg(\forall x_2)A_1^2(x_2, x_2)$$

Now, take as interpretation any domain with at least two members and let A_1^2 stand for the identity relation. Then the antecedent of (∇) is true and the consequent false.

In the case of axiom (A5), relaxation of the restriction that x_i not be free in $\mathscr{A}$ would lead to the following disaster. Let $\mathscr{A}$ and $\mathscr{B}$ both be $A_1^1(x_1)$. Thus, x_1 is free in $\mathscr{A}$. Consider the following pseudo-instance of axiom (A5):

$$(\nabla\nabla) \quad (\forall x_1)(A_1^1(x_1) \Rightarrow A_1^1(x_1)) \Rightarrow (A_1^1(x_1) \Rightarrow (\forall x_1)A_1^1(x_1))$$

*In talking about an *interpretation of K*, we need to specify only a domain and the interpretations of the symbols of K.

The antecedent of (∇∇) is logically valid. Take as domain the set of integers and let $A_1^1(x)$ mean that x *is even*. Then $(\forall x_1)A_1^1(x_1)$ is false. So any sequence $s = (s_1, s_2, \ldots)$ for which s_1 is even does not satisfy the consequent of (∇∇). Hence, (∇∇) is not true for this interpretation.

Examples of First-Order Theories

1. *Partial order.* Let K have a single predicate letter A_1^2 and no function letters and individual constants. We shall write $x_i < x_j$ instead of $A_1^2(x_i, x_j)$, and $x_i \nless x_j$ for $\neg(x_i < x_j)$. There are two proper axioms:

(a) $(\forall x_1)(x_1 \nless x_1)$ (irreflexivity)
(b) $(\forall x_1)(\forall x_2)(\forall x_3)(x_1 < x_2 \wedge x_2 < x_3 \Rightarrow x_1 < x_3)$ (transitivity)

A model of the theory is called a *partially ordered structure*.

2. *Group theory.* Let K have one predicate letter A_1^2, one function letter f_1^2, and one individual constant a_1. To conform with ordinary notation, we shall write $t = s$ instead of $A_1^2(t, s)$, $t + s$ instead of $f_1^2(t, s)$, and 0 instead of a_1. The proper axioms are:

(a) $(\forall x_1)(\forall x_2)(\forall x_3)(x_1 + (x_2 + x_3) = (x_1 + x_2) + x_3)$ (associativity)
(b) $(\forall x_1)(0 + x_1 = x_1)$ (identity)
(c) $(\forall x_1)(\exists x_2)(x_2 + x_1 = 0)$ (inverse)
(d) $(\forall x_1)(x_1 = x_1)$ (reflexivity of =)
(e) $(\forall x_1)(\forall x_2)(x_1 = x_2 \Rightarrow x_2 = x_1)$ (symmetry of =)
(f) $(\forall x_1)(\forall x_2)(\forall x_3)(x_1 = x_2 \wedge x_2 = x_3 \Rightarrow x_1 = x_3)$ (transitivity of =)
(g) $(\forall x_1)(\forall x_2)(\forall x_3)(x_2 = x_3 \Rightarrow$
$x_1 + x_2 = x_1 + x_3 \wedge x_2 + x_1 = x_3 + x_1)$ (substitutivity of =)

A model for this theory, in which the interpretation of = is the identity relation, is called a *group*. A group is said to be *abelian* if, in addition, the wf $(\forall x_1)(\forall x_2)(x_1 + x_2 = x_2 + x_1)$ is true.

The theories of partial order and of groups are both axiomatic. In general, any theory with a finite number of proper axioms is axiomatic, since it is obvious that one can effectively decide whether any given wf is a logical axiom.

4. PROPERTIES OF FIRST-ORDER THEORIES

All the results in this section refer to an arbitrary first-order theory K unless something is said to the contrary. Instead of writing $\vdash_K$, we shall sometimes simply write $\vdash$. Moreover, since we shall deal in this book with only first-order theories, from now on we shall refer to first-order theories simply as *theories*.

PROPOSITION 2.1 Every wf $\mathscr{A}$ of K that is an instance of a tautology is a theorem of K, and it may be proved using only axioms (A1)–(A3) and MP.

Proof $\mathscr{A}$ arises from a tautology W by substitution. By Proposition 1.13, there is a proof of W in L. In such a proof, make the same substitutions of wfs of

K for statement letters as were used in obtaining $\mathscr{A}$ from W, and, for all statement letters in the proof that do not occur in W, substitute an arbitrary wf of K. Then the resulting sequence of wfs is a proof of $\mathscr{A}$, and this proof uses only axiom schemas (A1)–(A3) and MP.

The application of Proposition 2.1 in a proof will be indicated by writing "Tautology".

PROPOSITION 2.2 Any first-order predicate calculus K is consistent.

Proof For each wf $\mathscr{A}$ of K, let $h[\mathscr{A}]$ be the expression obtained by erasing all the quantifiers and terms in $\mathscr{A}$ (together with the associated commas and parentheses). For example, $h[(\forall x_1)(A_1^2(x_1, x_2) \Rightarrow A_1^1(x_3))]$ is $A_1^2 \Rightarrow A_1^1$, and $h[\neg(\forall x_7)A_2^3(x_4, a_1, x_7) \Rightarrow A_3^1(x_4)]$ is $\neg A_2^3 \Rightarrow A_3^1$. Thus, $h[\mathscr{A}]$ is essentially a statement form, with the symbols A_k^n playing the role of statement letters. Clearly, $h[\neg\mathscr{A}] = \neg h[\mathscr{A}]$ and $h[\mathscr{A} \Rightarrow \mathscr{B}] = h[\mathscr{A}] \Rightarrow h[\mathscr{B}]$. Now, for every axiom $\mathscr{A}$ given by axiom schemas (A1)–(A5), $h[\mathscr{A}]$ is a tautology. This is clear for (A1)–(A3). An instance of (A4), $(\forall x_i)\mathscr{A}(x_i) \Rightarrow \mathscr{A}(t)$, is transformed by h into a tautology of the form $\mathscr{B} \Rightarrow \mathscr{B}$; and an instance of (A5), $(\forall x_i)(\mathscr{A} \Rightarrow \mathscr{B}) \Rightarrow (\mathscr{A} \Rightarrow (\forall x_i)\mathscr{B})$, is transformed into a tautology of the form $(\mathscr{D} \Rightarrow \mathscr{E}) \Rightarrow (\mathscr{D} \Rightarrow \mathscr{E})$. In addition, if $h[\mathscr{A}]$ and $h[\mathscr{A} \Rightarrow \mathscr{B}]$ are tautologies, then, by Proposition 1.1, $h[\mathscr{B}]$ is also a tautology; and, if $h[\mathscr{A}]$ is a tautology, so is $h[(\forall x_i)\mathscr{A}]$, which is the same as $h[\mathscr{A}]$. Hence, $h[\mathscr{A}]$ is a tautology whenever $\mathscr{A}$ is a theorem of K. If there were a wf $\mathscr{B}$ of K such that $\vdash_K \mathscr{B}$ and $\vdash_K \neg\mathscr{B}$, then both $h[\mathscr{B}]$ and $\neg h[\mathscr{B}]$ would be tautologies, which is impossible. Thus, K is consistent. (The transformation h amounts to interpreting K in a domain with a single element. All the theorems of K are true in such an interpretation, but no wf and its negation can be true in any interpretation.)

The Deduction Theorem (Proposition 1.8) for the propositional calculus cannot be carried over without modification to first-order theories. For example, for any wf $\mathscr{A}$, $\mathscr{A} \vdash_K (\forall x_1)\mathscr{A}$, but it is not always the case that $\vdash_K \mathscr{A} \Rightarrow (\forall x_1)\mathscr{A}$. Consider a domain containing at least two elements c and d. Let K be a predicate calculus and let $\mathscr{A}$ be $A_1^1(x_1)$. Interpret A_1^1 as a property that holds only for c. Then $A_1^1(x_1)$ is satisfied by any sequence $s = (s_1, s_2, \ldots)$ in which $s_1 = c$, but $(\forall x_1)A_1^1(x_1)$ is satisfied by no sequence at all. Hence, $A_1^1(x_1) \Rightarrow (\forall x_1)A_1^1(x_1)$ is not true in this interpretation, and so it is not logically valid. But it is easy to see that every theorem of a predicate calculus is logically valid (see Proposition 2.10).

A modified, but still useful, form of the Deduction Theorem may be derived, however. Let $\mathscr{A}$ be a wf in a set Γ of wfs, and assume that we are given a deduction $\mathscr{B}_1, \ldots, \mathscr{B}_n$ from Γ, together with justification for each step of the deduction. We shall say that $\mathscr{B}_i$ *depends upon* $\mathscr{A}$ in this proof if and only if (1) $\mathscr{B}_i$ is $\mathscr{A}$ and the justification for $\mathscr{B}_i$ is that it belongs to Γ or (2) $\mathscr{B}_i$ is justified as a direct consequence by MP or Gen of some preceding wfs of the sequence, where at least one of these preceding wfs depends upon $\mathscr{A}$.

Example $\mathscr{A}, (\forall x_1)\mathscr{A} \Rightarrow \mathscr{C} \vdash (\forall x_1)\mathscr{C}$

$(\mathscr{B}_1)$	$\mathscr{A}$	Hyp
$(\mathscr{B}_2)$	$(\forall x_1)\mathscr{A}$	$(\mathscr{B}_1)$, Gen
$(\mathscr{B}_3)$	$(\forall x_1)\mathscr{A} \Rightarrow \mathscr{C}$	Hyp
$(\mathscr{B}_4)$	$\mathscr{C}$	$(\mathscr{B}_2)$, $(\mathscr{B}_3)$, MP
$(\mathscr{B}_5)$	$(\forall x_1)\mathscr{C}$	$(\mathscr{B}_4)$, Gen

Here, $(\mathscr{B}_1)$ depends upon $\mathscr{A}$, $(\mathscr{B}_2)$ depends upon $\mathscr{A}$, $(\mathscr{B}_3)$ depends upon $(\forall x_1)\mathscr{A} \Rightarrow \mathscr{C}$, $(\mathscr{B}_4)$ depends upon $\mathscr{A}$ and $(\forall x_1)\mathscr{A} \Rightarrow \mathscr{C}$, and $(\mathscr{B}_5)$ depends upon $\mathscr{A}$ and $(\forall x_1)\mathscr{A} \Rightarrow \mathscr{C}$.

PROPOSITION 2.3 If $\mathscr{B}$ does not depend upon $\mathscr{A}$ in a deduction showing that $\Gamma, \mathscr{A} \vdash \mathscr{B}$, then $\Gamma \vdash \mathscr{B}$.

Proof Let $\mathscr{B}_1, \ldots, \mathscr{B}_n = \mathscr{B}$ be a deduction of $\mathscr{B}$ from Γ and $\mathscr{A}$, in which $\mathscr{B}$ does not depend upon $\mathscr{A}$. As an inductive hypothesis, let us assume that the proposition is true for all deductions of length less than n. If $\mathscr{B}$ belongs to Γ or is an axiom, then $\Gamma \vdash \mathscr{B}$. If $\mathscr{B}$ is a direct consequence of one or two preceding wfs, then, since $\mathscr{B}$ does not depend upon $\mathscr{A}$, neither do these preceding wfs. By the inductive hypothesis, these preceding wfs are deducible from Γ alone. Consequently, so is $\mathscr{B}$.

PROPOSITION 2.4 (DEDUCTION THEOREM) Assume that in some deduction showing that $\Gamma, \mathscr{A} \vdash \mathscr{B}$, no application of Gen to a wf that depends upon $\mathscr{A}$ has as its quantified variable a free variable of $\mathscr{A}$. Then $\Gamma \vdash \mathscr{A} \Rightarrow \mathscr{B}$.

Proof Let $\mathscr{B}_1, \ldots, \mathscr{B}_n = \mathscr{B}$ be a deduction of $\mathscr{B}$ from Γ and $\mathscr{A}$, satisfying the assumptions of our proposition. Let us show by induction that $\Gamma \vdash \mathscr{A} \Rightarrow \mathscr{B}_i$ for each $i \leqslant n$. If $\mathscr{B}_i$ is an axiom or belongs to Γ, then $\Gamma \vdash \mathscr{A} \Rightarrow \mathscr{B}_i$, since $\mathscr{B}_i \Rightarrow (\mathscr{A} \Rightarrow \mathscr{B}_i)$ is an axiom. If $\mathscr{B}_i$ is $\mathscr{A}$, then $\Gamma \vdash \mathscr{A} \Rightarrow \mathscr{B}_i$, since, by Proposition 2.1, $\vdash \mathscr{A} \Rightarrow \mathscr{A}$. If there exist j and k less than i such that $\mathscr{B}_k$ is $\mathscr{B}_j \Rightarrow \mathscr{B}_i$, then, by inductive hypothesis, $\Gamma \vdash \mathscr{A} \Rightarrow \mathscr{B}_j$ and $\Gamma \vdash \mathscr{A} \Rightarrow (\mathscr{B}_j \Rightarrow \mathscr{B}_i)$. Now, by axiom (A2), $\vdash (\mathscr{A} \Rightarrow (\mathscr{B}_j \Rightarrow \mathscr{B}_i)) \Rightarrow ((\mathscr{A} \Rightarrow \mathscr{B}_j) \Rightarrow (\mathscr{A} \Rightarrow \mathscr{B}_i))$. Hence, by MP twice, $\Gamma \vdash \mathscr{A} \Rightarrow \mathscr{B}_i$. Finally, suppose that there is some $j < i$ such that $\mathscr{B}_i$ is $(\forall x_k)\mathscr{B}_j$. By hypothesis, $\Gamma \vdash \mathscr{A} \Rightarrow \mathscr{B}_j$ and either $\mathscr{B}_j$ does not depend upon $\mathscr{A}$ or x_k is not a free variable of $\mathscr{A}$. If $\mathscr{B}_j$ does not depend upon $\mathscr{A}$, then, by Proposition 2.3, $\Gamma \vdash \mathscr{B}_j$ and, consequently, by Gen, $\Gamma \vdash (\forall x_k)\mathscr{B}_j$. Thus, $\Gamma \vdash \mathscr{B}_i$. Now, by axiom (A1), $\vdash \mathscr{B}_i \Rightarrow (\mathscr{A} \Rightarrow \mathscr{B}_i)$. So $\Gamma \vdash \mathscr{A} \Rightarrow \mathscr{B}_i$ by MP. If x_k is not a free variable of $\mathscr{A}$, then, by axiom (A5), $\vdash (\forall x_k)(\mathscr{A} \Rightarrow \mathscr{B}_j) \Rightarrow (\mathscr{A} \Rightarrow (\forall x_k)\mathscr{B}_j)$. Since $\Gamma \vdash \mathscr{A} \Rightarrow \mathscr{B}_j$, we have, by Gen, $\Gamma \vdash (\forall x_k)(\mathscr{A} \Rightarrow \mathscr{B}_j)$, and so, by MP, $\Gamma \vdash \mathscr{A} \Rightarrow (\forall x_k)\mathscr{B}_j$; that is, $\Gamma \vdash \mathscr{A} \Rightarrow \mathscr{B}_i$. This completes the induction, and our proposition is just the special case $i = n$.

The hypothesis of Proposition 2.4 is rather cumbersome; the following weaker corollaries often prove to be more useful.

COROLLARY 2.5 If a deduction $\Gamma, \mathscr{A} \vdash \mathscr{B}$, involves no application of Gen of which the quantified variable is free in $\mathscr{A}$, then $\Gamma \vdash \mathscr{A} \Rightarrow \mathscr{B}$.

COROLLARY 2.6 If $\mathscr{A}$ is a closed wf and $\Gamma, \mathscr{A} \vdash \mathscr{B}$, then $\Gamma \vdash \mathscr{A} \Rightarrow \mathscr{B}$.

In Propositions 2.3–2.6, the following additional conclusion can be drawn from the proofs. The new proof of $\Gamma \vdash \mathscr{A} \Rightarrow \mathscr{B}$ (in Proposition 2.3, of $\Gamma \vdash \mathscr{B}$) involves an application of Gen to a wf depending upon a wf $\mathscr{C}$ of Γ only if there is an application of Gen in the given proof of $\Gamma, \mathscr{A} \vdash \mathscr{B}$, that involves the same quantified variable and is applied to a wf that depends upon $\mathscr{C}$. (In the proof of Proposition 2.4, one should observe that $\mathscr{B}_j$ depends upon a premiss $\mathscr{C}$ of Γ in the original proof if and only if $\mathscr{A} \Rightarrow \mathscr{B}_j$ depends upon $\mathscr{C}$ in the new proof.)

This supplementary conclusion will be useful when we wish to apply the Deduction Theorem several times in a row to a given deduction—for example, to obtain $\Gamma \vdash \mathscr{D} \Rightarrow (\mathscr{A} \Rightarrow \mathscr{B})$ from $\Gamma, \mathscr{D}, \mathscr{A} \vdash \mathscr{B}$; from now on, it is to be considered an integral part of the statements of Propositions 2.3–2.6.

Example $\vdash (\forall x_1)(\forall x_2)\mathscr{A} \Rightarrow (\forall x_2)(\forall x_1)\mathscr{A}$

Proof

1. $(\forall x_1)(\forall x_2)\mathscr{A}$	Hyp
2. $(\forall x_1)(\forall x_2)\mathscr{A} \Rightarrow (\forall x_2)\mathscr{A}$	(A4)
3. $(\forall x_2)\mathscr{A}$	1, 2, MP
4. $(\forall x_2)\mathscr{A} \Rightarrow \mathscr{A}$	(A4)
5. $\mathscr{A}$	3, 4, MP
6. $(\forall x_1)\mathscr{A}$	5, Gen
7. $(\forall x_2)(\forall x_1)\mathscr{A}$	6, Gen

Thus, by 1–7, we have $(\forall x_1)(\forall x_2)\mathscr{A} \vdash (\forall x_2)(\forall x_1)\mathscr{A}$, where, in the deduction, no application of Gen has as a quantified variable a free variable of $(\forall x_1)(\forall x_2)\mathscr{A}$. Hence, by Corollary 2.5, $\vdash (\forall x_1)(\forall x_2)\mathscr{A} \Rightarrow (\forall x_2)(\forall x_1)\mathscr{A}$.

Exercises

2.26 Derive the following theorems.

(a) $\vdash (\forall x)(\mathscr{A} \Rightarrow \mathscr{B}) \Rightarrow ((\forall x)\mathscr{A} \Rightarrow (\forall x)\mathscr{B})$

(b) $\vdash (\forall x)(\mathscr{A} \Rightarrow \mathscr{B}) \Rightarrow ((\exists x)\mathscr{A} \Rightarrow (\exists x)\mathscr{B})$

(c) $\vdash (\forall x)(\mathscr{A} \wedge \mathscr{B}) \Leftrightarrow ((\forall x)\mathscr{A}) \wedge (\forall x)\mathscr{B}$

(d) $\vdash (\forall y_1) \ldots (\forall y_n)\mathscr{A} \Rightarrow \mathscr{A}$

(e) $\vdash \neg(\forall x)\mathscr{B} \Rightarrow (\exists x)\neg\mathscr{B}$

2.27[D] Let K be a first-order theory and let $K^{\#}$ be an axiomatic theory having the following axioms: (a) $(\forall y_1) \ldots (\forall y_n)\mathscr{A}$, where $\mathscr{A}$ is any axiom of K and $y_1, \ldots, y_n$ $(n \geqslant 0)$ are any variables (none at all when $n = 0$), and (b) $(\forall y_1) \ldots (\forall y_n)(\mathscr{A} \Rightarrow \mathscr{B}) \Rightarrow [(\forall y_1) \ldots (\forall y_n)\mathscr{A} \Rightarrow (\forall y_1) \ldots (\forall y_n)\mathscr{B}]$ where $\mathscr{A}$ and $\mathscr{B}$ are any wfs and $y_1, \ldots, y_n$ are any variables. Moreover, $K^{\#}$ has modus ponens as its *only* rule of inference. Show that $K^{\#}$ has the same theorem as K. (Thus, at the expense of adding more axioms, the generalization rule can be dispensed with.)

5. ADDITIONAL METATHEOREMS AND DERIVED RULES

For the sake of smoothness in working with particular theories later, we shall introduce various techniques for constructing proofs. In this section it is assumed that we are dealing with an arbitrary theory K.

Often one wants to obtain $\mathscr{A}(t)$ from $(\forall x)\mathscr{A}(x)$, where t is a term free for x in $\mathscr{A}(x)$. This is allowed by the following *derived rule*.

PARTICULARIZATION RULE A4 If t is free for x in $\mathscr{A}(x)$, then $(\forall x)\mathscr{A}(x) \vdash \mathscr{A}(t)$.

Proof From $(\forall x)\mathscr{A}(x)$ and the instance $(\forall x)\mathscr{A}(x) \Rightarrow \mathscr{A}(t)$ of axiom (A4), we obtain $\mathscr{A}(t)$ by modus ponens.

There is another very useful derived rule, which is essentially the contrapositive of Particularization Rule A4.

EXISTENTIAL RULE E4 Let t be a term that is free for x in a wf $\mathscr{A}(x, t)$, and let $\mathscr{A}(t, t)$ arise from $\mathscr{A}(x, t)$ by replacing all free occurrences of x by t.* Then, $\mathscr{A}(t, t) \vdash (\exists x)\mathscr{A}(x, t)$.

Proof It suffices to show that $\vdash \mathscr{A}(t, t) \Rightarrow (\exists x)\mathscr{A}(x, t)$. But, by axiom (A4), $\vdash (\forall x) \neg \mathscr{A}(x, t) \Rightarrow \neg \mathscr{A}(t, t)$. Hence, by the tautology, $(A \Rightarrow \neg B) \Rightarrow (B \Rightarrow \neg A)$ and, by MP, $\vdash \mathscr{A}(t, t) \Rightarrow \neg (\forall x) \neg \mathscr{A}(x, t)$, which, in abbreviated form, is $\vdash \mathscr{A}(t, t) \Rightarrow (\exists x)\mathscr{A}(x, t)$.

A special case of Rule E4 is $\mathscr{B}(t) \vdash (\exists x)\mathscr{B}(x)$, whenever t is free for x in $\mathscr{B}(x)$. In particular, when t is x itself, $\mathscr{B}(x) \vdash (\exists x)\mathscr{B}(x)$.

Example $\vdash (\forall x)\mathscr{A} \Rightarrow (\exists x)\mathscr{A}$

1. $(\forall x)\mathscr{A}$	Hyp
2. $\mathscr{A}$	1, Rule A4
3. $(\exists x)\mathscr{A}$	2, Rule E4
4. $(\forall x)\mathscr{A} \vdash (\exists x)\mathscr{A}$	1–3
5. $\vdash (\forall x)\mathscr{A} \Rightarrow (\exists x)\mathscr{A}$	1–4, Corollary 2.5

The following derived rules are extremely useful.

NEGATION: Elimination $\neg\neg\mathscr{A} \vdash \mathscr{A}$
Introduction $\mathscr{A} \vdash \neg\neg\mathscr{A}$

CONJUNCTION: Elimination $\mathscr{A} \wedge \mathscr{B} \vdash \mathscr{A}$
$\mathscr{A} \wedge \mathscr{B} \vdash \mathscr{B}$
$\neg(\mathscr{A} \wedge \mathscr{B}) \vdash \neg\mathscr{A} \vee \neg\mathscr{B}$
Introduction $\mathscr{A}, \mathscr{B} \vdash \mathscr{A} \wedge \mathscr{B}$

DISJUNCTION: Elimination $\mathscr{A} \Rightarrow \mathscr{C}; \mathscr{B} \Rightarrow \mathscr{C}, \mathscr{A} \vee \mathscr{B} \vdash \mathscr{C}$
$\neg(\mathscr{A} \vee \mathscr{B}) \vdash \neg\mathscr{A} \wedge \neg\mathscr{B}$
$\mathscr{A} \vee \mathscr{B}, \neg\mathscr{A} \vdash \mathscr{B}$
$\mathscr{A} \vee \mathscr{B}, \neg\mathscr{B} \vdash \mathscr{A}$
Introduction $\mathscr{A} \vdash \mathscr{A} \vee \mathscr{B}$
$\mathscr{B} \vdash \mathscr{A} \vee \mathscr{B}$

* $\mathscr{A}(x, t)$ may or may not contain occurrences of t.

CONDITIONAL: Elimination $\mathscr{A} \Rightarrow \mathscr{B}, \neg\mathscr{B} \vdash \neg\mathscr{A}$
$\neg(\mathscr{A} \Rightarrow \mathscr{B}) \vdash \mathscr{A}$
$\neg(\mathscr{A} \Rightarrow \mathscr{B}) \vdash \neg\mathscr{B}$

BICONDITIONAL: Elimination $\mathscr{A} \Leftrightarrow \mathscr{B}, \mathscr{A} \vdash \mathscr{B}$ $\quad$ $\mathscr{A} \Leftrightarrow \mathscr{B}, \neg\mathscr{A} \vdash \neg\mathscr{B}$
$\mathscr{A} \Leftrightarrow \mathscr{B}, \mathscr{B} \vdash \mathscr{A}$ $\quad$ $\mathscr{A} \Leftrightarrow \mathscr{B}, \neg\mathscr{B} \vdash \neg\mathscr{A}$
$\mathscr{A} \Leftrightarrow \mathscr{B} \vdash \mathscr{A} \Rightarrow \mathscr{B}$ $\quad$ $\mathscr{A} \Leftrightarrow \mathscr{B} \vdash \mathscr{B} \Rightarrow \mathscr{A}$
Introduction $\mathscr{A} \Rightarrow \mathscr{B}, \mathscr{B} \Rightarrow \mathscr{A} \vdash \mathscr{A} \Leftrightarrow \mathscr{B}$

PROOF BY CONTRADICTION: If a proof of $\Gamma, \neg\mathscr{A} \vdash \mathscr{C} \wedge \neg\mathscr{C}$ involves no application of Gen using a variable free in $\mathscr{A}$, then $\Gamma \vdash \mathscr{A}$. (Similarly, one obtains $\Gamma \vdash \neg\mathscr{A}$ from $\Gamma, \mathscr{A} \vdash \mathscr{C} \wedge \neg\mathscr{C}$.)

These derived rules will be designated in obvious ways. For example, the use of $\mathscr{A} \wedge \mathscr{B} \vdash \mathscr{A}$ will be indicated by "Conjunction Elimination".

Exercises

2.28 Justify the derived rules listed above.

2.29 Prove the following.

(a) $\vdash (\forall x)(\forall y)A_1^2(x, y) \Rightarrow (\forall x)A_1^2(x, x)$
(b) $\vdash [(\forall x)\mathscr{A}] \vee [(\forall x)\mathscr{B}] \Rightarrow (\forall x)(\mathscr{A} \vee \mathscr{B})$
(c) $\vdash \neg(\exists x)\mathscr{A} \Leftrightarrow (\forall x)\neg\mathscr{A}$
(d) $\vdash (\forall x)\mathscr{A} \Rightarrow (\forall x)(\mathscr{A} \vee \mathscr{B})$
(e) $\vdash (\forall x)(\forall y)(A_1^2(x, y) \Rightarrow \neg A_1^2(y, x)) \Rightarrow (\forall x)\neg A_1^2(x, x)$
(f) $\vdash [(\exists x)\mathscr{A} \Rightarrow (\forall x)\mathscr{B}] \Rightarrow (\forall x)(\mathscr{A} \Rightarrow \mathscr{B})$
(g) $\vdash (\forall x)(\mathscr{A} \vee \mathscr{B}) \Rightarrow [(\forall x)\mathscr{A}] \vee (\exists x)\mathscr{B}$
(h) $\vdash (\forall x)(A_1^2(x, x) \Rightarrow (\exists y)A_1^2(x, y))$
(i) $\vdash (\forall x)(\mathscr{A} \Rightarrow \mathscr{B}) \Rightarrow [(\forall x)\neg\mathscr{B} \Rightarrow (\forall x)\neg\mathscr{A}]$
(j) $\vdash (\exists y)[A_1^1(y) \Rightarrow (\forall y)A_1^1(y)]$

2.30 Assume that $\mathscr{A}$ and $\mathscr{B}$ are wfs and x is not free in $\mathscr{A}$. Prove the following.

(a) $\vdash \mathscr{A} \Leftrightarrow (\forall x)\mathscr{A}$
(b) $\vdash \mathscr{A} \Leftrightarrow (\exists x)\mathscr{A}$
(c) $\vdash (\mathscr{A} \Rightarrow (\forall x)\mathscr{B}) \Leftrightarrow (\forall x)(\mathscr{A} \Rightarrow \mathscr{B})$
(d) $\vdash ((\exists x)\mathscr{B} \Rightarrow \mathscr{A}) \Leftrightarrow (\forall x)(\mathscr{B} \Rightarrow \mathscr{A})$

We need a derived rule that will allow us to replace a part $\mathscr{B}$ of a wf $\mathscr{A}$ by a wf that is provably equivalent to $\mathscr{B}$. For this purpose, we first must prove the following auxiliary result.

LEMMA 2.7 For any wfs $\mathscr{A}$ and $\mathscr{B}$, $\vdash (\forall x)(\mathscr{A} \Leftrightarrow \mathscr{B}) \Rightarrow ((\forall x)\mathscr{A} \Leftrightarrow (\forall x)\mathscr{B})$.

Proof

1. $(\forall x)(\mathscr{A} \Leftrightarrow \mathscr{B})$	Hyp
2. $(\forall x)\mathscr{A}$	Hyp
3. $\mathscr{A} \Leftrightarrow \mathscr{B}$	1, Rule A4
4. $\mathscr{A}$	2, Rule A4

5. $\mathscr{B}$	3, 4, Biconditional Elimination
6. $(\forall x)\mathscr{B}$	5, Gen
7. $(\forall x)(\mathscr{A} \Leftrightarrow \mathscr{B}), (\forall x)\mathscr{A} \vdash (\forall x)\mathscr{B}$	1–6
8. $(\forall x)(\mathscr{A} \Leftrightarrow \mathscr{B}) \vdash (\forall x)\mathscr{A} \Rightarrow (\forall x)\mathscr{B}$	1–7, Corollary 2.5
9. $(\forall x)(\mathscr{A} \Leftrightarrow \mathscr{B}) \vdash (\forall x)\mathscr{B} \Rightarrow (\forall x)\mathscr{A}$	Proof like that of 8
10. $(\forall x)(\mathscr{A} \Leftrightarrow \mathscr{B}) \vdash (\forall x)\mathscr{A} \Leftrightarrow (\forall x)\mathscr{B}$	8, 9, Biconditional Introduction
11. $\vdash (\forall x)(\mathscr{A} \Leftrightarrow \mathscr{B}) \Rightarrow ((\forall x)\mathscr{A} \Leftrightarrow (\forall x)\mathscr{B})$	1–10, Corollary 2.5

PROPOSITION 2.8 If $\mathscr{B}$ is a subformula of $\mathscr{A}$, $\mathscr{A}'$ is the result of replacing zero or more occurrences of $\mathscr{B}$ in $\mathscr{A}$ by a wf $\mathscr{C}$, and every free variable of $\mathscr{B}$ or $\mathscr{C}$ that is also a bound variable of $\mathscr{A}$ occurs in the list $y_1, \ldots, y_k$, then:

(a) $\vdash [(\forall y_1) \ldots (\forall y_k)(\mathscr{B} \Leftrightarrow \mathscr{C})] \Rightarrow (\mathscr{A} \Leftrightarrow \mathscr{A}')$ (Equivalence Theorem)

(b) If $\vdash \mathscr{B} \Leftrightarrow \mathscr{C}$, then $\vdash \mathscr{A} \Leftrightarrow \mathscr{A}'$ (Replacement Theorem)

(c) If $\vdash \mathscr{B} \Leftrightarrow \mathscr{C}$ and $\vdash \mathscr{A}$, then $\vdash \mathscr{A}'$

Example (a) $\vdash (\forall x)(A_1^1(x) \Leftrightarrow A_2^1(x)) \Rightarrow [(\exists x)A_1^1(x) \Leftrightarrow (\exists x)A_2^1(x)]$

Proof (a) We use induction on the number of connectives and quantifiers in $\mathscr{A}$. Note that, if zero occurrences are replaced, $\mathscr{A}'$ is $\mathscr{A}$ and the wf to be proved is an instance of the tautology $B \Rightarrow (A \Leftrightarrow A)$. Note also that, if $\mathscr{B}$ is identical with $\mathscr{A}$ and this occurrence of $\mathscr{B}$ is replaced by $\mathscr{C}$, the wf to be proved $(\forall y_1) \ldots (\forall y_k)(\mathscr{B} \Leftrightarrow \mathscr{C}) \Rightarrow (\mathscr{B} \Leftrightarrow \mathscr{C})$ is derivable by Exercise 2.26(d). Thus, we may assume that $\mathscr{B}$ is a proper part of $\mathscr{A}$ and that at least one occurrence of $\mathscr{B}$ is replaced. Our inductive hypothesis is that the result holds for all wfs with fewer connectives and quantifiers than $\mathscr{A}$.

Case 1. $\mathscr{A}$ is an atomic wf. Then $\mathscr{B}$ cannot be a proper part of $\mathscr{A}$.

Case 2. $\mathscr{A}$ is $\neg\mathscr{D}$. Let $\mathscr{A}'$ be $\neg\mathscr{D}'$. By inductive hypothesis, $\vdash (\forall y_1) \ldots (\forall y_k)(\mathscr{B} \Leftrightarrow \mathscr{C}) \Rightarrow (\mathscr{D} \Leftrightarrow \mathscr{D}')$. Hence, by a suitable instance of the tautology $(C \Rightarrow (A \Leftrightarrow B)) \Rightarrow (C \Rightarrow (\neg A \Leftrightarrow \neg B))$ and MP, we obtain $\vdash (\forall y_1) \ldots (\forall y_k)(\mathscr{B} \Leftrightarrow \mathscr{C}) \Rightarrow (\mathscr{A} \Leftrightarrow \mathscr{A}')$.

Case 3. $\mathscr{A}$ is $\mathscr{D} \Rightarrow \mathscr{E}$. Let $\mathscr{A}'$ be $\mathscr{D}' \Rightarrow \mathscr{E}'$. By inductive hypothesis, $\vdash (\forall y_1) \ldots (\forall y_k)(\mathscr{B} \Leftrightarrow \mathscr{C}) \Rightarrow (\mathscr{D} \Leftrightarrow \mathscr{D}')$ and $\vdash (\forall y_1) \ldots (\forall y_k)(\mathscr{B} \Leftrightarrow \mathscr{C}) \Rightarrow (\mathscr{E} \Leftrightarrow \mathscr{E}')$. Using a suitable instance of the tautology

$$(A \Rightarrow (B \Leftrightarrow C)) \wedge (A \Rightarrow (D \Leftrightarrow E)) \Rightarrow (A \Rightarrow [(B \Rightarrow D) \Leftrightarrow (C \Rightarrow E)])$$

we obtain $\vdash (\forall y_1) \ldots (\forall y_k)(\mathscr{B} \Leftrightarrow \mathscr{C}) \Rightarrow (\mathscr{A} \Leftrightarrow \mathscr{A}')$.

Case 4. $\mathscr{A}$ is $(\forall x)\mathscr{D}$. Let $\mathscr{A}'$ be $(\forall x)\mathscr{D}'$. By inductive hypothesis, $\vdash (\forall y_1) \ldots (\forall y_k)(\mathscr{B} \Leftrightarrow \mathscr{C}) \Rightarrow (\mathscr{D} \Leftrightarrow \mathscr{D}')$. Now, x does not occur free in $(\forall y_1) \ldots (\forall y_k)(\mathscr{B} \Leftrightarrow \mathscr{C})$ because, if it did, it would be free in $\mathscr{B}$ or $\mathscr{C}$ and, since it is bound in $\mathscr{A}$, it would be one of $y_1, \ldots, y_k$ and x would not be free in $(\forall y_1) \ldots (\forall y_k)(\mathscr{B} \Leftrightarrow \mathscr{C})$. Hence, using axiom (A5), we obtain $\vdash (\forall y_1) \ldots (\forall y_k)(\mathscr{B} \Leftrightarrow \mathscr{C}) \Rightarrow (\forall x)(\mathscr{D} \Leftrightarrow \mathscr{D}')$. However, by Lemma 2.7, $\vdash (\forall x)(\mathscr{D} \Leftrightarrow \mathscr{D}') \Rightarrow ((\forall x)\mathscr{D} \Leftrightarrow (\forall x)\mathscr{D}')$. Then, by a suitable tautology and MP, $\vdash (\forall y_1) \ldots (\forall y_k)(\mathscr{B} \Leftrightarrow \mathscr{C}) \Rightarrow (\mathscr{A} \Leftrightarrow \mathscr{A}')$.

(b) From $\vdash \mathscr{B} \Leftrightarrow \mathscr{C}$, by several applications of Gen, we obtain $\vdash (\forall y_1) \ldots (\forall y_k)(\mathscr{B} \Leftrightarrow \mathscr{C})$. Then, by part (a) and MP, $\vdash \mathscr{A} \Leftrightarrow \mathscr{A}'$.

(c) Use part (b) and Biconditional Elimination.

Exercises

2.31 Prove the following.

(a) $\vdash (\exists x)\neg\mathscr{A} \Leftrightarrow \neg(\forall x)\mathscr{A}$

(b) $\vdash (\forall x)\mathscr{A} \Leftrightarrow \neg(\exists x)\neg\mathscr{A}$

(c) $\vdash (\exists x)(\mathscr{A} \Rightarrow \neg(\mathscr{B} \vee \mathscr{C})) \Leftrightarrow (\exists x)(\mathscr{A} \Rightarrow \neg\mathscr{B} \wedge \neg\mathscr{C})$

(d) $\vdash (\forall x)(\exists y)(\mathscr{A} \Rightarrow \mathscr{B}) \Leftrightarrow (\forall x)(\exists y)(\neg\mathscr{A} \vee \mathscr{B})$

2.32 Show by a counterexample that we cannot omit the quantifiers $(\forall y_1) \ldots (\forall y_k)$ in Proposition 2.8(a).

2.33 If $\mathscr{B}$ is obtained from $\mathscr{A}$ by erasing all quantifiers $(\forall x)$ or $(\exists x)$ whose scope does not contain x free, prove that $\vdash \mathscr{A} \Leftrightarrow \mathscr{B}$.

2.34 For each wf $\mathscr{A}$ below, find a wf $\mathscr{B}$ such that $\vdash \mathscr{B} \Leftrightarrow \neg\mathscr{A}$ and negation signs in $\mathscr{B}$ apply only to atomic wfs.

(a) $(\forall x)(\forall y)(\exists z)A_1^3(x, y, z)$

(b) $(\forall \varepsilon)(\varepsilon > 0 \Rightarrow (\exists \delta)(\delta > 0 \wedge (\forall x)(|x - c| < \delta \Rightarrow |f(x) - f(c)| < \varepsilon)))$

(c) $(\forall \varepsilon)(\varepsilon > 0 \Rightarrow (\exists n)(\forall m)(m > n \Rightarrow |a_m - b| < \varepsilon))$

2.35 Let $\mathscr{A}$ be a wf that does not contain $\Rightarrow$ and $\Leftrightarrow$. Exchange universal and existential quantifiers, and exchange $\wedge$ and $\vee$. The result $\mathscr{A}^*$ is called the *dual* of $\mathscr{A}$.

(a) In any predicate calculus, prove:

(i) $\vdash \mathscr{A}$ if and only if $\vdash \neg\mathscr{A}^*$,

(ii) $\vdash \mathscr{A} \Rightarrow \mathscr{B}$ if and only if $\vdash \mathscr{B}^* \Rightarrow \mathscr{A}^*$,

(iii) $\vdash \mathscr{A} \Leftrightarrow \mathscr{B}$ if and only if $\vdash \mathscr{A}^* \Leftrightarrow \mathscr{B}^*$, and

(iv) $\vdash (\exists x)(\mathscr{A} \vee \mathscr{B}) \Leftrightarrow [((\exists x)\mathscr{A}) \vee (\exists x)\mathscr{B}]$. [*Hint:* Use Exercise 2.26(c).]

(b) Show that the duality results of part (a), (i)–(iii), do not hold for arbitrary theories.

6. RULE C

It is very common in mathematics to reason in the following way: Assume that we have proved a wf of the form $(\exists x)\mathscr{A}(x)$. Then we say, let b be an object such that $\mathscr{A}(b)$. We continue the proof, finally arriving at a formula that does not involve the arbitrarily chosen element b.

For example, let us say that we wish to show that $(\exists x)(\mathscr{B}(x) \Rightarrow \mathscr{C}(x)), (\forall x)\mathscr{B}(x) \vdash (\exists x)\mathscr{C}(x)$.

1. $(\exists x)(\mathscr{B}(x) \Rightarrow \mathscr{C}(x))$		Hyp
2. $(\forall x)\mathscr{B}(x)$		Hyp
3. $\mathscr{B}(b) \Rightarrow \mathscr{C}(b)$	for some b	1
4. $\mathscr{B}(b)$		2, Rule A4
5. $\mathscr{C}(b)$		3, 4, MP
6. $(\exists x)\mathscr{C}(x)$		5, Rule E4

Such a proof seems to be perfectly legitimate on an intuitive basis. In fact, we can achieve the same result without making an arbitrary choice of an element b as in step 3. This can be done as follows:

1. $(\forall x)\mathscr{B}(x)$	Hyp
2. $(\forall x)\neg\mathscr{C}(x)$	Hyp
3. $\mathscr{B}(x)$	1, Rule A4
4. $\neg\mathscr{C}(x)$	2, Rule A4

5. $\neg(\mathscr{B}(x) \Rightarrow \mathscr{C}(x))$ 3, 4, Tautology $(A \wedge \neg B) \Rightarrow \neg(A \Rightarrow B)$, MP
6. $(\forall x)\neg(\mathscr{B}(x) \Rightarrow \mathscr{C}(x))$ 5, Gen
7. $(\forall x)\mathscr{B}(x), (\forall x)\neg\mathscr{C}(x) \vdash (\forall x)\neg(\mathscr{B}(x) \Rightarrow \mathscr{C}(x))$ 1–6
8. $(\forall x)\mathscr{B}(x) \vdash (\forall x)\neg\mathscr{C}(x) \Rightarrow (\forall x)\neg(\mathscr{B}(x) \Rightarrow \mathscr{C}(x))$ 1–7, Corollary 2.5
9. $(\forall x)\mathscr{B}(x) \vdash \neg(\forall x)\neg(\mathscr{B}(x) \Rightarrow \mathscr{C}(x)) \Rightarrow \neg(\forall x)\neg\mathscr{C}(x)$
8, Tautology $(A \Rightarrow B) \Rightarrow (\neg B \Rightarrow \neg A)$, MP
10. $(\forall x)\mathscr{B}(x) \vdash (\exists x)(\mathscr{B}(x) \Rightarrow \mathscr{C}(x)) \Rightarrow (\exists x)\mathscr{C}(x)$ Abbreviation of 9
11. $(\exists x)(\mathscr{B}(x) \Rightarrow \mathscr{C}(x)), (\forall x)\mathscr{B}(x) \vdash (\exists x)\mathscr{C}(x)$ 10, MP

In general, any wf that can be proved using a finite number of arbitrary choices can also be proved without such acts of choice. We shall call the rule that permits us to go from $(\exists x)\mathscr{A}(x)$ to $\mathscr{A}(b)$, *Rule C* ("C" for "choice"). More precisely, a Rule C deduction in a first-order theory K is defined in the following manner:

$\Gamma \vdash_C \mathscr{A}$ if and only if there is a sequence of wfs $\mathscr{B}_1, \ldots, \mathscr{B}_n$ such that $\mathscr{B}_n$ is $\mathscr{A}$ and the following four conditions hold:

1. For each $i \leqslant n$, either
   (a) $\mathscr{B}_i$ is an axiom of K, or
   (b) $\mathscr{B}_i$ is in Γ, or
   (c) $\mathscr{B}_i$ follows by MP or Gen from preceding wfs in the sequence, or
   (d) there is a preceding wf $(\exists x)\mathscr{C}(x)$ such that $\mathscr{B}_i$ is $\mathscr{C}(d)$, where d is a new individual constant (Rule C).
2. As axioms in condition 1(a), we can also use all logical axioms that involve the new individual constants already introduced by applications of condition 1(d).
3. No application of Gen is made using a variable that is free in some $(\exists x)\mathscr{C}(x)$ to which Rule C has been previously applied.
4. $\mathscr{A}$ contains none of the new individual constants introduced in any application of Rule C.

A word should be said about the reason for including condition 3. Without this clause, we could proceed as follows:

1. $(\forall x)(\exists y)A_1^2(x, y)$ Hyp
2. $(\exists y)A_1^2(x, y)$ 1, Rule A4
3. $A_1^2(x, b)$ 2, Rule C
4. $(\forall x)A_1^2(x, b)$ 3, Gen
5. $(\exists y)(\forall x)A_1^2(x, y)$ 4, Rule E4

However, there is an interpretation for which $(\forall x)(\exists y)A_1^2(x, y)$ is true but $(\exists y)(\forall x)A_1^2(x, y)$ is false (see Example 5 on page 53).

PROPOSITION 2.9 If $\Gamma \vdash_C \mathscr{A}$, then $\Gamma \vdash \mathscr{A}$. Moreover, from the following proof it is easy to verify that, if there is an application of Gen in the new proof of $\mathscr{A}$ from Γ using a certain variable and applied to a wf depending upon a certain wf of Γ, then there was such an application of Gen in the original proof.*

*The first formulation of a version of Rule C similar to that given here seems to be due to Rosser (1953).

Proof Let $(\exists y_1)\mathscr{C}_1(y_1)$, ..., $(\exists y_k)\mathscr{C}_k(y_k)$ be the wfs, in order of occurrence, to which Rule C is applied in the proof of $\Gamma \vdash_C \mathscr{A}$, and let $c_1, \ldots, c_k$ be the corresponding new individual constants. Then, $\Gamma, \mathscr{C}_1(c_1), \ldots, \mathscr{C}_k(c_k) \vdash \mathscr{A}$. Now, by condition 3 of the definition above, Corollary 2.5 is applicable, yielding $\Gamma, \mathscr{C}_1(c_k)$, $\ldots, \mathscr{C}_{k-1}(c_{k-1}) \vdash \mathscr{C}_k(c_k) \Rightarrow \mathscr{A}$. We replace c_k everywhere by a variable z that does not occur in the proof. Then

$\Gamma, \mathscr{C}_1(c_1), \ldots, \mathscr{C}_{k-1}(c_{k-1}) \vdash \mathscr{C}_k(z) \Rightarrow \mathscr{A}$	and, by Gen,
$\Gamma, \mathscr{C}_1(c_1), \ldots, \mathscr{C}_{k-1}(c_{k-1}) \vdash (\forall z)(\mathscr{C}_k(z) \Rightarrow \mathscr{A})$	Hence, by Exercise 2.30(d),
$\Gamma, \mathscr{C}_1(c_1), \ldots, \mathscr{C}_{k-1}(c_{k-1}) \vdash (\exists y_k)\mathscr{C}_k(y_k) \Rightarrow \mathscr{A}$	But,
$\Gamma, \mathscr{C}_1(c_1), \ldots, \mathscr{C}_{k-1}(c_{k-1}) \vdash (\exists y_k)\mathscr{C}_k(y_k)$	Hence,
$\Gamma, \mathscr{C}_1(c_1), \ldots, \mathscr{C}_{k-1}(c_{k-1}) \vdash \mathscr{A}$	

Repeating this argument, we can eliminate $\mathscr{C}_{k-1}(c_{k-1}), \ldots, \mathscr{C}_1(c_1)$ one after the other, finally obtaining $\Gamma \vdash \mathscr{A}$.

Example $\vdash (\forall x)(\mathscr{A}(x) \Rightarrow \mathscr{B}(x)) \Rightarrow ((\exists x)\mathscr{A}(x) \Rightarrow (\exists x)\mathscr{B}(x))$

1.	$(\forall x)(\mathscr{A}(x) \Rightarrow \mathscr{B}(x))$	Hyp
2.	$(\exists x)\mathscr{A}(x)$	Hyp
3.	$\mathscr{A}(b)$	2, Rule C
4.	$\mathscr{A}(b) \Rightarrow \mathscr{B}(b)$	1, Rule A4
5.	$\mathscr{B}(b)$	3, 4, MP
6.	$(\exists x)\mathscr{B}(x)$	5, Rule E4
7.	$(\forall x)(\mathscr{A}(x) \Rightarrow \mathscr{B}(x)), (\exists x)\mathscr{A}(x) \vdash_C (\exists x)\mathscr{B}(x)$	1–6
8.	$(\forall x)(\mathscr{A}(x) \Rightarrow \mathscr{B}(x)), (\exists x)\mathscr{A}(x) \vdash (\exists x)\mathscr{B}(x)$	7, Proposition 2.9
9.	$(\forall x)(\mathscr{A}(x) \Rightarrow \mathscr{B}(x)) \vdash (\exists x)\mathscr{A}(x) \Rightarrow (\exists x)\mathscr{B}(x)$	1–8, Corollary 2.5
10.	$\vdash (\forall x)(\mathscr{A}(x) \Rightarrow \mathscr{B}(x)) \Rightarrow ((\exists x)\mathscr{A}(x) \Rightarrow (\exists x)\mathscr{B}(x))$	1–9, Corollary 2.5

Exercises

Use Rule C and Proposition 2.9 to prove Exercises 2.36–2.43.

2.36 $\vdash (\exists x)(\mathscr{A}(x) \Rightarrow \mathscr{B}(x)) \Rightarrow ((\forall x)\mathscr{A}(x) \Rightarrow (\exists x)\mathscr{B}(x))$

2.37 $\vdash \neg(\exists y)(\forall x)(A_1^2(x, y) \Leftrightarrow \neg A_1^2(x, x))$

2.38 $\vdash [(\forall x)(A_1^1(x) \Rightarrow A_2^1(x) \lor A_3^1(x)) \land \neg(\forall x)(A_1^1(x) \Rightarrow A_2^1(x))] \Rightarrow (\exists x)(A_1^1(x) \land A_3^1(x))$

2.39 $\vdash [(\exists x)\mathscr{A}(x)] \land [(\forall x)\mathscr{B}(x)] \Rightarrow (\exists x)(\mathscr{A}(x) \land \mathscr{B}(x))$

2.40 $\vdash (\exists x)\mathscr{B}(x) \Rightarrow (\exists x)(\mathscr{A}(x) \lor \mathscr{B}(x))$

2.41 $\vdash (\exists x)(\exists y)\mathscr{A}(x, y) \Leftrightarrow (\exists y)(\exists x)\mathscr{A}(x, y)$

2.42 $\vdash (\exists x)(\forall y)\mathscr{A}(x, y) \Rightarrow (\forall y)(\exists x)\mathscr{A}(x, y)$

2.43 $\vdash (\exists x)(\mathscr{A}(x) \land \mathscr{B}(x)) \Rightarrow ((\exists x)\mathscr{A}(x)) \land (\exists x)\mathscr{B}(x)$

2.44 What is wrong with the following alleged derivations?

(a)

1.	$(\exists x)\mathscr{A}(x)$	Hyp
2.	$\mathscr{A}(b)$	1, Rule C
3.	$(\exists x)\mathscr{B}(x)$	Hyp
4.	$\mathscr{B}(b)$	3, Rule C
5.	$\mathscr{A}(b) \land \mathscr{B}(b)$	2, 4, Conjunction Introduction
6.	$(\exists x)(\mathscr{A}(x) \land \mathscr{B}(x))$	5, Rule E4
7.	$(\exists x)\mathscr{A}(x), (\exists x)\mathscr{B}(x) \vdash (\exists x)(\mathscr{A}(x) \land \mathscr{B}(x))$	1–6, Proposition 2.9

(b)	1. $(\exists x)(\mathscr{A}(x) \Rightarrow \mathscr{B}(x))$	Hyp
	2. $(\exists x)\mathscr{A}(x)$	Hyp
	3. $\mathscr{A}(b) \Rightarrow \mathscr{B}(b)$	1, Rule C
	4. $\mathscr{A}(b)$	2, Rule C
	5. $\mathscr{B}(b)$	3, 4, MP
	6. $(\exists x)\mathscr{B}(x)$	5, Rule E4
	7. $(\exists x)(\mathscr{A}(x) \Rightarrow \mathscr{B}(x)), (\exists x)\mathscr{A}(x) \vdash (\exists x)\mathscr{B}(x)$	1–6, Proposition 2.9

7. COMPLETENESS THEOREMS

PROPOSITION 2.10 Every theorem of a first-order predicate calculus is logically valid.

Proof Axioms (A1)–(A3) are logically valid, by property (VII) of the notion of truth (see page 49), and axioms (A4) and (A5) are logically valid by properties (X) and (XI). By properties (III) and (VI), the rules of inference MP and Gen preserve logical validity. Hence, every theorem of a predicate calculus is logically valid.

Proposition 2.10 establishes only half of the completeness result that we are seeking. The other half will follow from a much more general proposition established later. First we must prove a few preliminary lemmas.

If x_i and x_j are distinct, then $\mathscr{A}(x_i)$ and $\mathscr{A}(x_j)$ are said to be *similar* if and only if x_j is free for x_i in $\mathscr{A}(x_i)$ and $\mathscr{A}(x_i)$ has no free occurrences of x_j. It is assumed here that $\mathscr{A}(x_j)$ arises from $\mathscr{A}(x_i)$ by substituting x_j for all free occurrences of x_i. If $\mathscr{A}(x_i)$ and $\mathscr{A}(x_j)$ are similar, then x_i is free for x_j in $\mathscr{A}(x_j)$ and $\mathscr{A}(x_j)$ has no free occurrences of x_i. Thus, similarity is a symmetric relation. Intuitively, $\mathscr{A}(x_i)$ and $\mathscr{A}(x_j)$ are similar if and only if $\mathscr{A}(x_i)$ and $\mathscr{A}(x_j)$ are the same except that $\mathscr{A}(x_j)$ has free occurrences of x_j in exactly those places where $\mathscr{A}(x_i)$ has free occurrences of x_i.

LEMMA 2.11 If $\mathscr{A}(x_i)$ and $\mathscr{A}(x_j)$ are similar, then $\vdash (\forall x_i)\mathscr{A}(x_i) \Leftrightarrow (\forall x_j)\mathscr{A}(x_j)$.

Proof $\vdash (\forall x_i)\mathscr{A}(x_i) \Rightarrow \mathscr{A}(x_j)$ by axiom (A4). Then, by Gen, $\vdash (\forall x_j)((\forall x_i)\mathscr{A}(x_i) \Rightarrow \mathscr{A}(x_j))$, and so, by axiom (A5) and MP, $\vdash (\forall x_i)\mathscr{A}(x_i) \Rightarrow (\forall x_j)\mathscr{A}(x_j)$. Similarly, $\vdash (\forall x_j)\mathscr{A}(x_j) \Rightarrow (\forall x_i)\mathscr{A}(x_i)$. Hence, by Biconditional Introduction, $\vdash (\forall x_i)\mathscr{A}(x_i) \Leftrightarrow (\forall x_j)\mathscr{A}(x_j)$.

Exercises

2.45 If $\mathscr{A}(x_i)$ and $\mathscr{A}(x_j)$ are similar, prove that $\vdash (\exists x_i)\mathscr{A}(x_i) \Leftrightarrow (\exists x_j)\mathscr{A}(x_j)$.

2.46 *Change of bound variables.* If $\mathscr{B}(x)$ is similar to $\mathscr{B}(y)$, $(\forall x)\mathscr{B}(x)$ is a subformula of $\mathscr{A}$, and $\mathscr{A}'$ is the result of replacing one or more occurrences of $(\forall x)\mathscr{B}(x)$ in $\mathscr{A}$ by $(\forall y)\mathscr{B}(y)$, prove that $\vdash \mathscr{A} \Leftrightarrow \mathscr{A}'$.

LEMMA 2.12 If a closed wf $\neg\mathscr{A}$ of K is not provable in K, then the theory K′, obtained from K by adding $\mathscr{A}$ as an axiom, is consistent.

Proof Assume K′ is inconsistent. Then, for some wf $\mathscr{B}$, $\vdash_{K'} \mathscr{B}$ and $\vdash_{K'} \neg\mathscr{B}$. Now, $\vdash_{K'} \mathscr{B} \Rightarrow (\neg\mathscr{B} \Rightarrow \neg\mathscr{A})$ by Proposition 2.1. So, by two applications of MP, $\vdash_{K'} \neg\mathscr{A}$. Hence, $\mathscr{A} \vdash_K \neg\mathscr{A}$. Since $\mathscr{A}$ is closed, we have $\vdash_K \mathscr{A} \Rightarrow \neg\mathscr{A}$ by Corollary 2.6. However, by Proposition 2.1, $\vdash_K (\mathscr{A} \Rightarrow \neg\mathscr{A}) \Rightarrow \neg\mathscr{A}$. Hence, by MP, $\vdash_K \neg\mathscr{A}$, contradicting our hypothesis.

Similarly, if $\mathscr{A}$ is not provable in K, then the new theory obtained by adding $\neg\mathscr{A}$ as an axiom to K is consistent.

LEMMA 2.13 The set of expressions of a theory K is denumerable. Hence, the same is true of the set of terms, wfs, and closed wfs.

Proof First assign a distinct positive integer $g(u)$ to each symbol u as follows: $g(() = 3$, $g()) = 5$, $g(,) = 7$, $g(\neg) = 9$, $g(\Rightarrow) = 11$, $g(\forall) = 13$, $g(x_k) = 13 + 8k$, $g(a_k) = 7 + 8k$, $g(f_k^n) = 1 + 8(2^n 3^k)$, and $g(A_k^n) = 3 + 8(2^n 3^k)$. Then, to an expression $u_1 u_2 \dots u_r$ associate the number $2^{g(u_1)} 3^{g(u_2)} \dots p_r^{g(u_r)}$, where p_j is the jth prime number, starting with $p_0 = 2$. We can enumerate all expressions in the order of their associated numbers.

Moreover, if we can effectively tell whether any given symbol is a symbol of K, then this enumeration can be effectively carried out, and, in addition, we can effectively decide whether any given number is the number of an expression of K. The same holds true for terms, wfs, closed wfs, and so on. If K is also axiomatic—that is, if we can effectively decide whether any given wf is an axiom of K—then we can effectively enumerate the theorems of K as follows: starting with a list consisting of the first axiom of K in the given enumeration (according to the associated numbers) of the axioms, add all the direct consequences of this axiom by MP and by Gen used with only x_1 as quantified variable. Add the second axiom to this new list (if it is not already there) and write all new direct consequences of the wfs in this augmented list, this time with Gen used with only x_1 and x_2. If at the kth step, we add the kth axiom and restrict Gen to the variables $x_1, \dots, x_k$, we eventually obtain, in this manner, all theorems of K. However, in contradistinction to the case of expressions, wfs, terms, and so on, it turns out that there are theories K for which we cannot tell in advance whether any given wf of K will eventually appear in the list of theorems.

DEFINITIONS A theory K is said to be *complete* if, for any closed wf $\mathscr{A}$ of K, either $\vdash_K \mathscr{A}$ or $\vdash_K \neg\mathscr{A}$. A theory K′ is said to be an *extension* of a theory K if every theorem of K is a theorem of K′.

LEMMA 2.14 (LINDENBAUM'S LEMMA) If K is a consistent theory, then there is a consistent, complete extension of K.

Proof Let $\mathscr{B}_1, \mathscr{B}_2, \dots$ be an enumeration of all closed wfs of K by Lemma 2.13. Define a sequence $J_0, J_1, J_2, \dots$ of theories in the following way. J_0 is K. Assume J_n is defined, with $n \geqslant 0$. If it is not the case that $\vdash_{J_n} \neg\mathscr{B}_{n+1}$, then let J_{n+1} be obtained from J_n by adding $\mathscr{B}_{n+1}$ as an additional axiom. On the other hand, if $\vdash_{J_n} \neg\mathscr{B}_{n+1}$, let $J_{n+1} = J_n$. Let J be the theory obtained by taking as axioms all the axioms of all the J_i's. Clearly, J_{n+1} is an extension of J_n, and J is an extension of all the J_i's, including $J_0 = K$. To show that J is consistent, it suffices to prove that all the J_i's are consistent because a proof of a contradiction in J, involving as it does only a finite number of axioms, is also a proof of a contradiction in some J_n. We prove the consistency of the J_i's by induction. By hypothesis, $J_0 = K$ is consistent. Assume that J_i is consistent. If $J_{i+1} = J_i$, then J_{i+1} is consistent. If $J_i \neq J_{i+1}$ and, therefore, by the definition of J_{i+1}, $\neg\mathscr{B}_{i+1}$ is not provable in J_i, then, by Lemma 2.12, J_{i+1} is also consistent. Hence, J_{i+1} is consistent if J_i is, and, therefore, J is consistent. To prove the completeness of J, let $\mathscr{A}$ be any closed wf of K. Then $\mathscr{A} = \mathscr{B}_{j+1}$ for some $j \geqslant 0$. Now, either $\vdash_{J_j} \neg\mathscr{B}_{j+1}$ or $\vdash_{J_{j+1}} \mathscr{B}_{j+1}$, since, if not $\vdash_{J_j} \neg\mathscr{B}_{j+1}$, then $\mathscr{B}_{j+1}$ is added as an axiom in J_{j+1}. Therefore, either $\vdash_J \neg\mathscr{B}_{j+1}$ or $\vdash_J \mathscr{B}_{j+1}$. Thus, J is a consistent, complete extension of K.

Note that even if one can effectively determine whether any wf is an axiom of K, it may not be possible to do the same with (or even to effectively enumerate) the axioms of J; that is, J may not be axiomatic even if K is. This is due to the possibility of not being able to determine, at each step, whether or not $\neg\mathscr{B}_{n+1}$ is provable in J_n.

Exercises

2.47 Show that a theory K is complete if and only if, for any closed wfs $\mathscr{A}$ and $\mathscr{B}$ of K, if $\vdash_K \mathscr{A} \vee \mathscr{B}$, then $\vdash_K \mathscr{A}$ or $\vdash_K \mathscr{B}$.

2.48[D] Prove that every consistent, decidable theory has a consistent, decidable, complete extension.

DEFINITIONS A *closed term* is a term without variables.

A theory K is called a *scapegoat theory* if, for any wf $\mathscr{A}(x)$ that has x as its only free variable, there is a closed term t such that

$$\vdash_K (\exists x)\neg\mathscr{A}(x) \Rightarrow \neg\mathscr{A}(t)$$

LEMMA 2.15 Every consistent theory K has a consistent extension K′ such that K′ is a scapegoat theory and K′ contains denumerably many closed terms.

Proof Add to the symbols of K a denumerable set $\{b_1, b_2, \ldots\}$ of new individual constants. Call this new theory K_0. Its axioms are those of K plus those logical axioms that involve the symbols of K and the new constants. K_0 is consistent. For, if not, $\vdash_{K_0} \mathscr{A} \wedge \neg\mathscr{A}$ for some wf $\mathscr{A}$. Replace each b_i appearing in this proof by a variable that does not appear in the proof. This transforms axioms into axioms and preserves the correctness of the applications of the rules of inference. The final wf in the proof is still a contradiction, but now the proof does not involve any of the b_i's and therefore is a proof in K. This contradicts the consistency of K. Therefore, K_0 is consistent.

By Lemma 2.13, let $F_1(x_{i_1}), F_2(x_{i_2}), \ldots, F_k(x_{i_k}), \ldots$ be an enumeration of all wfs of K_0 that have one free variable. Choose a sequence $b_{j_1}, b_{j_2}, \ldots$ of some of the new individual constants such that b_{j_k} is not contained in any of the wfs $F_1(x_{i_1}), \ldots, F_k(x_{i_k})$

and such that b_{j_k} is different from each of $b_{j_1}, b_{j_2}, \ldots, b_{j_{k-1}}$. Consider the wf:

$$(S_k) \quad (\exists x_{i_k}) \neg F_k(x_{i_k}) \Rightarrow \neg F_k(b_{j_k})$$

Let K_n be the theory obtained by adding $(S_1), \ldots, (S_n)$ to the axioms of K_0, and let K_∞ be the theory obtained by adding all the (S_i)'s as axioms to K_0. Any proof in K_∞ contains only a finite number of the (S_i)'s and, therefore, will also be a proof in some K_n. Hence, if all the K_n's are consistent, so is K_∞. To demonstrate that all the K_n's are consistent, proceed by induction. We know that K_0 is consistent. Assume that K_{n-1} is consistent but that K_n is inconsistent ($n \geqslant 1$). Then, as we know, any wf is provable in K_n [by the tautology $\neg A \Rightarrow (A \Rightarrow B)$ and Proposition 2.1]. In particular, $\vdash_{K_n} \neg(S_n)$. Hence, $(S_n) \vdash_{K_{n-1}} \neg(S_n)$. Since (S_n) is closed, we have, by Corollary 2.6, $\vdash_{K_{n-1}} (S_n) \Rightarrow \neg(S_n)$. But, by the tautology $(A \Rightarrow \neg A) \Rightarrow \neg A$, Proposition 2.1, and MP, we then have $\vdash_{K_{n-1}} \neg(S_n)$; that is, $\vdash_{K_{n-1}} \neg[(\exists x_{i_n}) \neg F_n(x_{i_n}) \Rightarrow \neg F_n(b_{j_n})]$. Now, by Conditional Elimination, we obtain $\vdash_{K_{n-1}} (\exists x_{i_n}) \neg F_n(x_{i_n})$ and $\vdash_{K_{n-1}} \neg\neg F_n(b_{j_n})$ and then, by Negation Elimination, $\vdash_{K_{n-1}} F_n(b_{j_n})$. From the latter and the fact that b_{j_n} does not occur in $(S_0), \ldots, (S_{n-1})$, we conclude $\vdash_{K_{n-1}} F_n(x_p)$, where x_p is a variable that does not occur in the proof of $F_n(b_{j_n})$. (Simply replace in the proof all occurrences of b_{j_n} by x_p.) By Gen, $\vdash (\forall x_p) F_n(x_p)$ and then, by Lemma 2.11 and Biconditional Elimination, $\vdash_{K_{n-1}} (\forall x_{i_n}) F_n(x_{i_n})$. [We use the fact that $F_n(x_p)$ and $F_n(x_{i_n})$ are similar.] But, we already have $\vdash_{K_{n-1}} (\exists x_{i_n}) \neg F_n(x_{i_n})$, which is an abbreviation of $\vdash_{K_{n-1}} \neg(\forall x_{i_n}) \neg\neg F_n(x_{i_n})$, whence, by the Replacement Theorem, $\vdash_{K_{n-1}} \neg(\forall x_{i_n}) F_n(x_{i_n})$, contradicting the hypothesis that K_{n-1} is consistent. Hence, K_n must also be consistent. Thus, K_∞ is consistent, it is an extension of K, and it is clearly a scapegoat theory.

LEMMA 2.16 Let J be a consistent, complete, scapegoat theory. Then J has a model M whose domain is the set D of closed terms of J.

Proof For any individual constant a_i of J, let $(a_i)^M = a_i$. For any function letter f_k^n of J and for any closed terms $t_1, \ldots, t_n$ of J, let $(f_k^n)^M(t_1, \ldots, t_n) = f_k^n(t_1, \ldots, t_n)$. [Notice that $f_k^n(t_1, \ldots, t_n)$ is a closed term. Hence, $(f_k^n)^M$ is an n-ary operation on D.] For any predicate letter A_k^n of J and any closed terms $t_1, \ldots, t_n$ of J, let $(A_k^n)^M$ consist of all n-tuples $\langle t_1, \ldots, t_n \rangle$ of closed terms of J such that $\vdash_J A_k^n(t_1, \ldots, t_n)$. It now suffices to show that, for any closed wf $\mathscr{A}$ of J:

$$(\square) \quad \models_M \mathscr{A} \text{ if and only if } \vdash_J \mathscr{A}$$

[If this is established and $\mathscr{B}$ is any axiom of J, let $\mathscr{A}$ be the closure of $\mathscr{B}$. By Gen, $\vdash_J \mathscr{A}$. By $(\square)$, $\models_M \mathscr{A}$. By (VI) from page 49, $\models_M \mathscr{B}$. Hence, M would be a model of J.] The proof of $(\square)$ is by induction on the number r of connectives and quantifiers in $\mathscr{A}$. Assume that $(\square)$ holds for all closed wfs with fewer than r connectives and quantifiers.

Case 1. $\mathscr{A}$ is a closed atomic wf $A_k^n(t_1, \ldots, t_n)$. Then $(\square)$ is a direct consequence of the definition of $(A_k^n)^M$.

Case 2. $\mathscr{A}$ is $\neg\mathscr{B}$. If $\mathscr{A}$ is true for M, then $\mathscr{B}$ is false for M and so, by inductive hypothesis, not-$\vdash_J \mathscr{B}$. Since J is complete and $\mathscr{B}$ is closed, $\vdash_J \neg\mathscr{B}$—that is, $\vdash_J \mathscr{A}$.

On the other hand, if $\mathscr{A}$ is not true for M, then $\mathscr{B}$ is true for M. Hence, $\vdash_J \mathscr{B}$. Since J is consistent, not-$\vdash_J \neg\mathscr{B}$—that is, not-$\vdash_J \mathscr{A}$.

Case 3. $\mathscr{A}$ is $\mathscr{B} \Rightarrow \mathscr{C}$. Since $\mathscr{A}$ is closed, so are $\mathscr{B}$ and $\mathscr{C}$. If $\mathscr{A}$ is false for M, then $\mathscr{B}$ is true and $\mathscr{C}$ is false. Hence, by inductive hypothesis, $\vdash_J \mathscr{B}$ and not-$\vdash_J \mathscr{C}$. By the completeness of J, $\vdash_J \neg\mathscr{C}$. Therefore, by an instance of the tautology $B \Rightarrow (\neg C \Rightarrow \neg(B \Rightarrow C))$ and two applications of MP, $\vdash_J \neg(\mathscr{B} \Rightarrow \mathscr{C})$—that is, $\vdash_J \neg\mathscr{A}$—and so, by the consistency of J, not-$\vdash_J \mathscr{A}$. On the other hand, if not-$\vdash_J \mathscr{A}$, then, by the completeness of J, $\vdash_J \neg\mathscr{A}$. By Conditional Elimination, $\vdash_J \mathscr{B}$ and $\vdash_J \neg\mathscr{C}$. Hence, $\mathscr{B}$ is true for M. By the consistency of J, not-$\vdash_J \mathscr{C}$ and, therefore, $\mathscr{C}$ is false for M. Thus, $\mathscr{A}$ is false for M.

Case 4. $\mathscr{A}$ is $(\forall x_m)\mathscr{B}$.

Case 4a. $\mathscr{B}$ is a closed wf. By inductive hypothesis, $\models_M \mathscr{B}$ if and only if $\vdash_J \mathscr{B}$. By Exercise 2.30a, $\vdash_J \mathscr{B} \Leftrightarrow (\forall x_m)\mathscr{B}$. So, $\vdash_J \mathscr{B}$ if and only if $\vdash_J (\forall x_m)\mathscr{B}$ by Biconditional Elimination. Moreover, $\models_M \mathscr{B}$ if and only if $\models_M (\forall x_m)\mathscr{B}$ by property (VI) on page 49. Hence, $\models_M \mathscr{A}$ if and only if $\vdash_J \mathscr{A}$.

Case 4b. $\mathscr{B}$ is not a closed wf. Since $\mathscr{A}$ is closed, $\mathscr{B}$ has x_m as its only free variable—say, $\mathscr{B}$ is $F(x_m)$. Then $\mathscr{A}$ is $(\forall x_m)F(x_m)$.

(i) Assume $\models_M \mathscr{A}$ and not-$\vdash_J \mathscr{A}$. By the completeness of J, $\vdash_J \neg\mathscr{A}$—that is, $\vdash_J \neg(\forall x_m)F(x_m)$. Then, by Exercise 2.31(a) and Biconditional Elimination, $\vdash_J (\exists x_m)\neg F(x_m)$. Since J is a scapegoat theory, $\vdash_J \neg F(t)$ for some closed term t of J. But $\models_M \mathscr{A}$—that is, $\models_M (\forall x_m)F(x_m)$. Since $(\forall x_m)F(x_m) \Rightarrow F(t)$ is true for M by property (X) on page 50, $\models_M F(t)$. Hence, by inductive hypothesis, $\vdash_J F(t)$. This contradicts the consistency of J. Thus, if $\models_M \mathscr{A}$, then $\vdash_J \mathscr{A}$.

(ii) Assume $\vdash_J \mathscr{A}$ and not-$\models_M \mathscr{A}$. Thus, (*) $\vdash_J (\forall x_m)F(x_m)$ and (**) not-$\models_M (\forall x_m)F(x_m)$. By (**), some sequence of elements of the domain D does not satisfy $(\forall x_m)F(x_m)$. Hence, some sequence s does not satisfy $F(x_m)$. Let t be the ith component of s. Notice that $s^*(u) = u$ for all closed terms u of J [by the definition of $(a_i)^M$ and $(f_j)^M$]. Hence, by Lemma 2(a) on page 51, s does not satisfy $F(t)$. So, $F(t)$ is false for M. But, by (*) and Rule A4, $\vdash_J F(t)$, and so, by inductive hypothesis, $\models_M F(t)$. This contradiction shows that, if $\vdash_J \mathscr{A}$, then $\models_M \mathscr{A}$.

Now we can prove the fundamental theorem of quantification theory. By a *denumerable model* we mean a model in which the domain is denumerable.

PROPOSITION 2.17† Every consistent theory K has a denumerable model.

Proof By Lemma 2.15, K has a consistent extension K′ such that K′ is a scapegoat theory and has denumerably many closed terms. By Lindenbaum's Lemma, K′ has a consistent, complete extension J that has the same symbols as K′. Hence, J is also a scapegoat theory. By Lemma 2.16, J has a model M whose domain is the denumerable set of closed terms of J. Since J is an extension of K, M is a denumerable model of K.

† The proof given here is essentially due to Henkin (1949), as simplified by Hasenjaeger (1953). The result was originally proved by Gödel (1930). Other proofs have been published by Rasiowa and Sikorski (1951, 1952) and Beth (1951), using (Boolean) algebraic and topological methods, respectively. Still other proofs may be found in Hintikka (1955a, b) and in Beth (1959).

COROLLARY 2.18 Any logically valid wf $\mathscr{A}$ of a theory K is a theorem of K.

Proof We need consider only closed wfs $\mathscr{A}$, since a wf $\mathscr{B}$ is logically valid if and only if its closure is logically valid, and $\mathscr{B}$ is provable in K if and only if its closure is provable in K. So, let $\mathscr{A}$ be a logically valid closed wf of K. Assume that not-$\vdash_K \mathscr{A}$. By Lemma 2.12, if we add $\neg\mathscr{A}$ as a new axiom to K, the new theory K′ is consistent. Hence, by Proposition 2.17, K′ has a model M. Since $\neg\mathscr{A}$ is an axiom of K′, $\neg\mathscr{A}$ is true for M. But, since $\mathscr{A}$ is logically valid, $\mathscr{A}$ is true for M. Hence, $\mathscr{A}$ is both true and false for M, which is impossible [by (II) on page 49]. Thus, $\mathscr{A}$ must be a theorem of K.

COROLLARY 2.19 (GÖDEL'S COMPLETENESS THEOREM, 1930) In any predicate calculus, the theorems are precisely the logically valid wfs.

Proof The corollary is proved by Proposition 2.10 and Corollary 2.18. [Gödel's original proof runs along quite different lines. For a constructive proof of a related result, see Herbrand (1930, 1971), and for still other proofs, see Dreben (1952), Hintikka (1955a, b), Beth (1951), and Rasiowa and Sikorski (1951, 1952).]

COROLLARY 2.20 Let K be any theory.

(a) A wf $\mathscr{A}$ is true in every denumerable model of K if and only if $\vdash_K \mathscr{A}$. Hence, $\mathscr{A}$ is true in every model of K if and only if $\vdash_K \mathscr{A}$.
(b) If, in every model of K, every sequence that satisfies all wfs in a set Γ of wfs also satisfies a wf $\mathscr{B}$, then $\Gamma \vdash_K \mathscr{B}$.
(c) If a wf $\mathscr{B}$ of K is a logical consequence (see page 52) of a set Γ of wfs of K, then $\Gamma \vdash_K \mathscr{B}$.
(d) If a wf $\mathscr{B}$ of K is a logical consequence of a wf $\mathscr{A}$ of K, then $\mathscr{A} \vdash_K \mathscr{B}$.

Proof (a) We may assume $\mathscr{A}$ is closed. If not-$\vdash_K \mathscr{A}$, then the theory $K' = K + \{\neg\mathscr{A}\}$ is consistent.* Hence, K′ has a denumerable model M. However, $\neg\mathscr{A}$, being an axiom of K′, is true for M; and since M is also a model of K, $\mathscr{A}$ is true for M. Therefore, $\mathscr{A}$ is true and false for M, which is impossible.

(b) Consider the theory $K + \Gamma$. The wf $\mathscr{B}$ is true for every model of this theory. Hence, by part (a), $\vdash_{K+\Gamma} \mathscr{B}$. So, $\Gamma \vdash_K \mathscr{B}$.

Part (c) is a consequence of (b), and part (d) is a special case of (c).

Corollaries 2.18–2.20 show that the syntactical approach to quantification theory by means of first-order theories is equivalent to the semantical approach through the notions of interpretations, models, logical validity, and so on. For the propositional calculus, Corollary 1.14 demonstrated the analogous equivalence between the semantical notion (tautology) and the syntactical notion (theorem of L). Notice also that, in the propositional calculus, the completeness of the system L (see Proposition 1.13) led to a solution of the decision problem. However, for

*If K is a theory and Δ is a set of wfs of K, then $K + \Delta$ denotes the theory obtained from K by adding the wfs of Δ as additional axioms.

first-order theories, we cannot obtain a decision procedure for logical validity or, equivalently, for provability in first-order predicate calculi. We shall prove this and related results in Chapter 3, Section 6.

COROLLARY 2.21 (SKOLEM-LÖWENHEIM THEOREM, 1919, 1915) Any theory K that has a model has a denumerable model.

Proof If K has model, then K is consistent [by (II) on page 49]. Hence, by Proposition 2.17, K has a denumerable model.

The following stronger consequence of Proposition 2.17 is derivable.

COROLLARY 2.22[A] For any cardinal number $\alpha \geqslant \aleph_0$, any consistent theory K has a model of cardinality α.

Proof By Proposition 2.17, we know that K has a denumerable model. Therefore, it suffices to prove the following lemma.

LEMMA If α and β are two cardinal numbers such that $\alpha \leqslant \beta$ and if K has a model of cardinality α, then K has a model of cardinality β.

Proof Let M be a model of K with domain D of cardinality α. Let D' be a set of cardinality β that contains D. Extend the model M to an interpretation M′ that has D' as domain in the following way. Let c be a fixed element of D. We stipulate that the elements of $D' - D$ behave like c. For example, if B_j^n is the interpretation in M of the predicate letter A_j^n and $(B_j^n)'$ is the new interpretation in M′, then, for any $d_1, \ldots, d_n$ in D', $(B_j^n)'$ holds for $(d_1, \ldots, d_n)$ if and only if B_j^n holds for $(u_1, \ldots, u_n)$, where $u_i = d_i$ if $d_i \in D$ and $u_i = c$ if $d_i \in D' - D$. The interpretation of the function letters is extended in an analogous way, and the individual constants have the same interpretations as in M. It is an easy exercise to show, by induction on the number of connectives and quantifiers in a wf $\mathscr{A}$, that $\mathscr{A}$ is true for M′ if and only if it is true for M. Hence, M′ is a model of K of cardinality β.

Exercises

2.49 For any theory K, if $\Gamma \vdash_K \mathscr{A}$ and each wf in Γ is true for a model M of K, show that $\mathscr{A}$ is also true for M.

2.50 If a wf $\mathscr{A}$ without quantifiers is provable in a predicate calculus, prove that $\mathscr{A}$ is an instance of a tautology and hence, by Proposition 2.1, has a proof without quantifiers using only axioms (A1)–(A3) and MP. (*Hint:* If $\mathscr{A}$ were not a tautology, one could construct an interpretation, having the set of terms that occur in $\mathscr{A}$ as its domain, for which $\mathscr{A}$ is not true, contradicting Proposition 2.10.) Note that this implies the consistency of the predicate calculus and also provides a decision procedure for the provability of wfs without quantifiers.

2.51 Show that $\vdash_K \mathscr{A}$ if and only if there is a wf $\mathscr{C}$ that is the closure of the conjunction of some axioms of K such that $\mathscr{C} \Rightarrow \mathscr{A}$ is logically valid.

2.52 *Compactness.* If all finite subsets of the set of axioms of a theory K have models, prove that K has a model.

2.53[A] If, for some cardinal $\alpha \geqslant \aleph_0$, a wf $\mathscr{A}$ is true for every interpretation of cardinality α, prove that $\mathscr{A}$ is logically valid.

2.54[A] If a wf $\mathscr{A}$ is true for all interpretations of cardinality α, prove that $\mathscr{A}$ is true for all interpretations of cardinality $\leqslant \alpha$.

2.55 (a) For any wf $\mathscr{A}$, prove that there is only a finite number of interpretations of $\mathscr{A}$ on a given domain of finite cardinality k.

(b) For any wf $\mathscr{A}$, prove that there is an effective way of determining whether $\mathscr{A}$ is true for all interpretations with domain of some fixed finite cardinality k.

(c) Let a wf $\mathscr{A}$ be called *k-valid* if it is true for all interpretations that have a domain of k elements. Call $\mathscr{A}$ *precisely k-valid* if it is k-valid but not $(k+1)$-valid. Show that $(k+1)$-validity implies k-validity and give an example of a wf that is precisely k-valid. [See Hilbert and Bernays (1934, §4–5) and Wajsberg (1933).]

2.56 Show that the following wf is true for all finite domains but is false for some infinite domain:

$$((\forall x)(\forall y)(\forall z)[A_1^2(x,x) \wedge (A_1^2(x,y) \wedge A_1^2(y,z) \Rightarrow A_1^2(x,z)) \wedge (A_1^2(x,y) \vee A_1^2(y,x))])$$
$$\Rightarrow (\exists y)(\forall x)A_1^2(y,x)$$

2.57 Prove that there is no theory K whose models are exactly the interpretations with finite domains.

2.58 Let $\mathscr{A}$ be any wf that contains no quantifiers, function letters, or individual constants.

(a) Show that a closed *prenex* wf $(\forall x_1)\ldots(\forall x_n)(\exists y_1)\ldots(\exists y_m)\mathscr{A}$, with $m \geqslant 0$ and $n \geqslant 1$, is logically valid if and only if it is true for every interpretation with a domain of n objects.

(b) Prove that a closed prenex wf $(\exists y_1)\ldots(\exists y_m)\mathscr{A}$ is logically valid if and only if it is true for all interpretations with a domain of one element.

(c) Show that there is an effective procedure to determine the logical validity of all wfs of the forms given in parts (a) and (b).

2.59 Let K_1 and K_2 be theories having the same symbols. Assume that any interpretation of K_1 is a model of K_1 if and only if it is not a model of K_2. Prove that K_1 and K_2 are finitely axiomatizable; that is, there are finite sets of sentences Γ and Δ such that, for any sentence $\mathscr{A}$, $\vdash_{K_1} \mathscr{A}$ if and only if $\Gamma \vdash \mathscr{A}$, and $\vdash_{K_2} \mathscr{A}$ if and only if $\Delta \vdash \mathscr{A}$.*

2.60[D] A set Γ of sentences is called an *independent axiomatization* of a theory K if (a) all sentences in Γ are theorems of K, (b) $\Gamma \vdash \mathscr{A}$ for every theorem $\mathscr{A}$ of K, and (c) for every sentence $\mathscr{B}$ of Γ, it is not the case that $\Gamma - \{\mathscr{B}\} \vdash \mathscr{B}$.* Prove that every theory K has an independent axiomatization.

8. FIRST-ORDER THEORIES WITH EQUALITY

Let K be a theory that has as one of its predicate letters A_1^2. Let us write $t = s$ as an abbreviation for $A_1^2(t,s)$ and $t \neq s$ as an abbreviation for $\neg A_1^2(t,s)$. Then K is called a *first-order theory with equality* (or simply a *theory with equality*) if the following are theorems of K:

(A6)	$(\forall x_1)(x_1 = x_1)$	(reflexivity of equality)
(A7)	$x = y \Rightarrow (\mathscr{A}(x,x) \Rightarrow \mathscr{A}(x,y))$	(substitutivity of equality)

* Here, an expression $\Gamma \vdash \mathscr{A}$, without any subscript attached to $\vdash$, means that $\mathscr{A}$ is derivable from Γ using only logical axioms—that is, within the predicate calculus.

where x and y are any variables, $\mathscr{A}(x, x)$ is any wf, and $\mathscr{A}(x, y)$ arises from $\mathscr{A}(x, x)$ by replacing some, but not necessarily all, free occurrences of x by y, with the proviso that y is free for x in $\mathscr{A}(x, x)$. Thus, $\mathscr{A}(x, y)$ may or may not contain free occurrences of x.

The numbering (A6) and (A7) is a continuation of the numbering of the logical axioms on pages 55–56.

PROPOSITION 2.23 In any theory with equality,

(a) $\vdash t = t$ for any term t
(b) $\vdash x = y \Rightarrow y = x$
(c) $\vdash x = y \Rightarrow (y = z \Rightarrow x = z)$

Proof (a) By (A6), $\vdash (\forall x_1)(x_1 = x_1)$; hence, by Rule A4, $\vdash t = t$.

(b) Let $\mathscr{A}(x, x)$ be $x = x$ and $\mathscr{A}(x, y)$ be $y = x$. Then, by schema (A7), $\vdash x = y \Rightarrow (x = x \Rightarrow y = x)$. But, by part (a), $\vdash x = x$. So, by an instance of the tautology $(A \Rightarrow (B \Rightarrow C)) \Rightarrow (B \Rightarrow (A \Rightarrow C))$ and two applications of MP, we have $\vdash x = y \Rightarrow y = x$.

(c) Let $\mathscr{A}(y, y)$ be $y = z$ and $\mathscr{A}(y, x)$ be $x = z$. Then, by (A7) with x and y interchanged, $\vdash y = x \Rightarrow (y = z \Rightarrow x = z)$. But, by part (b), $\vdash x = y \Rightarrow y = x$. Hence, using an instance of the tautology $(A \Rightarrow B) \Rightarrow ((B \Rightarrow C) \Rightarrow (A \Rightarrow C))$ and two applications of MP, we obtain $\vdash x = y \Rightarrow (y = z \Rightarrow x = z)$.

Exercises

2.61 Show that (A6) and (A7) are true for any interpretation M in which $(A_1^2)^{\mathrm{M}}$ is the identity relation on the domain of the interpretation.

2.62 Prove the following in any theory with equality.

(a) $\vdash (\forall x)(\mathscr{B}(x) \Leftrightarrow (\exists y)(x = y \wedge \mathscr{B}(y)))$ if y does not occur in $\mathscr{B}(x)$
(b) $\vdash (\forall x)(\mathscr{B}(x) \Leftrightarrow (\forall y)(x = y \Rightarrow \mathscr{B}(y)))$ if y does not occur in $\mathscr{B}(x)$
(c) $\vdash (\forall x)(\exists y)\, x = y$

We can reduce schema (A7) for equality to a few simpler cases.

PROPOSITION 2.24 Let K be a theory for which (A6) holds and (A7) holds for all atomic wfs $\mathscr{A}(x, x)$. Then K is a theory with equality; that is, (A7) holds for all wfs $\mathscr{A}(x, x)$.

Proof We must prove (A7) for all wfs $\mathscr{A}(x, x)$. It holds for atomic wfs by assumption. Note that we have the results of Proposition 2.23, since its proof used (A7) only with atomic wfs. Proceeding by induction on the number n of connectives and quantifiers in $\mathscr{A}$, we assume that (A7) holds for all $k < n$.

Case 1. $\mathscr{A}(x, x)$ is $\neg \mathscr{B}(x, x)$. By inductive hypothesis, we have $\vdash y = x \Rightarrow (\mathscr{B}(x, y) \Rightarrow \mathscr{B}(x, x))$, since $\mathscr{B}(x, x)$ arises from $\mathscr{B}(x, y)$ by replacing some occurrences of y by x. Hence, by Proposition 2.23(b), instances of the tautologies $(A \Rightarrow B) \Rightarrow (\neg B \Rightarrow \neg A)$ and $(A \Rightarrow B) \Rightarrow ((B \Rightarrow C) \Rightarrow (A \Rightarrow C))$ and MP, we obtain $\vdash x = y \Rightarrow (\mathscr{A}(x, x) \Rightarrow \mathscr{A}(x, y))$.

Case 2. $\mathscr{A}(x, x)$ is $\mathscr{B}(x, x) \Rightarrow \mathscr{C}(x, x)$. By inductive hypothesis and Proposition

2.23(b), $\vdash x = y \Rightarrow (\mathscr{B}(x, y) \Rightarrow \mathscr{B}(x, x))$ and $\vdash x = y \Rightarrow (\mathscr{C}(x, x) \Rightarrow \mathscr{C}(x, y))$. Hence, by the tautology $(A \Rightarrow (B_1 \Rightarrow B)) \Rightarrow [(A \Rightarrow (C \Rightarrow C_1)) \Rightarrow (A \Rightarrow ((B \Rightarrow C) \Rightarrow (B_1 \Rightarrow C_1)))]$, we have $\vdash x = y \Rightarrow (\mathscr{A}(x, x) \Rightarrow \mathscr{A}(x, y))$.

Case 3. $\mathscr{A}(x, x)$ is $(\forall z)\mathscr{B}(x, x, z)$. By inductive hypothesis, $\vdash x = y \Rightarrow (\mathscr{B}(x, x, z) \Rightarrow \mathscr{B}(x, y, z))$. Now, by Gen and axiom (A5), $\vdash x = y \Rightarrow (\forall z)(\mathscr{B}(x, x, z) \Rightarrow \mathscr{B}(x, y, z))$. By Exercise 2.26(a), $\vdash (\forall z)(\mathscr{B}(x, x, z) \Rightarrow \mathscr{B}(x, y, z)) \Rightarrow [(\forall z)\mathscr{B}(x, x, z) \Rightarrow (\forall z)\mathscr{B}(x, y, z)]$, and so, by the tautology $(A \Rightarrow B) \Rightarrow ((B \Rightarrow C) \Rightarrow (A \Rightarrow C))$, $\vdash x = y \Rightarrow (\mathscr{A}(x, x) \Rightarrow \mathscr{A}(x, y))$.

The instances of (A7) can be still further reduced.

PROPOSITION 2.25 Let K be a theory in which (A6) holds and the following are true.

(a) Schema (A7) holds for all atomic wfs $\mathscr{A}(x, x)$ such that no function letters occur in $\mathscr{A}(x, x)$ and $\mathscr{A}(x, y)$ comes from $\mathscr{A}(x, x)$ by replacing exactly one occurrence of x by y.

(b) $\vdash x = y \Rightarrow f_j^n(z_1, \ldots, z_n) = f_j^n(w_1, \ldots, w_n)$, where f_j^n is any function letter of K; $z_1, \ldots, z_n$ are variables; and $f_j^n(w_1, \ldots, w_n)$ arises from $f_j^n(z_1, \ldots, z_n)$ by replacing exactly one occurrence of x by y.

Then K is a theory with equality.

Proof By repeated application, our assumptions can be extended to replacements of more than one occurrence of x by y. Also, Proposition 2.23 is still derivable. By Proposition 2.24, it suffices to prove (A7) for only atomic wfs. But, one can easily prove $\vdash (y_1 = z_1 \wedge \cdots \wedge y_n = z_n) \Rightarrow (\mathscr{A}(y_1, \ldots, y_n) \Rightarrow \mathscr{A}(z_1, \ldots, z_n))$ for all variables $y_1, \ldots, y_n, z_1, \ldots, z_n$, and any atomic wf $\mathscr{A}$ without function letters. Hence, the problem is reduced to showing that, if $t(x, x)$ is a term and $t(x, y)$ comes from $t(x, x)$ by replacing some occurrences of x by y, then $\vdash x = y \Rightarrow t(x, x) = t(x, y)$. But this can be proved, using part (b), by induction on the number of function letters in t, and we leave this as an exercise.

It is easy to see from Proposition 2.25 that, when K has only finitely many predicate and function letters, it is only necessary to verify (A7) for a finite list of special cases (in fact, n wfs for each A_j^n and n wfs for each f_j^n).

Exercises

2.63 Let K_1 be a theory that has only $=$ as a predicate letter and no function letters or individual constants. Let its proper axioms be $(\forall x_1)x_1 = x_1$, $(\forall x_1)(\forall x_2)(x_1 = x_2 \Rightarrow x_2 = x_1)$, and $(\forall x_1)(\forall x_2)(\forall x_3)(x_1 = x_2 \Rightarrow (x_2 = x_3 \Rightarrow x_1 = x_3))$. Show that K_1 is a theory with equality. [*Hint:* It suffices to prove that $\vdash x_1 = x_3 \Rightarrow (x_1 = x_2 \Rightarrow x_3 = x_2)$ and $\vdash x_2 = x_3 \Rightarrow (x_1 = x_2 \Rightarrow x_1 = x_3)$.] K_1 is called the *first-order theory of equality*.

2.64 Let K_2 be a theory that has only $=$ and $<$ as predicate letters and no function letters or individual constants. Let K_2 have the following proper axioms.

(a) $(\forall x_1)x_1 = x_1$
(b) $(\forall x_1)(\forall x_2)(x_1 = x_2 \Rightarrow x_2 = x_1)$
(c) $(\forall x_1)(\forall x_2)(\forall x_3)(x_1 = x_2 \Rightarrow (x_2 = x_3 \Rightarrow x_1 = x_3))$

(d) $(\forall x_1)(\exists x_2)(\exists x_3)(x_1 < x_2 \wedge x_3 < x_1)$
(e) $(\forall x_1)(\forall x_2)(\forall x_3)(x_1 < x_2 \wedge x_2 < x_3 \Rightarrow x_1 < x_3)$
(f) $(\forall x_1)(\forall x_2)(x_1 = x_2 \Rightarrow \neg x_1 < x_2)$
(g) $(\forall x_1)(\forall x_2)(x_1 < x_2 \vee x_1 = x_2 \vee x_2 < x_1)$
(h) $(\forall x_1)(\forall x_2)(x_1 < x_2 \Rightarrow (\exists x_3)(x_1 < x_3 \wedge x_3 < x_2))$

Using Proposition 2.25, show that K_2 is a theory with equality. K_2 is called the *theory of densely ordered sets with neither first nor last element.*

2.65 Let K be any theory with equality. Prove the following.

(a) $\vdash x_1 = y_1 \wedge \cdots \wedge x_n = y_n \Rightarrow t(x_1, \ldots, x_n) = t(y_1, \ldots, y_n)$, where $t(y_1, \ldots, y_n)$ arises from the term $t(x_1, \ldots, x_n)$ by substitution of $y_1, \ldots, y_n$ for $x_1, \ldots, x_n$, respectively.

(b) $\vdash x_1 = y_1 \wedge \cdots \wedge x_n = y_n \Rightarrow (\mathscr{A}(x_1, \ldots, x_n) \Leftrightarrow \mathscr{A}(y_1, \ldots, y_n))$, where $\mathscr{A}(y_1, \ldots, y_n)$ is obtained by substituting $y_1, \ldots, y_n$ for one or more free occurrences of $x_1, \ldots, x_n$, respectively, in the wf $\mathscr{A}(x_1, \ldots, x_n)$, and $y_1, \ldots, y_n$ are free for $x_1, \ldots, x_n$, respectively, in the wf $\mathscr{A}(x_1, \ldots, x_n)$.

Examples (In the literature, "elementary" is sometimes used instead of "first-order".)

1. *Elementary theory G of groups:* predicate letter $=$, function letter f_1^2, and individual constant a_1. We abbreviate $f_1^2(t, s)$ by $t + s$ and a_1 by 0. The proper axioms are the following.

(a) $x_1 + (x_2 + x_3) = (x_1 + x_2) + x_3$
(b) $x_1 + 0 = x_1$
(c) $(\forall x_1)(\exists x_2)x_1 + x_2 = 0$
(d) $x_1 = x_1$
(e) $x_1 = x_2 \Rightarrow x_2 = x_1$
(f) $x_1 = x_2 \Rightarrow (x_2 = x_3 \Rightarrow x_1 = x_3)$
(g) $x_1 = x_2 \Rightarrow (x_1 + x_3 = x_2 + x_3 \wedge x_3 + x_1 = x_3 + x_2)$

That G is a theory with equality follows easily from Proposition 2.25. If one adds to the axioms the following wf:

(h) $x_1 + x_2 = x_2 + x_1$

the new theory G_C is called the *elementary theory of abelian groups.*

2. *Elementary theory F of fields:* predicate letter $=$, function letters f_1^2 and f_2^2, and individual constants a_1 and a_2. Abbreviate $f_1^2(t, s)$ by $t + s$, $f_2^2(t, s)$ by $t \cdot s$, and a_1 and a_2 by 0 and 1. As proper axioms, take (a)–(h) of Example 1 plus the following.

(i) $x_1 = x_2 \Rightarrow (x_1 \cdot x_3 = x_2 \cdot x_3 \wedge x_3 \cdot x_1 = x_3 \cdot x_2)$
(j) $(x_1 \cdot x_2) \cdot x_3 = x_1 \cdot (x_2 \cdot x_3)$
(k) $x_1 \cdot (x_2 + x_3) = (x_1 \cdot x_2) + (x_1 \cdot x_3)$
(l) $x_1 \cdot x_2 = x_2 \cdot x_1$
(m) $x_1 \cdot 1 = x_1$
(n) $x_1 \neq 0 \Rightarrow (\exists x_2)x_1 \cdot x_2 = 1$
(o) $0 \neq 1$

F is a theory with equality. Axioms (a)–(m) define the elementary theory R_C of commutative rings with unit. If we add to F the predicate letter A_2^2, abbreviate $A_2^2(t, s)$ by $t < s$, and add axioms (e), (f), and (g) of Exercise 2.64, as well as $x_1 < x_2 \Rightarrow x_1 + x_3 < x_2 + x_3$ and $x_1 < x_2 \wedge 0 < x_3 \Rightarrow x_1 \cdot x_3 < x_2 \cdot x_3$, then the new theory $F_<$ is called the *elementary theory of ordered fields.*

Exercise

2.66 Show that the axioms (d)–(f) of equality (reflexivity, symmetry, transitivity) mentioned in Example 1 can be replaced by (d) and (f′): $x = y \Rightarrow (z = y \Rightarrow x = z)$.

One often encounters first-order theories K in which $=$ may be defined; that is, there is a wf $\mathscr{E}(x, y)$ with two free variables x and y, such that, if we abbreviate $\mathscr{E}(t, s)$ by $t = s$, then axioms (A6) and (A7) are provable in K. We make the convention that, if t and s are terms that are not free for x and y, respectively, in $\mathscr{E}(x, y)$, then, by suitable changes of bound variables (see Exercise 2.46), we replace $\mathscr{E}(x, y)$ by a logically equivalent wf $\mathscr{E}^*(x, y)$ such that t and s are free for x and y, respectively, in $\mathscr{E}^*(x, y)$; then $t = s$ is to be the abbreviation of $\mathscr{E}^*(t, s)$. Proposition 2.23 and analogues of Propositions 2.24 and 2.25 hold for such theories. There is no harm in extending the name *theory with equality* to cover such theories.

In theories with equality it is possible to define in the following way phrases that use the expression "There exists one and only one x such that ...".

DEFINITION $(\exists_1 x)\mathscr{A}(x)$ for $(\exists x)\mathscr{A}(x) \wedge (\forall x)(\forall y)(\mathscr{A}(x) \wedge \mathscr{A}(y) \Rightarrow x = y)$

In this definition, the new variable y is assumed to be the first variable that does not occur in $\mathscr{A}(x)$. A similar convention is to be made in all other definitions where new variables are introduced.

Exercise

2.67 In any theory with equality, prove the following.

(a) $\vdash (\forall x)(\exists_1 y)x = y$

(b) $\vdash (\exists_1 x)\mathscr{A}(x) \Leftrightarrow (\exists x)(\forall y)(x = y \Leftrightarrow \mathscr{A}(y))$

(c) $\vdash (\forall x)(\mathscr{A}(x) \Leftrightarrow \mathscr{B}(x)) \Rightarrow [(\exists_1 x)\mathscr{A}(x) \Leftrightarrow (\exists_1 x)\mathscr{B}(x)]$

(d) $\vdash (\exists_1 x)(\mathscr{A} \vee \mathscr{B}) \Rightarrow ((\exists_1 x)\mathscr{A}) \vee (\exists_1 x)\mathscr{B}$

(e) $\vdash (\exists_1 x)\mathscr{A}(x) \Leftrightarrow (\exists x)(\mathscr{A}(x) \wedge (\forall y)(\mathscr{A}(y) \Rightarrow y = x))$

In any model for a theory K with equality, the relation E in the model corresponding to the predicate letter $=$ is an equivalence relation (by Proposition 2.23). If this relation E is the identity relation in the domain of the model, then the model is said to be *normal.*

Any model M for K can be *contracted* to a normal model M′ for K by taking the domain D' of M′ to be the set of equivalence classes determined by the relation E in the domain D of M. For a predicate letter A_j^n with interpretation $(A_j^n)^*$ in M, we define the new interpretation $(A_j^n)'$ in M′ as follows: for any equivalence classes $[b_1], \ldots, [b_n]$ in D' determined by the elements $b_1, \ldots, b_n$ in D, $(A_j^n)'$ holds for $([b_1], \ldots, [b_n])$ if and only if $(A_j^n)^*$ holds for $(b_1, \ldots, b_n)$. Notice that it makes no difference which representatives $b_1, \ldots, b_n$ we select in the given equivalence classes because, by (A7), $\vdash x_1 = y_1 \wedge \cdots \wedge x_n = y_n \Rightarrow (A_j^n(x_1, \ldots, x_n) \Leftrightarrow A_j^n(y_1, \ldots, y_n))$. Likewise, if $(f_j^n)^*$ is the interpretation of f_j^n in M, then we define the new interpretation $(f_j^n)'$ in M′ as follows: for any equivalence classes $[b_1], \ldots, [b_n]$ in D', $(f_j^n)'([b_1], \ldots, [b_n]) = [(f_j^n)^*(b_1, \ldots, b_n)]$. Again note that this is independent of the choice of the representatives $b_1, \ldots, b_n$, since, by (A7), $\vdash x_1 = y_1 \wedge \cdots \wedge x_n = y_n \Rightarrow$

$f_j^n(x_1, \ldots, x_n) = f_j^n(y_1, \ldots, y_n)$. If c is the interpretation in M of an individual constant a_i, then we take the equivalence class $[c]$ to be the interpretation of a_i in M′. The relation E′ corresponding to $=$ in the model M′ is the identity relation in D': $E'([b_1], [b_2])$ if and only if $E(b_1, b_2)$—that is, if and only if $[b_1] = [b_2]$. Now one can easily prove by induction the following lemma: if $s = (b_1, b_2, \ldots)$ is a denumerable sequence of elements of D, and $s' = ([b_1], [b_2], \ldots)$ is the corresponding sequence of equivalence classes, then $\mathscr{A}$ is satisfied by s in M if and only if $\mathscr{A}$ is satisfied by s' in M′. It follows that, for any wf $\mathscr{A}$, $\mathscr{A}$ is true for M if and only if $\mathscr{A}$ is true for M′. Hence, because M is a model of K, M′ is a normal model of K.

PROPOSITION 2.26 (EXTENSION OF PROPOSITION 2.17) (Gödel, 1930) Any consistent theory with equality K has a finite or denumerable normal model.

Proof By Proposition 2.17, K has a denumerable model M. Hence, the contraction of M to a normal model yields a finite or denumerable normal model M′ because the set of equivalence classes in a set D has cardinality less than or equal to the cardinality of D.

COROLLARY 2.27 (EXTENSION OF THE SKOLEM-LÖWENHEIM THEOREM) Any theory with equality K that has an infinite normal model M has a denumerable normal model.

Proof Add to K the denumerably many new individual constants $b_1, b_2, \ldots$ together with the axioms $b_i \neq b_j$ for $i \neq j$. Then the new theory K′ is consistent. If K′ were inconsistent, there would be a proof in K′ of a contradiction $\mathscr{C} \wedge \neg\mathscr{C}$, where we may assume that $\mathscr{C}$ is a wf of K. But this proof uses only a finite number of the new axioms: $b_{i_1} \neq b_{j_1}, \ldots, b_{i_n} \neq b_{j_n}$. Now, M can be extended to a model of K plus the axioms $b_{i_1} \neq b_{j_1}, \ldots, b_{i_n} \neq b_{j_n}$ because, since M is an infinite normal model, we can choose interpretations of $b_{i_1}, b_{j_1}, \ldots, b_{i_n}, b_{j_n}$ so that the wfs $b_{i_1} \neq b_{j_1}, \ldots, b_{i_n} \neq b_{j_n}$ are true. But, since $\mathscr{C} \wedge \neg\mathscr{C}$ is derivable from these wfs and the axioms of K, it would follow that $\mathscr{C} \wedge \neg\mathscr{C}$ is true for M, which is impossible. Hence, K′ must be consistent. Now, by Proposition 2.26, K′ has a finite or denumerable normal model N. But, since the wfs $b_i \neq b_j$, for $i \neq j$, are axioms of K′, they are true for N. Hence, the elements in the domain of N that are the interpretations of $b_1, b_2, \ldots$ must be distinct, which implies that the domain of N is infinite and, therefore, denumerable.

Exercises

2.68 We define $(\exists_n x)\mathscr{A}(x)$ by induction on $n \geqslant 1$. The case $n = 1$ has already been taken care of. Let $(\exists_{n+1} x)\mathscr{A}(x)$ stand for $(\exists y)(\mathscr{A}(y) \wedge (\exists_n x)(x \neq y \wedge \mathscr{A}(x)))$.

(a) Show that $(\exists_n x)\mathscr{A}(x)$ asserts that there are exactly n objects for which $\mathscr{A}$ holds, in the sense that in any normal model for $(\exists_n x)\mathscr{A}(x)$ there are exactly n objects for which the property corresponding to $\mathscr{A}(x)$ holds.

(b) (i) For each positive integer n, write a wf $\mathscr{B}_n$ such that $\mathscr{B}_n$ holds in a normal model when and only when that model contains at least n elements. (ii) Prove that the theory K, whose axioms are those of the theory of equality K_1 (see Exercise 2.63) plus the axioms $\mathscr{B}_1, \mathscr{B}_2, \ldots$, is not finitely axiomatizable; that is, there is no theory K′ with a finite number of axioms such that K and K′ have the same theorems.

2.69 (a) Prove that, if a theory with equality K has arbitrarily large finite normal models, then it has a denumerable normal model.

(b) Prove that there is no theory with equality whose normal models are precisely all finite normal interpretations.

2.70 Prove that any predicate calculus with equality is consistent. [A predicate calculus with equality is assumed to have (A1)–(A7) as axioms.]

2.71^D Prove the independence of axioms (A1)–(A7) in any predicate calculus with equality.

2.72 If $\mathscr{A}$ is a wf that does not contain the $=$ symbol and $\mathscr{A}$ is provable in a predicate calculus with equality K, show that $\mathscr{A}$ is provable in K without using (A6) or (A7).

2.73^D Show that $=$ can be defined in any theory that has only a finite number of predicate letters and no function letters.

9. DEFINITIONS OF NEW FUNCTION LETTERS AND INDIVIDUAL CONSTANTS

In mathematics, once we have proved, for any $y_1, \ldots, y_n$, the existence of a unique object u that has a property $\mathscr{A}(u, y_1, \ldots, y_n)$, we often introduce a new function $f(y_1, \ldots, y_n)$ such that $\mathscr{A}(f(y_1, \ldots, y_n), y_1, \ldots, y_n)$ holds for all $y_1, \ldots, y_n$. In cases where we have proved the existence of a unique object u that satisfies a wf $\mathscr{A}(u)$, and $\mathscr{A}(u)$ contains u as its only free variable, then we introduce a new individual constant b. It is generally acknowledged that such definitions, though convenient, add nothing really new to the theory. This can be made precise in the following manner.

PROPOSITION 2.28 Let K be a theory with equality. Assume that $\vdash_K (\exists_1 u)\mathscr{A}(u, y_1, \ldots, y_n)$. Let K′ be the theory with equality obtained by adding to K a new function letter f of n arguments and the proper axiom $\mathscr{A}(f(y_1, \ldots, y_n), y_1, \ldots, y_n)$,* as well as all instances of axioms (A1)–(A7) that involve f. Then there is an effective transformation mapping each wf $\mathscr{B}$ of K′ into a wf $\mathscr{B}'$ of K such that:

(a) If f does not occur in $\mathscr{B}$, then $\mathscr{B}'$ is $\mathscr{B}$

(b) $(\neg\mathscr{B})'$ is $\neg(\mathscr{B}')$

(c) $(\mathscr{B} \Rightarrow \mathscr{C})'$ is $\mathscr{B}' \Rightarrow \mathscr{C}'$

(d) $((\forall x)\mathscr{B})'$ is $(\forall x)(\mathscr{B}')$

(e) $\vdash_{K'} \mathscr{B} \Leftrightarrow \mathscr{B}'$

(f) If $\vdash_{K'} \mathscr{B}$, then $\vdash_K \mathscr{B}'$

Hence, if $\mathscr{B}$ does not contain f and $\vdash_{K'} \mathscr{B}$, then $\vdash_K \mathscr{B}$.

Proof By a *simple f-term* we mean an expression $f(t_1, \ldots, t_n)$ in which $t_1, \ldots, t_n$ are terms that do not contain f. Given an atomic wf $\mathscr{B}$ of K′, let $\mathscr{B}^*$

*It is better to take this axiom in the form $(\forall u)(u = f(y_1, \ldots, y_n) \Rightarrow \mathscr{A}(u, y_1, \ldots, y_n))$, since $f(y_1, \ldots, y_n)$ might not be free for u in $\mathscr{A}(u, y_1, \ldots, y_n)$.

be the result of replacing the leftmost occurrence of a simple term $f(t_1,\ldots,t_n)$ in $\mathscr{B}$ by the first variable u not in $\mathscr{B}$. Call the wf $(\exists u)(\mathscr{A}(u,t_1,\ldots,t_n) \wedge \mathscr{B}^*)$ the *f-transform* of $\mathscr{B}$. If $\mathscr{B}$ does not contain f, let $\mathscr{B}$ be its own f-transform. Clearly, $\vdash_{K'} (\exists u)(\mathscr{A}(u,t_1,\ldots,t_n) \wedge \mathscr{B}^*) \Leftrightarrow \mathscr{B}$. [Here we use $\vdash_K (\exists_1 u)\mathscr{A}(u,y_1,\ldots,y_n)$ and the axiom $\mathscr{A}(f(y_1,\ldots,y_n),y_1,\ldots,y_n)$ of K′.] Since the f-transform $\mathscr{B}^\#$ of $\mathscr{B}$ contains one less f than $\mathscr{B}$ and $\vdash_{K'} \mathscr{B}^\# \Leftrightarrow \mathscr{B}$, if we take successive f-transforms, eventually we obtain a wf $\mathscr{B}'$ that does not contain f and such that $\vdash_{K'} \mathscr{B}' \Leftrightarrow \mathscr{B}$. Call $\mathscr{B}'$ the f-less transform of $\mathscr{B}$. Extend the definition to all wfs of K′ by letting $(\neg\mathscr{B})'$ be $\neg(\mathscr{B}')$, $(\mathscr{B} \Rightarrow \mathscr{C})'$ be $\mathscr{B}' \Rightarrow \mathscr{C}'$, and $((\forall x)\mathscr{B})'$ be $(\forall x)(\mathscr{B}')$. Properties (a)–(e) of Proposition 2.28 are then obvious. To prove property (f), it suffices, by property (e), to show that, if $\mathscr{B}$ does not contain f and $\vdash_{K'} \mathscr{B}$, then $\vdash_K \mathscr{B}$. We may assume that $\mathscr{B}$ is a closed wf, since a wf and its closure are deducible from each other.

Assume that M is a model of K. Let M_1 be the corresponding normal model of K (see page 78). We know that a wf is true for M if and only if it is true for M_1. Since $\vdash_K (\exists_1 u)\mathscr{A}(u,y_1,\ldots,y_n)$, then, for any $b_1, \ldots, b_n$ in the domain of M_1, there is a unique c in the domain of M_1 such that $\models_{M_1} \mathscr{A}[c,b_1,\ldots,b_n]$. If we define $f'(b_1,\ldots,b_n)$ to be c, then taking f' to be the interpretation of the function letter f, we obtain from M_1 a model M′ of K′. For, the logical axioms of K′ (including the equality axioms of K′) are true in any interpretation, and the axiom $\mathscr{A}(f(y_1,\ldots,y_n),y_1,\ldots,y_n)$ also holds in M′ by virtue of the definition of f'. Since the proper axioms of K′ do not contain f and since they are true for M_1, they are also true for M′. But, $\vdash_{K'} \mathscr{B}$. Therefore, $\mathscr{B}$ is true for M′, but since $\mathscr{B}$ does not contain f, $\mathscr{B}$ is true for M_1 and hence also for M. Thus, $\mathscr{B}$ is true for every model of K. Therefore, by Corollary 2.20(a), $\vdash_K \mathscr{B}$. [In the case where $\vdash_K (\exists_1 u)\mathscr{A}(u)$ and $\mathscr{A}(u)$ contains only u as a free variable, we form K′ by adding a new individual constant b and the axiom $\mathscr{A}(b)$. Then the analogue of Proposition 2.28 follows from practically the same proof as the one just given.]

Exercise

2.74 Find the f-less transforms of the following wfs.

(a) $(\forall x)(\exists y)(A_1^3(x,y,f(x,y,\ldots,y)) \Rightarrow f(y,x,\ldots,x) = x)$

(b) $A_1^1(f(y_1,\ldots,y_{n-1},f(y_1,\ldots,y_n))) \wedge (\exists x)A_1^2(x,f(y_1,\ldots,y_n))$

Note that Proposition 2.28 also applies when we have introduced several new symbols $f_1, \ldots, f_m$ because we can assume that we have added each f_i to the theory already obtained by the addition of $f_1, \ldots, f_{i-1}$; then m successive applications of Proposition 2.28 are necessary. In addition, the wf $\mathscr{B}'$ of K in Proposition 2.28 can be considered an f-free translation of $\mathscr{B}$ into the language of K.

Examples

1. In the elementary theory G of groups (see page 77) one can prove $(\exists_1 x_2)(x_1 + x_2 = 0)$. Then introduce a new function letter f of one argument, abbreviate $f(t)$ by $(-t)$, and add the new axiom $x_1 + (-x_1) = 0$. By Proposition 2.28, we now are not able to prove any wf of G that we could not prove before. Thus, the definition of $(-t)$ adds no really new power to the original theory.

2. In the elementary theory F of fields (see page 77) one can prove that $(\exists_1 x_2)((x_1 \neq 0 \wedge x_1 \cdot x_2 = 1) \vee (x_1 = 0 \wedge x_2 = 0))$. We then introduce a new function letter g of one argument, abbreviate $g(t)$ by t^{-1}, and introduce the axiom $(x_1 \neq 0 \wedge x_1 \cdot x_1^{-1} = 1) \vee (x_1 = 0 \wedge x_1^{-1} = 0)$, from which one can prove $x_1 \neq 0 \Rightarrow x_1 \cdot x_1{}^{-1} = 1$.

From Proposition 2.28, we can see that, in theories with equality, only predicate letters are needed; function letters and individual constants are dispensable. If f_j^n is a function letter, we can replace it by a new predicate letter A_k^{n+1} if we add the axiom $(\exists_1 u)A_k^{n+1}(u, y_1, \ldots, y_n)$. An individual constant is to be replaced by a new predicate letter A_k^1 if we add the axiom $(\exists_1 u)A_k^1(u)$.

Example In the elementary theory G of groups, we can replace + and 0 by predicate letters A_1^3 and A_1^1 if we add the axioms $(\forall x_1)(\forall x_2)(\exists_1 x_3)A_1^3(x_1, x_2, x_3)$ and $(\exists_1 x_1)A_1^1(x_1)$, and if we replace axioms (a), (b), (c), and (g) by the following:

(a′) $A_1^3(x_2, x_3, y_1) \wedge A_1^3(x_1, y_1, y_2) \wedge A_1^3(x_1, x_2, y_3) \wedge A_1^3(y_3, x_3, y_4) \Rightarrow y_2 = y_4$

(b′) $A_1^1(y_1) \wedge A_1^3(x_1, y_1, y_2) \Rightarrow y_2 = x_1$

(c′) $(\exists x_2)(\forall y_1)(\forall y_2)(A_1^1(y_1) \wedge A_1^3(x_1, x_2, y_2) \Rightarrow y_2 = y_1)$

(g′) $[x_1 = x_2 \wedge A_1^3(x_1, x_3, y_1) \wedge A_1^3(x_2, x_3, y_2) \wedge A_1^3(x_3, x_1, y_3) \wedge A_1^3(x_3, x_2, y_4)] \Rightarrow y_1 = y_2 \wedge y_3 = y_4$

Notice that the proof of Proposition 2.28 is highly nonconstructive, since it uses semantical notions (model, truth) and is based upon Corollary 2.20(a), which was proved in a nonconstructive way. Constructive, syntactical proofs have been given for Proposition 2.28 (see Kleene, 1952, § 74), but, in general, they are quite complex.

Descriptive phrases of the kind "the u such that $\mathscr{A}(u, y_1, \ldots, y_n)$" are very common in ordinary language and in mathematics. Such phrases are called *definite descriptions*. We let $\iota u(\mathscr{A}(u, y_1, \ldots, y_n))$ denote the unique object u such that $\mathscr{A}(u, y_1, \ldots, y_n)$ if there is such a unique object. If there is no such unique object, either we may let $\iota u(\mathscr{A}(u, y_1, \ldots, y_n))$ stand for some fixed object, or we may consider it meaningless. (For example, we may say that the phrases "the present king of France" and "the smallest integer" are meaningless, or we may arbitrarily make the convention that they denote 0.) There are various ways of incorporating these ι-terms in formalized theories, but since in most cases the same results are obtained by using new function letters as above, and since they all lead to theorems similar to Proposition 2.28, we shall not discuss them any further here. For details, see Hilbert and Bernays (1934) and Rosser (1939a, 1953).

10. PRENEX NORMAL FORMS

A wf $(Q_1 y_1) \ldots (Q_n y_n)\mathscr{A}$, where each $(Q_i y_i)$ is a universal or existential quantifier, y_i is different from y_j for $i \neq j$, and $\mathscr{A}$ contains no quantifiers, is said to be in *prenex normal form*. (We include the case $n = 0$ when there are no quantifiers at all.) We shall prove that, for every wf, we can construct an equivalent prenex normal form.

LEMMA 2.29 In any theory, if y is not free in $\mathscr{D}$, and $\mathscr{C}(x)$ and $\mathscr{C}(y)$ are similar, then the following hold.

(a) $\vdash ((\forall x)\mathscr{C}(x) \Rightarrow \mathscr{D}) \Leftrightarrow (\exists y)(\mathscr{C}(y) \Rightarrow \mathscr{D})$
(b) $\vdash ((\exists x)\mathscr{C}(x) \Rightarrow \mathscr{D}) \Leftrightarrow (\forall y)(\mathscr{C}(y) \Rightarrow \mathscr{D})$
(c) $\vdash (\mathscr{D} \Rightarrow (\forall x)\mathscr{C}(x)) \Leftrightarrow (\forall y)(\mathscr{D} \Rightarrow \mathscr{C}(y))$
(d) $\vdash (\mathscr{D} \Rightarrow (\exists x)\mathscr{C}(x)) \Leftrightarrow (\exists y)(\mathscr{D} \Rightarrow \mathscr{C}(y))$
(e) $\vdash \neg(\forall x)\mathscr{C} \Leftrightarrow (\exists x)\neg\mathscr{C}$
(f) $\vdash \neg(\exists x)\mathscr{C} \Leftrightarrow (\forall x)\neg\mathscr{C}$

Proof For part (a):

1. $(\forall x)\mathscr{C}(x) \Rightarrow \mathscr{D}$	Hyp
2. $\neg(\exists y)(\mathscr{C}(y) \Rightarrow \mathscr{D})$	Hyp
3. $\neg\neg(\forall y)\neg(\mathscr{C}(y) \Rightarrow \mathscr{D})$	2, Abbreviation
4. $(\forall y)\neg(\mathscr{C}(y) \Rightarrow \mathscr{D})$	3, Negation Elimination
5. $(\forall y)(\mathscr{C}(y) \wedge \neg\mathscr{D})$	4, Tautology, Proposition 2.8(c)
6. $\mathscr{C}(y) \wedge \neg\mathscr{D}$	5, Rule A4
7. $\mathscr{C}(y)$	6, Conjunction Elimination
8. $(\forall y)\mathscr{C}(y)$	7, Gen
9. $(\forall x)\mathscr{C}(x)$	8, Lemma 2.11, Biconditional Elimination
10. $\mathscr{D}$	1, 9, MP
11. $\neg\mathscr{D}$	6, Conjunction Elimination
12. $\mathscr{D} \wedge \neg\mathscr{D}$	10, 11, Conjunction Introduction
13. $(\forall x)\mathscr{C}(x) \Rightarrow \mathscr{D}, \neg(\exists y)(\mathscr{C}(y) \Rightarrow \mathscr{D}) \vdash \mathscr{D} \wedge \neg\mathscr{D}$	1–12
14. $(\forall x)\mathscr{C}(x) \Rightarrow \mathscr{D} \vdash (\exists y)(\mathscr{C}(y) \Rightarrow \mathscr{D})$	1–13, Proof by Contradiction
15. $\vdash ((\forall x)\mathscr{C}(x) \Rightarrow \mathscr{D}) \Rightarrow (\exists y)(\mathscr{C}(y) \Rightarrow \mathscr{D})$	1–14, Corollary 2.5

The converse is proven in the following manner:

1. $(\exists y)(\mathscr{C}(y) \Rightarrow \mathscr{D})$	Hyp
2. $(\forall x)\mathscr{C}(x)$	Hyp
3. $\mathscr{C}(b) \Rightarrow \mathscr{D}$	1, Rule C
4. $\mathscr{C}(b)$	2, Rule A4
5. $\mathscr{D}$	3, 4, MP
6. $(\exists y)(\mathscr{C}(y) \Rightarrow \mathscr{D}), (\forall x)\mathscr{C}(x) \vdash_C \mathscr{D}$	1–5
7. $(\exists y)(\mathscr{C}(y) \Rightarrow \mathscr{D}), (\forall x)\mathscr{C}(x) \vdash \mathscr{D}$	6, Proposition 2.9
8. $\vdash (\exists y)(\mathscr{C}(y) \Rightarrow \mathscr{D}) \Rightarrow ((\forall x)\mathscr{C}(x) \Rightarrow \mathscr{D})$	1–7, Corollary 2.5 twice

Part (a) follows from the two proofs above by Biconditional Introduction. Parts (b)–(f) are proved easily and left as an exercise. [Part (f) is trivial, and part (e) appeared as Exercise 2.31(a); parts (c) and (d) follow easily from parts (b) and (a), respectively.]

Lemma 2.29 allows us to move interior quantifiers to the front of a wf. This is the essential process in the proof of the following proposition.

PROPOSITION 2.30 There is an effective procedure for transforming any wf $\mathscr{A}$ into a wf $\mathscr{B}$ in prenex normal form such that $\vdash \mathscr{A} \Leftrightarrow \mathscr{B}$.

Proof We describe the procedure by induction on the number k of occurrences of connectives and quantifiers in $\mathscr{A}$. [By Exercise 2.30(a, b), we may assume that the quantified variables in the prefix that we shall obtain are distinct.] If $k = 0$, $\mathscr{B}$ is $\mathscr{A}$ itself. Assume that we can find a corresponding $\mathscr{B}$ for all wfs with $k < n$, and assume that $\mathscr{A}$ has n occurrences of connectives and quantifiers.

Case 1. If $\mathscr{A}$ is $\neg\mathscr{C}$, then, by inductive hypothesis, we can construct a wf $\mathscr{D}$ in prenex normal form such that $\vdash \mathscr{C} \Leftrightarrow \mathscr{D}$. Hence, $\vdash \neg\mathscr{C} \Leftrightarrow \neg\mathscr{D}$—that is, $\vdash \mathscr{A} \Leftrightarrow \neg\mathscr{D}$; but, applying parts (e) and (f) of Lemma 2.29 and the Replacement Theorem [Proposition 2.8(b)], we can find a wf $\mathscr{B}$ in prenex normal form such that $\vdash \neg\mathscr{D} \Leftrightarrow \mathscr{B}$. Hence, $\vdash \mathscr{A} \Leftrightarrow \mathscr{B}$.

Case 2. If $\mathscr{A}$ is $\mathscr{C} \Rightarrow \mathscr{E}$, then, by inductive hypothesis, we can find wfs $\mathscr{C}_1$ and $\mathscr{E}_1$ in prenex normal form such that $\vdash \mathscr{C} \Leftrightarrow \mathscr{C}_1$ and $\vdash \mathscr{E} \Leftrightarrow \mathscr{E}_1$. Hence, by a suitable tautology, $\vdash (\mathscr{C} \Rightarrow \mathscr{E}) \Leftrightarrow (\mathscr{C}_1 \Rightarrow \mathscr{E}_1)$; that is, $\vdash \mathscr{A} \Leftrightarrow (\mathscr{C}_1 \Rightarrow \mathscr{E}_1)$. Now, applying parts (a)–(d) of Lemma 2.29 and the Replacement Theorem, we can move the quantifiers in the prefixes of $\mathscr{C}_1$ and $\mathscr{E}_1$ to the front, obtaining a wf $\mathscr{B}$ in prenex normal form such that $\vdash \mathscr{A} \Leftrightarrow \mathscr{B}$.

Case 3. $\mathscr{A}$ is $(\forall x)\mathscr{C}$. By inductive hypothesis, there is a wf $\mathscr{C}_1$ in prenex normal form such that $\vdash \mathscr{C} \Leftrightarrow \mathscr{C}_1$; hence, $\vdash \mathscr{A} \Leftrightarrow (\forall x)\mathscr{C}_1$. But $(\forall x)\mathscr{C}_1$ is in prenex normal form.

Examples

1. Let $\mathscr{A}$ be $(\forall x)(A_1^1(x) \Rightarrow (\forall y)(A_2^2(x, y) \Rightarrow \neg(\forall z)A_3^2(y, z)))$.
   By part (e) of Lemma 2.29: $(\forall x)(A_1^1(x) \Rightarrow (\forall y)[A_2^2(x, y) \Rightarrow (\exists z)\neg A_3^2(y, z)])$.
   By part (d): $(\forall x)(A_1^1(x) \Rightarrow (\forall y)(\exists u)[A_2^2(x, y) \Rightarrow \neg A_3^2(y, u)])$.
   By part (c): $(\forall x)(\forall v)(A_1^1(x) \Rightarrow (\exists u)[A_2^2(x, v) \Rightarrow \neg A_3^2(v, u)])$.
   By part (d): $(\forall x)(\forall v)(\exists w)(A_1^1(x) \Rightarrow (A_2^2(x, v) \Rightarrow \neg A_3^2(v, w)))$.
   Changing bound variables: $(\forall x)(\forall y)(\exists z)(A_1^1(x) \Rightarrow (A_2^2(x, y) \Rightarrow \neg A_3^2(y, z)))$.
2. Let $\mathscr{A}$ be $A_1^2(x, y) \Rightarrow (\exists y)[A_1^1(y) \Rightarrow ([(\exists x)A_1^1(x)] \Rightarrow A_2^1(y))]$.
   By part (b): $A_1^2(x, y) \Rightarrow (\exists y)(A_1^1(y) \Rightarrow (\forall u)[A_1^1(u) \Rightarrow A_2^1(y)])$.
   By part (c): $A_1^2(x, y) \Rightarrow (\exists y)(\forall v)(A_1^1(y) \Rightarrow [A_1^1(v) \Rightarrow A_2^1(y)])$.
   By part (d): $(\exists w)(A_1^2(x, y) \Rightarrow (\forall v)[A_1^1(w) \Rightarrow (A_1^1(v) \Rightarrow A_2^1(w))])$.
   By part (c): $(\exists w)(\forall z)(A_1^2(x, y) \Rightarrow [A_1^1(w) \Rightarrow (A_1^1(z) \Rightarrow A_2^1(w))])$.

Exercise

2.75 Find a prenex normal form equivalent to the following wfs.
(a) $[(\forall x)(A_1^1(x) \Rightarrow A_1^2(x, y))] \Rightarrow ([(\exists y)A_1^1(y)] \Rightarrow (\exists z)A_1^2(y, z))$
(b) $(\exists x)A_1^2(x, y) \Rightarrow (A_1^1(x) \Rightarrow \neg(\exists u)A_1^2(x, u))$

A predicate calculus in which there are no function letters or individual constants and in which, for any positive integer n, there are infinitely many predicate letters with n arguments is called a *pure predicate calculus*. For pure predicate calculi, we can find a very simple prenex normal form theorem. A wf in prenex normal form such that all existential quantifiers precede all universal quantifiers is said to be in *Skolem normal form*.

PROPOSITION 2.31 In a pure predicate calculus, there is an effective process assigning to each wf $\mathscr{A}$ another wf $\mathscr{B}$ in Skolem normal form such that $\vdash \mathscr{A}$ if and only if $\vdash \mathscr{B}$ (or, equivalently, by Gödel's Completeness Theorem, such that $\mathscr{A}$ is logically valid if and only if $\mathscr{B}$ is logically valid).

Proof First we may assume that $\mathscr{A}$ is a closed wf, since a wf is provable if and only if its closure is provable. By Proposition 2.30, we may also assume that $\mathscr{A}$ is in prenex normal form. Let the *rank* r of $\mathscr{A}$ be the number of universal quantifiers in $\mathscr{A}$ that precede existential quantifiers. By induction on the rank, we shall describe the process for finding Skolem normal forms. Clearly, when the rank is 0, we already have the Skolem normal form. Let us assume that we can construct Skolem normal forms when the rank is less than r, and let r be the rank of $\mathscr{A}$. $\mathscr{A}$ can be written as follows: $(\exists y_1)\ldots(\exists y_n)(\forall u)\mathscr{B}(y_1,\ldots,y_n,u)$, where $\mathscr{B}(y_1,\ldots,y_n,u)$ has only $y_1,\ldots,y_n$, u as its free variables. Let A_j^{n+1} be the first predicate letter of $n+1$ arguments that does not occur in $\mathscr{A}$. Construct the wf

$$(\mathscr{A}_1)\quad (\exists y_1)\ldots(\exists y_n)([(\forall u)(\mathscr{B}(y_1,\ldots,y_n,u)\Rightarrow A_j^{n+1}(y_1,\ldots,y_n,u))] \Rightarrow (\forall u)A_j^{n+1}(y_1,\ldots,y_n,u))$$

Let us show that $\vdash \mathscr{A}$ if and only if $\vdash \mathscr{A}_1$. Assume $\vdash \mathscr{A}_1$. In the proof of $\mathscr{A}_1$, replace all occurrences of $A_j^{n+1}(z_1,\ldots,z_n,w)$ by $\mathscr{B}^*(z_1,\ldots,z_n,w)$, where $\mathscr{B}^*$ is obtained from $\mathscr{B}$ by replacing all bound variables having free occurrences in the proof by new variables not occurring in the proof. The result is a proof of the wf

$$(\exists y_1)\ldots(\exists y_n)(((\forall u)(\mathscr{B}(y_1,\ldots,y_n,u)\Rightarrow \mathscr{B}^*(y_1,\ldots,y_n,u)))\Rightarrow(\forall u)\mathscr{B}^*(y_1,\ldots,y_n,u))$$

[$\mathscr{B}$ was replaced by $\mathscr{B}^*$ so that applications of axiom (A4) would remain applications of the same axiom.] Now, by changing the bound variables back again, we see that

$$\vdash (\exists y_1)\ldots(\exists y_n)[(\forall u)(\mathscr{B}(y_1,\ldots,y_n,u)\Rightarrow \mathscr{B}(y_1,\ldots,y_n,u))\Rightarrow(\forall u)\mathscr{B}(y_1,\ldots,y_n,u)]$$

Since $\vdash (\forall u)(\mathscr{B}(y_1,\ldots,y_n,u)\Rightarrow\mathscr{B}(y_1,\ldots,y_n,u))$, we obtain, by the Replacement Theorem, $\vdash (\exists y_1)\ldots(\exists y_n)(\forall u)\mathscr{B}(y_1,\ldots,y_n,u)$—that is, $\vdash \mathscr{A}$. Conversely, assume $\vdash \mathscr{A}$. By Rule C, we obtain $(\forall u)\mathscr{B}(b_1,\ldots,b_n,u)$. But, $\vdash (\forall u)\mathscr{D}\Rightarrow((\forall u)(\mathscr{D}\Rightarrow\mathscr{F})\Rightarrow(\forall u)\mathscr{F})$ [see Exercise 2.26(a)] for any wfs $\mathscr{D}$, $\mathscr{F}$. Hence, $\vdash_C (\forall u)(\mathscr{B}(b_1,\ldots,b_n,u)\Rightarrow A^{n+1}(b_1,\ldots,b_n,u))\Rightarrow(\forall u)A^{n+1}(b_1,\ldots,b_n,u)$. So, by Rule E4, $\vdash_C (\exists y_1)\ldots(\exists y_n)([(\forall u)(\mathscr{B}(y_1,\ldots,y_n,u)\Rightarrow A^{n+1}(y_1,\ldots,y_n,u))]\Rightarrow(\forall u)A^{n+1}(y_1,\ldots,y_n,u))$—that is, $\vdash_C \mathscr{A}_1$. By Proposition 2.9, $\vdash \mathscr{A}_1$. A prenex normal form of $\mathscr{A}_1$ has the form $\mathscr{A}_2$: $(\exists y_1)\ldots(\exists y_n)(\exists u)(Q_1 z_1)\ldots(Q_s z_s)(\forall v)\mathscr{G}$, where $\mathscr{G}$ has no quantifiers and $(Q_1 z_1)\ldots(Q_s z_s)$ is the prefix of $\mathscr{B}$. [In deriving the prenex normal form, first, by Lemma 2.29(a), we pull out the first $(\forall u)$, which changes to $(\exists u)$; then we pull out of the first conditional the quantifiers in the prefix of $\mathscr{B}$. By Lemma 2.29(b), this changes existential and universal quantifiers, but then we again pull these out of the second conditional of $\mathscr{A}_1$, which brings the prefix back to its original form. Finally, by Lemma 2.29(c), we bring the second $(\forall u)$ out to the prefix, changing it to a new variable $(\forall v)$.] Clearly,

$\mathscr{A}_2$ has rank one less than the rank of $\mathscr{A}$, and, by Proposition 2.30, $\vdash \mathscr{A}_1 \Leftrightarrow \mathscr{A}_2$; but, $\vdash \mathscr{A}$ if and only if $\vdash \mathscr{A}_1$. Hence, $\vdash \mathscr{A}$ if and only if $\vdash \mathscr{A}_2$. By inductive hypothesis, we can find a Skolem normal form for $\mathscr{A}_2$, which is also a Skolem normal form for $\mathscr{A}$.

Example

$\mathscr{A}$: $(\forall x)(\forall y)(\exists z)\mathscr{C}(x, y, z)$, where $\mathscr{C}$ contains no quantifiers

$\mathscr{A}_1$: $(\forall x)((\forall y)(\exists z)\mathscr{C}(x, y, z) \Rightarrow A_j^1(x)) \Rightarrow (\forall x)A_j^1(x)$, where A_j^1 is not in $\mathscr{C}$

We obtain the prenex normal form of $\mathscr{A}_1$:

$(\exists x)([(\forall y)(\exists z)\mathscr{C}(x, y, z) \Rightarrow A_j^1(x)] \Rightarrow (\forall x)A_j^1(x))$	2.29(a)
$(\exists x)((\exists y)[(\exists z)\mathscr{C}(x, y, z) \Rightarrow A_j^1(x)] \Rightarrow (\forall x)A_j^1(x))$	2.29(a)
$(\exists x)((\exists y)(\forall z)[\mathscr{C}(x, y, z) \Rightarrow A_j^1(x)] \Rightarrow (\forall x)A_j^1(x))$	2.29(b)
$(\exists x)(\forall y)[(\forall z)(\mathscr{C}(x, y, z) \Rightarrow A_j^1(x)) \Rightarrow (\forall x)A_j^1(x)]$	2.29(b)
$(\exists x)(\forall y)(\exists z)([\mathscr{C}(x, y, z) \Rightarrow A_j^1(x)] \Rightarrow (\forall x)A_j^1(x))$	2.29(a)
$(\exists x)(\forall y)(\exists z)(\forall v)[(\mathscr{C}(x, y, z) \Rightarrow A_j^1(x)) \Rightarrow A_j^1(v)]$	2.29(c)

We repeat this process again: let $\mathscr{D}(x, y, z, v)$ be $(\mathscr{C}(x, y, z) \Rightarrow A_j^1(x)) \Rightarrow A_j^1(v)$. Let A_k^2 not occur in $\mathscr{D}$. Form:

$(\exists x)([(\forall y)[(\exists z)(\forall v)(\mathscr{D}(x, y, z, v)) \Rightarrow A_k^2(x, y)]] \Rightarrow (\forall y)A_k^2(x, y))$	
$(\exists x)(\exists y)[[(\exists z)(\forall v)(\mathscr{D}(x, y, z, v)) \Rightarrow A_k^2(x, y)] \Rightarrow (\forall y)A_k^2(x, y)]$	2.29(a)
$(\exists x)(\exists y)(\exists z)(\forall v)([\mathscr{D}(x, y, z, v) \Rightarrow A_k^2(x, y)] \Rightarrow (\forall y)A_k^2(x, y))$	2.29(a, b)
$(\exists x)(\exists y)(\exists z)(\forall v)(\forall w)([\mathscr{D}(x, y, z, v) \Rightarrow A_k^2(x, y)] \Rightarrow A_k^2(x, w))$	2.29(c)

Thus, a Skolem normal form of $\mathscr{A}$ is:

$$(\exists x)(\exists y)(\exists z)(\forall v)(\forall w)([((\mathscr{C}(x, y, z) \Rightarrow A_j^1(x)) \Rightarrow A_j^1(v)) \Rightarrow A_k^2(x, y)] \Rightarrow A_k^2(x, w))$$

Exercises

2.76 Find Skolem normal forms for the following wfs.

(a) $\neg(\exists x)A_1^1(x) \Rightarrow (\forall u)(\exists y)(\forall x)A_1^3(u, x, y)$

(b) $(\forall x)(\exists y)(\forall u)(\exists v)A_1^4(x, y, u, v)$

2.77 Show that there is an effective process that gives, for each wf $\mathscr{A}$ of a pure predicate calculus, another wf $\mathscr{B}$ of this calculus of the form $(\forall y_1)\ldots(\forall y_n)(\exists z_1)\ldots(\exists z_m)\mathscr{C}$, such that $\mathscr{C}$ is quantifier-free, $n, m \geqslant 0$, and $\mathscr{A}$ is satisfiable if and only if $\mathscr{B}$ is satisfiable. [*Hint:* Apply Proposition 2.31 to $\neg\mathscr{A}$.]

2.78 Find a Skolem normal form $\mathscr{B}$ for $(\forall x)(\exists y)A_1^2(x, y)$ and show that it is not the case that $\vdash \mathscr{B} \Leftrightarrow (\forall x)(\exists y)A_1^2(x, y)$. Hence, a Skolem normal form for a wf $\mathscr{A}$ is not necessarily logically equivalent to $\mathscr{A}$, in contradistinction to the prenex normal form given by Proposition 2.30.

11. ISOMORPHISM OF INTERPRETATIONS. CATEGORICITY OF THEORIES

We shall say that an interpretation M of the wfs of some first-order theory K is *isomorphic* with an interpretation M′ of K if and only if there is a one–one correspondence g (called an isomorphism) of the domain D of M with the domain D' of M′ such that:

1. For any predicate letter A_j^n of K and for any $b_1, \ldots, b_n$ in D, $\models_M A_j^n[b_1, \ldots, b_n]$ if and only if $\models_{M'} A_j^n[g(b_1), \ldots, g(b_n)]$.
2. For any function letter f_j^n of K and for any $b_1, \ldots, b_n$ in D, $g((f_j^n)^M(b_1, \ldots, b_n)) = (f_j^n)^{M'}(g(b_1), \ldots, g(b_n)$.
3. For any individual constant a_j of K, $g((a_j)^M) = (a_j)^{M'}$.

The notation $M_1 \approx M_2$ will be used to indicate that M_1 is isomorphic with M_2. Notice that, if $M_1 \approx M_2$, the domains of M_1 and M_2 must be of the same cardinality.

PROPOSITION 2.32 If g is an isomorphism of M with M′ then (a) for any wf $\mathscr{A}$ of K, any sequence $s = (b_1, b_2, \ldots)$ of elements of D, and the corresponding sequence $g(s) = (g(b_1), g(b_2), \ldots)$, s satisfies $\mathscr{A}$ if and only if $g(s)$ satisfies $\mathscr{A}$; (b) hence, $\models_M \mathscr{A}$ if and only if $\models_{M'} \mathscr{A}$.

Proof Part (b) follows directly from part (a). The proof of part (a) is a simple induction on the number of connectives and quantifiers in $\mathscr{A}$ and is left as an exercise.

We see from Proposition 2.32 that isomorphic interpretations have the same "structure" and, thus, differ in no essential way.

Exercises

2.79 Prove that if M is an interpretation with domain D, and D' is a set that has the same cardinality as D, then one can define an interpretation M′ with domain D' such that M is isomorphic with M′.

2.80 Prove that: (a) M is isomorphic with M. (b) If M is isomorphic with M′, then M′ is isomorphic with M. (c) If M is isomorphic with M′ and M′ is isomorphic with M″, then M is isomorphic with M″.

A theory K with equality is said to be $\mathfrak{m}$-*categorical*, where $\mathfrak{m}$ is a cardinal number, if and only if (1) any two normal models of K of cardinality $\mathfrak{m}$ are isomorphic, and (2) K has at least one normal model of cardinality $\mathfrak{m}$ (see Loś, 1954c).

Examples

1. Let K^2 be the theory of equality K_1 (see page 76) to which we have added the axiom (E2): $(\exists x_1)(\exists x_2)(x_1 \neq x_2 \wedge (\forall x_3)(x_3 = x_1 \vee x_3 = x_2))$. Then K^2 is 2-categorical. Every normal model of K^2 has exactly two elements. More generally, define (En) to be:

$$(\exists x_1)\ldots(\exists x_n)\left(\bigwedge_{1 \leqslant i < j \leqslant n} x_i \neq x_j \wedge (\forall x_{n+1})(x_{n+1} = x_1 \vee \cdots \vee x_{n+1} = x_n)\right)$$

where $\bigwedge_{1 \leqslant i < j \leqslant n} x_i \neq x_j$ is the conjunction of all wfs $x_i \neq x_j$ with $1 \leqslant i < j \leqslant n$. Then, if K^n is obtained from K_1 by adding (En) as an axiom, K^n is n-categorical, and every normal model of K^n has exactly n elements.

2. Theory K_2 (see page 76) of densely ordered sets with neither first nor last

element is $\aleph_0$-categorical (see Kamke, 1950, p. 71: every denumerable normal model of K_2 is isomorphic with the model consisting of the set of rational numbers under their natural ordering). But one can prove that K_2 is not $\mathfrak{m}$-categorical for any $\mathfrak{m}$ different from $\aleph_0$.

Exercises

2.81[A] Find a first-order theory with equality that is not $\aleph_0$-categorical but is $\mathfrak{m}$-categorical for all $\mathfrak{m} > \aleph_0$. [*Hint:* Consider the theory G_c of abelian groups (see page 77). For each integer n, let nx stand for the term $\underbrace{(x + x) + \cdots + x}_{n\text{ times}}$. Add to G_c the new axioms $(\mathscr{B}_n)$: $(\forall x)(\exists_1 y)(ny = x)$ for all $n \geqslant 2$. The new theory is the theory of uniquely divisible abelian groups. Its normal models are essentially vector spaces over the field of rational numbers. However, any two vector spaces over the rationals of the same nondenumerable cardinality are isomorphic, and there are denumerable vector spaces over the rationals that are not isomorphic (see Bourbaki, 1947).]

2.82[A] Find a theory with equality that is $\mathfrak{m}$-categorical for all infinite cardinals $\mathfrak{m}$. [*Hint:* Add to the theory G_c of abelian groups the axiom $(\forall x_1)(2x_1 = 0)$. The normal models of the new theory are just the vector spaces over the field of integers modulo 2. Any two such vector spaces of the same cardinality are isomorphic (see Bourbaki, 1947).]

2.83 Show that the theorems of the theory K^n in Example 1 above are precisely the set of all wfs of K^n that are true in all normal models of cardinality n.

2.84[A] Find two nonisomorphic densely ordered sets of cardinality $2^{\aleph_0}$ with neither first nor last element. (This shows that the theory K_2 of Example 2 is not $2^{\aleph_0}$-categorical.)

Is there a theory with equality that is $\mathfrak{m}$-categorical for some noncountable cardinal $\mathfrak{m}$ but not $\mathfrak{n}$-categorical for some other noncountable cardinal $\mathfrak{n}$? In Example 2 we found a theory that is only $\aleph_0$-categorical; in Exercise 2.81 we found a theory that is $\mathfrak{m}$-categorical for all infinite $\mathfrak{m} > \aleph_0$ but not $\aleph_0$-categorical, and in Exercise 2.82, a theory that is $\mathfrak{m}$-categorical for all infinite $\mathfrak{m}$. The elementary theory G of groups is not $\mathfrak{m}$-categorical for any infinite $\mathfrak{m}$. The problem is whether these four cases exhaust all the possibilities. That this is so has been proved by M. D. Morley (1965).

12. GENERALIZED FIRST-ORDER THEORIES. COMPLETENESS AND DECIDABILITY*

If, in the definition of the notion of first-order theory, we allow a noncountable number of predicate letters, function letters, and individual constants, and possibly a noncountable number of axioms, we arrive at the notion of a generalized first-order theory. First-order theories are special cases of generalized first-order theories. The reader may easily check that all the results for first-order theories, through Lemma 2.12, hold also for generalized first-order theories without any changes in the proofs. Lemma 2.13 becomes Lemma 2.13′: if the set of symbols of a generalized

* Presupposed in parts of this section is a slender acquaintance with ordinal and cardinal numbers (see Chapter 4, or Kamke, 1950, or Sierpinski, 1958).

theory K has cardinality $\aleph_\alpha$, then the set of expressions of K also can be well-ordered and has cardinality $\aleph_\alpha$. (First, order the expressions by their length, which is some positive integer, and then stipulate that if e_1 and e_2 are two distinct expressions of the same length k, and j is the first place in which they differ, then e_1 "precedes" e_2 if the jth symbol of e_1 precedes the jth symbol of e_2 according to the given well-ordering of the symbols of K.) Now, under the same assumption as for Lemma 2.13′, Lindenbaum's Lemma 2.14′ can be proved for generalized theories much as before, except that all the enumerations (of the wfs $\mathscr{B}_i$ and of the theories J_i) are transfinite, and the proof that J is consistent and complete uses transfinite induction. The analogue of Henkin's theorem 2.17 runs as follows.

PROPOSITION 2.33 If the set of symbols of a consistent generalized theory K has cardinality $\aleph_\alpha$, then K has a model of cardinality $\aleph_\alpha$.

Proof The original proof of Lemma 2.15 is modified in the following way. Add $\aleph_\alpha$ new individual constants $b_1, b_2, \ldots, b_\lambda, \ldots$. As before, the new theory K_0 is consistent. Let $F_1(x_{i_1}), \ldots, F(x_{i_\lambda}), \ldots$ $(\lambda < \omega_\alpha)$ be a sequence consisting of all wfs of K_0 with exactly one free variable. Let (S_λ) be $(\exists x_{i_\lambda}) \neg F_\lambda(x_{i_\lambda}) \Rightarrow \neg F_\lambda(b_{j_\lambda})$, where the sequence $b_{j_1}, b_{j_2}, \ldots, b_{j_\lambda}, \ldots$ of distinct constants is chosen so that b_{j_λ} does not occur in $F_\beta(x_{i_\beta})$ for $\beta \leqslant \lambda$. The new theory K′, obtained by adding all the wfs (S_λ) as axioms, is consistent by a transfinite induction analogous to the inductive proof in Lemma 2.15. K′ is a scapegoat theory which is an extension of K and contains $\aleph_\alpha$ closed terms. By the extended Lindenbaum Lemma 2.14′, K′ can be extended to a consistent, complete scapegoat theory J with $\aleph_\alpha$ closed terms. The same proof as in Lemma 2.16 provides a model M of J of cardinality $\aleph_\alpha$.

COROLLARY 2.34

(a) If the set of symbols of a consistent generalized theory K with equality has cardinality $\aleph_\alpha$, then K has a normal model of cardinality $\leqslant \aleph_\alpha$.
(b) If, in addition, K has an infinite normal model (or if K has arbitrarily large finite normal models), then K has a normal model of any cardinality $\aleph_\beta \geqslant \aleph_\alpha$.
(c) In particular, if K is an ordinary theory with equality (i.e., $\aleph_\alpha = \aleph_0$) and K has an infinite normal model (or if K has arbitrarily large finite normal models), then K has a normal model of any cardinality $\aleph_\beta (\beta \geqslant 0)$.

Proof (a) The model guaranteed by Proposition 2.33 can be contracted to a normal model (see page 78) consisting of equivalence classes in a set of cardinality $\aleph_\alpha$. Such a set of equivalence classes has cardinality $\leqslant \aleph_\alpha$.

(b) Assume $\aleph_\beta \geqslant \aleph_\alpha$. Let $b_1, b_2, \ldots$ be a set of new individual constants of cardinality $\aleph_\beta$, and add the axioms $b_\lambda \neq b_\mu$ for $\lambda \neq \mu$. As in the proof of Corollary 2.27, this new theory is consistent and so, by part (a), has a normal model of cardinality $\leqslant \aleph_\beta$ (since the new theory has $\aleph_\beta$ symbols). But, because of the axioms $b_\lambda \neq b_\mu$, the normal model has exactly $\aleph_\beta$ elements.

(c) Part (c) is a special case of part (b).

Exercise

2.85 If the set of symbols of a predicate calculus K with equality has cardinality $\aleph_\alpha$, prove that there is an extension K′ of K (with the same symbols as K) such that K′ has a normal model of cardinality $\aleph_\alpha$ but K′ has no normal model of cardinality $<\aleph_\alpha$.

From Lemma 2.12 and Corollary 2.34(a, b), it follows easily that, if a generalized first-order theory K with equality has $\aleph_\alpha$ symbols, is $\aleph_\beta$-categorical for some $\beta \geqslant \alpha$, and has no finite models, then K is complete, in the sense that, for any closed wf $\mathscr{A}$, either $\vdash_K \mathscr{A}$ or $\vdash_K \neg\mathscr{A}$ (Vaught, 1954). If not-$\vdash_K \mathscr{A}$ and not-$\vdash_K \neg\mathscr{A}$, then the theories $K' = K + \{\neg\mathscr{A}\}$ and $K'' = K + \{\mathscr{A}\}$ are consistent by Lemma 2.12, and so, by Corollary 2.34(a), there are normal models M_1 and M_2 of K′ and K″, respectively, of cardinality $\leqslant\aleph_\alpha$. Since K has no finite models, M_1 and M_2 are infinite. Hence, by Corollary 2.34(b), there are normal models N_1 and N_2 of K′ and K″, respectively, of cardinality $\aleph_\beta$. By the $\aleph_\beta$-categoricity of K, N_1 and N_2 must be isomorphic. But, since $\neg\mathscr{A}$ is true in N_1 and $\mathscr{A}$ is true in N_2, this is impossible. Therefore, either $\vdash_K \mathscr{A}$ or $\vdash_K \neg\mathscr{A}$.

In particular, if K is an ordinary first-order theory with equality that has no finite models and is $\aleph_\beta$-categorical for some $\beta \geqslant 0$, then K is complete. As an example, consider the theory K_2 of densely ordered sets with neither first nor last element (see page 76, Example 2). K_2 has no finite models and is $\aleph_0$-categorical.

If an ordinary first-order theory K is axiomatic (i.e., one can effectively decide whether any wf is an axiom) and complete, then K is decidable; that is, there is an effective procedure to determine whether any given wf is a theorem. To see this, remember (see page 68) that if a theory is axiomatic, one can effectively enumerate the theorems. Any wf $\mathscr{A}$ is provable if and only if its closure is provable. Hence, we may confine our attention to closed wfs $\mathscr{A}$. Since K is complete, either $\mathscr{A}$ is a theorem or $\neg\mathscr{A}$ is a theorem, and, therefore, one or the other will eventually turn up in our enumeration of the theorems. This provides an effective test for theoremhood. Notice that if K is inconsistent, then every wf is a theorem and there is an obvious decision procedure; if K is consistent, then not both $\mathscr{A}$ and $\neg\mathscr{A}$ can show up as theorems and we need only wait until one or the other appears.

If an ordinary axiomatic theory K with equality has no finite models and is $\aleph_\beta$-categorical for some $\beta \geqslant 0$, then, by what we have proved, K is decidable. In particular, the theory K_2 mentioned above is decidable.

In certain cases, there is a more direct method of proving completeness or decidability. Let us take as an example the theory K_2 of densely ordered sets with neither first nor last element. Langford (1927) has given the following procedure for K_2. Consider any closed wf $\mathscr{A}$. By Proposition 2.30, we can assume that $\mathscr{A}$ is in prenex normal form $(Qy_1)\ldots(Qy_n)\mathscr{B}$, where $\mathscr{B}$ contains no quantifiers. If (Qy_n) is $(\forall y_n)$, replace $(\forall y_n)\mathscr{B}$ by $\neg(\exists y_n)\neg\mathscr{B}$. In all cases, then, we have, at the right side of the wf, $(\exists y_n)\mathscr{C}$, where $\mathscr{C}$ has no quantifiers. Any negation $x \neq y$ can be replaced by $x < y \vee y < x$, and $x \nless y$ can be replaced by $x = y \vee y < x$. Hence, all negation signs may be eliminated from $\mathscr{C}$. We can now put $\mathscr{C}$ into disjunctive normal form—that is, a disjunction of conjunctions of atomic wfs (see page 25, Exercise 1.41). Now $(\exists y_n)(\mathscr{C}_1 \vee \mathscr{C}_2 \vee \cdots \vee \mathscr{C}_k)$ is equivalent to $(\exists y_n)\mathscr{C}_1 \vee (\exists y_n)\mathscr{C}_2 \vee \cdots \vee$

$(\exists y_n)\mathscr{C}_k$. Consider each $(\exists y_n)\mathscr{C}_i$ separately. $\mathscr{C}_i$ is a conjunction of atomic wfs of the form $t < s$ and $t = s$. If $\mathscr{C}_i$ does not contain y_n, just erase $(\exists y_n)$. Note that, if a wf $\mathscr{D}$ does not contain y_n, then $(\exists y_n)(\mathscr{D} \wedge \mathscr{E})$ may be replaced by $\mathscr{D} \wedge (\exists y_n)\mathscr{E}$. Hence, we are reduced to the consideration of $(\exists y_n)\mathscr{F}$, where $\mathscr{F}$ is a conjunction of atomic wfs, each of which contains y_n. Now, if one of the conjuncts is $y_n = z$ for some z different from y_n, replace in $\mathscr{F}$ all occurrences of y_n by z and erase $(\exists y_n)$. If we have $y_n = y_n$ alone, then just erase $(\exists y_n)$. If we have $y_n = y_n$ as one conjunct among others, erase $y_n = y_n$. If $\mathscr{F}$ has a conjunct $y_n < y_n$, replace all of $(\exists y_n)\mathscr{F}$ by $y_n < y_n$. If $\mathscr{F}$ consists of $y_n < z_1 \wedge \cdots \wedge y_n < z_j$, or if $\mathscr{F}$ consists of $u_1 < y_n \wedge \cdots \wedge u_m < y_n$, where $z_1, \ldots, z_j, u_1, \ldots, u_m$ are different from y_n, replace $(\exists y_n)\mathscr{F}$ by $y_n = y_n$. If $\mathscr{F}$ consists of $y_n < z_1 \wedge \cdots \wedge y_n < z_j \wedge u_1 < y_n \wedge \cdots \wedge u_m < y_n$, replace $(\exists y_n)\mathscr{F}$ by the conjunction of all the wfs $u_i < z_p$ for $1 \leqslant i \leqslant m$ and $1 \leqslant p \leqslant j$. This exhausts all possibilities, and, in every case, we have replaced $(\exists y_n)\mathscr{C}$ by a wf $\mathscr{R}$ containing no quantifiers; that is, we have eliminated the quantifier $(\exists y_n)$. We are left with $(Qy_1)\ldots(Qy_{n-1})\mathscr{S}$ where $\mathscr{S}$ contains no quantifiers. Now we apply the same procedure successively to $(Qy_{n-1}), \ldots, (Qy_1)$. Finally, we are left with a wf without quantifiers built up out of wfs of the form $x = x$ and $x < x$. Now, if we replace $x = x$ by $x = x \Rightarrow x = x$ and $x < x$ by $\neg(x = x \Rightarrow x = x)$, then the result is either an instance of a tautology or the negation of such an instance. Hence, by Proposition 2.1, either the result or its negation is provable. Now, one can easily check that all the replacements we have made in this whole reduction process applied to $\mathscr{A}$ have been replacements of wfs $\mathscr{T}$ by other wfs $\mathscr{U}$ such that $\vdash_K \mathscr{T} \Leftrightarrow \mathscr{U}$. Hence, by Proposition 2.8(c), if our final result is provable, then so is the original wf $\mathscr{A}$, and, if the negation of our result is provable, so is $\neg\mathscr{A}$. Thus, K_2 is complete and decidable.

The method used in this proof, the successive elimination of existential quantifiers, has been applied to other theories. It yields a decision procedure (see Hilbert & Bernays, 1934, §5) for the elementary theory K_1 of equality (see page 76). It has been applied by Tarski (1951) to prove the completeness and decidability of elementary algebra (i.e., of the elementary theory of real-closed fields; see van der Waerden, 1949) and by Szmielew (1955) to prove the decidability of the elementary theory of abelian groups. For more details and examples, see Chang and Keisler (1973, § 1.5).

Exercises

2.86 (Henkin, 1955b) If an ordinary theory K with equality is finitely axiomatizable and $\aleph_\alpha$-categorical for some α, prove that K is decidable.

2.87 (a) Prove the decidability of the theory K_1 of equality.
(b) Give an example of a theory with equality that is $\aleph_\alpha$-categorical for some α but is incomplete.

MATHEMATICAL APPLICATIONS

1. Let F be the elementary theory of fields (see page 77). We let n stand for the term $\underbrace{1 + 1 + \cdots + 1}_{n\text{ times}}$. Then the assertion that a field has characteristic p

can be expressed by the wf $\mathscr{C}_p$: $p = 0$. A field has characteristic 0 if and only if it does not have characteristic p for any prime p. Then for any closed wf $\mathscr{A}$ of F that holds for all fields of characteristic 0, there is a prime number q such that $\mathscr{A}$ holds for all fields of characteristic $\geqslant q$. To see this, notice that if F′ is obtained from F by adding as axioms $\neg\mathscr{C}_2, \neg\mathscr{C}_3, \ldots, \neg\mathscr{C}_p, \ldots$ (for all primes p), the normal models of F′ are the fields of characteristic 0. Hence, by Corollary 2.20(a), noting that if $\mathscr{A}$ holds in all normal models of F′, it holds in all models of F′, $\vdash_{\mathrm{F}'} \mathscr{A}$; but then, for some finite number of the new axioms $\neg\mathscr{C}_{q_1}, \neg\mathscr{C}_{q_2}, \ldots, \neg\mathscr{C}_{q_n}$, we have $\neg\mathscr{C}_{q_1}, \ldots, \neg\mathscr{C}_{q_n} \vdash_{\mathrm{F}} \mathscr{A}$. Let q be a prime greater than all $q_1, \ldots, q_n$. In every field of characteristic $\geqslant q$, the wfs $\neg\mathscr{C}_{q_1}, \neg\mathscr{C}_{q_2}, \ldots, \neg\mathscr{C}_{q_n}$ are true; hence, $\mathscr{A}$ is also true. [Other applications in algebra may be found in A. Robinson (1951) and Cherlin (1976).]

2. A *graph* may be considered as a set partially ordered by a symmetric binary relation R (i.e., the relation that holds between any two vertices if and only if they are connected by an edge). Call a graph *k*-colorable if and only if the graph can be divided into k disjoint (possibly empty) sets such that no two elements in the same set are in the relation R. (Intuitively, these k sets correspond to k colors, each color being painted on the points in the corresponding set, with the proviso that two points connected by an edge are painted different colors.) Notice that any subgraph of a k-colorable graph is also k-colorable. Now, we can show that if every finite subgraph of a graph $\mathscr{G}$ is k-colorable and if $\mathscr{G}$ can be well-ordered, then the whole graph $\mathscr{G}$ is k-colorable. To prove this, construct the following generalized theory K with equality (Beth, 1953). There are two binary predicate letters $A_1^2(=)$ and A_2^2 (corresponding to the relation R on $\mathscr{G}$); there are k monadic predicate letters $A_1^1, \ldots, A_k^1$ (corresponding to the k subsets into which we hope to divide the graph); and there are individual constants a_c, one for each element c of the graph $\mathscr{G}$. We have as proper axioms, in addition to the usual assumptions (A6) and (A7) for equality, the following wfs:

(I) $\neg A_2^2(x, x)$ (irreflexivity of R)
(II) $A_2^2(x, y) \Rightarrow A_2^2(y, x)$ (symmetry of R)
(III) $(\forall x)(A_1^1(x) \vee A_2^1(x) \vee \cdots \vee A_k^1(x))$ (division into k classes)
(IV) $(\forall x)\neg(A_i^1(x) \wedge A_j^1(x))$ for $1 \leqslant i < j \leqslant k$ (disjointness of the k classes)
(V) For $1 \leqslant i \leqslant k$, $(\forall x)(\forall y)(A_i^1(x) \wedge A_i^1(y) \Rightarrow \neg A_2^2(x, y))$ (two elements in the same class are not in the relation R)
(VI) For any two distinct elements b, c of $\mathscr{G}$, $a_b \neq a_c$.
(VII) If $\mathrm{R}(b, c)$ holds in $\mathscr{G}$, $A_2^2(a_b, a_c)$.

Now, any finite set of these axioms involves only a finite number of the individual constants $a_{c_1}, \ldots, a_{c_n}$, and since the corresponding subgraph $\{c_1, \ldots, c_n\}$ is, by assumption, k-colorable, the given finite set of axioms has a model and is, therefore, consistent. Since any finite set of axioms is consistent, K is consistent. By Corollary 2.34(a), K has a normal model of cardinality $\leqslant$ the cardinality of the graph $\mathscr{G}$. This model is a k-colorable graph and, by (VI)–(VII), has $\mathscr{G}$ as a subgraph. Hence, $\mathscr{G}$ is also k-colorable. [Compare this proof with a standard mathematical proof of the same result by Bruijn and Erdos (1951). Generally, use of the method above replaces complicated applications of Tychonoff's Theorem or König's Unendlichkeit's Lemma.]

Exercises

2.88[A] (Loś, 1954b) A group B is said to be orderable if there exists a binary relation R on B that totally orders B such that, if $x \mathrm{R} y$, then $(x + z)\mathrm{R}(y + z)$ and $(z + x)\mathrm{R}(z + y)$. Show, by a method similar to that used in example 2 above, that a group B is orderable if and only if every finitely generated subgroup is orderable (if we assume that the set B can be well-ordered).

2.89[A] Set up a theory for algebraically closed fields of characteristic $p\,(\geqslant 0)$ by adding to the theory F of fields the new axioms P_n, where P_n states that every nonconstant polynomial of degree $\leqslant n$ has a root, as well as axioms to determine the characteristic. Show that every wf of F that holds for one algebraically closed field of characteristic 0 holds for all of them. [*Hint:* This theory is $\aleph_\beta$-categorical for $\beta > 0$, is axiomatizable, and has no finite models. See A. Robinson (1952).]

2.90 By ordinary mathematical reasoning, solve the *finite marriage problem*. Given a finite set M of m men and a set N of women such that each man knows only a finite number of women and, for $1 \leqslant k \leqslant m$, any subset of M having k elements is acquainted with at least k women of N (i.e., there are at least k women in N acquainted with at least one of the k given men), then it is possible to marry (monogamously) all the men of M to women in N so that every man is married to a woman with whom he is acquainted. [*Hint*: Halmos and Vaughn (1950): $m = 1$ is trivial. For $m > 1$, use induction, considering the cases: (I) for all k with $1 \leqslant k < m$, every set of k men knows at least $k + 1$ women, and (II) for some k with $1 \leqslant k < m$, there is a set of k men knowing exactly k women.] Extend this result to the infinite case—that is, when M is infinite and well-orderable and the assumptions above hold for all finite k. [*Hint:* Construct an appropriate generalized first-order theory, analogous to that in the second mathematical application just listed, and use Corollary 2.34(a).]

2.91 Prove that there is no generalized theory K with equality, having one predicate letter $<$ in addition to $=$, such that the normal models of K are exactly those interpretations in which the interpretation of $<$ is a well-ordering of the domain of the interpretation.

Let $\mathscr{A}$ be a wf in prenex normal form. If $\mathscr{A}$ is not closed, form its closure instead. Suppose, for example, $\mathscr{A}$ is $(\exists y_1)(\forall y_2)(\forall y_3)(\exists y_4)(\exists y_5)(\forall y_6)\mathscr{B}(y_1, y_2, y_3, y_4, y_5, y_6)$, where $\mathscr{B}$ contains no quantifiers. Erase $(\exists y_1)$ and replace y_1 in $\mathscr{B}$ by a new individual constant b_1: $(\forall y_2)(\forall y_3)(\exists y_4)(\exists y_5)(\forall y_6)\mathscr{B}(b_1, y_2, y_3, y_4, y_5, y_6)$. Erase $(\forall y_2)$ and $(\forall y_3)$, obtaining $(\exists y_4)(\exists y_5)(\forall y_6)\mathscr{B}(b_1, y_2, y_3, y_4, y_5, y_6)$. Now erase $(\exists y_4)$ and replace y_4 in $\mathscr{B}$ by $g(y_2, y_3)$, where g is a new function letter: $(\exists y_5)(\forall y_6)\mathscr{B}(b_1, y_2, y_3, g(y_2, y_3), y_5, y_6)$. Erase $(\exists y_5)$ and replace y_5 by $h(y_2, y_3)$, where h is a new function letter: $(\forall y_6)\mathscr{B}(b_1, y_2, y_3, g(y_2, y_3), h(y_2, y_3), y_6)$. Finally, erase $(\forall y_6)$. The resulting wf $\mathscr{B}(b_1, y_2, y_3, g(y_2, y_3), h(y_2, y_3), y_6)$ contains no quantifiers and will be denoted by $\mathscr{A}^*$. Thus, by introducing new function letters and individual constants, we can eliminate the quantifiers from a wf.

Examples

1. If $\mathscr{A}$ is $(\forall y_1)(\exists y_2)(\forall y_3)(\forall y_4)(\exists y_5)\mathscr{B}(y_1, y_2, y_3, y_4, y_5)$, where $\mathscr{B}$ is quantifier-free, then $\mathscr{A}^*$ may be taken to be $\mathscr{B}(y_1, g(y_1), y_3, y_4, h(y_1, y_3, y_4))$.
2. If $\mathscr{A}$ is $(\exists y_1)(\exists y_2)(\forall y_3)(\forall y_4)(\exists y_5)\mathscr{B}(y_1, y_2, y_3, y_4, y_5)$, where $\mathscr{B}$ is quantifier-free, $\mathscr{A}^*$ may be taken to be $\mathscr{B}(b, c, y_3, y_4, g(y_3, y_4))$.

Notice that $\mathscr{A}^* \vdash \mathscr{A}$, since we can put the quantifiers back on by several applications of Gen and Rule E4. [To be more precise, in the process of obtaining

$\mathscr{A}^*$, we drop all quantifiers and, for each existentially quantified variable y_i, we substitute a term $g(z_1, \ldots, z_k)$, where g is a new function letter and $z_1, \ldots, z_k$ are the variables that were universally quantified in the prefix preceding $(\exists y_i)$. If there are no such variables $z_1, \ldots, z_k$, we replace y_i by a new individual constant.]

PROPOSITION 2.35 (SECOND ε-THEOREM) (Rasiowa, 1956; Hilbert & Bernays, 1939) Let K be a generalized theory. Replace each axiom $\mathscr{A}$ of K by $\mathscr{A}^*$. (The new function letters and individual constants introduced for one wf are to be different from those introduced for another wf.) Let K* be the generalized theory with the proper axioms $\mathscr{A}^*$. Then:

(a) If $\mathscr{C}$ is a wf of K and $\vdash_{K^*} \mathscr{C}$, then $\vdash_K \mathscr{C}$.
(b) K is consistent if and only if K* is consistent.

Proof (a) Let $\mathscr{C}$ be a wf of K such that $\vdash_{K^*} \mathscr{C}$. Consider the ordinary theory K′ whose axioms $\mathscr{A}_1, \ldots, \mathscr{A}_n$ are such that $\mathscr{A}_1^*, \ldots, \mathscr{A}_n^*$ are the axioms used in the proof of $\mathscr{C}$. Let K′* be the theory whose axioms are $\mathscr{A}_1^*, \ldots, \mathscr{A}_n^*$. Assume that M is a denumerable model of K′. We may assume that the domain of M is the set P of positive integers (see Exercise 2.79). Let $\mathscr{A}$ be any axiom of K′. For example, suppose that $\mathscr{A}$ has the form $(\exists y_1)(\forall y_2)(\forall y_3)(\exists y_4)\mathscr{B}(y_1, y_2, y_3, y_4)$, where $\mathscr{B}$ is quantifier-free. $\mathscr{A}^*$ has the form $\mathscr{B}(b, y_2, y_3, g(y_2, y_3))$. Extend the model M step by step in the following way (note that the domain always remains the set P): since $\mathscr{A}$ is true for M, $(\exists y_1)(\forall y_2)(\forall y_3)(\exists y_4)\mathscr{B}(y_1, y_2, y_3, y_4)$ is true for M. Let the interpretation b^* of b be the least positive integer y_1 such that $(\forall y_2)(\forall y_3)(\exists y_4)\mathscr{B}(y_1, y_2, y_3, y_4)$ is true in the model. Hence, $(\exists y_4)\mathscr{B}(b, y_2, y_3, y_4)$ is true in this extended model. For any positive integers y_2 and y_3, let the interpretation of $g(y_2, y_3)$ be the least positive integer y_4 such that $\mathscr{B}(b, y_2, y_3, y_4)$ is true in the extended model. Hence, $\mathscr{B}(b, y_2, y_3, g(y_2, y_3))$ is true in the extended model. If we do this for all the axioms $\mathscr{A}$ of K′, we obtain a model M* of K′*. Since $\vdash_{K'^*} \mathscr{C}$, $\mathscr{C}$ is true for M*. Since M* differs from M only in having interpretations of the new individual constants and function letters, and since $\mathscr{C}$ does not contain any of these constants or function letters, $\mathscr{C}$ is true for M. Thus, $\mathscr{C}$ is true in every denumerable model of K′. Hence, $\vdash_{K'} \mathscr{C}$ by Corollary 2.20(a). Since the axioms of K′ are axioms of K, we have $\vdash_K \mathscr{C}$. [For a constructive proof of an equivalent result, see Hilbert and Bernays (1939).]

(b) Clearly, K* is an extension of K, since $\mathscr{A}^* \vdash \mathscr{A}$. Hence, if K* is consistent, so is K. Conversely, assume K is consistent. Let $\mathscr{C}$ be any wf of K. If K* is inconsistent, $\vdash_{K^*} \mathscr{C} \wedge \neg\mathscr{C}$. By part (a), $\vdash_K \mathscr{C} \wedge \neg\mathscr{C}$, contradicting the consistency of K.

Let us use the term *Generalized Completeness Theorem* for the proposition that every consistent generalized theory has a model. If we assume that every set can be well-ordered (or, equivalently, the axiom of choice), then the Generalized Completeness Theorem is a consequence of Proposition 2.33.

By the Maximal Ideal Theorem (MI) we mean the proposition that every Boolean algebra has a maximal ideal. This is equivalent to the Boolean Representation Theorem, which states that every Boolean algebra is isomorphic to a Boolean algebra of sets. [Compare Stone (1936). For the theory of Boolean algebras, see Sikorski (1960).] The only known proof of the MI Theorem uses the axiom of choice, but it is a remarkable fact that the MI Theorem is equivalent to the Generalized Completeness Theorem, and this equivalence can be proved without using the axiom of choice.

PROPOSITION 2.36 (Loś, 1954a; Rasiowa & Sikorski, 1951, 1952) The Generalized Completeness Theorem is equivalent to the Maximal Ideal Theorem.

Proof (a) Assume the Generalized Completeness Theorem. Let B be a Boolean algebra. Construct a generalized theory K with equality having the binary function letters $\cup$ and $\cap$, the singulary function letter f_1^1 [we denote $f_1^1(t)$ by $\bar{t}$], predicate letters $=$ and A_1^1, and, for each element b in B, an individual constant a_b. As axioms, we take the usual axioms for a Boolean algebra (see Sikorski, 1960), axioms (A6) and (A7) for equality, a complete description of B (i.e., if b, c, d, e, and b_1 are in B, the axioms $a_b \neq a_c$ if $b \neq c$; $a_b \cup a_c = a_d$ if $b \cup c = d$ in B; $a_b \cap a_c = a_e$ if $b \cap c = e$ in B; and $\bar{a}_b = a_{b_1}$ if $\bar{b} = b_1$ in B, where $\bar{b}$ denotes the complement of b), and axioms asserting that A_1^1 determines a maximal ideal [i.e., $A_1^1(x \cap \bar{x})$, $A_1^1(x) \wedge A_1^1(y) \Rightarrow A_1^1(x \cup y)$, $A_1^1(x) \Rightarrow A_1^1(x \cap y)$, $A_1^1(x) \vee A_1^1(\bar{x})$, and $\neg A_1^1(x \cup \bar{x})$]. Now K is consistent, for, if there is a proof in K of a contradiction, this proof contains only a finite number of the symbols $a_b, a_c, \ldots$—say, $a_{b_1}, \ldots, a_{b_n}$. The elements $b_1, \ldots, b_n$ generate a finite subalgebra B′ of B. Every finite Boolean algebra clearly has a maximal ideal. Hence, B′ is a model for the wfs that occur in the proof of the contradiction, and therefore the contradiction is true in B′, which is impossible. Thus, K is consistent and, by the Generalized Completeness Theorem, K has a model, which is a Boolean algebra A with a maximal ideal I. But B is a subalgebra of A and I $\cap$ B is a maximal ideal in B.

(b) Assume the Maximal Ideal Theorem. Let K be a consistent generalized theory. For each axiom $\mathscr{A}$ of K, form the wf $\mathscr{A}^*$ obtained by constructing a prenex normal form for $\mathscr{A}$ and then eliminating the quantifiers through the addition of new individual constants and function letters. Let K′ be a new theory having the wfs $\mathscr{A}^*$, plus all instances of tautologies, as its axioms, such that its wfs contain no quantifiers and its rules of inference are modus ponens and a rule of substitution for variables (namely, substitution of terms for variables). Now K′ is consistent, since the theorems of K′ are also theorems of the consistent theory K* of Proposition 2.35. Let B be the Lindenbaum algebra determined by K′ (i.e., for any wfs $\mathscr{A}$ and $\mathscr{B}$, let $\mathscr{A}$ Eq $\mathscr{B}$ mean that $\vdash_{K'} \mathscr{A} \Leftrightarrow \mathscr{B}$; Eq is an equivalence relation; let $[\mathscr{A}]$ be the equivalence class of $\mathscr{A}$; define $[\mathscr{A}] \cup [\mathscr{B}] = [\mathscr{A} \vee \mathscr{B}]$, $[\mathscr{A}] \cap [\mathscr{B}] = [\mathscr{A} \wedge \mathscr{B}]$, $\overline{[\mathscr{A}]} = [\neg \mathscr{A}]$; under these operations, the set of equivalence classes is a Boolean algebra, called the Lindenbaum algebra of K′). By the Maximal Ideal Theorem, let I be a maximal ideal in B. Define a model M of K′ having the set of terms of K′ as its domain; the individual constants and function letters are their own interpretations, and, for any predicate letter A_j^n, we say that $A_j^n(t_1, \ldots, t_n)$ is true in M if and only if $[A_j^n(t_1, \ldots, t_n)]$ is not in I. One can show easily that a wf $\mathscr{A}$ of K′ is true in M if and only if $[\mathscr{A}]$ is not in I. But, for any theorem $\mathscr{B}$ of K′, $[\mathscr{B}] = 1$, which is not in I. Hence, M is a model for K′. For any axiom $\mathscr{A}$ of K, every substitution instance of $\mathscr{A}^*(y_1, \ldots, y_n)$ is a theorem in K′; therefore, $\mathscr{A}^*(y_1, \ldots, y_n)$ is true for all $y_1, \ldots, y_n$ in the model. It follows easily, by reversing the process through which $\mathscr{A}^*$ arose from $\mathscr{A}$, that $\mathscr{A}$ is true in the model. Hence, M is a model for K.

The Maximal Ideal Theorem (and, therefore, also the Generalized Completeness Theorem) turns out to be strictly weaker than the axiom of choice (see Halpern, 1964).

Exercises

2.92 Show that the Generalized Completeness Theorem implies that every set can be totally ordered (and, therefore, that the axiom of choice holds for any set of nonempty disjoint finite sets).

2.93 In the proof of Proposition 2.36(b), show that if K is an ordinary first-order theory, then the Lindenbaum algebra B is countable and the Maximal Ideal Theorem need not be assumed in the proof.

The natural algebraic structures corresponding to the propositional calculus are Boolean algebras (see Exercise 1.59 and Rosenbloom, 1950, chaps. 1 and 2). For first-order theories, the presence of quantifiers introduces more algebraic structure. For example, if K is a first-order theory, then, in the corresponding Lindenbaum algebra B, $[(\exists x)\mathscr{A}(x)] = \sum_t [\mathscr{A}(t)]$, where $\sum_t$ indicates the least upper bound in B, and t ranges over all terms of K that are free for x in $\mathscr{A}(x)$. Two types of algebraic structures have been proposed to serve as algebraic counterparts of quantification theory. The first, cylindrical algebras, have been studied extensively by Tarski, Thompson, Henkin, Monk, and others (see Henkin, Monk & Tarski, 1971). The other approach is the theory of polyadic algebras, invented and developed by Halmos (1962).

13. ELEMENTARY EQUIVALENCE. ELEMENTARY EXTENSIONS

Two interpretations M_1 and M_2 of a generalized first-order predicate calculus K are said to be *elementarily equivalent* (written $M_1 \equiv M_2$) if the sentences of K true for M_1 are the same as the sentences true for M_2. Intuitively, $M_1 \equiv M_2$ if and only if M_1 and M_2 cannot be distinguished by means of the language of K.† Of course, K is a generalized predicate calculus and may have nondenumerably many symbols.

Clearly, (1) $M \equiv M$; (2) if $M_1 \equiv M_2$, then $M_2 \equiv M_1$; and (3) if $M_1 \equiv M_2$ and $M_2 \equiv M_3$, then $M_1 \equiv M_3$.

Two models of a complete theory K must be elementarily equivalent, since the sentences true in these models are precisely the sentences provable in K. This applies, for example, to any two densely ordered sets without first or last elements (see page 90).

We already know, by Proposition 2.32(a), that isomorphic models are elemen-

†Notice that for M to be a model of a predicate calculus K nothing more is required than that the interpretations provided by M consist of only interpretations of the symbols of K. M is then automatically a model of K, since the only axioms of K are logical axioms.

tarily equivalent. The converse, however, is not true. Consider, for example, any complete theory K that has an infinite normal model. By Corollary 2.34(b), K has normal models of any infinite cardinal $\aleph_\alpha$. If we take two normal models of K of different cardinality, they are elementarily equivalent but not isomorphic. A concrete example is the complete theory K_2 of densely ordered sets that have neither first nor last element. The rational numbers and the real numbers, under their natural orderings, are elementarily equivalent models of K_2 but are not isomorphic.

Exercises

2.94 Let K_∞, the theory of infinite sets, consist of the pure theory K_1 of equality plus the axioms $\mathscr{B}_n$, where $\mathscr{B}_n$ asserts that there are at least n elements. Show that any two models of K_∞ are elementarily equivalent (see Exercises 2.63 and 2.87(a)].

2.95[D] If M_1 and M_2 are elementarily equivalent normal models and M_1 is finite, prove that M_1 and M_2 are isomorphic.

2.96 Let K be a theory with equality having $\aleph_\alpha$ symbols.

(a) Prove that there are at most $2^{\aleph_\alpha}$ models of K, no two of which are elementarily equivalent.

(b) Prove that there are at most $2^{\aleph_\gamma}$ mutually nonisomorphic models of K of cardinality $\aleph_\beta$, where γ is the maximum of α and β.

2.97 Let M be any infinite normal model of a theory with equality K having $\aleph_\alpha$ symbols. Prove that, for any cardinal $\aleph_\gamma \geqslant \aleph_\alpha$, there is a normal model M′ of K of cardinality $\aleph_\alpha$ such that $M \equiv M'$.

A model M_2 of a predicate calculus K is said to be an *extension* of a model M_1 of K (written $M_1 \subseteq M_2$)* if the following conditions hold:

1. The domain D_1 of M_1 is a subset of the domain D_2 of M_2.
2. For any individual constant c of K, $c^{M_2} = c^{M_1}$, where c^{M_2} and c^{M_1} are the interpretations of c in M_2 and M_1, respectively.
3. For any function letter f_j^n of K and any $a_1, \ldots, a_n$ in D_1, $(f_j^n)^{M_2}(a_1, \ldots, a_n) = (f_j^n)^{M_1}(a_1, \ldots, a_n)$.
4. For any predicate letter A_j^n of K and any $a_1, \ldots, a_n$ in D_1, $\models_{M_1} A_j^n[a_1, \ldots, a_n]$ if and only if $\models_{M_2} A_j^n[a_1, \ldots, a_n]$.

When $M_1 \subseteq M_2$, one also says that M_1 is a *substructure* (or *submodel*) of M_2.

Examples

1. If K contains only the predicate letters = and <, then the set of rational numbers under its natural ordering is an extension of the set of integers under its natural ordering.

2. If K is the predicate calculus in the language of field theory (with the predicate letter =, function letters + and ×, and individual constants 0 and 1), then the field of real numbers is an extension of the field of rational numbers, the field of rational numbers is an extension of the ring of integers, and the ring of integers is an extension of the "semiring" of nonnegative integers. For any fields F_1 and F_2, $F_1 \subseteq F_2$ if and only if F_1 is a subfield of F_2 in the usual algebraic sense.

*The reader will have no occasion to confuse this use of $\subseteq$ with that for the inclusion relation.

Exercises

2.98 Prove: (a) $M \subseteq M$; (b) if $M_1 \subseteq M_2$ and $M_2 \subseteq M_3$, then $M_1 \subseteq M_3$; and (c) if $M_1 \subseteq M_2$ and $M_2 \subseteq M_1$, then $M_1 = M_2$.

2.99 Assume $M_1 \subseteq M_2$.

(a) Let $\mathscr{B}(x_1,\dots,x_n)$ be a wf of the form $(\forall y_1)\dots(\forall y_m)\mathscr{A}(x_1,\dots,x_n,y_1,\dots,y_m)$, where $\mathscr{A}$ contains no quantifiers. Show that, for any $a_1, \dots, a_n$ in the domain of M_1, if $\models_{M_2} \mathscr{B}[a_1,\dots,a_n]$, then $\models_{M_1} \mathscr{B}[a_1,\dots,a_n]$. In particular, any sentence $(\forall y_1)\dots(\forall y_m)\mathscr{A}(y_1,\dots,y_m)$, where $\mathscr{A}$ contains no quantifiers, is true in M_1 if it is true in M_2.

(b) Let $\mathscr{B}(x_1,\dots,x_n)$ be a wf of the form $(\exists y_1)\dots(\exists y_m)\mathscr{A}(x_1,\dots,x_n,y_1,\dots,y_m)$, where $\mathscr{A}$ has no quantifiers. Show that, for any $a_1, \dots, a_n$ in the domain of M_1, if $\models_{M_1} \mathscr{B}[a_1,\dots,a_n]$, then $\models_{M_2} \mathscr{B}[a_1,\dots,a_n]$. In particular, any sentence $(\exists y_1)\dots(\exists y_m)\mathscr{A}(y_1,\dots,y_m)$, where $\mathscr{A}$ contains no quantifiers, is true in M_2 if it is true in M_1.

2.100 (a) Let K be the predicate calculus of the language of field theory. Find a model M of K and a nonempty subset X of the domain D of M such that there is no substructure of M having domain X.

(b) If K is a predicate calculus with no individual constants or function letters, show that, if M is a model of K and X is a subset of the domain D of M, then there is one and only one substructure of M having domain X.

(c) Let K be any predicate calculus. Let M be any model of K and let X be any subset of the domain D of M. Let Y be the intersection of the domains of all submodels M′ of M such that $X \subseteq D_{M'}$, the domain of M′. Show that there is one and only one submodel of M having domain Y. (This submodel is called the *submodel generated* by X.)

A somewhat stronger relation between interpretations than "extension" is useful in model theory. Let M_1 and M_2 be models of some predicate calculus K. We say that M_2 is an *elementary extension* of M_1 (written $M_1 \leqslant_e M_2$) if

(1) $M_1 \subseteq M_2$, and

(2) For any wf $\mathscr{A}(y_1,\dots,y_n)$ of K and for any $a_1, \dots, a_n$ in the domain D_1 of M_1,

$$\models_{M_1} \mathscr{A}[a_1,\dots,a_n] \text{ if and only if } \models_{M_2} \mathscr{A}[a_1,\dots,a_n]$$

(In particular, for any sentence $\mathscr{A}$ of K, $\mathscr{A}$ is true for M_1 if and only if $\mathscr{A}$ is true for M_2.) When $M_1 \leqslant_e M_2$, we shall also say that M_1 is an *elementary substructure* (or *elementary submodel*) of M_2.

It is obvious that, if $M_1 \leqslant_e M_2$, then $M_1 \subseteq M_2$ and $M_1 \equiv M_2$. The converse is not true, as the following example shows. Let K be the first-order theory of groups (see page 77). K has the predicate letter $=$, function letter $+$, and individual constant 0. Let I be the group of integers, and $2I$ the group of even integers. Then $2I \subseteq I$ and $I \simeq 2I$. [The function g such that $g(x) = 2x$ for all x in I is an isomorphism of I with $2I$.] Since $I \simeq 2I$, $I \equiv 2I$. Consider the wf $\mathscr{A}(y)$: $(\exists x)(x + x = y)$. Then $\models_I \mathscr{A}[2]$, but not-$\models_{2I} \mathscr{A}[2]$. Thus, I is not an elementary extension of $2I$. (This example shows the stronger result that even assuming $M_1 \subseteq M_2$ and $M_1 \simeq M_2$ does not imply $M_1 \leqslant_e M_2$.)

The following theorem provides an easy method for showing that $M_1 \leqslant_e M_2$.

PROPOSITION 2.37 (Tarski & Vaught, 1957) Let $M_1 \subseteq M_2$. Assume the following condition:

(\$) For every wf $\mathscr{B}(x_1, \ldots, x_n)$ of the form $(\exists y)\mathscr{A}(x_1, \ldots, x_n, y)$ and for all $a_1, \ldots, a_n$ in the domain D_1 of M_1, if $\models_{M_2} \mathscr{B}[a_1, \ldots, a_n]$, then there is some b in D_1 such that $\models_{M_2} \mathscr{A}[a_1, \ldots, a_n, b]$.
Then $M_1 \leqslant_e M_2$.

Proof Let us prove: $(*)$ $\models_{M_1} \mathscr{C}[a_1, \ldots, a_k]$ if and only if $\models_{M_2} \mathscr{C}[a_1, \ldots, a_k]$ for any wf $\mathscr{C}(x_1, \ldots, x_k)$ and any $a_1, \ldots, a_k$ in D_1. The proof is by induction on the number m of connectives and quantifiers in $\mathscr{C}$. If $m = 0$, then $(*)$ follows from clause 4 of the definition of $M_1 \subseteq M_2$. Now assume that $(*)$ holds true for all wfs having fewer than m connectives and quantifiers.

Case 1. $\mathscr{C}$ is $\neg\mathscr{D}$. By inductive hypothesis, $\models_{M_1} \mathscr{D}[a_1, \ldots, a_k]$ if and only if $\models_{M_2} \mathscr{D}[a_1, \ldots, a_k]$. Using the fact that not-$\models_{M_1} \mathscr{D}[a_1, \ldots, a_k]$ if and only if $\models_{M_1} \neg\mathscr{D}[a_1, \ldots, a_k]$, and similarly for M_2, we obtain $(*)$.

Case 2. $\mathscr{C}$ is $\mathscr{D} \Rightarrow \mathscr{E}$. By inductive hypothesis, $\models_{M_1} \mathscr{D}[a_1, \ldots, a_k]$ if and only if $\models_{M_2} \mathscr{D}[a_1, \ldots, a_k]$, and similarly for $\mathscr{E}$. $(*)$ then follows easily.

Case 3. $\mathscr{C}$ is $(\exists y)\mathscr{A}(x_1, \ldots, x_k, y)$. By inductive hypothesis, $(**)$ $\models_{M_1} \mathscr{A}[a_1, \ldots, a_k, b]$ if and only if $\models_{M_2} \mathscr{A}[a_1, \ldots, a_k, b]$ for any $a_1, \ldots, a_k, b$ in D_1.

Case 3a. Assume $\models_{M_1} (\exists y)\mathscr{A}(x_1, \ldots, x_k, y)[a_1, \ldots, a_k]$ for some $a_1, \ldots, a_k$ in D_1. Then, $\models_{M_1} \mathscr{A}[a_1, \ldots, a_k, b]$ for some b in D_1. So, by $(**)$, $\models_{M_2} \mathscr{A}[a_1, \ldots, a_k, b]$. Hence, $\models_{M_2} (\exists y)\mathscr{A}(x_1, \ldots, x_k, y)[a_1, \ldots, a_k]$.

Case 3b. Assume $\models_{M_2} (\exists y)\mathscr{A}(x_1, \ldots, x_k, y)[a_1, \ldots, a_k]$ for some $a_1, \ldots, a_k$ in D_1. By assumption (\$), there exist b in D_1 such that $\models_{M_2} \mathscr{A}[a_1, \ldots, a_k, b]$. Hence, by $(**)$, $\models_{M_1} \mathscr{A}[a_1, \ldots, a_k, b]$ and, therefore, $\models_{M_1} (\exists y)\mathscr{A}(x_1, \ldots, x_k, y)[a_1, \ldots, a_k]$.

This completes the induction proof, since any wf is logically equivalent to a wf that can be built up from atomic wfs by forming negations, conditionals, and existential quantifications.

Exercises

2.101 Prove: (a) $M \leqslant_e M$; (b) if $M_1 \leqslant_e M_2$ and $M_2 \leqslant_e M_3$, then $M_1 \leqslant_e M_3$; and (c) if $M_1 \leqslant_e M$, $M_2 \leqslant_e M$, and $M_1 \subseteq M_2$, then $M_1 \leqslant_e M_2$.

2.102 Let K be the theory of totally ordered sets with equality [axioms (a)–(c) and (e)–(g) of Exercise 2.64]. Let M_1 and M_2 be the models for K with domains the set of positive integers and the set of nonnegative integers, respectively (under the natural orderings $<$ in both cases). Prove that $M_1 \subseteq M_2$ and $M_1 \simeq M_2$, but $M_1 \nleqslant_e M_2$.

Let M be a model of a theory K. Extend K to a theory K′ by adding a new individual constant a_d for every member d of the domain of M. We can extend M to a model of K′ by taking d as the interpretation of a_d. By the *diagram* of M we mean the set of all true sentences of M of the forms $A_j^n(a_{d_1}, \ldots, a_{d_n})$, $\neg A_j^n(a_{d_1}, \ldots, a_{d_n})$, and $f_j^n(a_{d_1}, \ldots, a_{d_n}) = a_{d_m}$. In particular, $a_{d_1} \neq a_{d_2}$ belongs to the diagram if $d_1 \neq d_2$. By the *complete diagram* of M we mean the set of all sentences of K′ that are true for M.

Clearly, any model M′ of the complete diagram of M determines an elementary extension M″ of M,* and vice versa.

* The elementary extension M″ of M is obtained from M′ by forgetting about the interpretations of the a_d's.

Exercises

2.103 (a) Let M be a denumerable normal model of an ordinary theory K with equality such that every element of the domain of M is the interpretation of some closed term of K.

(i) Show that, if $M \subseteq M'$ and $M \equiv M'$, then $M \leqslant_e M'$.

(ii) Prove that there is a denumerable normal elementary extension M′ of M such that M and M′ are not isomorphic.

(b) Let K be a predicate calculus with equality having two function letters $+$ and $\times$ and two individual constants 0 and 1. Let M be the standard model of arithmetic, with domain the set of natural numbers, and $+$, $\times$, 0, and 1 having their ordinary meaning. Prove that M has a proper denumerable extension that is not isomorphic to M; that is, there is a denumerable nonstandard model of arithmetic.

PROPOSITION 2.38 (UPWARD LÖWENHEIM-SKOLEM-TARSKI THEOREM) Let K be a theory with equality having $\aleph_\alpha$ symbols, and let M be a normal model of K with domain of cardinality $\aleph_\beta$. Let γ be the maximum of α and β. Then, for any $\delta \geqslant \gamma$, there is a model M′ of cardinality $\aleph_\delta$ such that $M \neq M'$ and $M \leqslant_e M'$.

Proof Add to the complete diagram of M a set of cardinality $\aleph_\delta$ of new individual constants b_τ, together with axioms $b_\tau \neq b_\rho$ for distinct τ and ρ and axioms $b_\tau \neq a_d$ for all individual constants a_d corresponding to members d of the domain of M. This new theory K′ is consistent, since M can be used as a model for any finite number of axioms of K′. (If $b_{\tau_1}, \ldots, b_{\tau_k}, a_{d_1}, \ldots, a_{d_m}$ are the new individual constants in these axioms, interpret $b_{\tau_1}, \ldots, b_{\tau_k}$ as distinct elements of the domain of M different from $d_1, \ldots, d_m$.) Hence, by Corollary 2.34(a), K′ has a normal model M′ of cardinality $\aleph_\delta$ such that $M \subsetneqq M'$ and $M \leqslant_e M'$.

PROPOSITION 2.39 (DOWNWARD LÖWENHEIM-SKOLEM-TARSKI THEOREM) Let K be a theory with $\aleph_\alpha$ symbols, and let M be a model of K of cardinality $\aleph_\gamma \geqslant \aleph_\alpha$. Assume A is a subset of the domain D of M having cardinality $\mathfrak{n}$, and assume $\aleph_\beta$ is such that $\aleph_\gamma \geqslant \aleph_\beta \geqslant \max(\aleph_\alpha, \mathfrak{n})$. Then there is an elementary submodel M′ of M of cardinality $\aleph_\beta$ and with domain $D' \supseteq A$.

Proof Since $\mathfrak{n} \leqslant \aleph_\beta \leqslant \aleph_\gamma$, we can add $\aleph_\beta$ elements of D to A to obtain a larger set B of cardinality $\aleph_\beta$. Consider any subset C of D having cardinality $\aleph_\beta$. For every wf $\mathscr{A}(y_1, \ldots, y_n, z)$ of K and any $a_1, \ldots, a_n$ in C such that $\models_M (\exists z)\mathscr{A}(y_1, \ldots, y_n, z)[a_1, \ldots, a_n]$, add to C the first element b of D (with respect to some fixed well-ordering of D) such that $\models_M \mathscr{A}[a_1, \ldots, a_n, b]$. Denote the so-enlarged set by $C^\#$. Since K has $\aleph_\alpha$ symbols, there are $\aleph_\alpha$ wfs. Since $\aleph_\alpha \leqslant \aleph_\beta$, there are at most $\aleph_\beta$ new elements in $C^\#$, and, therefore, the cardinality of $C^\#$ is $\aleph_\beta$. Form by induction a sequence of sets $C_0, C_1, \ldots$ by setting $C_0 = B$ and $C_{n+1} = C_n^\#$. Let $D' = \bigcup_{n \in \omega} C_n$. Then the cardinality of D' is $\aleph_\beta$. In addition, D' is closed under all the functions $(f_j^n)^M$. {Assume $a_1, \ldots, a_n$ in D'. We may assume $a_1, \ldots, a_n$ in C_k for some k. Now $\models_M (\exists z)(f_j^n(x_1, \ldots, x_n) = z)[a_1, \ldots, a_n]$. Hence, $(f_j^n)^M(a_1, \ldots, a_n)$, being the first and only member b of D such that $\models_M (f_j^n(x_1, \ldots, x_n) = z)[a_1, \ldots, a_n, b]$, must belong to

$C_k^\# = C_{k+1} \subseteq D'$.} Similarly, all interpretations $(a_j)^M$ of individual constants are in D'. Hence, D' determines a substructure M′ of M. To show that $M' \leqslant_e M$, consider any wf $\mathscr{A}(y_1, \ldots, y_n, z)$ and any $a_1, \ldots, a_n$ in D' such that $\models_M (\exists z)\mathscr{A}(y_1, \ldots, y_n, z)$ $[a_1, \ldots, a_n]$. There exists C_k such that $a_1, \ldots, a_n$ are in C_k. Let b be the first element of D such that $\models_M \mathscr{A}[a_1, \ldots, a_n, b]$. Then $b \in C_k^\# = C_{k+1} \subseteq D'$. So, by the Tarski-Vaught Theorem (Proposition 2.37), $M' \leqslant_e M$.

14. ULTRAPOWERS. NONSTANDARD ANALYSIS

By a *filter*† on a nonempty set A we mean a set $\mathscr{F}$ of subsets of A such that

1. $A \in \mathscr{F}$
2. $B \in \mathscr{F} \wedge C \in \mathscr{F} \Rightarrow B \cap C \in \mathscr{F}$
3. $B \in \mathscr{F} \wedge B \subseteq C \Rightarrow C \in \mathscr{F}$

Examples

1. $\mathscr{F} = \{A\}$ is a filter on A.
2. $\mathscr{F} = \mathscr{P}(A)$ is a filter on A. It is said to be *improper* and every other filter on A is said to be *proper*.
3. Let $B \subseteq A$. The set $\mathscr{F}_B = \{C \mid B \subseteq C \subseteq A\}$ is a filter on A. $\mathscr{F}_B$ consists of all subsets of A that include B. Any filter of the form $\mathscr{F}_B$ is called a *principal* filter. In particular, $\mathscr{F}_A = \{A\}$ and $\mathscr{P}(A) = \mathscr{F}_0$ are principal filters. (Remember that 0 denotes the empty set.)

Exercises

2.104 Show that a filter $\mathscr{F}$ on A is proper if and only if $0 \notin \mathscr{F}$.

2.105 Show that a filter $\mathscr{F}$ on A is a principal filter if and only if the intersection of all sets in $\mathscr{F}$ is a member of $\mathscr{F}$.

2.106 Prove that every finite filter is a principal filter. In particular, any filter on a finite set A is a principal filter.

2.107 Let A be infinite and let $\mathscr{F}$ be the set of all subsets of A that are complements of finite sets: $\mathscr{F} = \{C \mid (\exists W)(C = A - W \wedge \text{Fin}(W))\}$. Show that $\mathscr{F}$ is a nonprincipal filter on A.

2.108 Assume A has cardinality $\aleph_\beta$. Let $\aleph_\alpha \leqslant \aleph_\beta$. Let $\mathscr{F}$ be the set of all subsets of A whose complements have cardinality $< \aleph_\alpha$. Show that $\mathscr{F}$ is a nonprincipal filter on A.

2.109 A collection $\mathscr{G}$ of sets is said to have the *finite intersection property* if $B_1 \cap B_2 \cap \cdots \cap B_k \neq 0$ for any sets $B_1, \ldots, B_k$ in $\mathscr{G}$. If $\mathscr{G}$ is a collection of subsets of A having the finite intersection property, and $\mathscr{F} = \{C \mid (\exists B)(B \in \mathscr{G} \wedge B \subseteq C \subseteq A)\}$, show that $\mathscr{F}$ is a *proper* filter on A.

DEFINITION A filter $\mathscr{F}$ on a set A is called an *ultrafilter* on A if $\mathscr{F}$ is a maximal proper filter on A; that is, $\mathscr{F}$ is a proper filter on A and there is no proper filter $\mathscr{G}$ on A such that $\mathscr{F} \subset \mathscr{G}$.

†The notion of a filter is related to that of an ideal. A collection $\mathscr{F} \subseteq \mathscr{P}(A)$ is a filter on A if and only if the set $\mathscr{G} = \{A - B \mid B \in \mathscr{F}\}$ of complements of sets in $\mathscr{F}$ is an ideal in the Boolean algebra $\mathscr{P}(A)$. Remember that $\mathscr{P}(A)$ denotes the set of all subsets of A.

Example Let $a \in A$. The principal filter $\mathscr{F}_a = \{B | a \in B \wedge B \subseteq A\}$ is an ultrafilter on A. Assume that $\mathscr{G}$ is a filter on A such that $\mathscr{F}_a \subset \mathscr{G}$. Let $C \in \mathscr{G} - \mathscr{F}_a$. Then $C \subseteq A$ and $a \notin C$. Hence, $a \in A - C$. Thus, $A - C \in \mathscr{F}_a \subset \mathscr{G}$. Since $\mathscr{G}$ is a filter and C and $A - C$ are both in $\mathscr{G}$, then $0 = C \cap (A - C) \in \mathscr{G}$. Hence, $\mathscr{G}$ is not a proper filter.

Exercises

2.110 Let $\mathscr{F}$ be a proper filter on A, and assume that $B \subseteq A$ and $A - B \notin \mathscr{F}$. Prove that there is a proper filter $\mathscr{F}' \supseteq \mathscr{F}$ such that $B \in \mathscr{F}'$.

2.111 Let $\mathscr{F}$ be a proper filter on A. Prove that $\mathscr{F}$ is an ultrafilter on A if and only if, for every $B \subseteq A$, either $B \in \mathscr{F}$ or $A - B \in \mathscr{F}$.

2.112 Let $\mathscr{F}$ be a proper filter on A. Show that $\mathscr{F}$ is an ultrafilter on A if and only if, for all B and C in $\mathscr{P}(A)$, if $B \notin \mathscr{F}$ and $C \notin \mathscr{F}$, then $B \cup C \notin \mathscr{F}$.

2.113 (a) Show that every principal ultrafilter on A is of the form $\mathscr{F}_a = \{B | a \in B \wedge B \subseteq A\}$, where $a \in A$.

(b) Show that a nonprincipal ultrafilter on A contains no finite sets.

2.114 Let $\mathscr{F}$ be a filter on a set A and let $\mathscr{I}$ be the corresponding ideal: $B \in \mathscr{I}$ if and only if $A - B \in \mathscr{F}$. Prove that $\mathscr{F}$ is an ultrafilter on A if and only if $\mathscr{I}$ is a maximal ideal.

2.115 Let X be a chain of proper filters on a set A; that is, for any B and C in X, either $B \subseteq C$ or $C \subseteq B$. Prove that the union $\cup(X) = \{a | (\exists B)(B \in X \wedge a \in B\}$ is a proper filter on A and $B \subseteq \cup(X)$ for all B in X.

PROPOSITION 2.40 (ULTRAFILTER THEOREM) Every filter $\mathscr{F}$ on a set A can be extended to an ultrafilter on A.†

Proof Let $\mathscr{F}$ be a filter on A. Let $\mathscr{I}$ be the corresponding ideal: $B \in \mathscr{I}$ if and only if $A - B \in \mathscr{F}$. By Proposition 2.36, every ideal can be extended to a maximal ideal. In particular, $\mathscr{I}$ can be extended to a maximal ideal $\mathscr{J}$. If we let $\mathscr{G} = \{B | A - B \in \mathscr{J}\}$, then $\mathscr{G}$ is easily seen to be an ultrafilter, and $\mathscr{F} \subseteq \mathscr{G}$.

The existence of an ultrafilter including $\mathscr{F}$ can be proved easily on the basis of Zorn's Lemma. (In fact, consider the set X of all proper filters $\mathscr{F}'$ such that $\mathscr{F} \subseteq \mathscr{F}'$. X is partially ordered by $\subset$ and any $\subset$-chain in X has an upper bound in X—namely, by Exercise 2.115, the union of all filters in the chain. Hence, by Zorn's Lemma, there is a maximal element $\mathscr{F}^*$ in X, which is the required ultrafilter.) However, Zorn's Lemma is equivalent to the axiom of choice, which is a stronger assumption than the Generalized Completeness Theorem.

COROLLARY 2.41 If A is an infinite set, there exists a nonprincipal ultrafilter on A.

Proof Let $\mathscr{F}$ be the filter on A consisting of all complements $A - B$ of finite subsets B of A (see Exercise 2.107). By Proposition 2.40, there is an ultrafilter $\mathscr{G} \supseteq \mathscr{F}$. Assume $\mathscr{G}$ is a principal ultrafilter. By Exercise 2.113(a), $\mathscr{G} = \mathscr{F}_a$ for some $a \in A$. Then $A - \{a\} \in \mathscr{F} \subseteq \mathscr{G}$. Also, $\{a\} \in \mathscr{G}$. Hence, $0 = \{a\} \cap (A - \{a\}) \in \mathscr{G}$, contradicting the fact that an ultrafilter is proper.

† We assume the Generalized Completeness Theorem.

REDUCED DIRECT PRODUCTS

We shall now study an important way of constructing models. Let K be any predicate calculus with equality. Let J be a nonempty set and, for each j in J, let M_j be some normal model of K. In other words, consider a function F assigning to each j in J some normal model. We denote $F(j)$ by M_j.

Let $\mathscr{F}$ be an ultrafilter on J. For each j in J, let D_j denote the domain of the model M_j. By the Cartesian product $\prod_{j \in J} D_j$ we mean the set of all functions f with domain J such that $f(j) \in D_j$ for all j in J. If $f \in \prod_{j \in J} D_j$, we shall refer to $f(j)$ as the jth *component* of f. Let us define a binary relation $=_{\mathscr{F}}$ in $\prod_{j \in J} D_j$ as follows:

$$f =_{\mathscr{F}} g \text{ if and only if } \{j \mid f(j) = g(j)\} \in \mathscr{F}$$

If we think of the sets in $\mathscr{F}$ as being "large" sets, then, borrowing a phrase from measure theory, we read $f =_{\mathscr{F}} g$ as "$f(j) = g(j)$ almost everywhere".

It is easy to see that $=_{\mathscr{F}}$ is an equivalence relation: (1) $f =_{\mathscr{F}} f$; (2) if $f =_{\mathscr{F}} g$, then $g =_{\mathscr{F}} f$; and (3) if $f =_{\mathscr{F}} g$ and $g =_{\mathscr{F}} h$, then $f =_{\mathscr{F}} h$. For the proof of (3), observe that $\{j \mid f(j) = g(j)\} \cap \{j \mid g(j) = h(j)\} \subseteq \{j \mid f(j) = h(j)\}$. If $\{j \mid f(j) = g(j)\}$ and $\{j \mid g(j) = h(j)\}$ are in $\mathscr{F}$, then so is their intersection and, therefore, also $\{j \mid f(j) = h(j)\}$.

On the basis of the equivalence relation $=_{\mathscr{F}}$, we can divide $\prod_{j \in J} D_j$ into equivalence classes: for any f in $\prod_{j \in J} D_j$, we define its equivalence class $f_{\mathscr{F}}$ as $\{g \mid f =_{\mathscr{F}} g\}$. Clearly, (1) $f \in f_{\mathscr{F}}$; (2) $f_{\mathscr{F}} = h_{\mathscr{F}}$ if and only if $f =_{\mathscr{F}} h$; and (3) if $f_{\mathscr{F}} \neq h_{\mathscr{F}}$, then $f_{\mathscr{F}} \cap h_{\mathscr{F}} = 0$. We denote the set of all equivalence class $f_{\mathscr{F}}$ by $\prod_{j \in J} D_j/\mathscr{F}$. Intuitively, $\prod_{j \in J} D_j/\mathscr{F}$ is obtained from $\prod_{j \in J} D_j$ by identifying (or merging) elements of $\prod_{j \in J} D_j$ that are equal almost everywhere.

Now we shall define a model M of K with domain $\prod_{j \in J} D_j/\mathscr{F}$.

1. Let c be any individual constant of K and let c_j be the interpretation of c in M_j. Then the interpretation of c in M will be $f_{\mathscr{F}}$, where f is the function such that $f(j) = c_j$ for all j in J. We denote f by $\{c_j\}_{j \in J}$.

2. Let f_k^n be any function letter of K and let A_k^n be any predicate letter of K. Their interpretations $(f_k^n)^M$ and $(A_k^n)^M$ are defined in the following manner. Let $(g_1)_{\mathscr{F}}, \ldots, (g_n)_{\mathscr{F}}$ be any members of $\prod_{j \in J} D_j/\mathscr{F}$.

(a) $(f_k^n)^M((g_1)_{\mathscr{F}}, \ldots, (g_n)_{\mathscr{F}}) = h_{\mathscr{F}}$, where $h(j) = (f_k^n)^{M_j}(g_1(j), \ldots, g_n(j))$ for all j in J.

(b) $(A_k^n)^M((g_1)_{\mathscr{F}}, \ldots, (g_n)_{\mathscr{F}})$ holds if and only if $\{j \mid \models_{M_j} A_k^n[g_1(j), \ldots, g_n(j)]\} \in \mathscr{F}$.

Intuitively, $(f_k^n)^M$ is calculated component-wise, and $(A_k^n)^M$ holds if and only if it holds in almost all components. Definitions (a) and (b) have to be shown to be independent of the choice of the representatives $g_1, \ldots, g_n$ in the equivalence classes $(g_1)_{\mathscr{F}}, \ldots, (g_n)_{\mathscr{F}}$:

If $g_1 =_{\mathscr{F}} g_1^*, \ldots, g_n =_{\mathscr{F}} g_n^*$ and $h^*(j) = (f_k^n)^{M_j}(g_1^*(j), \ldots, g_n^*(j))$, then (i) $h_{\mathscr{F}} =_{\mathscr{F}} h_{\mathscr{F}}^*$ and (ii) $\{j \mid \models_{M_j} A_k^n[g_1(j), \ldots, g_n(j)]\} \in \mathscr{F}$ if and only if $\{j \mid \models_{M_j} A_k^n[g_1^*(j), \ldots, g_n^*(j)]\} \in \mathscr{F}$.

Part (i) follows from the inclusion:

$$\begin{aligned}&\{j \mid g_1(j) = g_1^*(j)\} \cap \cdots \cap \{j \mid g_n(j) = g_n^*(j)\} \\ &\qquad \subseteq \{j \mid (f_k^n)^{M_j}(g_1(j), \ldots, g_n(j)) = (f_k^n)^{M_j}(g_1^*(j), \ldots, g_n^*(j))\}\end{aligned}$$

Part (ii) follows from the inclusions:

$$\{j|g_1(j) = g_1^*(j)\} \cap \cdots \cap \{j|g_n(j) = g_n^*(j)\}$$
$$\subseteq \{j| \models_{M_j} A_k^n[g_1(j), \ldots, g_n(j)] \text{ if and only if } \models_{M_j} A_k^n[g_1^*(j), \ldots, g_n^*(j)]\}$$

and

$$\{j| \models_{M_j} A_k^n[g_1(j), \ldots, g_n(j)]\} \cap \{j| \models_{M_j} A_k^n[g_1(j), \ldots, g_n(j)] \text{ if and only if } \models_{M_j} A_k^n[g_1^*(j), \ldots, g_n^*(j)]\} \subseteq \{j| \models_{M_j} A_k^n[g_1^*(j), \ldots, g_n^*(j)]\}$$

In the case of the equality relation $=$, which is an abbreviation for A_1^2,

$$(A_1^2)^{M}(g_{\mathscr{F}}, h_{\mathscr{F}}) \quad \text{if and only if} \quad \{j| \models_{M_j} A_1^2[g(j), h(j)]\} \in \mathscr{F}$$
$$\text{if and only if} \quad \{j|g(j) = h(j)\} \in \mathscr{F}$$
$$\text{if and only if} \quad g =_{\mathscr{F}} h$$

that is, if and only if $g_{\mathscr{F}} = h_{\mathscr{F}}$. Hence, the interpretation $(A_1^2)^M$ is the identity relation and the model M is normal.

The model M just defined will be denoted $\prod_{j \in J} M_j/\mathscr{F}$ and will be called a *reduced direct product*. When $\mathscr{F}$ is an ultrafilter, $\prod_{j \in J} M_j/\mathscr{F}$ is called an *ultraproduct*. When $\mathscr{F}$ is an ultrafilter and all the M_j's are the same model N, then $\prod_{j \in J} M_j/\mathscr{F}$ is denoted $N^J/\mathscr{F}$ and is called an *ultrapower*.

Examples

1. Choose a fixed element r of the index set J, and let $\mathscr{F}$ be the principal ultrafilter $\mathscr{F}_r = \{B|r \in B \wedge B \subseteq J\}$. Then for any f, g in $\prod_{j \in J} D_j$, $f =_{\mathscr{F}} g$ if and only if $\{j|f(j) = g(j)\} \in \mathscr{F}$—that is, if and only if $f(r) = g(r)$. Hence, a member of $\prod_{j \in J} D_j/\mathscr{F}$ consists of all f in $\prod_{j \in J} D_j$ that have the same rth component. For any predicate letter A_k^n of K and any $g_1, \ldots, g_n$ in $\prod_{j \in J} D_j$, $\models_M A_k^n[(g_1)_{\mathscr{F}}, \ldots, (g_n)_{\mathscr{F}}]$ if and only if $\{j| \models_{M_j} A_k^n[g_1(j), \ldots, g_n(j)]\} \in \mathscr{F}$—that is, if and only if $\models_{M_r} A_k^n[g_1(r), \ldots, g_n(r)]$. Hence, it is easy to verify that the function φ: $\prod_{j \in J} D_j/\mathscr{F} \to D_r$, defined by $\varphi(g_{\mathscr{F}}) = g(r)$, is an isomorphism of $\prod_{j \in J} M_j/\mathscr{F}$ with M_r. Thus, when $\mathscr{F}$ is a principal ultrafilter, the ultraproduct $\prod_{j \in J} M_j/\mathscr{F}$ is essentially the same as one of its components and yields nothing new.

2. Let $\mathscr{F}$ be the filter $\{J\}$. Then, for any f, g in $\prod_{j \in J} D_j$, $f =_{\mathscr{F}} g$ if and only if $\{j|f(j) = g(j)\} \in \mathscr{F}$—that is, if and only if $f(j) = g(j)$ for all j in J, or if and only if $f = g$. Thus, every member of $\prod_{j \in J} D_j/\mathscr{F}$ is a singleton $\{g\}$ for some g in $\prod_{j \in J} D_j$. Moreover, $(f_k^n)^M((g_1)_{\mathscr{F}}, \ldots, (g_n)_{\mathscr{F}}) = \{g\}$, where g is such that $g(j) = (f_k^n)^{M_j}(g_1(j), \ldots, g_n(j))$ for all $j \in J$. $\models_M A_k^n[(g_1)_{\mathscr{F}}, \ldots, (g_n)_{\mathscr{F}}]$ if and only if $\models_{M_j} A_k^n[g_1(j), \ldots, g_n(j)]$ for all j. Hence, $\prod_{j \in J} M_j/\mathscr{F}$ is, in this case, essentially the same as the ordinary "direct product" $\prod_{j \in J} M_j$, in which the operations and relations are defined component-wise.

3. Let $\mathscr{F}$ be the improper filter $\mathscr{P}(J)$. Then, for any f, g in $\prod_{j \in J} D_j$, $f =_{\mathscr{F}} g$ if and only if $\{j|f(j) = g(j)\} \in \mathscr{F}$—that is, if and only if $\{j|f(j) = g(j)\} \in \mathscr{P}(J)$. Thus, $f =_{\mathscr{F}} g$ for all f and g, and $\prod_{j \in J} D_j/\mathscr{F}$ consists of only one element. For any predicate

letter A_k^n, $\models_M A_k^n[f_{\mathscr{F}},\ldots,f_{\mathscr{F}}]$ if and only if $\{j \mid \models_{M_j} A_k^n[f(j),\ldots,f(j)]\} \in \mathscr{P}(J)$; that is, every atomic wf is true.

The basic theorem on ultraproducts is due to Łoś (1955b).

PROPOSITION 2.42 (ŁOŚ'S THEOREM) Let $\mathscr{F}$ be an ultrafilter on a set J and let $M = \prod_{j\in J} M_j/\mathscr{F}$ be an ultraproduct.

(a) Let $s = ((g_1)_{\mathscr{F}}, (g_2)_{\mathscr{F}}, \ldots)$ be a denumerable sequence of elements of $\prod_{j\in J} D_j/\mathscr{F}$. For each j in J, let s_j be the denumerable sequence $(g_1(j), g_2(j), \ldots)$ in D_j. Then, for any wf $\mathscr{A}$ of K, s satisfies $\mathscr{A}$ in M if and only if $\{j \mid s_j \text{ satisfies } \mathscr{A} \text{ in } M_j\} \in \mathscr{F}$.

(b) For any sentence $\mathscr{A}$ of K, $\mathscr{A}$ is true in $\prod_{j\in J} M_j/\mathscr{F}$ if and only if $\{j \mid \models_{M_j} \mathscr{A}\} \in \mathscr{F}$. [Thus, part (b) asserts that a sentence $\mathscr{A}$ is true in an ultraproduct if and only if it is true in almost all components.]

Proof (a) We shall use induction on the number m of connectives and quantifiers in $\mathscr{A}$. We can reduce the case $m = 0$ to the following subcases:* (i) $A_k^n(x_{i_1},\ldots,x_{i_n})$; (ii) $x_l = f_k^n(x_{i_1},\ldots,x_{i_n})$; and (iii) $x_l = a_k$. For part (i), s satisfies $A_k^n(x_{i_1},\ldots,x_{i_n})$ if and only if $\models_M A_k^n[(g_{i_1})_{\mathscr{F}},\ldots,(g_{i_n})_{\mathscr{F}}]$, which is equivalent to $\{j \mid \models_{M_j} A_k^n[g_{i_1}(j),\ldots, g_{i_n}(j)]\} \in \mathscr{F}$; that is, $\{j \mid s_j \text{ satisfies } A_k^n(x_{i_1},\ldots,x_{i_n}) \text{ in } M_j\} \in \mathscr{F}$. Parts (ii) and (iii) are handled in similar fashion.

Now let us assume the result holds for all wfs that have fewer than m connectives and quantifiers.

Case 1. $\mathscr{A}$ is $\neg\mathscr{B}$. By inductive hypothesis, s satisfies $\mathscr{B}$ in M if and only if $\{j \mid s_j \text{ satisfies } \mathscr{B} \text{ in } M_j\} \in \mathscr{F}$. s satisfies $\neg\mathscr{B}$ in M if and only if $\{j \mid s_j \text{ satisfies } \mathscr{B} \text{ in } M_j\} \notin \mathscr{F}$. But, since $\mathscr{F}$ is an ultrafilter, the last condition is equivalent, by Exercise 2.111, to $\{j \mid s_j \text{ satisfies } \neg\mathscr{B} \text{ in } M_j\} \in \mathscr{F}$.

Case 2. $\mathscr{A}$ is $\mathscr{B} \wedge \mathscr{C}$. By inductive hypothesis, s satisfies $\mathscr{B}$ in M if and only if $\{j \mid s_j \text{ satisfies } \mathscr{B} \text{ in } M_j\} \in \mathscr{F}$; s satisfies $\mathscr{C}$ in M if and only if $\{j \mid s_j \text{ satisfies } \mathscr{C} \text{ in } M_j\} \in \mathscr{F}$. Therefore, s satisfies $\mathscr{B} \wedge \mathscr{C}$ if and only if both of the indicated sets belong to $\mathscr{F}$. But this is equivalent to their intersection belonging to $\mathscr{F}$, which, in turn, is equivalent to $\{j \mid s_j \text{ satisfies } \mathscr{B} \wedge \mathscr{C} \text{ in } M_j\} \in \mathscr{F}$.

Case 3. $\mathscr{A}$ is $(\exists x_i)\mathscr{B}$. Assume s satisfies $(\exists x_i)\mathscr{B}$. Then there exists h in $\prod_{j\in J} D_j$ such that s' satisfies $\mathscr{B}$ in M, where s' is the same as s except that $h_{\mathscr{F}}$ is the ith component of s'. By inductive hypothesis, s' satisfies $\mathscr{B}$ in M if and only if $\{j \mid s'_j \text{ satisfies } \mathscr{B} \text{ in } M_j\} \in \mathscr{F}$. Hence, $\{j \mid s_j \text{ satisfies } (\exists x_i)\mathscr{B} \text{ in } M_j\} \in \mathscr{F}$, since, if s'_j satisfies $\mathscr{B}$ in M_j, then s_j satisfies $(\exists x_i)\mathscr{B}$ in M_j.

Conversely, assume $W = \{j \mid s_j \text{ satisfies } (\exists x_i)\mathscr{B} \text{ in } M_j\} \in \mathscr{F}$. For each j in W, choose some s'_j such that s'_j is the same as s_j except in at most the ith component and s'_j satisfies $\mathscr{B}$. Now define h in $\prod_{j\in J} D_j$ as follows: for j in W, let $h(j)$ be the ith component of s'_j, and, for $j \notin W$, choose $h(j)$ to be an arbitrary element of D_j. Let s'' be the same as s except that its ith component is $h_{\mathscr{F}}$. Then $W \subseteq \{j \mid s''_j \text{ satisfies } \mathscr{B}$ in

* A wf $A_k^n(t_1,\ldots,t_n)$ can be replaced by $(\forall u_1)\ldots(\forall u_n)(u_1 = t_1 \wedge \cdots \wedge u_n = t_n \Rightarrow A_k^n(u_1,\ldots,u_n))$, and a wf $x = f_k^n(t_1,\ldots,t_n)$ can be replaced by $(\forall z_1)\ldots(\forall z_n)(z_1 = t_1 \wedge \cdots \wedge z_n = t_n \Rightarrow x = f_k^n(z_1,\ldots,z_n))$. In this way, every wf is equivalent to a wf built up from wfs of the forms (i)–(iii) by applying connectives and quantifiers.

$M_j\} \in \mathscr{F}$. Hence, by the inductive hypothesis, s'' satisfies $\mathscr{B}$ in M. Therefore, s satisfies $(\exists x_i)\mathscr{B}$ in M.

(b) This follows from part (a) by noting that a sentence $\mathscr{A}$ is true in a model if and only if some sequence satisfies $\mathscr{A}$.

COROLLARY 2.43 If M is a model and $\mathscr{F}$ is an ultrafilter on J, and if M* is the ultrapower $M^J/\mathscr{F}$, then $M^* \equiv M$.

Proof Let $\mathscr{A}$ be any sentence. Then, by Proposition 2.42(b), $\mathscr{A}$ is true in M* if and only if $\{j|\mathscr{A}$ is true in M$\} \in \mathscr{F}$. If $\mathscr{A}$ is true in M, $\{j|\mathscr{A}$ is true in M$\} = J \in \mathscr{F}$. If $\mathscr{A}$ is false in M, $\{j|\mathscr{A}$ is true in M$\} = 0 \notin \mathscr{F}$.

Corollary 2.43 can be strengthened considerably. For each c in the domain D of M, let $c^{\#}$ stand for the constant function such that $c^{\#}(j) = c$ for all j in J. Define the function ψ such that, for each c in D, $\psi(c) = (c^{\#})_{\mathscr{F}} \in D^J/\mathscr{F}$, and denote the range of ψ by $M^{\#}$. $M^{\#}$ obviously contains the interpretations in M* of the individual constants. Moreover, $M^{\#}$ is closed under the operations $(f_k^n)^{M^*}$; for $(f_k^n)^{M^*}((c_1^{\#})_{\mathscr{F}}, \ldots, (c_n^{\#})_{\mathscr{F}})$ is $h_{\mathscr{F}}$, where $h(j) = (f_k^n)^M(c_1, \ldots, c_n)$ for all j in J, and $(f_k^n)^M(c_1, \ldots, c_n)$ is a fixed element b of D. So $h_{\mathscr{F}} = (b^{\#})_{\mathscr{F}} \in M^{\#}$. Thus, $M^{\#}$ is a substructure of M*.

COROLLARY 2.44 ψ is an isomorphism of M with $M^{\#}$, and $M^{\#} \leqslant_e M^*$.

Proof (a) By definition of $M^{\#}$, the range of ψ is $M^{\#}$.

(b) ψ is one–one. (For any c, d in D, $(c^{\#})_{\mathscr{F}} = (d^{\#})_{\mathscr{F}}$ if and only if $c^{\#} =_{\mathscr{F}} d^{\#}$, which is equivalent to $\{j|c^{\#}(j) = d^{\#}(j)\} \in \mathscr{F}$; that is, $\{j|c = d\} \in \mathscr{F}$. If $c \neq d$, $\{j|c = d\} = 0 \notin \mathscr{F}$ and, therefore, $\psi(c) \neq \psi(d)$.)

(c) For any $c_1, \ldots, c_n$ in D, $(f_k^n)^{M^*}(\psi(c_1), \ldots, \psi(c_n)) = (f_k^n)^{M^*}((c_1^{\#})_{\mathscr{F}}, \ldots, (c_n^{\#})_{\mathscr{F}}) = h_{\mathscr{F}}$, where $h(j) = (f_k^n)^M(c_1^{\#}(j), \ldots, c_n^{\#}(j)) = (f_k^n)^M(c_1, \ldots, c_n)$. Thus, $h_{\mathscr{F}} = ((f_k^n)^M(c_1, \ldots, c_n))^{\#}/\mathscr{F} = \psi((f_k^n)^M(c_1, \ldots, c_n))$.

(d) $\models_{M^*} A_k^n[\psi(c_1), \ldots, \psi(c_n)]$ if and only if $\{j|\models_M A_k^n(\psi(c_1)(j), \ldots, \psi(c_n)(j))\} \in \mathscr{F}$, which is equivalent to $\{j|\models_M A_k^n(c_1, \ldots, c_n)\} \in \mathscr{F}$; that is, $\models_M A_k^n[c_1, \ldots, c_n]$. Thus, ψ is an isomorphism of M with $M^{\#}$.

To see that $M^{\#} \leqslant_e M^*$, let $\mathscr{A}$ be any wf and $(c_1^{\#})_{\mathscr{F}}, \ldots, (c_n^{\#})_{\mathscr{F}} \in M^{\#}$. Then, by Proposition 2.42(a), $\models_{M^*} \mathscr{A}[(c_1^{\#})_{\mathscr{F}}, \ldots, (c_n^{\#})_{\mathscr{F}}]$ if and only if $\{j|\models_M \mathscr{A}[c_1^{\#}(j), \ldots, c_n^{\#}(j)]\} \in \mathscr{F}$, which is equivalent to $\{j|\models_M \mathscr{A}[c_1, \ldots, c_n]\} \in \mathscr{F}$, which, in turn, is equivalent to $\models_M \mathscr{A}[c_1, \ldots, c_n]$—that is, to $\models_{M^*} \mathscr{A}[(c_1^{\#})_{\mathscr{F}}, \ldots, (c_n^{\#})_{\mathscr{F}}]$, since ψ is an isomorphism of M with $M^{\#}$.

Exercises

2.116 (The Compactness Theorem again; see Exercise 2.52) If all finite subsets of a set of sentences Γ have a model, prove that Γ has a model.

2.117 (a) A class $\mathscr{W}$ of models of a predicate calculus K is called *elementary* if there is a set Γ of sentences of K such that $\mathscr{W}$ is the class of all models of Γ. Prove that $\mathscr{W}$ is elementary if and only if $\mathscr{W}$ is closed under elementary equivalence and the formation of ultraproducts.

(b) A class $\mathscr{W}$ of models of a predicate calculus K will be called *sentential* if there is a sentence $\mathscr{A}$ of K such that $\mathscr{W}$ is the class of all models of $\mathscr{A}$. Prove that a class

$\mathscr{W}$ is sentential if and only if both $\mathscr{W}$ and its complement $\overline{\mathscr{W}}$(all models of K not in $\mathscr{W}$) are closed with respect to elementary equivalence and ultraproducts.

(c) Prove that the theory K of fields of characteristic 0 (see page 92) is axiomatizable but not finitely axiomatizable.

NONSTANDARD ANALYSIS

From the invention of the calculus until relatively recent times the idea of *infinitesimals* has been an intuitively meaningful tool for finding new results in analysis. The fact that there was no rigorous foundation for infinitesimals was a source of embarrassment and led mathematicians to discard them in favor of the rigorous limit ideas of Cauchy and Weierstrass. However, about 25 years ago, Abraham Robinson discovered that it was possible to resurrect infinitesimals in an entirely legitimate and precise way. This can be done by constructing models that are elementarily equivalent to, but not isomorphic to, the ordered field of real numbers. Such models can be produced either by using Proposition 2.33 or as ultrapowers. We shall sketch here the method based on ultrapowers.

Let R be the set of real numbers. Let K be a generalized predicate calculus with equality having the following symbols:

1. For each real number r, there is an individual constant a_r.
2. For every n-ary operation φ on R, there is a function letter f_φ.
3. For every n-ary relation Φ on R, there is a predicate letter A_Φ.

We can think of R as forming the domain of a model $\mathscr{R}$ for K; we simply let $(a_r)^{\mathscr{R}} = r$, $(f_\varphi)^{\mathscr{R}} = \varphi$, and $(A_\Phi)^{\mathscr{R}} = \Phi$.

Let $\mathscr{F}$ be a nonprincipal ultrafilter on the set ω of natural numbers. We can then form the ultrapower $\mathscr{R}^* = \mathscr{R}^\omega/\mathscr{F}$. We denote the domain $R^\omega/\mathscr{F}$ of $\mathscr{R}^*$ by R^*. By Corollary 2.43, $\mathscr{R}^* \equiv \mathscr{R}$ and, therefore, $\mathscr{R}^*$ has all the properties formalizable in K that $\mathscr{R}$ possesses. Moreover, by Corollary 2.44, $\mathscr{R}^*$ has an elementary submodel $\mathscr{R}^\#$, which is an isomorphic image of $\mathscr{R}$. The domain $R^\#$ of $\mathscr{R}^\#$ consists of all elements $(c^\#)_{\mathscr{F}}$ corresponding to the constant function $c^\#(i) = c$ for all i in ω. We shall sometimes refer to the members of $R^\#$ also as *real* numbers; the elements of $R^* - R^\#$ will be called *nonstandard reals.*

That there exist nonstandard reals can be shown by explicitly exhibiting one. Let $\iota(j) = j$ for all j in ω. Then $\iota_{\mathscr{F}} \in R^*$. However, $(c^\#)_{\mathscr{F}} < \iota_{\mathscr{F}}$ for all c in R, by virtue of Loś's Theorem and the fact that $\{j | c^\#(j) < \iota(j)\} = \{j | c < j\}$, being the set of all natural numbers greater than a fixed real number, is the complement of a finite set and is, therefore, in the nonprincipal ultrafilter $\mathscr{F}$. $\iota_{\mathscr{F}}$ is an "infinitely large" nonstandard real. [The relation $<$ used in the assertion $(c^\#)_{\mathscr{F}} < \iota_{\mathscr{F}}$ is the relation on the ultrapower $\mathscr{R}^*$ corresponding to the predicate letter $<$ of K. We use the symbol $<$ instead of $(<)^{\mathscr{R}^*}$ in order to avoid excessive notation, and we shall often do the same with other relations and functions, such as $u + v$, $u \times v$, and $|u|$.]

Since $\mathscr{R}^*$ possesses all the properties of $\mathscr{R}$ formalizable in K, $\mathscr{R}^*$ is an ordered field having the real number field $\mathscr{R}^\#$ as a proper subfield. [$\mathscr{R}^*$ is non-Archimedean: the element $\iota_{\mathscr{F}}$ defined above is greater than all the natural numbers $(n^\#)_{\mathscr{F}}$ of $\mathscr{R}^*$.] Let R_1, the set of "finite" elements of R^*, contain those elements z such that $|z| < u$ for some real number u in $R^\#$. (R_1 is easily seen to form a subring of R^*.) Let R_0, the set of "infinitesimals" of R^*, contain those elements z such that $|z| < u$ for all positive real numbers u in $R^{\#}$. The reciprocal $1/\iota_{\mathscr{F}}$ is an infinitesimal. (It is not difficult to verify that R_0 is an ideal in the ring R_1. In fact, since $x \in R_1 - R_0$ implies that $1/x \in R_1 - R_0$, it can be easily proved that R_0 is a maximal ideal in R_1.)

Exercises

2.118 Prove that the cardinality of R^* is $2^{\aleph_0}$.
2.119 Prove that the set R_0 of infinitesimals is closed under the operations of $+$, $-$, and $\times$.
2.120 Prove that, if $x \in R_1$ and $y \in R_0$, then $xy \in R_0$.
2.121 Prove that, if $x \in R_1 - R_0$, then $1/x \in R_1 - R_0$.

Let $x \in R_1$. Let $A = \{u | u \in R^{\#} \wedge u < x\}$ and $B = \{u | u \in R^{\#} \wedge u \geqslant x\}$. Then (A, B) is a "cut" and, therefore, determines a unique real number r such that (1) $(\forall x)(x \in A \Rightarrow x \leqslant r)$ and (2) $(\forall x)(x \in B \Rightarrow x \geqslant r)$.† The difference $x - r$ is an infinitesimal. [*Proof:* Assume $x - r$ is not an infinitesimal. Then $|x - r| > r_1$ for some positive real number r_1. *Case 1:* $x > r$. Then $x - r > r_1$. So, $x > r + r_1 > r$. But then $r + r_1 \in A$, contradicting condition (1). *Case 2:* $x < r$. Then $r - x > r_1$, and so, $r > r - r_1 > x$. Thus, $r - r_1 \in B$, contradicting condition (2).] The real number r such that $x - r$ is an infinitesimal is called the *standard part* of x and is denoted $\mathrm{st}(x)$. Note that, if x is itself a real number, then $\mathrm{st}(x) = x$. We shall use the notation $x \approx y$ to mean $\mathrm{st}(x) = \mathrm{st}(y)$. Clearly, $x \approx y$ if and only if $x - y$ is an infinitesimal. If $x \approx y$, we say that x and y are *infinitely close.*

Exercises

2.122 If $x \in R_1$, show that there is a unique real number r such that $x - r$ is an infinitesimal. [It is necessary to check this to ensure that $\mathrm{st}(x)$ is well-defined.]
2.123 If x and y are in R_1, prove the following.
(a) $\mathrm{st}(x + y) = \mathrm{st}(x) + \mathrm{st}(y)$
(b) $\mathrm{st}(xy) = \mathrm{st}(x)\mathrm{st}(y)$
(c) $\mathrm{st}(-x) = -\mathrm{st}(x) \wedge \mathrm{st}(y - x) = \mathrm{st}(y) - \mathrm{st}(x)$
(d) $x \geqslant 0 \Rightarrow \mathrm{st}(x) \geqslant 0$
(e) $x \leqslant y \Rightarrow \mathrm{st}(x) \leqslant \mathrm{st}(y)$

The set ω of natural numbers is a subset of the real numbers. Therefore, in the theory K there is a predicate letter N corresponding to the property $x \in \omega$. Hence, in R^*, there is a set ω^* of elements satisfying the wf $N(x)$. An element $f_{\mathscr{F}}$ of R^* satisfies $N(x)$ if and only if $\{j | f(j) \in \omega\} \in \mathscr{F}$. In particular, the elements $n^{\#}_{\mathscr{F}}$, for $n \in \omega$, are the "standard" members of ω^*, whereas $\iota_{\mathscr{F}}$, for example, is a "nonstandard" natural number in R^*.

Many of the properties of the real number system can be studied from the viewpoint of nonstandard analysis. For example, if s is an ordinary sequence of real numbers and c is a real number, one ordinarily says that $\lim s_n = c$ if

$$(\&) \quad (\forall \varepsilon)(\varepsilon > 0 \Rightarrow (\exists n)(n \in \omega \wedge (\forall k)(k \in \omega \wedge k \geqslant n \Rightarrow |s_n - c| < \varepsilon)))$$

Since $s \in R^{\omega}$, s is a relation and, therefore, the theory K contains a predicate letter $S(n, x)$ corresponding to the relation $s_n = x$. Hence, R^* will have a relation of all

† See Mendelson (1973, chap. 5).

pairs (n, x) satisfying $S(n, x)$. Since $\mathscr{R}^* \equiv \mathscr{R}$, this relation will be a function that is an extension of the given sequence to the larger domain ω^*. Then we have the following result.

PROPOSITION 2.45 Let s be a sequence of real numbers and c a real number. Let s^* denote the function from ω^* into R^* corresponding to s in $\mathscr{R}^*$. Then $\lim s_n = c$ if and only if $s^*(n) \approx c$ for all n in $\omega^* - \omega$. [The latter condition can be paraphrased by saying that $s^*(n)$ is infinitely close to c when n is infinitely large.]

Proof Assume $\lim s_n = c$. Consider any positive real ε. By (&), there is a natural number n_0 such that $(\forall k)(k \in \omega \wedge k \geqslant n_0 \Rightarrow |s_n - c| < \varepsilon)$ holds in $\mathscr{R}$. Hence, the corresponding sentence $(\forall k)(k \in \omega^* \wedge k \geqslant n_0 \Rightarrow |s^*(n) - c| < \varepsilon)$ holds in $\mathscr{R}^*$. For any $n \in \omega^* - \omega$, $n > n_0$ and, therefore, $|s^*(n) - c| < \varepsilon$. Since this holds for all positive real ε, $s^*(n) - c$ is an infinitesimal.

Conversely, assume $s^*(n) \approx c$ for all $n \in \omega^* - \omega$. Take any positive real ε. Fix some n_1 in $\omega^* - \omega$. Then $(\forall k)(k \geqslant n_1 \Rightarrow |s^*(k) - c| < \varepsilon)$. So, the sentence $(\exists n)(n \in \omega \wedge (\forall k)(k \in \omega \wedge k \geqslant n \Rightarrow |s_k - c| < \varepsilon)$ is true for $\mathscr{R}^*$ and, therefore, also for $\mathscr{R}$. So, there must be a natural number n_0 such that $(\forall k)(k \in \omega \wedge k \geqslant n_0 \Rightarrow |s_k - c| < \varepsilon)$. Since ε was an arbitrary positive real, we have proved $\lim s_n = c$.

Exercise

2.124 Using Proposition 2.45, prove the following limit theorems for the real number system: let s and u be sequences of real numbers, and c_1 and c_2 real numbers such that $\lim s_n = c_1$ and $\lim u_n = c_2$. Then:

(a) $\lim (s_n + u_n) = c_1 + c_2$
(b) $\lim (s_n u_n) = c_1 c_2$
(c) If $c_2 \neq 0$ and all $u_n \neq 0$, $\lim (s_n/u_n) = c_1/c_2$.

Let us now consider another important notion of analysis, continuity. Let B be a set of real numbers, let $c \in B$, and let f be a function defined on B and taking real values. One says that f is continuous at c if

(&&) $(\forall \varepsilon)(\varepsilon > 0 \Rightarrow (\exists \delta)(\delta > 0 \wedge (\forall x)(x \in B \wedge |x - c| < \delta \Rightarrow |f(x) - f(c)| < \varepsilon)))$

PROPOSITION 2.46 Let f be a real-valued function defined on a set B of real numbers. Let $c \in B$. Let B^* be the subset of R^* corresponding to B, and let f^* be the function corresponding to f.† Then f is continuous at c if and only if $(\forall x)(x \in B^* \wedge x \approx c \Rightarrow f^*(x) \approx f(c))$.

Exercises

2.125 Prove Proposition 2.46.

2.126 Assume f and g are real-valued functions defined on a set B of real numbers, and assume that f and g are continuous at a point c in B. Using Proposition 2.46, prove the following.

†To be more precise, f is represented in the theory K by a predicate letter A_f, where $A_f(x, y)$ corresponds to the relation $f(x) = y$. Then the corresponding relation A_f^* in R^* determines a function f^* with domain B^*.

(a) $f + g$ is continuous at c.
(b) $f \cdot g$ is continuous at c.

2.127 Let f be a real-valued function defined on a set B of real numbers and continuous at a point c in B, and let g be a real-valued function defined on a set A of real numbers containing the image of B under f. Assume that g is continuous at the point $f(c)$. Prove, by Proposition 2.46, that the composition $g \circ f$ is continuous at c.

2.128 (a) Let $C \subseteq R$. C is said to be *closed* if $(\forall x)((\forall \varepsilon)[\varepsilon > 0 \Rightarrow (\exists y)(y \in C \wedge |x - y| < \varepsilon)] \Rightarrow x \in C)$. Show that C is closed if and only if every real number that is infinitely close to a member of C^* is in C.

(b) Let $C \subseteq R$. C is said to be *open* if $(\forall x)(x \in C \Rightarrow (\exists \delta)(\delta > 0 \wedge (\forall y)(|y - x| < \delta \Rightarrow y \in C)))$. Show that C is open if and only if every nonstandard real number that is infinitely close to a member of C is a member of C^*.

Many standard theorems of analysis turn out to have much simpler proofs within nonstandard analysis. Even stronger results can be obtained by starting with a theory K that has symbols, not only for the elements, operations, and relations on R, but also for sets of subsets of R, sets of sets of subsets of R, and so on. In this way, the methods of nonstandard analysis can be applied to all areas of modern analysis, sometimes with original and striking results. For further development and applications, see A. Robinson (1966), Luxemburg (1969), Bernstein (1973), Stroyan and Luxemburg (1976), and Davis (1977a). A calculus textbook based on nonstandard analysis has been written by H. J. Keisler (1976) and has been used in some experimental undergraduate courses.

Exercises

2.129 A real-valued function f defined on a closed interval $[a, b] = \{x | a \leqslant x \leqslant b\}$ is said to be *uniformly continuous* if

$$(\forall \varepsilon)(\varepsilon > 0 \Rightarrow (\exists \delta)(\delta > 0 \wedge (\forall x)(\forall y)(a \leqslant x \leqslant b$$
$$\wedge\ a \leqslant y \leqslant b \wedge |x - y| < \delta \Rightarrow |f(x) - f(y)| < \varepsilon)))$$

Prove that f is uniformly continuous if and only if, for all x and y in $[a, b]^*$, $x \approx y \Rightarrow f^*(x) \approx f^*(y)$.

2.130 Prove, using nonstandard methods, that any function continuous on $[a, b]$ is uniformly continuous on $[a, b]$.

2.131 *Bolzano–Weierstrass Theorem.* A real number c is said to be a *limit point* of a set A of real numbers if $(\forall \varepsilon)(\varepsilon > 0 \Rightarrow (\exists u)(u \in A \wedge |c - u| < \varepsilon))$. Let s be a bounded sequence of reals; that is, there is a number b such that $|s_n| < b$ for all n in ω. Prove that the set of terms of s (i.e., the range of the function $s \in R^{\omega}$) has a limit point.

15. SEMANTIC TREES

Remember that a wf is logically valid if and only if it is true for all interpretations. Since there are uncountably many interpretations, there is no simple, direct way to determine logical validity. Gödel's Completeness Theorem (Corollary 2.19) showed that logical validity is equivalent to derivability in a predicate calculus. But, to find out whether a wf is provable in a predicate calculus, we have only a very clumsy method: start generating the theorems and watch to see whether the given wf ever appears. Our aim here is to outline a more intuitive and usable approach in the case of wfs without function letters. Throughout this section, we assume that no function letters occur in our wfs.

A closed wf is logically valid if and only if its negation is not satisfiable. We shall now explain a simple procedure for trying to determine satisfiability of a closed wf $\mathscr{A}$. Our purpose is either to show that $\mathscr{A}$ is not satisfiable or to find a model for $\mathscr{A}$.

We shall construct a figure in the shape of an inverted tree. Start with the wf $\mathscr{A}$ at the top (the "root" of the tree). We apply certain rules for writing wfs below those already obtained. These rules replace complicated formulas by simpler ones in a way that corresponds to the meaning of the connectives and quantifiers.

RULES

NEGATION:

$$\begin{array}{c} \neg\neg\mathscr{B} \\ \downarrow \\ \mathscr{B} \end{array} \qquad \begin{array}{c} \neg(\mathscr{A} \vee \mathscr{B}) \\ \downarrow \\ \neg\mathscr{A} \\ \neg\mathscr{B} \end{array} \qquad \begin{array}{c} \neg(\mathscr{A} \Rightarrow \mathscr{B}) \\ \downarrow \\ \mathscr{A} \\ \neg\mathscr{B} \end{array} \qquad \begin{array}{c} \neg(\forall x)\mathscr{B} \\ \downarrow \\ (\exists x)\neg\mathscr{B} \end{array} \qquad \begin{array}{c} \neg(\exists x)\mathscr{B} \\ \downarrow \\ (\forall x)\neg\mathscr{B} \end{array}$$

$$\begin{array}{ccc} & \neg(\mathscr{A} \wedge \mathscr{B}) & \\ \swarrow & & \searrow \\ \neg\mathscr{A} & & \neg\mathscr{B} \end{array} \qquad \begin{array}{ccc} & \neg(\mathscr{A} \Leftrightarrow \mathscr{B}) & \\ \swarrow & & \searrow \\ \mathscr{A} & & \neg\mathscr{A} \\ \neg\mathscr{B} & & \mathscr{B} \end{array}$$

CONJUNCTION:

$$\begin{array}{c} \mathscr{A} \wedge \mathscr{B} \\ \downarrow \\ \mathscr{A} \\ \mathscr{B} \end{array}$$

DISJUNCTION:

$$\begin{array}{ccc} & \mathscr{A} \vee \mathscr{B} & \\ \swarrow & & \searrow \\ \mathscr{A} & & \mathscr{B} \end{array}$$

CONDITIONAL:

$$\begin{array}{ccc} & \mathscr{A} \Rightarrow \mathscr{B} & \\ \swarrow & & \searrow \\ \neg\mathscr{A} & & \mathscr{B} \end{array}$$

BICONDITIONAL:

$$\begin{array}{ccc} & \mathscr{A} \Leftrightarrow \mathscr{B} & \\ \swarrow & & \searrow \\ \mathscr{A} & & \neg\mathscr{A} \\ \mathscr{B} & & \neg\mathscr{B} \end{array}$$

UNIVERSAL QUANTIFIER:

$$\begin{array}{c} (\forall x)\mathscr{B}(x) \\ \downarrow \\ \mathscr{B}(a) \end{array}$$

(Rule U)
[Here, a is any individual constant already present.]

EXISTENTIAL QUANTIFIER:

$$\begin{array}{c} (\exists x)\mathscr{B}(x) \\ \downarrow \\ \mathscr{B}(b) \end{array}$$

[b is a new individual constant not already in the figure.]

REMARK Some of the rules require a fork or branching. This occurs when the given wf implies that one of two possible situations holds.

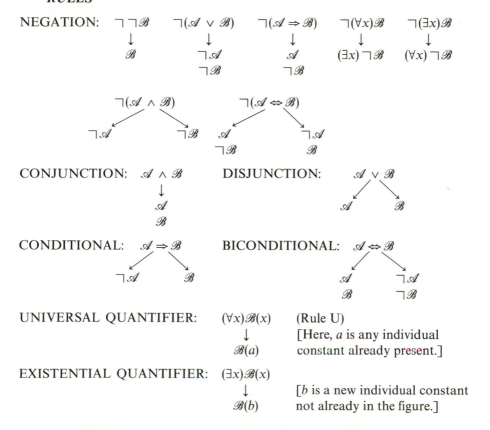

A *branch* is a sequence of wfs starting at the top and proceeding down the figure by applications of the rules. When a wf and its negation appear in a branch, that branch becomes *closed* and no further rules are applied to the wfs in that branch. Closure of a branch will be indicated by a large cross ×.

Inspection of the rules shows that, when a rule is applied to a formula, the usefulness of that formula has been exhausted (the formula will be said to be *discharged*) and that formula need never be subject to a rule again, except in the case of a universally quantified formula. In the latter case, whenever a new individual constant appears in the branch below the formula, Rule U can again be applied with that new variable. *In addition, if no further rule applications are possible along a branch and no individual constant occurs in that branch, then we must introduce a new individual constant for use in possible applications of Rule U along that branch.* (The idea behind this requirement is that, if we are trying to build a model, we must introduce symbols for objects that can serve as elements of the domain of that model.)

BASIC PRINCIPLE OF SEMANTIC TREES If all branches become closed, the original wf is unsatisfiable. If, however, a branch remains unclosed, that branch can be used to construct a model in which the original wf is true; the domain of the model consists of the individual constants that appear in that branch.

We shall discuss the justification of this principle later on. First we shall give examples of its use.

Examples

1. To prove that $(\forall x)\mathscr{B}(x) \Rightarrow \mathscr{B}(a)$ is logically valid, we build a semantic tree starting from its negation.

(i)	$\neg((\forall x)\mathscr{B}(x) \Rightarrow \mathscr{B}(a))$	
(ii)	$(\forall x)\mathscr{B}(x)$	(i)
(iii)	$\neg\mathscr{B}(a)$	(i)
(iv)	$\mathscr{B}(a)$	(ii)
	×	

The number to the right of a given wf indicates the number of the line of the wf from which the given wf is derived. Since the only branch in this tree is closed, $\neg((\forall x)\mathscr{B}(x) \Rightarrow \mathscr{B}(a))$ is unsatisfiable and, therefore, $(\forall x)\mathscr{B}(x) \Rightarrow \mathscr{B}(a)$ is logically valid.

2.

(i)	$\neg[(\forall x)(\mathscr{B}(x) \Rightarrow \mathscr{C}(x)) \Rightarrow ((\forall x)\mathscr{B}(x) \Rightarrow (\forall x)\mathscr{C}(x))]$		
(ii)	$(\forall x)(\mathscr{B}(x) \Rightarrow \mathscr{C}(x))$		(i)
(iii)	$\neg((\forall x)\mathscr{B}(x) \Rightarrow (\forall x)\mathscr{C}(x))$		(i)
(iv)	$(\forall x)\mathscr{B}(x)$		(iii)
(v)	$\neg(\forall x)\mathscr{C}(x)$		(iii)
(vi)	$(\exists x)\neg\mathscr{C}(x)$		(v)
(vii)	$\neg\mathscr{C}(b)$		(vi)
(viii)	$\mathscr{B}(b)$		(iv)
(ix)	$\mathscr{B}(b) \Rightarrow \mathscr{C}(b)$		(ii)
	↙	↘	
(x)	$\neg\mathscr{B}(b)$	$\mathscr{C}(b)$	(ix)
	×	×	

Since both branches are closed, the original wf (i) is unsatisfiable and, therefore, $(\forall x)(\mathscr{B}(x) \Rightarrow \mathscr{C}(x)) \Rightarrow ((\forall x)\mathscr{B}(x) \Rightarrow (\forall x)\mathscr{C}(x))$ is logically valid.

3.

(i)	$\neg[(\exists x)A_1^1(x) \Rightarrow (\forall x)A_1^1(x)]$	
(ii)	$(\exists x)A_1^1(x)$	(i)
(iii)	$\neg(\forall x)A_1^1(x)$	(i)
(iv)	$A_1^1(b)$	(ii)
(v)	$(\exists x)\neg A_1^1(x)$	(iii)
(vi)	$\neg A_1^1(c)$	(v)

No further applications of rules are possible and there is still an open branch. Define a model M with domain $\{b, c\}$ such that the interpretation of A_1^1 holds for b but not for c. Thus, $(\exists x)A_1^1(x)$ is true in M but $(\forall x)A_1^1(x)$ is false in M. Hence, $(\exists x)A_1^1(x) \Rightarrow (\forall x)A_1^1(x)$ is false in M and is not logically valid.

4.

(i)	$\neg[(\exists y)(\forall x)\mathscr{B}(x, y) \Rightarrow (\forall x)(\exists y)\mathscr{B}(x, y)]$	
(ii)	$(\exists y)(\forall x)\mathscr{B}(x, y)$	(i)
(iii)	$\neg(\forall x)(\exists y)\mathscr{B}(x, y)$	(i)
(iv)	$(\forall x)\mathscr{B}(x, b)$	(ii)
(v)	$(\exists x)\neg(\exists y)\mathscr{B}(x, y)$	(iii)
(vi)	$\mathscr{B}(b, b)$	(iv)
(vii)	$\neg(\exists y)\mathscr{B}(c, y)$	(v)
(viii)	$\mathscr{B}(c, b)$	(iv)
(ix)	$(\forall y)\neg\mathscr{B}(c, y)$	(vii)
(x)	$\neg\mathscr{B}(c, b)$	(ix)
	×	

Hence, $(\exists y)(\forall x)\mathscr{B}(x, y) \Rightarrow (\forall x)(\exists y)\mathscr{B}(x, y)$ is logically valid.

Notice that, in the last tree, step (vi) served no purpose but was required by our method of constructing trees. We should be a little more precise in describing that method. At each step, we apply the appropriate rule to each undischarged wf, except to universally quantified wfs. Then, to every universally quantified wf on a given branch, we apply Rule U with every individual constant that has appeared on that branch since the last step. In every application of a rule to a given wf, we write the resulting wf(s) below the branch that contains that wf.

5.

(i)	$\neg[(\forall x)\mathscr{B}(x) \Rightarrow (\exists x)\mathscr{B}(x)]$	
(ii)	$(\forall x)\mathscr{B}(x)$	(i)
(iii)	$\neg(\exists x)\mathscr{B}(x)$	(i)
(iv)	$(\forall x)\neg\mathscr{B}(x)$	(iii)
(v)	$\mathscr{B}(a_1)$	(ii)*
(vi)	$\neg\mathscr{B}(a_1)$	(iv)
	×	

Hence, $(\forall x)\mathscr{B}(x) \Rightarrow (\exists x)\mathscr{B}(x)$ is logically valid.

6.

(i)	$\neg[(\forall x)\neg A_1^2(x, x) \Rightarrow (\exists x)(\forall y)\neg A_1^2(x, y)]$	
(ii)	$(\forall x)\neg A_1^2(x, x)$	(i)
(iii)	$\neg(\exists x)(\forall y)\neg A_1^2(x, y)$	(i)
(iv)	$(\forall x)\neg(\forall y)\neg A_1^2(x, y)$	(iii)
(v)	$\neg A_1^2(a_1, a_1)$	(ii)*

* Here, we must introduce a new individual constant a_1 for use with Rule U, since, otherwise, the branch would end and would not contain any individual constants.

(vi)	$\neg(\forall y)\neg A_1^2(a_1, y)$	(iv)
(vii)	$(\exists y)\neg\neg A_1^2(a_1, y)$	(vi)
(viii)	$\neg\neg A_1^2(a_1, a_2)$	(vii)
(ix)	$A_1^2(a_1, a_2)$	(viii)
(x)	$\neg A_1^2(a_2, a_2)$	(ii)
(xi)	$\neg(\forall y)\neg A_1^2(a_2, y)$	(iv)
(xii)	$(\exists y)\neg\neg A_1^2(a_2, y)$	(xi)
(xiii)	$\neg\neg A_1^2(a_2, a_3)$	(xii)
(xiv)	$A_1^2(a_2, a_3)$	(xiii)

We can see that the branch will never end and that we will obtain a sequence of constants a_1, a_2,... with wfs $A_1^2(a_n, a_{n+1})$ and $\neg A_1^2(a_n, a_n)$. Thus, we construct a model M with domain $\{a_1, a_2, \ldots\}$ and we define $(A_1^2)^{\mathrm{M}}$ to contain only the pairs $\langle a_n, a_{n+1}\rangle$. Then, $(\forall x)\neg A_1^2(x, x)$ is true in M, whereas $(\exists x)(\forall y)\neg A_1^2(x, y)$ is false in M. Hence, $(\forall x)\neg A_1^2(x, x) \Rightarrow (\exists x)(\forall y)\neg A_1^2(x, y)$ is not logically valid.

Exercise

2.132 Use semantic trees to determine whether the following wfs are logically valid.

(a) $(\forall x)(A_1^1(x) \vee A_2^1(x)) \Rightarrow ((\forall x)A_1^1(x)) \vee (\forall x)A_2^1(x)$
(b) $((\forall x)\mathscr{A}(x)) \wedge (\forall x)\mathscr{B}(x) \Rightarrow (\forall x)(\mathscr{A}(x) \wedge \mathscr{B}(x))$
(c) $(\forall x)(\mathscr{A}(x) \wedge \mathscr{B}(x)) \Rightarrow ((\forall x)\mathscr{A}(x)) \wedge (\forall x)\mathscr{B}(x)$
(d) $(\exists x)(A_1^1(x) \Rightarrow A_2^1(x)) \Rightarrow ((\exists x)A_1^1(x) \Rightarrow (\exists x)A_2^1(x))$
(e) $(\exists x)(\exists y)A_1^2(x, y) \Rightarrow (\exists z)A_1^2(z, z)$
(f) $((\forall x)A_1^1(x)) \vee (\forall x)A_2^1(x) \Rightarrow (\forall x)(A_1^1(x) \vee A_2^1(x))$
(g) The wfs of Exercises 2.23, 2.29(a, e, j), 2.37, and 2.38.
(h) The wfs of Exercise 2.19(a, b, g).

PROPOSITION 2.47 Assume that Γ is a set of closed wfs that satisfy the following closure conditions: (a) if $\neg\neg\mathscr{B}$ is in Γ, then $\mathscr{B}$ is in Γ; (b) if $\neg(\mathscr{B} \vee \mathscr{C})$ is in Γ, then $\neg\mathscr{B}$ and $\neg\mathscr{C}$ are in Γ; (c) if $\neg(\mathscr{B} \Rightarrow \mathscr{C})$ is in Γ, then $\mathscr{B}$ and $\neg\mathscr{C}$ are in Γ; (d) if $\neg(\forall x)\mathscr{B}$ is in Γ, then $(\exists x)\neg\mathscr{B}$ is in Γ; (e) if $\neg(\exists x)\mathscr{B}$ is in Γ, then $(\forall x)\neg\mathscr{B}$ is in Γ; (f) if $\neg(\mathscr{B} \wedge \mathscr{C})$ is in Γ, then at least one of $\neg\mathscr{B}$ and $\neg\mathscr{C}$ is in Γ; (g) if $\neg(\mathscr{B} \Leftrightarrow \mathscr{C})$ is in Γ, then either $\mathscr{B}$ and $\neg\mathscr{C}$ are in Γ or $\neg\mathscr{B}$ and $\mathscr{C}$ are in Γ; (h) if $\mathscr{B} \wedge \mathscr{C}$ is in Γ, then so are $\mathscr{B}$ and $\mathscr{C}$; (i) if $\mathscr{B} \vee \mathscr{C}$ is in Γ, then at least one of $\mathscr{B}$ and $\mathscr{C}$ is in Γ; (j) if $\mathscr{B} \Rightarrow \mathscr{C}$ is in Γ, then at least one of $\neg\mathscr{B}$ and $\mathscr{C}$ is in Γ; (k) if $\mathscr{B} \Leftrightarrow \mathscr{C}$ is in Γ, then either $\mathscr{B}$ and $\mathscr{C}$ are in Γ or $\neg\mathscr{B}$ and $\neg\mathscr{C}$ are in Γ; (l) if $(\forall x)\mathscr{B}(x)$ is in Γ, then $\mathscr{B}(b)$ is in Γ, where b is any individual constant that occurs in some wf of Γ; and (m) if $(\exists x)\mathscr{B}(x)$ is in Γ, then $\mathscr{B}(b)$ is in Γ for some individual constant b. If no wf and its negation both belong to Γ and some wfs in Γ contain individual constants, then there is a model for Γ whose domain is the set D of individual constants that occur in wfs of Γ.

Proof Define a model M with domain D by specifying that the interpretation of any predicate letter A_k^n in Γ contains an n-tuple $\langle b_1, \ldots, b_n\rangle$ if and only if $A_k^n(b_1, \ldots, b_n)$ is in Γ. By induction on the number of connectives and quantifiers in any closed wf $\mathscr{E}$, it is easy to prove: (a) if $\mathscr{E}$ is in Γ, $\mathscr{E}$ is true in M; and (b) if $\neg\mathscr{E}$ is in Γ, $\mathscr{E}$ is false in M. Hence, M is a model for Γ.

If a branch of a semantic tree remains open, the set Γ of wfs of that branch satisfies the hypotheses of Proposition 2.47. It follows that, if a branch of a semantic tree remains open, then the set Γ of wfs of that branch has a model M whose domain is the set of individual constants that appear in that branch. This yields half of the Basic Principle of Semantic Trees.

PROPOSITION 2.48 If all the branches of a semantic tree are closed, then the wf $\mathscr{A}$ at the root of the tree is unsatisfiable.

Proof From the derivation rules it is clear that, if a sequence of wfs starts at $\mathscr{A}$ and continues down the tree through applications of the rules, and if the wfs in that sequence are simultaneously satisfiable in some model M, then that sequence can be extended by another application of a rule so that the added wf (or wfs) also would be true in M. Otherwise, the sequence would form an unclosed branch, contrary to our hypothesis. Assume now that $\mathscr{A}$ is satisfiable in a model M. Then, starting with $\mathscr{A}$, we could construct an infinite branch in which all the wfs are true in M. (In the case of a branching rule, if there are two ways to extend the sequence, we choose the left-hand wf.) Therefore, this branch would not be closed, contrary to our hypothesis. Hence, $\mathscr{A}$ is unsatisfiable.

This completes the proof of the Basic Principle of Semantic Trees. Notice that this principle does not yield a decision procedure for logical validity. If a closed wf $\mathscr{B}$ is not logically valid, the semantic tree of $\neg\mathscr{B}$ may (and often does) contain an infinite unclosed branch. At any stage of the construction of this tree, we have no general procedure for deciding whether or not, at some later stage, all branches of the tree will have become closed. Thus, we have no general way of knowing whether $\mathscr{B}$ is unsatisfiable.

For the sake of brevity, our exposition has been loose and imprecise. A clear and masterful study of semantic trees and related matters can be found in Smullyan (1968).

CHAPTER THREE

Formal Number Theory

1. AN AXIOM SYSTEM

Together with geometry, the theory of numbers is the most immediately intuitive of all branches of mathematics. It is not surprising then that attempts to formalize mathematics and to establish a rigorous foundation for mathematics should begin with number theory. The first semiaxiomatic presentation of this subject was given by Dedekind in 1879 and has come to be known as Peano's Postulates.* It can be formulated as follows:

(P1) 0 is a natural number.

(P2) If x is a natural number, there is another natural number denoted by x' (and called the *successor* of x).†

(P3) $0 \neq x'$ for any natural number x.

(P4) If $x' = y'$, then $x = y$.

(P5) If Q is a property that may or may not hold for natural numbers, and if (I) 0 has the property Q and (II) whenever a natural number x has the property Q, then x' has the property Q, then all natural numbers have the property Q (Principle of Induction).

These axioms, together with a certain amount of set theory, can be used to develop not only number theory but also the theory of rational, real, and complex numbers (see Mendelson, 1973). However, the axioms involve certain intuitive notions, such as "property", that prevent this system from being a rigorous formalization. We therefore shall build a first-order theory S that is based upon Peano's Postulates and seems to be adequate for the proofs of all the basic results of elementary number theory.

The first-order theory S has a single predicate letter A_1^2. As usual, we shall write $t = s$ for $A_1^2(t, s)$. S has one individual constant a_1. We shall use 0 as an alternative notation for a_1. Finally, S has three function letters f_1^1, f_1^2, and f_2^2. We shall write (t') instead of $f_1^1(t)$; $(t + s)$ instead of $f_1^2(t, s)$; and $(t \cdot s)$ instead of $f_2^2(t, s)$. However,

* For historical information, see Wang (1957).

† The intuitive meaning of x' is $x + 1$.

we shall write t', $t + s$, and $t \cdot s$ instead of (t'), $(t + s)$, and $(t \cdot s)$ whenever this will cause no confusion.

The proper axioms of S are:

(S1) $x_1 = x_2 \Rightarrow (x_1 = x_3 \Rightarrow x_2 = x_3)$
(S2) $x_1 = x_2 \Rightarrow x_1' = x_2'$
(S3) $0 \neq x_1'$
(S4) $x_1' = x_2' \Rightarrow x_1 = x_2$
(S5) $x_1 + 0 = x_1$
(S6) $x_1 + x_2' = (x_1 + x_2)'$
(S7) $x_1 \cdot 0 = 0$
(S8) $x_1 \cdot (x_2') = (x_1 \cdot x_2) + x_1$
(S9) For any wf $\mathscr{A}(x)$ of S, $\mathscr{A}(0) \Rightarrow ((\forall x)(\mathscr{A}(x) \Rightarrow \mathscr{A}(x')) \Rightarrow (\forall x)\mathscr{A}(x))$.

Notice that axioms (S1)–(S8) are particular wfs, whereas (S9) is an axiom schema providing an infinite number of axioms. However, (S9), which we shall call the *Principle of Mathematical Induction*, cannot fully correspond to Peano's Postulate (P5), since the latter refers intuitively to the $2^{\aleph_0}$ properties of natural numbers, whereas (S9) can take care of only the denumerable number of properties defined by wfs of S.

Axioms (S3) and (S4) correspond to the Peano Postulates (P3) and (P4), respectively. Peano's axioms (P1) and (P2) are taken care of by the presence of 0 as an individual constant and f_1^1 as a function letter. Our axioms (S1) and (S2) furnish some needed properties of equality; they would have been assumed as intuitively obvious by Dedekind and Peano. Axioms (S5)–(S8) are the recursion equations for addition and multiplication. Dedekind and Peano did not have to assume them because they allowed the use of intuitive set theory, from which the existence of operations + and · satisfying (S5)–(S8) is deducible (see Mendelson, 1973, theorems 3.1 and 5.1).

Any theory that has the same theorems as S is often referred to in the literature as *Peano Arithmetic*, or simply PA.

From (S9), by MP, we can obtain the *Induction Rule*: from $\mathscr{A}(0)$ and $(\forall x)(\mathscr{A}(x) \Rightarrow \mathscr{A}(x'))$, we can derive $(\forall x)\mathscr{A}(x)$.

It will be our immediate aim to establish the usual rules of equality; that is, we shall show that the properties (A6) and (A7) of equality (see page 74) are derivable in S and, hence, that S is a first-order theory with equality.

First, for convenience and brevity in carrying out proofs, we cite some immediate, trivial consequences of the axioms.

LEMMA 3.1 For any terms t, s, r of S, the following wfs are theorems.

(S1′) $t = r \Rightarrow (t = s \Rightarrow r = s)$
(S2′) $t = r \Rightarrow t' = r'$
(S3′) $0 \neq t'$
(S4′) $t' = r' \Rightarrow t = r$
(S5′) $t + 0 = t$
(S6′) $t + r' = (t + r)'$

(S7′) $t \cdot 0 = 0$
(S8′) $t \cdot r' = (t \cdot r) + t$

Proof (S1′)–(S8′) follow from (S1)–(S8), respectively, by first forming the closure by means of Gen and then applying Rule A4 with the appropriate terms t, r, s.

PROPOSITION 3.2 For any terms t, r, s, the following wfs are theorems of S.

(a) $t = t$
(b) $t = r \Rightarrow r = t$
(c) $t = r \Rightarrow (r = s \Rightarrow t = s)$
(d) $r = t \Rightarrow (s = t \Rightarrow r = s)$
(e) $t = r \Rightarrow t + s = r + s$
(f) $t = 0 + t$
(g) $t' + r = (t + r)'$
(h) $t + r = r + t$
(i) $t = r \Rightarrow s + t = s + r$
(j) $(t + r) + s = t + (r + s)$
(k) $t = r \Rightarrow t \cdot s = r \cdot s$
(l) $0 \cdot t = 0$
(m) $t' \cdot r = t \cdot r + r$
(n) $t \cdot r = r \cdot t$
(o) $t = r \Rightarrow s \cdot t = s \cdot r$

Proof

(a)	1. $t + 0 = t$	(S5′)
	2. $(t + 0 = t) \Rightarrow (t + 0 = t \Rightarrow t = t)$	(S1′)
	3. $t + 0 = t \Rightarrow t = t$	1, 2, MP
	4. $t = t$	1, 3, MP
(b)	1. $t = r \Rightarrow (t = t \Rightarrow r = t)$	(S1′)
	2. $t = t \Rightarrow (t = r \Rightarrow r = t)$	1, Tautology
	3. $t = r \Rightarrow r = t$	2, part (a), MP
(c)	1. $r = t \Rightarrow (r = s \Rightarrow t = s)$	(S1′)
	2. $t = r \Rightarrow r = t$	Part (b)
	3. $t = r \Rightarrow (r = s \Rightarrow t = s)$	1, 2, Tautology
(d)	1. $r = t \Rightarrow (t = s \Rightarrow r = s)$	Part (c)
	2. $t = s \Rightarrow (r = t \Rightarrow r = s)$	1, Tautology
	3. $s = t \Rightarrow t = s$	Part (b)
	4. $s = t \Rightarrow (r = t \Rightarrow r = s)$	2, 3, Tautology
	5. $r = t \Rightarrow (s = t \Rightarrow r = s)$	4, Tautology

(e) Apply the Induction Rule to $\mathscr{A}(z)$: $x = y \Rightarrow (x + z = y + z)$.

(i)	1. $x + 0 = x$	(S5′)
	2. $y + 0 = y$	(S5′)

3. $x = y$	Hyp
4. $x + 0 = y$	1, 3, part (c)
5. $x + 0 = y + 0$	2, 4, part (d)
6. $x = y \Rightarrow x + 0 = y + 0$	1–5, Deduction Theorem
Thus, $\vdash \mathscr{A}(0)$	
(ii) 1. $x = y \Rightarrow x + z = y + z$	Hyp
2. $x = y$	Hyp
3. $x + z' = (x + z)'$	(S6′)
4. $y + z' = (y + z)'$	(S6′)
5. $x + z = y + z$	1, 2, MP
6. $(x + z)' = (y + z)'$	5, (S2′)
7. $x + z' = (y + z)'$	3, 6, part (c)
8. $x + z' = y + z'$	4, 7, part (d)
9. $(x = y \Rightarrow (x + z = y + z))$ $\Rightarrow (x = y \Rightarrow (x + z' = y + z'))$	1–8, Deduction Theorem
Thus, $\vdash \mathscr{A}(z) \Rightarrow \mathscr{A}(z')$	

Hence, $\vdash (\forall z)\mathscr{A}(z)$ by the Induction Rule, from (i), (ii), and Gen. Therefore, by Gen and Rule A4, $\vdash t = r \Rightarrow t + s = r + s$.

(f) Let $\mathscr{A}(x)$ be $x = 0 + x$.

(i) $0 = 0 + 0$, by (S5′) and part (b); i.e., $\vdash \mathscr{A}(0)$

(ii) 1. $x = 0 + x$	Hyp
2. $0 + x' = (0 + x)'$	(S6′)
3. $x' = (0 + x)'$	1, (S2′)
4. $x' = 0 + x'$	2, 3, part (d)
5. $x = 0 + x \Rightarrow x' = 0 + x'$	1–4, Deduction Theorem
Thus, $\vdash \mathscr{A}(x) \Rightarrow \mathscr{A}(x')$	

By (i) and (ii), Gen, and the Induction Rule, $\vdash (\forall x)(x = 0 + x)$. So, by Rule A4, $\vdash t = 0 + t$.

(g) Let $\mathscr{A}(y)$ be $x' + y = (x + y)'$.

(i) 1. $x' + 0 = x'$	(S5′)
2. $x + 0 = x$	(S5′)
3. $(x + 0)' = x'$	2, (S2′)
4. $x' + 0 = (x + 0)'$	1, 3, part (d)
Thus, $\vdash \mathscr{A}(0)$	
(ii) 1. $x' + y = (x + y)'$	Hyp
2. $x' + y' = (x' + y)'$	(S6′)
3. $(x' + y)' = (x + y)''$	1, (S2′)
4. $x' + y' = (x + y)''$	2, 3, part (c)
5. $(x + y') = (x + y)'$	(S6′)
6. $(x + y')' = (x + y)''$	5, (S2′)
7. $x' + y' = (x + y')'$	4, 6 part (d)
8. $x' + y = (x + y)'$ $\Rightarrow x' + y' = (x + y')'$	1–7, Deduction Theorem
Thus, $\vdash \mathscr{A}(y) \Rightarrow \mathscr{A}(y')$	

So, by (i), (ii), Gen, and the Induction Rule, $\vdash (\forall y)(x' + y = (x + y)')$ and, then by Gen and Rule A4, $\vdash t' + r = (t + r)'$.

(h) Let $\mathscr{A}(y)$ be $x + y = y + x$.

(i)	1. $x + 0 = x$	(S5′)
	2. $x = 0 + x$	Part (f)
	3. $x + 0 = 0 + x$	1, 2, part (c)
	i.e., $\vdash \mathscr{A}(0)$	
(ii)	1. $x + y = y + x$	Hyp
	2. $x + y' = (x + y)'$	(S6′)
	3. $y' + x = (y + x)'$	Part (g)
	4. $(x + y)' = (y + x)'$	1, (S2′)
	5. $x + y' = (y + x)'$	2, 4, part (c)
	6. $x + y' = y' + x$	3, 5, part (d)
	7. $x + y = y + x \Rightarrow x + y' = y' + x$	1–6, Deduction Theorem
	Thus, $\vdash \mathscr{A}(y) \Rightarrow \mathscr{A}(y')$	

So, by (i), (ii), Gen, and the Induction Rule, $\vdash (\forall y)(x + y = y + x)$ and, then by Gen and Rule A4, $\vdash t + r = r + t$.

(i)	1. $t = r \Rightarrow t + s = r + s$	Part (e)
	2. $t + s = s + t$	Part (h)
	3. $r + s = s + r$	Part (h)
	4. $t = r$	Hyp
	5. $t + s = r + s$	1, 4, MP
	6. $s + t = r + s$	2, 5, (S1′)
	7. $s + t = s + r$	3, 6, part (c)
	8. $t = r \Rightarrow s + t = s + r$	1–7, Deduction Theorem

(j) Let $\mathscr{A}(z)$ be $(x + y) + z = x + (y + z)$.

(i)	1. $(x + y) + 0 = x + y$	(S5′)
	2. $y + 0 = y$	(S5′)
	3. $x + (y + 0) = x + y$	2, part (i)
	4. $(x + y) + 0 = x + (y + 0)$	1, 3, part (d)
	i.e., $\vdash \mathscr{A}(0)$	
(ii)	1. $(x + y) + z = x + (y + z)$	Hyp
	2. $(x + y) + z' = ((x + y) + z)'$	(S6′)
	3. $((x + y) + z)' = (x + (y + z))'$	1, (S2′)
	4. $(x + y) + z' = (x + (y + z))'$	2, 3, part (c)
	5. $y + z' = (y + z)'$	(S6′)
	6. $x + (y + z') = x + (y + z)'$	5, part (i)
	7. $x + (y + z)' = (x + (y + z))'$	(S6′)
	8. $x + (y + z') = (x + (y + z))'$	6, 7, part (d)
	9. $(x + y) + z' = x + (y + z')$	4, 8, part (d)
	10. $(x + y) + z = x + (y + z) \Rightarrow (x + y) + z' = x + (y + z')$	1–9, Deduction Theorem
	Thus, $\vdash \mathscr{A}(z) \Rightarrow \mathscr{A}(z')$	

By (i), (ii), Gen, and the Induction Rule, $\vdash (\forall z)((x + y) + z = x + (y + z))$ and then, by Gen and Rule A4, $\vdash (t + r) + s = t + (r + s)$.

Parts (k)–(o) are left as exercises.

COROLLARY 3.3 S is a theory with equality; that is, we have (A6): $\vdash x_1 = x_1$, and (A7): $\vdash x = y \Rightarrow (\mathscr{A}(x, x) \Rightarrow \mathscr{A}(x, y))$, where $\mathscr{A}(x, y)$ comes from $\mathscr{A}(x, x)$ by replacing one or more occurrences of x by y, with the proviso that y is free for x in $\mathscr{A}(x, x)$.

Proof By Proposition 2.25, this reduces to Proposition 3.2(a–e, i, k, o), and (S2′).

Notice that the interpretation in which

(a) the set of nonnegative integers is the domain
(b) the integer 0 is the interpretation of the symbol 0
(c) the successor operation (addition of 1) is the interpretation of the ′ function (i.e., of f_1^1)
(d) ordinary addition and multiplication are the interpretations of $+$ and $\cdot$
(e) the interpretation of the predicate letter $=$ is the identity relation

is a normal model for S. This model is called the *standard interpretation* or *standard model*. Any normal model for S that is not isomorphic to the standard model will be called a *nonstandard model* for S.

If we recognize the standard interpretation to be a model for S, then, of course, S is consistent. However, semantic methods, involving as they do a certain amount of set-theoretic reasoning, are regarded by some as too precarious to serve as a basis for consistency proofs; likewise, we have not proved in a rigorous way that the axioms of S are true under the standard interpretation, but we have taken it as intuitively obvious. For these and other reasons, when the consistency of S enters into the argument of a proof, it is common practice to take the statement of the consistency of S as an explicit, unproved assumption.

Some important additional properties of addition and multiplication are covered by the following result.

PROPOSITION 3.4 For any terms t, r, s, the following wfs are theorems of S.

(a) $t \cdot (r + s) = (t \cdot r) + (t \cdot s)$ (distributivity)
(b) $(r + s) \cdot t = (r \cdot t) + (s \cdot t)$ (distributivity)
(c) $(t \cdot r) \cdot s = t \cdot (r \cdot s)$ (associativity of $\cdot$)
(d) $t + s = r + s \Rightarrow t = r$ (cancellation law for $+$)

Proof

(a) Prove $\vdash x \cdot (y + z) = (x \cdot y) + (x + z)$ by induction on z.
(b) Prove from part (a) by Proposition 3.2(n).
(c) Prove $\vdash (x \cdot y) \cdot z = x \cdot (y \cdot z)$ by induction on z.

(d) Prove $\vdash x + z = y + z \Rightarrow x = y$ by induction on z. This requires, for the first time, use of (S4′).

The terms $0, 0', 0'', 0''', \ldots$ we shall call *numerals* and denote by $\overline{0}, \overline{1}, \overline{2}, \overline{3}, \ldots$. More precisely, $\overline{0}$ is 0 and, for any natural number n, $\overline{n+1}$ is $(\overline{n})'$. In general, if n is a natural number, $\overline{n}$ stands for the corresponding numeral $0'''^{\cdots\prime}$—that is, for 0 followed by n strokes. The numerals can be defined recursively by stating that 0 is a numeral and, if u is a numeral, then u' is also a numeral.

PROPOSITION 3.5

(a) $\vdash t + \overline{1} = t'$
(b) $\vdash t \cdot \overline{1} = t$
(c) $\vdash t \cdot \overline{2} = t + t$
(d) $\vdash t + s = 0 \Rightarrow t = 0 \wedge s = 0$
(e) $\vdash t \neq 0 \Rightarrow (s \cdot t = 0 \Rightarrow s = 0)$
(f) $\vdash t + s = \overline{1} \Rightarrow (t = 0 \wedge s = \overline{1}) \vee (t = \overline{1} \wedge s = 0)$
(g) $\vdash t \cdot s = \overline{1} \Rightarrow (t = \overline{1} \wedge s = \overline{1})$
(h) $\vdash t \neq 0 \Rightarrow (\exists y)(t = y')$
(i) $\vdash s \neq 0 \Rightarrow (t \cdot s = r \cdot s \Rightarrow t = r)$
(j) $\vdash t \neq 0 \Rightarrow (t \neq \overline{1} \Rightarrow (\exists y)(t = y''))$

Proof

(a)	1. $t + 0' = (t + 0)'$	(S6′)
	2. $t + 0 = t$	(S5′)
	3. $(t + 0)' = t'$	2, (S2′)
	4. $t + 0' = t'$	1, 3, Proposition 3.2(c)
	5. $t + \overline{1} = t'$	4, abbreviation
(b)	1. $t \cdot 0' = t \cdot 0 + t$	(S8′)
	2. $t \cdot 0 = 0$	(S7′)
	3. $(t \cdot 0) + t = 0 + t$	2, Proposition 3.2(e)
	4. $t \cdot 0' = 0 + t$	1, 3, Proposition 3.2(c)
	5. $0 + t = t$	Proposition 3.2(f, b)
	6. $t \cdot 0' = t$	4, 5, Proposition 3.2(c)
	7. $t \cdot \overline{1} = t$	6, abbreviation
(c)	1. $t \cdot \overline{1}' = (t \cdot \overline{1}) + t$	(S8′)
	2. $t \cdot \overline{1} = t$	Part (b)
	3. $(t \cdot \overline{1}) + t = t + t$	2, Proposition 3.2(e)
	4. $t \cdot \overline{1}' = t + t$	1, 3, Proposition 3.2(c)
	5. $t \cdot \overline{2} = t + t$	4, abbreviation

(d) Let $\mathscr{A}(y)$ be $x + y = 0 \Rightarrow x = 0 \wedge y = 0$. It is easy to prove that $\vdash \mathscr{A}(0)$. Also, since $\vdash (x + y)' \neq 0$ by (S3′) and Proposition 3.2(b), then, by (S6′), it follows that $\vdash x + y' \neq 0$. Hence, $\vdash \mathscr{A}(y')$ by the tautology $\neg A \Rightarrow (A \Rightarrow B)$. So, $\vdash \mathscr{A}(y) \Rightarrow \mathscr{A}(y')$ by the tautology $A \Rightarrow (B \Rightarrow A)$. Thus, by the Induction Rule, $\vdash (\forall y)\mathscr{A}(y)$ and, then, by Gen and Rule A4, we obtain part (d).

(e) The proof is similar to that for part (d) and is left as an exercise.

(f) By induction on y in $x + y = \bar{1} \Rightarrow ((x = 0 \wedge y = \bar{1}) \vee (x = \bar{1} \wedge y = 0))$.

(g) By induction on y in $x \cdot y = \bar{1} \Rightarrow (x = \bar{1} \wedge y = \bar{1})$.

(h) Perform induction on x in $x \neq 0 \Rightarrow (\exists w)(x = w')$.

(i) Let $\mathscr{A}(y)$ be $(\forall x)(z \neq 0 \Rightarrow (x \cdot z = y \cdot z \Rightarrow x = y))$.

(i)

1.	$z \neq 0$	Hyp
2.	$x \cdot z = 0 \cdot z$	Hyp
3.	$0 \cdot z = 0$	Proposition 3.2(l)
4.	$x \cdot z = 0$	2, 3, Proposition 3.2(c)
5.	$x = 0$	1, 4, part (e)
6.	$z \neq 0$ $\Rightarrow (x \cdot z = 0 \cdot z \Rightarrow x = 0)$	1–5, Deduction Theorem
7.	$(\forall x)(z \neq 0$ $\Rightarrow (x \cdot z = 0 \cdot z \Rightarrow x = 0))$	6, Gen
	Thus, $\vdash \mathscr{A}(0)$	

(ii)

1.	$(\forall x)(z \neq 0$ $\Rightarrow (x \cdot z = y \cdot z \Rightarrow x = y))$	Hyp ($\mathscr{A}(y)$)
2.	$z \neq 0$	Hyp
3.	$x \cdot z = y' \cdot z$	Hyp
4.	$y' \neq 0$	(S3′), Proposition 3.2(b)
5.	$y' \cdot z \neq 0$	2, 4, part (e) and a tautology
6.	$x \cdot z \neq 0$	3, 5, (S1′) and tautologies
7.	$x \neq 0$	6, (S7′), Proposition 3.2(o, n), (S1′), and tautologies
8.	$(\exists w)(x = w')$	7, part (h)
9.	$x = b'$	8, Rule C
10.	$b' \cdot z = y' \cdot z$	3, 9, equality law (A7)
11.	$b \cdot z + z = y \cdot z + z$	10, Proposition 3.2(m, d)
12.	$b \cdot z = y \cdot z$	11, Proposition 3.4(d)
13.	$z \neq 0$ $\Rightarrow ((b \cdot z = y \cdot z) \Rightarrow (b = y))$	1, Rule A4
14.	$b \cdot z = y \cdot z \Rightarrow b = y$	2, 13, MP
15.	$b = y$	12, 14, MP
16.	$b' = y'$	15, (S2′)
17.	$x = y'$	9, 16, Proposition 3.2(c)
18.	$\mathscr{A}(y), z \neq 0,$ $x \cdot z = y' \cdot z \vdash x = y'$	1–17, Proposition 2.9
19.	$\mathscr{A}(y) \vdash z \neq 0$ $\Rightarrow (x \cdot z = y' \cdot z \Rightarrow x = y')$	18, Deduction Theorem twice
20.	$\mathscr{A}(y) \vdash (\forall x)(z \neq 0$ $\Rightarrow (x \cdot z = y' \cdot z \Rightarrow x = y'))$	19, Gen
21.	$\vdash \mathscr{A}(y) \Rightarrow \mathscr{A}(y')$	20, Deduction Theorem

Hence, by (i), (ii), Gen, and the Induction Rule, we obtain $\vdash (\forall y)\mathscr{A}(y)$, and, then, by Gen and Rule A4, we have the desired result.

(j) This is left as an exercise.

PROPOSITION 3.6

(a) Let m and n be any natural numbers.
  (i) If $m \neq n$, then $\vdash \bar{m} \neq \bar{n}$.
  (ii) $\vdash \overline{m+n} = \bar{m} + \bar{n}$ and $\vdash \overline{m \cdot n} = \bar{m} \cdot \bar{n}$
(b) Any model for S is infinite.
(c) For any cardinal number $\aleph_\beta$, S has a normal model of cardinality $\aleph_\beta$.

Proof

(a)(i) Assume $m \neq n$. Now, either $m < n$ or $n < m$; say, $m < n$.

1. $\bar{m} = \bar{n}$ — Hyp
2. $0\underbrace{'' \ldots '}_{m \text{ times}} = 0\underbrace{''' \ldots '}_{n \text{ times}}$ — 1 is an abbreviation of 2
3. Apply (S4′) m times in a row. Then $0 = 0\underbrace{'' \ldots '}_{(n-m)\text{times}}$. Let t be $\overline{(n-m-1)}$. Since $n > m$, $n - m - 1 \geqslant 0$. Thus, $0 = t'$.
4. $0 \neq t'$ — (S3′)
5. $0 = t' \wedge 0 \neq t'$ — 3, 4, Conjunction Introduction
6. $\bar{m} = \bar{n} \vdash (0 = t' \wedge 0 \neq t')$ — 1–5
7. $\vdash \bar{m} \neq \bar{n}$ — 1–6, Proof by Contradiction

A similar proof holds in the case when $n < m$. (A more rigorous proof can be given by induction in the metalanguage with respect to n.)

(ii) We use induction in the metalanguage. First, $\overline{m+0}$ is $\bar{m}$. Hence, $\vdash \overline{m+0} = \bar{m} + \bar{0}$ by (S5′). Now assume $\vdash \overline{m+n} = \bar{m} + \bar{n}$. Therefore, $\vdash (\overline{m+n})' = \bar{m} + (\bar{n})'$ by (S2′) and (S6′). But $\overline{m+(n+1)}$ is $(\overline{m+n})'$ and $\overline{n+1}$ is $(\bar{n})'$. Hence, $\vdash \overline{m+(n+1)} = \bar{m} + \overline{n+1}$. The proof that $\vdash \overline{m \cdot n} = \bar{m} \cdot \bar{n}$ is left as an exercise.

(b) By part (a) (i), in a model for S, the objects corresponding to the numerals must be distinct. But there are denumerably many numerals.

(c) This follows from Corollary 2.34(a) and the fact that the standard model is an infinite normal model.

An order relation can be introduced by definition in S.

DEFINITIONS

$t < s$ for $(\exists w)(w \neq 0 \wedge w + t = s)$
$t \leqslant s$ for $t < s \vee t = s$
$t > s$ for $s < t$
$t \geqslant s$ for $s \leqslant t$
$t \nless s$ for $\neg(t < s)$, and so on

In the first definition, to be precise, we can choose w to be the first variable not in t or s.

PROPOSITION 3.7 For any terms t, r, s, the following wfs are theorems.

(a) $t \nless t$
(b) $t < s \Rightarrow (s < r \Rightarrow t < r)$
(c) $t < s \Rightarrow s \nless t$
(d) $t < s \Leftrightarrow (t + r < s + r)$

(e) $t \leqslant t$
(f) $t \leqslant s \Rightarrow (s \leqslant r \Rightarrow t \leqslant r)$
(g) $t \leqslant s \Leftrightarrow (t + r \leqslant s + r)$
(h) $t \leqslant s \Rightarrow (s < r \Rightarrow t < r)$
(i) $0 \leqslant t$
(j) $0 < t'$
(k) $t < r \Leftrightarrow t' \leqslant r$
(l) $t \leqslant r \Leftrightarrow t < r'$
(m) $t < t'$
(n) $(0 < \overline{1}), (\overline{1} < \overline{2}), (\overline{2} < \overline{3}), \ldots$
(o) $t \neq r \Rightarrow (t < r \vee r < t)$
(o′) $t = r \vee t < r \vee r < t$
(p) $t \leqslant r \vee r \leqslant t$
(q) $t + r \geqslant t$
(r) $r \neq 0 \Rightarrow t + r > t$
(s) $r \neq 0 \Rightarrow t \cdot r \geqslant t$
(t) $r \neq 0 \Leftrightarrow r > 0$
(u) $r > 0 \Rightarrow (t > 0 \Rightarrow r \cdot t > 0)$
(v) $r \neq 0 \Rightarrow (t > 1 \Rightarrow t \cdot r > r)$
(w) $r \neq 0 \Rightarrow (t < s \Leftrightarrow t \cdot r < s \cdot r)$
(x) $r \neq 0 \Rightarrow (t \leqslant s \Leftrightarrow t \cdot r \leqslant s \cdot r)$
(y) $t \nless 0$
(z) $t \leqslant r \wedge r \leqslant t \Rightarrow t = r$

Proof

(a) By Proposition 3.4(d)

(b)

1.	$t < s$	Hyp
2.	$s < r$	Hyp
3.	$(\exists w)(w \neq 0 \wedge w + t = s)$	1, Definition
4.	$(\exists v)(v \neq 0 \wedge v + s = r)$	2, Definition
5.	$b \neq 0 \wedge b + t = s$	3, Rule C
6.	$c \neq 0 \wedge c + s = r$	4, Rule C
7.	$b + t = s$	5, Conjunction Elimination
8.	$c + s = r$	6, Conjunction Elimination
9.	$c + (b + t) = r$	7, 8, Proposition 3.2(i, c)
10.	$(c + b) + t = r$	9, Proposition 3.2(j, c)
11.	$b \neq 0$	5, Conjunction Elimination
12.	$c + b \neq 0$	11, Proposition 3.5(d)
13.	$c + b \neq 0 \wedge (c + b) + t = r$	10, 12, Conjunction Introduction
14.	$(\exists u)(u \neq 0 \wedge u + t = r)$	13, Rule E4
15.	$t < r$	14, Definition
16.	$\vdash t < s \Rightarrow (s < r \Rightarrow t < r)$	1–15, Deduction Theorem, Proposition 2.9

Parts (c)–(z) are left as exercises. These theorems are not arranged in any special order, though, generally, they can be proved more or less directly from preceding ones in the list.

PROPOSITION 3.8

(a) For any natural number k, $\vdash x = 0 \vee \cdots \vee x = \overline{k} \Leftrightarrow x \leqslant \overline{k}$.
(a′) For any natural number k and any wf $\mathscr{A}$,
$\vdash \mathscr{A}(0) \wedge \mathscr{A}(\overline{1}) \wedge \cdots \wedge \mathscr{A}(\overline{k}) \Leftrightarrow (\forall x)(x \leqslant \overline{k} \Rightarrow \mathscr{A}(x))$.
(b) For any natural number $k > 0$, $\vdash x = 0 \vee \cdots \vee x = (\overline{k - 1}) \Leftrightarrow x < \overline{k}$.
(b′) For any natural number $k > 0$ and any wf $\mathscr{A}$,
$\vdash \mathscr{A}(0) \wedge \mathscr{A}(\overline{1}) \wedge \cdots \wedge \mathscr{A}(\overline{k - 1}) \Leftrightarrow (\forall x)(x < \overline{k} \Rightarrow \mathscr{A}(x))$.
(c) $\vdash ((\forall x)(x < y \Rightarrow \mathscr{A}(x)) \wedge (\forall x)(x \geqslant y \Rightarrow \mathscr{B}(x))) \Rightarrow (\forall x)(\mathscr{A}(x) \vee \mathscr{B}(x))$

Proof (a) We prove $\vdash x = 0 \vee \cdots \vee x = \bar{k} \Leftrightarrow x \leqslant \bar{k}$ by induction in the metalanguage on k. The case for $k = 0$, $\vdash x = 0 \Leftrightarrow x \leqslant 0$, is obvious from the definitions and Proposition 3.7. Assume $\vdash x = 0 \vee \cdots \vee x = \bar{k} \Leftrightarrow x \leqslant \bar{k}$. Now, assume $x = 0 \vee \cdots \vee x = \bar{k} \vee x = \overline{k+1}$. But, $x = \overline{k+1} \Rightarrow x \leqslant \overline{k+1}$; also, $x = 0 \vee \cdots \vee x = \bar{k} \Rightarrow x \leqslant \bar{k}$ and $x \leqslant \bar{k} \Rightarrow x \leqslant \overline{k+1}$. Hence, $x = 0 \vee \cdots \vee x = \overline{k+1} \Rightarrow x \leqslant \overline{k+1}$. On the other hand, assume $x \leqslant \overline{k+1}$. Then $x = \overline{k+1} \vee x < \overline{k+1}$. If $x = \overline{k+1}$, then $x = 0 \vee \cdots \vee x = \overline{k+1}$. If $x < \overline{k+1}$, then, since $\overline{k+1}$ is $(\bar{k})'$, we have $x \leqslant \bar{k}$ by Proposition 3.7(l). By inductive hypothesis, $x = 0 \vee \cdots \vee x = \bar{k}$; so, $x = 0 \vee \cdots \vee x = \overline{k+1}$. (This proof has been given in an informal manner that we shall generally use from now on. In particular, the Deduction Theorem, the eliminability of Rule C, the Replacement Theorem, and various tautologies will be applied without being explicitly mentioned.)

Parts (a′), (b), and (b′) follow easily from part (a). Part (c) follows almost immediately from Proposition 3.7(o), using obvious tautologies.

There are several stronger forms of the induction principle that we can prove at this point.

PROPOSITION 3.9

(a) *Complete induction.* $\vdash (\forall x)((\forall z)(z < x \Rightarrow \mathscr{A}(z)) \Rightarrow \mathscr{A}(x)) \Rightarrow (\forall x)\mathscr{A}(x)$ (In ordinary language, consider a property P such that, for any x, if P holds for all natural numbers less than x, then P holds for x also. Then P holds for all natural numbers.)

(b) *Least-number principle.* $\vdash (\exists x)\mathscr{A}(x) \Rightarrow (\exists y)(\mathscr{A}(y) \wedge (\forall z)(z < y \Rightarrow \neg\mathscr{A}(z)))$ (If a property P holds for some natural number, then there is a least number satisfying P.)

Proof

(a) Let $\mathscr{B}(x)$ be $(\forall z)(z \leqslant x \Rightarrow \mathscr{A}(z))$.

(i)	1. $(\forall x)((\forall z)(z < x \Rightarrow \mathscr{A}(z)) \Rightarrow \mathscr{A}(x))$	Hyp
	2. $(\forall z)(z < 0 \Rightarrow \mathscr{A}(z)) \Rightarrow \mathscr{A}(0)$	1, Rule A4
	3. $z \not< 0$	Proposition 3.7(y)
	4. $(\forall z)(z < 0 \Rightarrow \mathscr{A}(z))$	3, Tautology, Gen
	5. $\mathscr{A}(0)$	2, 4, MP
	6. $(\forall z)(z \leqslant 0 \Rightarrow \mathscr{A}(z))$ i.e., $\mathscr{B}(0)$	5, Proposition 3.8(a′)
	7. $(\forall x)((\forall z)(z < x \Rightarrow \mathscr{A}(z)) \Rightarrow \mathscr{A}(x)) \vdash \mathscr{B}(0)$	1–6
(ii)	1. $(\forall x)((\forall z)(z < x \Rightarrow \mathscr{A}(z)) \Rightarrow \mathscr{A}(x))$	Hyp
	2. $\mathscr{B}(x)$, i.e., $(\forall z)(z \leqslant x \Rightarrow \mathscr{A}(z))$	Hyp
	3. $(\forall z)(z < x' \Rightarrow \mathscr{A}(z))$	2, Proposition 3.7(l)
	4. $(\forall z)(z < x' \Rightarrow \mathscr{A}(z)) \Rightarrow \mathscr{A}(x')$	1, Rule A4
	5. $\mathscr{A}(x')$	3, 4, MP
	6. $z \leqslant x' \Rightarrow z < x' \vee z = x'$	Definition, Tautology
	7. $z < x' \Rightarrow \mathscr{A}(z)$	3, Rule A4

8. $z = x' \Rightarrow \mathscr{A}(z)$	5, axiom (A7)
9. $(\forall z)(z \leqslant x' \Rightarrow \mathscr{A}(z))$ i.e., $\mathscr{B}(x')$	6, 7, 8, Tautology, Gen
10. $(\forall x)((\forall z)(z < x \Rightarrow \mathscr{A}(z)) \Rightarrow \mathscr{A}(x)) \vdash (\forall x)(\mathscr{B}(x) \Rightarrow \mathscr{B}(x'))$	1–9, Deduction Theorem, Gen

By (i), (ii), and the Induction Rule, we obtain $\mathscr{C} \vdash (\forall x)\mathscr{B}(x)$; that is, $\mathscr{C} \vdash (\forall x)(\forall z)(z \leqslant x \Rightarrow \mathscr{A}(z))$, where $\mathscr{C}$ is $(\forall x)((\forall z)(z < x \Rightarrow \mathscr{A}(z)) \Rightarrow \mathscr{A}(x))$. Hence, by Rule A4 twice, $\mathscr{C} \vdash x \leqslant x \Rightarrow \mathscr{A}(x)$; but, $\vdash x \leqslant x$. So, $\mathscr{C} \vdash \mathscr{A}(x)$ and, by Gen and the Deduction Theorem, $\vdash \mathscr{C} \Rightarrow (\forall x)\mathscr{A}(x)$.

(b)

1. $\neg(\exists y)(\mathscr{A}(y) \wedge (\forall z)(z < y \Rightarrow \neg\mathscr{A}(z)))$	Hyp
2. $(\forall y)\neg(\mathscr{A}(y) \wedge (\forall z)(z < y \Rightarrow \neg\mathscr{A}(z)))$	1, Tautology
3. $(\forall y)((\forall z)(z < y \Rightarrow \neg\mathscr{A}(z)) \Rightarrow \neg\mathscr{A}(y))$	2, Tautology
4. $(\forall y)\neg\mathscr{A}(y)$	3, part (a) with $\neg\mathscr{A}$ instead of $\mathscr{A}$
5. $\neg(\exists y)\mathscr{A}(y)$	4, Tautology
6. $\neg(\exists x)\mathscr{A}(x)$	5, change of bound variable
7. $\neg(\exists y)(\mathscr{A}(y) \wedge (\forall z)(z < y \Rightarrow \neg\mathscr{A}(z))) \Rightarrow \neg(\exists x)\mathscr{A}(x)$	1–6, Deduction Theorem
8. $(\exists x)\mathscr{A}(x) \Rightarrow (\exists y)(\mathscr{A}(y) \wedge (\forall z)(z < y \Rightarrow \neg\mathscr{A}(z)))$	7, Tautology

Exercise

3.1 *Method of infinite descent.* Prove $\vdash (\forall x)(\mathscr{A}(x) \Rightarrow (\exists y)(y < x \wedge \mathscr{A}(y))) \Rightarrow (\forall x)\neg\mathscr{A}(x)$.

Another important notion in number theory is divisibility, which we now define.

DEFINITION $t|s$ for $(\exists z)(s = t \cdot z)$. Here, z is the first variable not in t or s.

PROPOSITION 3.10 The following wfs are theorems.

(a) $t\|t$	(e) $s \neq 0 \wedge t\|s \Rightarrow t \leqslant s$
(b) $\overline{1}\|t$	(f) $t\|s \wedge s\|t \Rightarrow s = t$
(c) $t\|0$	(g) $t\|s \Rightarrow t\|(r \cdot s)$
(d) $t\|s \wedge s\|r \Rightarrow t\|r$	(h) $t\|s \wedge t\|r \Rightarrow t\|(s + r)$

Proof

(a) $t = t \cdot \overline{1}$. Hence, $t|t$.
(b) $t = \overline{1} \cdot t$. Hence, $\overline{1}|t$.
(c) $0 = t \cdot 0$. Hence, $t|0$.
(d) If $s = t \cdot z$ and $r = s \cdot w$, then $r = t \cdot (z \cdot w)$.
(e) If $s \neq 0$ and $t|s$, then $s = t \cdot z$ for some z. If $z = 0$, then $s = 0$. Hence, $z \neq 0$. So, $z = u'$ for some u. $s = t \cdot (u') = t \cdot u + t \geqslant t$.
(f)–(h). These proofs are left as exercises.

Exercises

3.2 Prove $\vdash t|\overline{1} \Rightarrow t = \overline{1}$.

3.3 Prove $\vdash (t|s \wedge t|s') \Rightarrow t = \overline{1}$.

It will be useful for later purposes to prove the existence of a unique quotient and remainder upon division of one number x by another nonzero number y.

PROPOSITION 3.11 $\vdash y \neq 0 \Rightarrow (\exists u)(\exists v)[x = y \cdot u + v \wedge v < y \wedge (\forall u_1)(\forall v_1)((x = y \cdot u_1 + v_1 \wedge v_1 < y) \Rightarrow u = u_1 \wedge v = v_1)]$

Proof Let $\mathscr{A}(x)$ be $y \neq 0 \Rightarrow (\exists u)(\exists v)(x = y \cdot u + v \wedge v < y)$.

(i)	1.	$y \neq 0$	Hyp
	2.	$0 = y \cdot 0 + 0$	(S5′), (S7′)
	3.	$0 < y$	1, Proposition 3.7(t)
	4.	$0 = y \cdot 0 + 0 \wedge 0 < y$	2, 3, Conjunction Introduction
	5.	$(\exists y)(\exists v)(0 = y \cdot u + v \wedge v < y)$	4, Rule E4 twice
	6.	$y \neq 0 \Rightarrow (\exists u)(\exists v)(0 = y \cdot u + v \wedge v < y)$	1–5, Deduction Theorem
(ii)	1.	$\mathscr{A}(x)$, i.e., $y \neq 0 \Rightarrow (\exists u)(\exists v)(x = y \cdot u + v \wedge v < y)$	Hyp
	2.	$y \neq 0$	Hyp
	3.	$(\exists u)(\exists v)(x = y \cdot u + v \wedge v < y)$	1, 2, MP
	4.	$x = y \cdot a + b \wedge b < y$	3, Rule C twice
	5.	$b < y$	4, Conjunction Elimination
	6.	$b' \leqslant y$	5, Proposition 3.7(k)
	7.	$b' < y \vee b' = y$	6, Definition
	8.	$b' < y \Rightarrow (x' = y \cdot a + b' \wedge b' < y)$	4, (S6′), Tautology
	9.	$b' < y \Rightarrow (\exists u)(\exists v)(x' = y \cdot u + v \wedge v < y)$	8, Rule E4, Deduction Theorem
	10.	$b' = y \Rightarrow x' = y \cdot a + y \cdot \overline{1}$	4, (S6′), Proposition 3.5(b)
	11.	$b' = y \Rightarrow (x' = y \cdot (a + \overline{1}) + 0 \wedge 0 < y)$	10, Proposition 3.4, 2, Proposition 3.7(t), (S5′)
	12.	$b' = y \Rightarrow (\exists u)(\exists v)(x' = y \cdot u + v \wedge v < y)$	11, Rule E4, Deduction Theorem
	13.	$(\exists u)(\exists v)(x' = y \cdot u + v \wedge v < y)$	7, 9, 12, Disjunction Elimination
	14.	$\mathscr{A}(x) \Rightarrow (y \neq 0 \Rightarrow (\exists u)(\exists v)(x' = y \cdot u + v \wedge v < y))$	1–13, Deduction Theorem
		Thus, $\mathscr{A}(x) \Rightarrow \mathscr{A}(x')$	

By (i), (ii), Gen, and the Induction Rule, $\vdash (\forall x)\mathscr{A}(x)$. This establishes the existence of a quotient u and a remainder v. To prove uniqueness, proceed as follows. Assume $y \neq 0$. Assume $x = y \cdot u + v \wedge v < y$ and $x = y \cdot u_1 + v_1 \wedge v_1 < y$. Now, $u = u_1$ or $u < u_1$ or $u_1 < u$. If $u = u_1$, then $v = v_1$ by Proposition 3.4(d). If $u < u_1$, then

$u_1 = u + w$ for some $w \neq 0$. Then $y \cdot u + v = y \cdot (u + w) + v_1 = y \cdot u + y \cdot w + v_1$. Hence, $v = y \cdot w + v_1$. Since $w \neq 0$, $y \cdot w \geqslant y$. So, $v = y \cdot w + v_1 \geqslant y$, contradicting $v < y$. Hence, $u \nless u_1$. Similarly, $u_1 \nless u$. Thus, $u = u_1$ and so, $v = v_1$.

From this point on, one can generally translate into S and prove the results from any text on elementary number theory. There are certain number-theoretic functions, such as x^y and $x!$, that we have to be able to define in S, and this we shall do later in this chapter. (In most cases, by suitable paraphrasing, one can get along without explicitly defining these functions, but, after a short time, this leads to unwieldy complications.) Some standard results of number theory, such as Dirichlet's Theorem, are proved with the aid of the theory of complex variables, and it is often not known whether elementary proofs (or proofs in S) can be given for such theorems. The statement of some results in number theory involves nonelementary concepts, such as the logarithmic function, and, except in cases where an equivalent elementary formula can be obtained, cannot even be formulated in S. More information about the strength and expressive powers of S will be revealed in the sequel. For example, it will be shown later that there are closed wfs that are neither provable nor disprovable in S, if S is consistent; hence there is a wf that is true under the standard interpretation but is not provable in S. We shall also see that this incompleteness of S cannot be attributed to omission of some essential axiom but has deeper underlying causes that apply to other theories as well.

Exercises

3.4 Show that the Induction Principle (S9) is independent of the other axioms of S.

3.5^D (a) Show that there exist nonstandard models for S of any cardinality $\aleph_\alpha$.

(b) Ehrenfeucht (1958) has shown the existence of at least $2^{\aleph_0}$ mutually nonisomorphic models of S of cardinality $\aleph_\alpha$. Prove the special case that there are $2^{\aleph_0}$ mutually nonisomorphic denumerable models of S.

3.6^D Give a standard mathematical proof of the categoricity of Peano's Postulates, in the sense that any two "models" are isomorphic. Explain why this proof does not apply to the first-order theory S.

3.7^D (Presburger, 1929) If we eliminate from S the function letter f_2^2 for multiplication and the axioms (S7) and (S8), show that the new system S_+ is complete and decidable (in the sense of Chapter 1, page 28).

3.8 (a) Show that every closed atomic wf $t = s$ of S is *decidable*—that is, either $\vdash_S t = s$ or $\vdash_S t \neq s$.

(b) Show that every closed wf of S without quantifiers is decidable.

2. NUMBER-THEORETIC FUNCTIONS AND RELATIONS

A *number-theoretic function* is a function whose arguments and values are natural numbers. Addition and multiplication are familiar examples of number-theoretic functions of two arguments. A *number-theoretic relation* is a relation whose arguments are natural numbers. For example, $=$ and $<$ are binary number-theoretic relations, and the expression $x + y < z$ determines a number-theoretic relation of three arguments. Number-theoretic functions and relations are intuitive and are not bound up with any formal system.

Let K be any theory that has the same symbols as S. A number-theoretic relation R of n arguments is said to be *expressible in K* if and only if there is a wf $\mathscr{A}(x_1, \ldots, x_n)$ of K with n free variables such that, for any natural numbers $k_1, \ldots, k_n$, the following hold:

1. If $R(k_1, \ldots, k_n)$ is true, then $\vdash_K \mathscr{A}(\overline{k}_1, \ldots, \overline{k}_n)$.
2. If $R(k_1, \ldots, k_n)$ is false, then $\vdash_K \neg \mathscr{A}(\overline{k}_1, \ldots, \overline{k}_n)$.

For example, the number-theoretic relation of equality is expressed in S by the wf $x_1 = x_2$. In fact, if $k_1 = k_2$, then $\overline{k}_1$ is the same term as $\overline{k}_2$ and so, by Proposition 3.2(a), $\vdash_S \overline{k}_1 = \overline{k}_2$. Moreover, if $k_1 \neq k_2$, then, by Proposition 3.6(a), $\vdash_S \overline{k}_1 \neq \overline{k}_2$.

Likewise, the relation "less than" is expressed in S by the wf $x_1 < x_2$. Recall that $x_1 < x_2$ is $(\exists x_3)(x_3 \neq 0 \wedge x_3 + x_1 = x_2)$. If $k_1 < k_2$, then there is some nonzero number n such that $k_2 = n + k_1$. Now, by Proposition 3.6(a)(ii), $\vdash_S \overline{k}_2 = \overline{n} + \overline{k}_1$. Also, by (S3′), since $n \neq 0$, $\vdash_S \overline{n} \neq 0$. Hence, by Rule E4, one can prove in S the wf $(\exists w)(w \neq 0 \wedge w + \overline{k}_1 = \overline{k}_2)$; that is, $\vdash_S \overline{k}_1 < \overline{k}_2$. Now, if $k_1 \nless k_2$, then $k_2 < k_1$ or $k_2 = k_1$. If $k_2 < k_1$, then, as we have just seen, $\vdash_S \overline{k}_2 < \overline{k}_1$. If $k_2 = k_1$, then $\vdash_S \overline{k}_2 = \overline{k}_1$. In either case, $\vdash_S \overline{k}_2 \leqslant \overline{k}_1$ and then, by Proposition 3.7(a, c), $\vdash_S \overline{k}_1 \nless \overline{k}_2$.

Observe that, if a relation is expressible in a theory K, then it is expressible in any extension of K.

Exercises

3.9 Show that the negation, disjunction, and conjunction of relations that are expressible in K are also expressible in K.
3.10 Show that the relation $x + y = z$ is expressible in S.

Let K be any theory with equality that has the same symbols as S. A number-theoretic function f of n arguments is said to be *representable in K* if and only if there is a wf $\mathscr{A}(x_1, \ldots, x_{n+1})$ of K with the free variables $x_1, \ldots, x_{n+1}$ such that, for any natural numbers $k_1, \ldots, k_n, m$, the following hold:

1. If $f(k_1, \ldots, k_n) = m$, then $\vdash_K \mathscr{A}(\overline{k}_1, \ldots, \overline{k}_n, \overline{m})$.
2. $\vdash_K (\exists_1 x_{n+1}) \mathscr{A}(\overline{k}_1, \ldots, \overline{k}_n, x_{n+1})$

If, in this definition, we replace condition 2 by

2′. $\vdash_K (\exists_1 x_{n+1}) \mathscr{A}(x_1, \ldots, x_n, x_{n+1})$

then the function f is said to be *strongly representable in K*. Notice that 2′ implies 2, by Gen and Rule A4. Hence, every strongly representable function is representable. (The converse is also true; see Exercise 3.32).

Observe that a function representable (strongly representable) in a theory K is representable (strongly representable) in any extension of K.

Examples In these examples, let K be any theory with equality that has the same symbols as S.

1. The zero function, $Z(x) = 0$, is strongly representable in the theory with equality K by the wf $x_1 = x_1 \wedge x_2 = 0$. For any k_1 and k_2, if $Z(k_1) = k_2$, then

$k_2 = 0$ and $\vdash_K \bar{k}_1 = \bar{k}_1 \wedge 0 = 0$; that is, condition 1 holds. Also, $\vdash_K (\exists_1 x_2)(x_1 = x_1 \wedge x_2 = 0)$. Thus, condition 2′ holds.

2. The successor function, $N(x) = x + 1$, is strongly representable in the theory with equality K by the wf $x_2 = (x_1)'$. For any k_1 and k_2, if $N(k_1) = k_2$, then $k_2 = k_1 + 1$; hence, $\bar{k}_2$ is $(\bar{k}_1)'$. Then $\vdash_K \bar{k}_2 = (\bar{k}_1)'$. It is easy to verify that $\vdash_K (\exists_1 x_2)(x_2 = (x_1)')$.

3. The projection function, $U_j^n(x_1, \ldots, x_n) = x_j$, is strongly representable in the theory with equality K by the wf $x_1 = x_1 \wedge x_2 = x_2 \wedge \cdots \wedge x_n = x_n \wedge x_{n+1} = x_j$. If $U_j^n(k_1, \ldots, k_n) = m$, then $m = k_j$. Hence, $\vdash_K \bar{k}_1 = \bar{k}_1 \wedge \bar{k}_2 = \bar{k}_2 \wedge \cdots \wedge \bar{k}_n = \bar{k}_n \wedge \bar{m} = \bar{k}_j$. Thus, condition 1 holds. Also, $\vdash_K (\exists_1 x_{n+1})(x_1 = x_1 \wedge x_2 = x_2 \wedge \cdots \wedge x_n = x_n \wedge x_{n+1} = x_j)$; that is, condition 2′ holds.

4. Assume that the functions $g(x_1, \ldots, x_m)$, $h_1(x_1, \ldots, x_n), \ldots, h_m(x_1, \ldots, x_n)$ are (strongly) representable in the theory with equality K by the wfs $\mathscr{B}(x_1, \ldots, x_{m+1})$, $\mathscr{A}_1(x_1, \ldots, x_{n+1}), \ldots, \mathscr{A}_m(x_1, \ldots, x_{n+1})$, respectively. Define a new function f by the equation

$$f(x_1, \ldots, x_n) = g(h_1(x_1, \ldots, x_n), \ldots, h_m(x_1, \ldots, x_n))$$

f is said to be obtained from $g, h_1, \ldots, h_m$ by *substitution*. Then f is also (strongly) representable in K by the following wf $\mathscr{C}(x_1, \ldots, x_{n+1})$:

$$(\exists y_1)\ldots(\exists y_m)(\mathscr{A}_1(x_1, \ldots, x_n, y_1) \wedge \cdots \wedge \mathscr{A}_m(x_1, \ldots, x_n, y_m) \wedge \mathscr{B}(y_1, \ldots, y_m, x_{n+1}))$$

To prove condition 1, let $f(k_1, \ldots, k_n) = p$. Let $h_j(k_1, \ldots, k_n) = r_j$ for $1 \leqslant j \leqslant m$; then $g(r_1, \ldots, r_m) = p$. By our assumption that $\mathscr{B}, \mathscr{A}_1, \ldots, \mathscr{A}_m$ represent $g, h_1, \ldots, h_m$, we have $\vdash_K \mathscr{A}_j(\bar{k}_1, \ldots, \bar{k}_n, \bar{r}_j)$ for $1 \leqslant j \leqslant m$, and $\vdash_K \mathscr{B}(\bar{r}_1, \ldots, \bar{r}_m, \bar{p})$. So, by Conjunction Introduction, $\vdash_K \mathscr{A}_1(\bar{k}_1, \ldots, \bar{k}_n, \bar{r}_1) \wedge \cdots \wedge \mathscr{A}_m(\bar{k}_1, \ldots, \bar{k}_n, \bar{r}_m) \wedge \mathscr{B}(\bar{r}_1, \ldots, \bar{r}_m, \bar{p})$. Hence, by Rule E4, $\vdash_K \mathscr{C}(\bar{k}_1, \ldots, \bar{k}_n, \bar{p})$; that is, condition 1 holds. We shall prove condition 2′ in the case of strong representability; the proof of condition 2 in the case of representability is similar. Assume

(▲) $(\exists y_1)\ldots(\exists y_m)(\mathscr{A}_1(x_1, \ldots, x_n, y_1) \wedge \cdots \wedge \mathscr{A}_m(x_1, \ldots, x_n, y_m) \wedge \mathscr{B}(y_1, \ldots, y_m, u))$

and

(■) $(\exists y_1)\ldots(\exists y_m)(\mathscr{A}_1(x_1, \ldots, x_n, y_1) \wedge \cdots \wedge \mathscr{A}_m(x_1, \ldots, x_n, y_m) \wedge \mathscr{B}(y_1, \ldots, y_m, v))$

By (▲), using Rule C m times,

$$\mathscr{A}_1(x_1, \ldots, x_n, b_1) \wedge \cdots \wedge \mathscr{A}_m(x_1, \ldots, x_n, b_m) \wedge \mathscr{B}(b_1, \ldots, b_m, u)$$

By (■), using Rule C again,

$$\mathscr{A}_1(x_1, \ldots, x_n, c_1) \wedge \cdots \wedge \mathscr{A}_m(x_1, \ldots, x_n, c_m) \wedge \mathscr{B}(c_1, \ldots, c_m, v)$$

Since $\vdash_K (\exists_1 x_{n+1})\mathscr{A}_j(x_1, \ldots, x_n, x_{n+1})$, we obtain, from $\mathscr{A}_j(x_1, \ldots, x_n, b_j)$ and $\mathscr{A}_j(x_1, \ldots, x_n, c_j)$, that $b_j = c_j$. From $\mathscr{B}(b_1, \ldots, b_m, u)$ and $b_1 = c_1, \ldots, b_m = c_m$, we have $\mathscr{B}(c_1, \ldots, c_m, u)$. Hence, from $\vdash_K (\exists_1 x_{m+1})\mathscr{B}(x_1, \ldots, x_m, x_{m+1})$ and $\mathscr{B}(c_1, \ldots, c_m, v)$, we obtain $u = v$. We have shown $\vdash_K \mathscr{C}(x_1, \ldots, x_n, u) \wedge \mathscr{C}(x_1, \ldots, x_n, v) \Rightarrow u = v$. It is also easy to show that $\vdash_K (\exists x_{n+1})\mathscr{C}(x_1, \ldots, x_{n+1})$. Hence, $\vdash_K (\exists_1 x_{n+1})\mathscr{C}(x_1, \ldots, x_{n+1})$—that is, condition 2′ holds.

Exercises

3.11 Let K be a theory with equality that has the same symbols as S. Show that the following functions are strongly representable in K.

(a) $Z_n(x_1,\dots,x_n) = 0$ [*Hint:* $Z_n(x_1,\dots,x_n) = Z(U_1^n(x_1,\dots,x_n))$.]

(b) $C_k^n(x_1,\dots,x_n) = k$, where k is a fixed natural number [*Hint:* By part (a), we have C_0^n. Assume C_k is strongly representable. Then $C_{k+1}^n(x_1,\dots,x_n) = N(C_k^n(x_1,\dots,x_n))$.]

3.12 Prove that addition and multiplication are strongly representable in S.

If R is a relation of n arguments, then the characteristic function C_R is defined as follows:

$$C_R(x_1,\dots,x_n) = \begin{cases} 0 & \text{if } R(x_1,\dots,x_n) \text{ is true} \\ 1 & \text{if } R(x_1,\dots,x_n) \text{ is false} \end{cases}$$

PROPOSITION 3.12 Let K be a theory with equality having the same symbols as S such that $\vdash_K 0 \neq \bar{1}$. Then, a number-theoretic relation R is expressible in K if and only if C_R is (strongly) representable in K.

Proof If R is expressible in K by a wf $\mathscr{A}(x_1,\dots,x_n)$, then it is easy to verify that C_R is strongly representable in K by the wf $(\mathscr{A}(x_1,\dots,x_n) \wedge x_{n+1} = 0) \vee (\neg\mathscr{A}(x_1,\dots,x_n) \wedge x_{n+1} = \bar{1})$. Conversely, if C_R is representable in K by a wf $\mathscr{B}(x_1,\dots,x_n,x_{n+1})$, then it is easy to show that R is expressible in K by the wf $\mathscr{B}(x_1,\dots,x_n,0)$. [Here, we must use the assumption that $\vdash_K 0 \neq \bar{1}$.]

Exercises

3.13 The *graph* of a function $f(x_1,\dots,x_n)$ is the relation $f(x_1,\dots,x_n) = x_{n+1}$. Show that $f(x_1,\dots,x_n)$ is representable in S if and only if its graph is expressible in S.

3.14 If R_1 and R_2 are relations of n arguments, prove that $C_{\text{not}-R_1} = 1 - C_{R_1}$, $C_{(R_1 \text{ or } R_2)} = C_{R_1} \cdot C_{R_2}$, and $C_{(R_1 \,\&\, R_2)} = C_{R_1} + C_{R_2} - C_{R_1} \cdot C_{R_2}$.

3.15 Show that $f(x_1,\dots,x_n)$ is representable in a theory with equality K if and only if there is a wf $\mathscr{A}(x_1,\dots,x_{n+1})$ such that, for any $k_1, \dots, k_n, m$, if $f(k_1,\dots,k_n) = m$, then $\vdash_K (\forall x_{n+1})(\mathscr{A}(\bar{k}_1,\dots,\bar{k}_n,x_{n+1}) \Leftrightarrow x_{n+1} = \bar{m})$.

3. PRIMITIVE RECURSIVE AND RECURSIVE FUNCTIONS

The study of representability of functions in S leads to a class of number-theoretic functions that turn out to be of great importance in mathematical logic.

DEFINITION

1. The following functions are called *initial functions*.
   (I) *The zero function.* $Z(x) = 0$ for all x.
   (II) *The successor function.* $N(x) = x + 1$ for all x.
   (III) *The projection functions.* $U_i^n(x_1,\dots,x_n) = x_i$ for all $x_1, \dots, x_n$.
2. The following are rules for obtaining new functions from given functions.

(IV) *Substitution*

$$f(x_1,\ldots,x_n) = g(h_1(x_1,\ldots,x_n),\ldots,h_m(x_1,\ldots,x_n))$$

f is said to be obtained by substitution from the functions $g(y_1,\ldots,y_m)$, $h_1(x_1,\ldots,x_n)$, $\ldots$, $h_m(x_1,\ldots,x_n)$.

(V) *Recursion*

$$f(x_1,\ldots,x_n,0) = g(x_1,\ldots,x_n)$$

$$f(x_1,\ldots,x_n,y+1) = h(x_1,\ldots,x_n,y,f(x_1,\ldots,x_n,y))$$

Here, we allow $n = 0$, in which case we have

$$f(0) = k \qquad \text{where } k \text{ is a fixed natural number}$$

$$f(y+1) = h(y,f(y))$$

We shall say that f is obtained from g and h (or, in the case $n = 0$, from h alone) by recursion. The *parameters* of the recursion are x_1, $\ldots$, x_n. Notice that f is well-defined: $f(x_1,\ldots,x_n,0)$ is given by the first equation, and if we already know $f(x_1,\ldots,x_n,y)$, then we can obtain $f(x_1,\ldots,x_n,y+1)$ by the second equation.

(VI) *μ-Operator.* Assume that $g(x_1,\ldots,x_n,y)$ is a function such that for any x_1, $\ldots$, x_n there is at least one y such that $g(x_1,\ldots,x_n,y) = 0$. We denote by $\mu y(g(x_1,\ldots,x_n,y) = 0)$ the least number y such that $g(x_1,\ldots,x_n,y) = 0$. In general, for any relation $\mathrm{R}(x_1,\ldots,x_n,y)$, we denote by $\mu y\mathrm{R}(x_1,\ldots,x_n,y)$ the least y such that $\mathrm{R}(x_1,\ldots,x_n,y)$ is true, if there is any y at all such that $\mathrm{R}(x_1,\ldots,x_n,y)$ holds. Let $f(x_1,\ldots,x_n) = \mu y(g(x_1,\ldots,x_n,y) = 0)$. Then f is said to be obtained from g by means of the μ-operator, if the given assumption about g holds: for any x_1, $\ldots$, x_n, there is at least one y such that $g(x_1,\ldots,x_n,y) = 0$.

3. A function f is said to be *primitive recursive* if and only if it can be obtained from the initial functions by any finite number of substitutions (IV) and recursions (V)—that is, if there is a finite sequence of functions f_0, $\ldots$, f_n such that $f_n = f$ and, for $0 \leqslant i \leqslant n$, either f_i is an initial function or f_i comes from preceding functions in the sequence by an application of rule (IV) (substitution) or rule (V) (recursion).
4. A function f is said to be *recursive* if and only if it can be obtained from the initial functions by any finite number of applications of substitution (IV), recursion (V), and the μ-operator (VI). This differs from the definition above of primitive recursive functions only in the addition of possible applications of the μ-operator, rule (VI). Hence, every primitive recursive function is recursive. We shall see later that the converse is false.

We shall show that the class of recursive functions is identical with the class of functions representable in S. (In the literature, the phrase "general recursive" is sometimes used instead of "recursive".)

First, let us prove that we can add "dummy variables" to and also permute and identify variables in any primitive recursive or recursive function, obtaining a function of the same type.

PROPOSITION 3.13 Let $g(y_1, \ldots, y_k)$ be primitive recursive (or recursive). Let $x_1, \ldots, x_n$ be distinct variables, and, for $1 \leqslant i \leqslant k$, let z_i be one of $x_1, \ldots, x_n$. Then the function f such that $f(x_1, \ldots, x_n) = g(z_1, \ldots, z_k)$ is primitive recursive (or recursive).

Proof Let $z_i = x_{j_i}$ (where $1 \leqslant j_i \leqslant n$). Then $z_i = U_{j_i}^n(x_1, \ldots, x_n)$. Thus,

$$f(x_1, \ldots, x_n) = g(U_{j_1}^n(x_1, \ldots, x_n), U_{j_2}^n(x_1, \ldots, x_n), \ldots, U_{j_k}^n(x_1, \ldots, x_n))$$

and therefore f is primitive recursive (or recursive), since it arises from g, $U_{j_1}^n, \ldots, U_{j_k}^n$ by substitution.

Examples

1. *Adding dummy variables.* If $g(x_1, x_3)$ is primitive recursive and if $f(x_1, x_2, x_3) = g(x_1, x_3)$, then $f(x_1, x_2, x_3)$ is also primitive recursive. In Proposition 3.13, let $z_1 = x_1$ and $z_2 = x_3$. [The new variable x_2 is called a "dummy variable", since its value has no influence on the value of $f(x_1, x_2, x_3)$.]
2. *Permuting variables.* If $g(x_1, x_2)$ is primitive recursive and if $f(x_1, x_2) = g(x_2, x_1)$, then $f(x_1, x_2)$ is also primitive recursive. In Proposition 3.13, let $z_1 = x_2$ and $z_2 = x_1$.
3. *Identifying variables.* If $g(x_1, x_2, x_3)$ is primitive recursive and if $f(x_1, x_2) = g(x_1, x_2, x_1)$, then $f(x_1, x_2)$ is primitive recursive. In Proposition 3.13, let $n = 2$ and let $z_1 = x_1$, $z_2 = x_2$, and $z_3 = x_1$.

COROLLARY 3.14

(a) The zero function $Z_n(x_1, \ldots, x_n) = 0$ is primitive recursive.
(b) The constant function $C_k^n(x_1, \ldots, x_n) = k$, where k is some fixed natural number, is primitive recursive.
(c) The Substitution Rule (IV) can be extended to the case where each h_i may be a function of some but not all of the variables. Likewise, in the Recursion Rule (V), the function g may not involve all of the variables $x_1, \ldots, x_n$, and h may not involve all of the variables $x_1, \ldots, x_n$, y, or $f(x_1, \ldots, x_n, y)$.

Proof

(a) In Proposition 3.13, let g be the zero function Z; then $k = 1$. Take z_1 to be x_1.
(b) For $k = 0$, this is part (a). Assume true for k. Then $C_{k+1}^n(x_1, \ldots, x_n) = N(C_k^n(x_1, \ldots, x_n))$.
(c) By Proposition 3.13, any variables among $x_1, \ldots, x_n$ not present in a function can be added as "dummy variables". For example, if $h(x_1, x_3)$ is given as primitive recursive (or recursive), then $h\#(x_1, x_2, x_3) =$

$h(x_1, x_3) = h(U_1^3(x_1, x_2, x_3), U_3^3(x_1, x_2, x_3))$ is also primitive recursive (or recursive).

PROPOSITION 3.15 The following functions are primitive recursive.

(a) $x + y$
(b) $x \cdot y$
(c) x^y
(d) $\delta(x) = \begin{cases} x - 1 & \text{if } x > 0 \\ 0 & \text{if } x = 0 \end{cases}$
(e) $x \dot{-} y = \begin{cases} x - y & \text{if } x \geqslant y \\ 0 & \text{if } x < y \end{cases}$
(f) $|x - y| = \begin{cases} x - y & \text{if } x \geqslant y \\ y - x & \text{if } x < y \end{cases}$
(g) $\mathrm{sg}(x) = \begin{cases} 0 & \text{if } x = 0 \\ 1 & \text{if } x \neq 0 \end{cases}$
(h) $\overline{\mathrm{sg}}(x) = \begin{cases} 1 & \text{if } x = 0 \\ 0 & \text{if } x \neq 0 \end{cases}$
(i) $x!$
(j) $\min(x, y) =$ minimum of x and y
(k) $\min(x_1, \ldots, x_n)$
(l) $\max(x, y) =$ maximum of x and y
(m) $\max(x_1, \ldots, x_n)$
(n) $\mathrm{rm}(x, y) =$ remainder upon division of y by x
(o) $\mathrm{qt}(x, y) =$ quotient upon division of y by x

Proof

(a) Recursion Rule (V)

$$x + 0 = x \qquad\qquad f(x, 0) = U_1^1(x)$$
$$x + (y + 1) = N(x + y) \quad \text{or} \quad f(x, y + 1) = N(f(x, y))$$

(b)
$$x \cdot 0 = 0 \qquad\qquad g(x, 0) = Z(x)$$
$$x \cdot (y + 1) = (x \cdot y) + x \quad \text{or} \quad g(x, y + 1) = f(g(x, y), x),$$
where f is the addition function

(c)
$$x^0 = 1$$
$$x^{y+1} = (x^y) \cdot x$$

(d)
$$\delta(0) = 0$$
$$\delta(y + 1) = y$$

(e)
$$x \dot{-} 0 = x$$
$$x \dot{-} (y + 1) = \delta(x \dot{-} y)$$

(f) $|x - y| = (x \dot{-} y) + (y \dot{-} x)$ (substitution)

(g)
$$\mathrm{sg}(0) = 0$$
$$\mathrm{sg}(y + 1) = 1$$

(h) $\overline{\mathrm{sg}}(x) = 1 \dot{-} \mathrm{sg}(x)$

(i)
$$0! = 1$$
$$(y + 1)! = (y!) \cdot (y + 1)$$

(j) $\min(x, y) = x \dot{-} (x \dot{-} y)$
(k) Assume $\min(x_1, \ldots, x_n)$ already shown primitive recursive.

$$\min(x_1, \ldots, x_n, x_{n+1}) = \min(\min(x_1, \ldots, x_n), x_{n+1})$$

(l) $\max(x, y) = y + (x \dot{-} y)$
(m) $\max(x_1, \ldots, x_{n+1}) = \max(\max(x_1, \ldots, x_n), x_{n+1})$
(n) $\mathrm{rm}(x, 0) = 0$
$\mathrm{rm}(x, y+1) = N(\mathrm{rm}(x, y)) \cdot \mathrm{sg}(|x - N(\mathrm{rm}(x, y))|)$
(o) $\mathrm{qt}(x, 0) = 0$
$\mathrm{qt}(x, y+1) = \mathrm{qt}(x, y) + \overline{\mathrm{sg}}(|x - N(\mathrm{rm}(x, y))|)$

DEFINITIONS

$$\sum_{y<z} f(x_1, \ldots, x_n, y) = \begin{cases} 0 & \text{if } z = 0 \\ f(x_1, \ldots, x_n, 0) + \cdots + f(x_1, \ldots, x_n, z-1) & \text{if } z > 0 \end{cases}$$

$$\sum_{y \leqslant z} f(x_1, \ldots, x_n, y) = \sum_{y<z+1} f(x_1, \ldots, x_n, y)$$

$$\prod_{y<z} f(x_1, \ldots, x_n, y) = \begin{cases} 1 & \text{if } z = 0 \\ f(x_1, \ldots, x_n, 0) \cdot \ldots \cdot f(x_1, \ldots, x_n, z-1) & \text{if } z > 0 \end{cases}$$

$$\prod_{y \leqslant z} f(x_1, \ldots, x_n, y) = \prod_{y<z+1} f(x_1, \ldots, x_n, y)$$

These *bounded* sums and products are functions of $x_1, \ldots, x_n, z$. We can also define doubly bounded sums and products in terms of the ones already given; for example,

$$\sum_{u<y<v} f(x_1, \ldots, x_n, y) = f(x_1, \ldots, x_n, u+1) + \cdots + f(x_1, \ldots, x_n, v-1)$$

$$= \sum_{y<(v \dot{-} u) \dot{-} 1} f(x_1, \ldots, x_n, y + u + 1)$$

PROPOSITION 3.16 If $f(x_1, \ldots, x_n, y)$ is primitive recursive (or recursive), then all the bounded sums and products defined above are also primitive recursive (or recursive).

Proof Let $g(x_1, \ldots, x_n, z) = \sum_{y<z} f(x_1, \ldots, x_n, y)$. Then, we have the following recursion:

$$g(x_1, \ldots, x_n, 0) = 0$$

$$g(x_1, \ldots, x_n, z+1) = g(x_1, \ldots, x_n, z) + f(x_1, \ldots, x_n, z)$$

If $h(x_1, \ldots, x_n, z) = \sum_{y \leqslant z} f(x_1, \ldots, x_n, y)$, then

$$h(x_1, \ldots, x_n, z) = g(x_1, \ldots, x_n, z+1) \qquad \text{(substitution)}$$

The proofs for bounded products and doubly bounded sums and products are left as exercises.

Example Let $\tau(x)$ be the number of divisors of x, if $x > 0$; let $\tau(0) = 1$. Then $\tau(x)$ is primitive recursive, since

$$\tau(x) = \sum_{y \leqslant x} \overline{\text{sg}}(\text{rm}(y, x))$$

Given expressions for number-theoretic relations, we can apply the connectives of the propositional calculus to them to obtain new relations. We shall use the same symbols ($\neg$, $\wedge$, $\vee$, $\Rightarrow$, $\Leftrightarrow$) for them here, except where confusion may arise between these symbols as they occur in our intuitive metalanguage and as they occur in first-order theories. For example, if $R_1(x_1, \ldots, x_n)$ and $R_2(x_1, \ldots, x_n)$ are relations, then $R_1(x_1, \ldots, x_n) \vee R_2(x_1, \ldots, x_n)$ is a new relation that holds for $x_1, \ldots, x_n$ when and only when $R_1(x_1, \ldots, x_n)$ holds or $R_2(x_1, \ldots, x_n)$ holds. We shall use $(\forall y)_{y<z} R(x_1, \ldots, x_n, y)$ to express the relation: for all y, if y is less than z, then $R(x_1, \ldots, x_n, y)$ holds. We shall use $(\forall y)_{y \leqslant z}$, $(\exists y)_{y<z}$, and $(\exists y)_{y \leqslant z}$ in an analogous way; for example, $(\exists y)_{y<z} R(x_1, \ldots, x_n, y)$ means that there is some $y < z$ such that $R(x_1, \ldots, x_n, y)$ holds. We shall call $(\forall y)_{y<z}$, $(\forall y)_{y \leqslant z}$, $(\exists y)_{y<z}$, and $(\exists y)_{y \leqslant z}$ *bounded quantifiers*. In addition, we define a *bounded μ-operator*:

$$\mu y_{y<z} R(x_1, \ldots, x_n, y) = \begin{cases} \text{the least } y < z \text{ for which } R(x_1, \ldots, x_n, y) \\ \quad \text{holds if there is such a } y \\ z \text{ otherwise} \end{cases}$$

(The value z is chosen in the second case because it is more convenient in later proofs; this choice has no intuitive significance.)

A relation $R(x_1, \ldots, x_n)$ is said to be primitive recursive (or recursive) if and only if its characteristic function $C_R(x_1, \ldots, x_n)$ is primitive recursive (or recursive). In particular, a set A of natural numbers is primitive recursive (or recursive) if and only if its characteristic function $C_A(x)$ is primitive recursive (or recursive).

Examples

1. The relation $x_1 = x_2$ is primitive recursive. Its characteristic function is $\text{sg}(|x_1 - x_2|)$, which is primitive recursive, by Proposition 3.15(f, g).
2. The relation $x_1 < x_2$ is primitive recursive, since its characteristic function is $\overline{\text{sg}}(x_2 \dot{-} x_1)$, which is primitive recursive, by Proposition 3.15(e, h).
3. The relation $x_1 | x_2$ is primitive recursive, since its characteristic function is $\text{sg}(\text{rm}(x_1, x_2))$.
4. The relation $\text{Pr}(x)$, x is a prime, is primitive recursive, since $C_{\text{Pr}}(x) = \text{sg}((\tau(x) \dot{-} 2) + \overline{\text{sg}}(|x - 1|) + \overline{\text{sg}}(|x - 0|))$. Remember that x is a prime if and only if it has exactly two divisors and is not equal to 0 or 1.

PROPOSITION 3.17 Relations obtained from primitive recursive (or recursive) relations by means of the propositional connectives and the bounded quantifiers are also primitive recursive (or recursive). Also, application of the bounded μ-operators $\mu y_{y<z}$ or $\mu y_{y \leqslant z}$ leads from primitive recursive (or recursive) relations to primitive recursive (or recursive) functions.

Proof Assume $R_1(x_1, \ldots, x_n)$ and $R_2(x_1, \ldots, x_n)$ are primitive recursive (or recursive) relations. Then the characteristic functions C_{R_1} and C_{R_2} are primitive recursive (or recursive). But $C_{\neg R_1}(x_1, \ldots, x_n) = 1 \dot{-} C_{R_1}(x_1, \ldots, x_n)$; hence, $\neg R_1$ is primitive recursive (or recursive). Also, $C_{R_1 \vee R_2}(x_1, \ldots, x_n) = C_{R_1}(x_1, \ldots, x_n) \cdot$

$C_{R_2}(x_1, \ldots, x_n)$; so, $R_1 \vee R_2$ is primitive recursive (or recursive). Since all the propositional connectives are definable in terms of $\neg$ and $\vee$, this takes care of them. Now, assume $R(x_1, \ldots, x_n, y)$ is primitive recursive (or recursive). If $Q(x_1, \ldots, x_n, z)$ is the relation $(\exists y)_{y<z} R(x_1, \ldots, x_n, y)$, then it is easy to verify that $C_Q(x_1, \ldots, x_n, z) = \prod_{y<z} C_R(x_1, \ldots, x_n, y)$, which, by Proposition 3.16, is primitive recursive (or recursive). The bounded quantifier $(\exists y)_{y \leqslant z}$ is equivalent to $(\exists y)_{y<z+1}$, which is obtainable from $(\exists y)_{y<z}$ by substitution. Also, $(\forall y)_{y<z}$ is equivalent to $\neg(\exists y)_{y<z}\neg$, and $(\forall y)_{y \leqslant z}$ is equivalent to $\neg(\exists y)_{y \leqslant z}\neg$. Doubly bounded quantifiers, such as $(\exists y)_{u<y<v}$ can be defined by substitution in the bounded quantifiers already mentioned. Finally, $\prod_{u \leqslant y} C_R(x_1, \ldots, x_n, u)$ has the value 1 for all y such that $R(x_1, \ldots, x_n, u)$ is false for all $u \leqslant y$; it has the value 0 as soon as there is some $u \leqslant y$ such that $R(x_1, \ldots, x_n, u)$ holds. Hence, $\sum_{y<z}(\prod_{u \leqslant y} C_R(x_1, \ldots, x_n, u))$ counts the number of integers from 0 up to but not including the first $y < z$ such that $R(x_1, \ldots, x_n, y)$ holds and is z if there is no such y; thus, it is equal to $\mu y_{y<z} R(x_1, \ldots, x_n, y)$ and so the latter function is primitive recursive (or recursive) by Proposition 3.16.

Examples

1. Let $p(x)$ be the xth prime number in ascending order. Thus, $p(0) = 2$, $p(1) = 3$, $p(2) = 5$, and so on. We shall write p_x instead of $p(x)$. Then p_x is a primitive recursive function. In fact,

$$p_0 = 2$$

$$p_{x+1} = \mu y_{y \leqslant (p_x)!+1}(p_x < y \wedge \Pr(y))$$

Notice that the relation $u < y \wedge \Pr(y)$ is primitive recursive. Hence, by Proposition 3.17, the function $\mu y_{y \leqslant v}(u < y \wedge \Pr(y))$ is a primitive recursive function $g(u, v)$. If we substitute the primitive recursive functions z and $(z)! + 1$ for u and v, respectively, in $g(u, v)$, we obtain the primitive recursive function

$$h(z) = \mu y_{y \leqslant z!+1}(z < y \wedge \Pr(y))$$

and the right-hand side of the second equation above is $h(p_x)$; hence, we have an application of the Recursion Rule (V). The bound $(p_x)! + 1$ on the first prime after p_x is obtained from Euclid's proof of the infinitude of primes (see Exercise 3.23).

2. Every positive integer x has a unique factorization into prime powers: $x = p_0^{a_0} p_1^{a_1} \ldots p_k^{a_k}$. Let us denote by $(x)_j$ the exponent a_j in this factorization. If $x = 1$, $(x)_j = 0$ for all j. If $x = 0$, we arbitrarily let $(x)_j = 0$ for all j. Then the function $(x)_j$ is primitive recursive, since $(x)_j = \mu y_{y<x}(p_j^y | x \wedge \neg(p_j^{y+1} | x))$.

3. For $x > 0$, let $\ell h(x)$ be the number of nonzero exponents in the factorization of x into powers of primes or, equivalently, the number of distinct primes that divide x. Let $\ell h(0) = 0$. Then ℓh is primitive recursive. To see this, let $R(x, y)$ be the primitive recursive relation $\Pr(y) \wedge y | x \wedge x \neq 0$. Then $\ell h(x) = \sum_{y \leqslant x} \overline{\text{sg}}(C_R(x, y))$. Observe that this yields the special cases $\ell h(0) = \ell h(1) = 0$.

4. If the number $x = 2^{a_0} 3^{a_1} \ldots p_k^{a_k}$ is used to "represent" or "encode" the sequence of positive integers $a_0, a_1, \ldots, a_k$, and $y = 2^{b_0} 3^{b_1} \ldots p_m^{b_m}$ "represents" the sequence of positive integers $b_0, b_1, \ldots, b_m$, then the number

$$x * y = 2^{a_0}3^{a_1} \cdots p_k{}^{a_k} p_{k+1}{}^{b_0} p_{k+2}{}^{b_1} \cdots p_{k+1+m}{}^{b_m}$$

"represents" the new sequence $a_0, a_1, \ldots, a_k, b_0, b_1, \ldots, b_m$ obtained by juxtaposing the two sequences. Note that $k + 1 = \ell h(x)$, $m + 1 = \ell h(y)$, and $b_j = (y)_j$. Hence,

$$x * y = x \cdot \prod_{j < \ell h(y)} (p_{\ell h(x)+j})^{(y)_j}$$

and, thus, $*$ is a primitive recursive function, which is called the *juxtaposition* function. It is not difficult to show that $x * (y * z) = (x * y) * z$ as long as $y \neq 0$ (which will be the only case of interest to us). Therefore, there is no harm in omitting parentheses when writing two or more applications of $*$. Also observe that $x * 0 = x * 1 = x$.

Exercises

3.16 Assume that $R(x_1, \ldots, x_n, y)$ is a primitive recursive (or recursive) relation. Using Proposition 3.17, prove the following.

(a) $(\exists y)_{u<y<v} R(x_1, \ldots, x_n, y)$, $(\exists y)_{u \leqslant y \leqslant v} R(x_1, \ldots, x_n, y)$, and $(\exists y)_{u \leqslant y < v} R(x_1, \ldots, x_n, y)$ are primitive recursive (or recursive) relations.

(b) $(\mu y)_{u<y<v} R(x_1, \ldots, x_n, y)$, $(\mu y)_{u \leqslant y \leqslant v} R(x_1, \ldots, x_n, y)$, and $(\mu y)_{u \leqslant y < v} R(x_1, \ldots, x_n, y)$ are primitive recursive (or recursive) functions.

3.17 Show that the intersection, union, and complement of primitive recursive (or recursive) sets are also primitive recursive (or recursive). Prove that every finite set is primitive recursive.

3.18 Prove that a function $f(x_1, \ldots, x_n)$ is recursive if and only if its representing relation $f(x_1, \ldots, x_n) = y$ is a recursive relation.

3.19 Let $[\sqrt{n}]$ denote the greatest integer $\leqslant \sqrt{n}$, and let $\Pi(n)$ denote the number of primes $\leqslant n$. Show that $[\sqrt{n}]$ and $\Pi(n)$ are primitive recursive.

3.20 Let e be the base of the natural logarithms. Show that $[ne]$, the greatest integer $\leqslant ne$, is a primitive recursive function of n.

3.21 Let $RP(y, z)$ hold if and only if y and z are relatively prime; that is, y and z have no common factor greater than 1. Let $\varphi(n)$ be the number of positive integers $\leqslant n$ that are relatively prime to n. Prove that RP and φ are primitive recursive.

3.22 Show that, in the definition of the primitive recursive functions, one need not assume that $Z(x) = 0$ is one of the initial functions.

3.23 Prove that $p_{k+1} \leqslant (p_0 p_1 \cdots p_k) + 1$. Hence, $p_{k+1} \leqslant p_k! + 1$.

For use in the further study of recursive functions, we prove the following theorem on definition by cases.

PROPOSITION 3.18 Let

$$f(x_1, \ldots, x_n) = \begin{cases} g_1(x_1, \ldots, x_n) & \text{if } R_1(x_1, \ldots, x_n) \text{ holds} \\ g_2(x_1, \ldots, x_n) & \text{if } R_2(x_1, \ldots, x_n) \text{ holds} \\ \vdots & \\ g_k(x_1, \ldots, x_n) & \text{if } R_k(x_1, \ldots, x_n) \text{ holds} \end{cases}$$

If the functions $g_1, \ldots, g_k$ and the relations $R_1, \ldots, R_k$ are primitive recursive (or recursive), and if, for any $x_1, \ldots, x_n$, exactly one of the relations $R_1(x_1, \ldots, x_n), \ldots, R_k(x_1, \ldots, x_n)$ is true, then f is primitive recursive (or recursive).

Proof $f(x_1, \ldots, x_n) = g_1(x_1, \ldots, x_n) \cdot \overline{\text{sg}}(C_{R_1}(x_1, \ldots, x_n)) + \cdots + g_k(x_1, \ldots, x_n) \cdot \overline{\text{sg}}(C_{R_k}(x_1, \ldots, x_n))$

Exercises

3.24 Show that in Proposition 3.18 it is not necessary to assume that R_k is primitive recursive (or recursive).

3.25 Let

$$f(x) = \begin{cases} x^2 & \text{if } x \text{ is even} \\ x + 1 & \text{if } x \text{ is odd} \end{cases}$$

Prove that f is primitive recursive.

3.26 Let

$$h(x) = \begin{cases} 2 & \text{if Fermat's Last Theorem is true} \\ 1 & \text{if Fermat's Last Theorem is false} \end{cases}$$

Is h primitive recursive?

It is often important to have available a primitive recursive one–one correspondence between the set of ordered pairs of natural numbers and the set of natural numbers. We shall enumerate the pairs as follows:

$$\overbrace{(0,0)},\quad \overbrace{(0,1),(1,0),(1,1)}\quad \overbrace{(0,2),(2,0),(1,2),(2,1),(2,2)},\ldots$$

After we have enumerated all the pairs having components $\leqslant k$, we then add a new group of all the new pairs involving components $\leqslant k + 1$ in the following order: $(0, k+1), (k+1, 0), (1, k+1), (k+1, 1), \ldots, (k, k+1), (k+1, k), (k+1, k+1)$. Now, if $x < y$, then (x, y) occurs before (y, x) and both are in the $(y+1)$th group. (Note that we start from 1 in counting groups.) The first y groups contain y^2 pairs, and (x, y) is the $(2x+1)$th pair in the $(y+1)$th group. Hence, (x, y) is the $(y^2 + 2x + 1)$th pair in the ordering, and (y, x) is the $(y^2 + 2x + 2)$th pair. On the other hand, if $x = y$, (x, y) is the $((x+1)^2)$th pair. This justifies the following definition, in which $\sigma^2(x, y)$ denotes the place of the pair (x, y) in the above enumeration, with $(0, 0)$ considered to be in the 0th place:

$$\sigma^2(x, y) = (\text{sg}(x \dot{-} y)) \cdot (x^2 + 2y + 1) + (\overline{\text{sg}}(x \dot{-} y)) \cdot (y^2 + 2x)$$

Clearly σ^2 is primitive recursive.

Let us define inverse functions σ_1^2 and σ_2^2 such that $\sigma_1^2(\sigma^2(x, y)) = x$, $\sigma_2^2(\sigma^2(x, y)) = y$, and $\sigma^2(\sigma_1^2(z), \sigma_2^2(z)) = z$. Thus, $\sigma_1^2(z)$ and $\sigma_2^2(z)$ are the first and second components, respectively, of the zth ordered pair in the given enumeration. Note first that $\sigma_1^2(0) = 0$, $\sigma_2^2(0) = 0$,

$$\sigma_1^2(n+1) = \begin{cases} \sigma_2^2(n) & \text{if } \sigma_1^2(n) < \sigma_2^2(n) \\ \sigma_2^2(n) + 1 & \text{if } \sigma_1^2(n) > \sigma_2^2(n) \\ 0 & \text{if } \sigma_1^2(n) = \sigma_2^2(n) \end{cases}$$

and

$$\sigma_2^2(n+1) = \begin{cases} \sigma_1^2(n) & \text{if } \sigma_1^2(n) \neq \sigma_2^2(n) \\ \sigma_1^2(n)+1 & \text{if } \sigma_1^2(n) = \sigma_2^2(n) \end{cases}$$

Hence,

$$\begin{aligned} \sigma_1^2(n+1) &= \sigma_2^2(n)\cdot(\mathrm{sg}(\sigma_2^2(n) \dot{-} \sigma_1^2(n))) + (\sigma_2^2(n)+1)\cdot(\mathrm{sg}(\sigma_1^2(n) \dot{-} \sigma_2^2(n))) \\ &= \varphi(\sigma_1^2(n), \sigma_2^2(n)) \\ \sigma_2^2(n+1) &= \mathrm{sg}(|\sigma_1^2(n) - \sigma_2^2(n)|)\cdot\sigma_1^2(n) + \overline{\mathrm{sg}}(|\sigma_1^2(n) - \sigma_2^2(n)|)\cdot(\sigma_1^2(n)+1) \\ &= \psi(\sigma_1^2(n), \sigma_2^2(n)) \end{aligned}$$

where φ and ψ are primitive recursive functions. Thus, σ_1^2 and σ_2^2 are defined recursively at the same time. We can show that σ_1^2 and σ_2^2 are primitive recursive in the following devious way: let $h(u) = 2^{\sigma_1^2(u)}3^{\sigma_2^2(u)}$. Now, h is primitive recursive, since $h(0) = 2^{\sigma_1^2(0)}3^{\sigma_2^2(0)} = 2^0 \cdot 3^0 = 1$, and $h(n+1) = 2^{\sigma_1^2(n+1)} \cdot 3^{\sigma_2^2(n+1)} = 2^{\varphi(\sigma_1^2(n), \sigma_2^2(n))}3^{\psi(\sigma_1^2(n), \sigma_2^2(n))} = 2^{\varphi((h(n))_0, (h(n))_1)}3^{\psi((h(n))_0, (h(n))_1)}$. Remembering that the function $(x)_i$ is primitive recursive (see Example 2 on page 138), we conclude by Recursion Rule (V) that h is primitive recursive. But $\sigma_1^2(x) = (h(x))_0$ and $\sigma_2^2(x) = (h(x))_1$; by substitution, σ_1^2 and σ_2^2 are primitive recursive.

One–one primitive recursive correspondences between all n-tuples of natural numbers and all natural numbers can be defined step by step, using induction on n. For $n = 2$, it has already been established. Assume that, for $n = k$, we have primitive recursive functions $\sigma^k(x_1, \ldots, x_k)$, $\sigma_1^k(x)$, $\ldots$, $\sigma_k^k(x)$ such that $\sigma_i^k(\sigma^k(x_1, \ldots, x_n)) = x_i$ for $1 \leqslant i \leqslant k$, and $\sigma^k(\sigma_1^k(x), \ldots, \sigma_k^k(x)) = x$. Now, for $n = k+1$, define $\sigma^{k+1}(x_1, \ldots, x_k, x_{k+1}) = \sigma^2(\sigma^k(x_1, \ldots, x_k), x_{k+1})$, $\sigma_i^{k+1}(x) = \sigma_i^k(\sigma_1^2(x))$ for $1 \leqslant i \leqslant k$, and $\sigma_{k+1}^{k+1}(x) = \sigma_2^2(x)$. Then σ^{k+1}, σ_1^{k+1}, $\ldots$, σ_{k+1}^{k+1} are all primitive recursive, and we leave it as an exercise to verify that $\sigma_i^{k+1}(\sigma^{k+1}(x_1, \ldots, x_{k+1})) = x_i$ for $1 \leqslant i \leqslant k+1$, and $\sigma^{k+1}(\sigma_1^{k+1}(x), \ldots, \sigma_{k+1}^{k+1}(x)) = x$.

It is often convenient to define functions by a recursion in which the value of $f(x_1, \ldots, x_n, y+1)$ depends not only upon $f(x_1, \ldots, x_n, y)$ but also upon several or all values of $f(x_1, \ldots, x_n, u)$, with $u \leqslant y$. This type of recursion is called a course-of-values recursion. Let $f\#(x_1, \ldots, x_n, y) = \prod_{u<y} p_u^{f(x_1, \ldots, x_n, u)}$. Note that f can be obtained from $f\#$ as follows: $f(x_1, \ldots, x_n, y) = (f\#(x_1, \ldots, x_n, y+1))_y$.

PROPOSITION 3.19 If $h(x_1, \ldots, x_n, y, z)$ is primitive recursive (or recursive) and $f(x_1, \ldots, x_n, y) = h(x_1, \ldots, x_n, y, f\#(x_1, \ldots, x_n. y))$, then f is primitive recursive (or recursive).

Proof

$$\begin{aligned} f\#(x_1, \ldots, x_n, 0) &= 1 \\ f\#(x_1, \ldots, x_n, y+1) &= f\#(x_1, \ldots, x_n, y) \cdot p_y^{f(x_1, x_2, \ldots, x_n, y)} \\ &= f\#(x_1, \ldots, x_n, y) \cdot (p_y)^{h(x_1, \ldots, x_n, y, f\#(x_1, \ldots, x_n, y))} \end{aligned}$$

Thus, by the Recursion Rule (V), $f\#$ is primitive recursive (or recursive), and $f(x_1, \ldots, x_n, y) = (f\#(x_1, \ldots, x_n, y+1))_y$.

Example The Fibonacci sequence is defined as follows: $f(0) = 1, f(1) = 1$, and $f(k+2) = f(k) + f(k+1)$ for $k \geqslant 0$. Then f is primitive recursive, since

$$f(k) = \overline{\text{sg}}(k) + \overline{\text{sg}}(|k-1|) + ((f\#(k))_{k \dot{-} 1} + (f\#(k))_{k \dot{-} 2}) \cdot \text{sg}(k \dot{-} 1)$$

The function

$$h(y,z) = \overline{\text{sg}}(y) + \overline{\text{sg}}(|y-1|) + ((z)_{y \dot{-} 1} + (z)_{y \dot{-} 2}) \cdot \text{sg}(y \dot{-} 1)$$

is primitive recursive, and

$$f(k) = h(k, f\#(k))$$

Exercise

3.27 Let $g(0) = 2$, $g(1) = 4$, and $g(k+2) = 3g(k+1) \dot{-} (2g(k)+1)$. Show that g is primitive recursive.

COROLLARY 3.20 If $\text{H}(x_1, \ldots, x_n, y, z)$ is a primitive recursive (or recursive) relation and $\text{R}(x_1, \ldots, x_n, y)$ holds if and only if $\text{H}(x_1, \ldots, x_n, y, (C_{\text{R}})\#(x_1, \ldots, x_n, y))$, where C_{R} is the characteristic function of R, then R is primitive recursive (or recursive).

Proof $C_{\text{R}}(x_1, \ldots, x_n, y) = C_{\text{H}}(x_1, \ldots, x_n, y, (C_{\text{R}})\#(x_1, \ldots, x_n, y))$, where the characteristic function C_{H} of H is primitive recursive (or recursive). Hence, by Proposition 3.19, C_{R} is primitive recursive (or recursive) and, therefore, so is the relation R.

Proposition 3.19 and Corollary 3.20 will be drawn upon heavily in the sequel. They are applicable whenever the value of a function or relation for y is defined in terms of values for arguments less than y. Notice in this connection that $\text{R}(x_1, \ldots, x_n, u)$ is equivalent to $C_{\text{R}}(x_1, \ldots, x_n, u) = 0$, which, in turn, for $u < y$, is equivalent to $((C_{\text{R}})\#(x_1, \ldots, x_n, y))_u = 0$.

Exercises

3.28 Prove that the set of recursive functions is denumerable.
3.29 If $f_0, f_1, f_2, \ldots$ is an enumeration of all primitive recursive functions (or all recursive functions) of one variable, prove that the function $f_x(y)$ is not primitive recursive (or recursive).

PROPOSITION 3.21 (GÖDEL'S β-FUNCTION) Let $\beta(x_1, x_2, x_3) = \text{rm}(1 + (x_3+1) \cdot x_2, x_1)$. Then β is primitive recursive, by Proposition 3.15(n). Also, $\beta(x_1, x_2, x_3)$ is strongly representable in S by the wf $\text{Bt}(x_1, x_2, x_3, x_4)$:

$$(\exists w)(x_1 = (1 + (x_3+1) \cdot x_2) \cdot w + x_4 \wedge x_4 < 1 + (x_3+1) \cdot x_2)$$

Proof By Proposition 3.11, $\vdash (\exists_1 x_4)\text{Bt}(x_1, x_2, x_3, x_4)$. Assume $\beta(k_1, k_2, k_3) = k_4$. Then $k_1 = (1 + (k_3+1) \cdot k_2) \cdot k + k_4$ for some k, and $k_4 < 1 + (k_3+1) \cdot k_2$. So, $\vdash \overline{k_1} = (\overline{1} + (\overline{k_3} + \overline{1}) \cdot \overline{k_2}) \cdot \overline{k} + \overline{k_4}$, by Proposition 3.6(a); and $\vdash \overline{k_4} < \overline{1} + (\overline{k_3} + \overline{1}) \cdot \overline{k_2}$ by the expressibility of $<$ and Proposition 3.6(a). Hence,

$\vdash \overline{k_1} = (\overline{1} + (\overline{k_3} + \overline{1})\cdot\overline{k_2})\cdot\overline{k} + \overline{k_4} \wedge \overline{k_4} < \overline{1} + (\overline{k_3} + \overline{1})\overline{k_2}$ from which, by Rule E4, $\vdash \mathrm{Bt}(\overline{k_1}, \overline{k_2}, \overline{k_3}, \overline{k_4})$. Thus, Bt strongly represents β in S.

PROPOSITION 3.22 For any sequence of natural numbers $k_0, k_1, \ldots, k_n$, there exist natural numbers b and c such that $\beta(b, c, i) = k_i$ for $0 \leqslant i \leqslant n$.

Proof Let $j = \max(n, k_0, k_1, \ldots, k_n)$ and let $c = j!$. Consider the numbers $u_i = 1 + (i + 1)c$ for $0 \leqslant i \leqslant n$; they have no factors in common other than 1. If p were a prime dividing both $1 + (i + 1)c$ and $1 + (m + 1)c$ with $0 \leqslant i < m \leqslant n$, then p would divide their difference $(m - i)c$; now, p does not divide c, since, in that case, p would divide both $(i + 1)c$ and $1 + (i + 1)c$ and so would divide 1, which is impossible. Hence, p also does not divide $(m - i)$; for, $m - i \leqslant n \leqslant j$, and so, $m - i$ divides $j! = c$; if p divided $m - i$, then p would divide c. Hence, p does not divide $(m - i)c$, which yields a contradiction. Thus, the numbers u_i, $0 \leqslant i \leqslant n$, are relatively prime in pairs. Also, for $0 \leqslant i \leqslant n$, $k_i \leqslant j \leqslant j! = c < 1 + (i + 1)c = u_i$; that is, $k_i < u_i$. Now, by the Chinese Remainder Theorem (see Exercise 3.33), there is a number $b < u_0 u_1 \cdots u_n$ such that $\mathrm{rm}(u_i, b) = k_i$ for $0 \leqslant i \leqslant n$. But, $\beta(b, c, i) = \mathrm{rm}(1 + (i + 1)c, b) = \mathrm{rm}(u_i, b) = k_i$.

Propositions 3.21 and 3.22 enable us to express within S assertions about finite sequences of natural numbers, and this ability is crucial in part of the proof of the following fundamental theorem.

PROPOSITION 3.23 Every recursive function is representable in S.

Proof The initial functions Z, N, and U_i^n are representable in S, by Examples 1–3 on pages 130–131. The Substitution Rule (IV) does not lead out of the class of representable functions, by Example 4 on page 131.

For the Recursion Rule (V), assume that $g(x_1, \ldots, x_n)$ and $h(x_1, \ldots, x_n, y, z)$ are representable in S by wfs $\mathscr{A}(x_1, \ldots, x_n, x_{n+1})$ and $\mathscr{B}(x_1, \ldots, x_{n+3})$, respectively, and let

$$\text{(I)} \qquad \begin{aligned} f(x_1, \ldots, x_n, 0) &= g(x_1, \ldots, x_n) \\ f(x_1, \ldots, x_n, y + 1) &= h(x_1, \ldots, x_n, y, f(x_1, \ldots, x_n, y)) \end{aligned}$$

Now, $f(x_1, \ldots, x_n, y) = z$ if and only if there is a finite sequence of numbers $b_0, \ldots, b_y$ such that $b_0 = g(x_1, \ldots, x_n)$, $b_{w+1} = h(x_1, \ldots, x_n, w, b_w)$ for $w + 1 \leqslant y$, and $b_y = z$. But, by Proposition 3.22, reference to finite sequences can be paraphrased in terms of the function β, and, by Proposition 3.21, β is representable in S by the wf $\mathrm{Bt}(x_1, x_2, x_3, x_4)$.

We shall show that $f(x_1, \ldots, x_n, x_{n+1})$ is representable in S by the following wf $\mathscr{C}(x_1, \ldots, x_{n+2})$:

$$(\exists u)(\exists v)[((\exists w)(\mathrm{Bt}(u, v, 0, w) \wedge \mathscr{A}(x_1, \ldots, x_n, w))) \wedge \mathrm{Bt}(u, v, x_{n+1}, x_{n+2}) \wedge$$
$$(\forall w)(w < x_{n+1} \Rightarrow (\exists y)(\exists z)(\mathrm{Bt}(u, v, w, y) \wedge \mathrm{Bt}(u, v, w', z) \wedge \mathscr{B}(x_1, \ldots, x_n, w, y, z)))]$$

(i) First, assume that $f(k_1, \ldots, k_n, p) = m$. We wish to show that $\vdash \mathscr{C}(\overline{k_1}, \ldots, \overline{k_n}, \overline{p}, \overline{m})$. If $p = 0$, then $m = g(k_1, \ldots, k_n)$. Consider the sequence consisting of m alone. By Proposition 3.22, there exist b and c such that $\beta(b, c, 0) = m$. Hence, by Proposition 3.21,

$$(\Xi) \quad \vdash \text{Bt}(\bar{b}, \bar{c}, 0, \bar{m})$$

Also, since $m = g(k_1, \ldots, k_n)$, we have $\vdash \mathscr{A}(\bar{k}_1, \ldots, \bar{k}_n, \bar{m})$. Hence, by Rule E4,

$$(\Xi\Xi) \quad \vdash (\exists w)(\text{Bt}(\bar{b}, \bar{c}, 0, w) \wedge \mathscr{A}(\bar{k}_1, \ldots, \bar{k}_n, w))$$

In addition, since $\vdash w \not< 0$, a tautology and Gen yield

$$(\Xi\Xi\Xi) \quad (\forall w)(w < 0 \Rightarrow (\exists y)(\exists z)(\text{Bt}(\bar{b}, \bar{c}, w, y) \wedge \text{Bt}(\bar{b}, \bar{c}, w', z) \wedge \mathscr{B}(\bar{k}_1, \ldots, \bar{k}_n, w, y, z))$$

Applying Rule E4 to the conjunction of (Ξ), $(\Xi\Xi)$, and $(\Xi\Xi\Xi)$, we obtain $\vdash \mathscr{C}(\bar{k}_1, \ldots, \bar{k}_n, 0, \bar{m})$. Now, for $p > 0$, $f(k_1, \ldots, k_n, p)$ is calculated from the equations (I) in $p + 1$ steps. Let $r_i = f(k_1, \ldots, k_n, i)$. For the sequence of numbers $r_0, r_1, \ldots, r_p$, there are, by Proposition 3.22, numbers b and c such that $\beta(b, c, i) = r_i$ for $0 \leqslant i \leqslant p$. Hence, by Proposition 3.21, $\vdash \text{Bt}(\bar{b}, \bar{c}, \bar{i}, \bar{r}_i)$. In particular, $\beta(b, c, 0) = r_0 = f(k_1, \ldots, k_n, 0) = g(k_1, \ldots, k_n)$. Therefore, $\vdash \text{Bt}(\bar{b}, \bar{c}, 0, \bar{r}_0) \wedge \mathscr{A}(\bar{k}_1, \ldots, \bar{k}_n, \bar{r}_0)$, and, by Rule E4, (1) $\vdash (\exists w)(\text{Bt}(\bar{b}, \bar{c}, 0\, w) \wedge \mathscr{A}(\bar{k}_1, \ldots, \bar{k}_n, w))$. Since $r_p = f(k_1, \ldots, k_n, p) = m$, $\beta(b, c, p) = m$; hence, (2) $\vdash \text{Bt}(\bar{b}, \bar{c}, \bar{p}, \bar{m})$. For $0 \leqslant i \leqslant p - 1$, $\beta(b, c, i) = r_i = f(k_1, \ldots, k_n, i)$, and $\beta(b, c, i + 1) = r_{i+1} = f(k_1, \ldots, k_n, i + 1) = h(k_1, \ldots, k_n, i, f(k_1, \ldots, k_n, i)) = h(k_1, \ldots, k_n, i, r_i)$. Hence, $\vdash \text{Bt}(\bar{b}, \bar{c}, \bar{i}, \bar{r}_i) \wedge \text{Bt}(\bar{b}, \bar{c}, \bar{i}', \bar{r}_{i+1}) \wedge \mathscr{B}(\bar{k}_1, \ldots, \bar{k}_n, \bar{i}, \bar{r}_i, \bar{r}_{i+1})$. By Rule E4, $\vdash (\exists y)(\exists z)(\text{Bt}(\bar{b}, \bar{c}, \bar{i}, y) \wedge \text{Bt}(\bar{b}, \bar{c}, \bar{i}', z) \wedge \mathscr{B}(\bar{k}_1, \ldots, \bar{k}_n, \bar{i}, y, z))$. So, by Proposition 3.8(b′), we have (3),

$$\vdash (\forall w)(w < \bar{p} \Rightarrow (\exists y)(\exists z)(\text{Bt}(\bar{b}, \bar{c}, w, y) \wedge \text{Bt}(\bar{b}, \bar{c}, w', z) \wedge \mathscr{B}(\bar{k}_1, \ldots, \bar{k}_n, w, y, z)))$$

Then, applying Rule E4 twice to the conjunction of (1), (2), and (3), we obtain $\vdash \mathscr{C}(\bar{k}_1, \ldots, \bar{k}_n, \bar{p}, \bar{m})$. Thus, we have verified clause 1 of the definition of representability in S (see page 130).

(ii) We must show that $\vdash (\exists_1 x_{n+2})\mathscr{C}(\bar{k}_1, \ldots, \bar{k}_n, \bar{p}, x_{n+2})$. The proof is by induction on p in the metalanguage. Notice that, by what we have proved above, it suffices to prove only uniqueness. The case for $p = 0$ is left as an easy exercise. Assume $\vdash (\exists_1 x_{n+2})\mathscr{C}(\bar{k}_1, \ldots, \bar{k}_n, \bar{p}, x_{n+2})$. Let $\alpha = g(k_1, \ldots, k_n)$, $\beta = f(k_1, \ldots, k_n, p)$, and $\gamma = f(k_1, \ldots, k_n, p + 1) = h(k_1, \ldots, k_n, p, \beta)$. Then,

(1) $\vdash \mathscr{B}(\bar{k}_1, \ldots, \bar{k}_n, \bar{p}, \bar{\beta}, \bar{\gamma})$
(2) $\vdash \mathscr{A}(\bar{k}_1, \ldots, \bar{k}_n, \bar{\alpha})$
(3) $\vdash \mathscr{C}(\bar{k}_1, \ldots, \bar{k}_n, \bar{p}, \bar{\beta})$
(4) $\vdash \mathscr{C}(\bar{k}_1, \ldots, \bar{k}_n, \overline{p + 1}, \bar{\gamma})$
(5) $\vdash (\exists_1 x_{n+2})\mathscr{C}(\bar{k}_1, \ldots, \bar{k}_n, \bar{p}, x_{n+2})$

Assume

(6) $\mathscr{C}(\bar{k}_1, \ldots, \bar{k}_n, \overline{p + 1}, x_{n+2})$

We must prove $x_{n+2} = \bar{\gamma}$. From (6), by Rule C,

(a) $(\exists w)(\text{Bt}(b, c, 0, w) \wedge \mathscr{A}(\bar{k}_1, \ldots, \bar{k}_n, w))$
(b) $\text{Bt}(b, c, \overline{p + 1}, x_{n+2})$
(c) $(\forall w)(w < \overline{p + 1} \Rightarrow (\exists y)(\exists z)(\text{Bt}(b, c, w, y) \wedge \text{Bt}(b, c, w', z)$
$\wedge \mathscr{B}(\bar{k}_1, \ldots, \bar{k}_n, w, y, z))$

From (c),

(d) $(\forall w)(w < \bar{p} \Rightarrow (\exists y)(\exists z)(\mathrm{Bt}(b, c, w, y) \wedge \mathrm{Bt}(b, c, w', z) \wedge \mathscr{B}(\bar{k}_1, \ldots, \bar{k}_n, w, y, z))$

From (c) by Rule A4 and Rule C,

(e) $\mathrm{Bt}(b, c, \bar{p}, d) \wedge \mathrm{Bt}(b, c, \overline{p+1}, e) \wedge \mathscr{B}(\bar{k}_1, \ldots, \bar{k}_n, \bar{p}, d, e)$

From (a), (d), and (e),

(f) $\mathscr{C}(\bar{k}_1, \ldots, \bar{k}_n, \bar{p}, d)$

From (f), (5), and (3),

(g) $d = \bar{\beta}$

From (e) and (g),

(h) $\mathscr{B}(\bar{k}_1, \ldots, \bar{k}_n, \bar{p}, \bar{\beta}, e)$

Since $\mathscr{B}$ represents h, we obtain from (1) and (h),

(i) $\bar{\gamma} = e$

From (e) and (i),

(j) $\mathrm{Bt}(b, c, \overline{p+1}, \bar{\gamma})$

From (b), (j), and Proposition 3.21,

(k) $x_{n+2} = \bar{\gamma}$

This completes the induction.

The μ-Operator (VI). Let us assume that, for any $x_1, \ldots, x_n$, there is some y such that $g(x_1, \ldots, x_n, y) = 0$, and let us assume g is representable in S by a wf $\mathscr{D}(x_1, \ldots, x_{n+2})$. Let $f(x_1, \ldots, x_n) = \mu y(g(x_1, \ldots, x_n, y) = 0)$. Then we shall show that f is representable in S by the wf $\mathscr{E}(x_1, \ldots, x_{n+1})$:

$$\mathscr{D}(x_1, \ldots, x_{n+1}, 0) \wedge (\forall y)(y < x_{n+1} \Rightarrow \neg\mathscr{D}(x_1, \ldots, x_n, y, 0))$$

Assume $f(k_1, \ldots, k_n) = m$. Then $g(k_1, \ldots, k_n, m) = 0$ and, for $k < m$, $g(k_1, \ldots, k_n, k) \neq 0$. So, $\vdash \mathscr{D}(\bar{k}_1, \ldots, \bar{k}_n, \bar{m}, 0)$, and, for $k < m$, $\vdash \neg\mathscr{D}(\bar{k}_1, \ldots, \bar{k}_n, \bar{k}, 0)$. By Proposition 3.8(b′), $\vdash (\forall y)(y < \bar{m} \Rightarrow \neg\mathscr{D}(\bar{k}_1, \ldots, \bar{k}_n, y, 0))$. Hence, $\vdash \mathscr{E}(\bar{k}_1, \ldots, \bar{k}_n, \bar{m})$. We must also show: $\vdash (\exists_1 x_{n+1})\mathscr{E}(\bar{k}_1, \ldots, \bar{k}_n, x_{n+1})$. It suffices to prove the uniqueness. Assume $\mathscr{D}(\bar{k}_1, \ldots, \bar{k}_n, u, 0) \wedge (\forall y)(y < u \Rightarrow \neg\mathscr{D}(\bar{k}_1, \ldots, \bar{k}_n, y, 0))$. Now, $\vdash \bar{m} < u \vee \bar{m} = u \vee u < \bar{m}$. Since $\vdash \mathscr{D}(\bar{k}_1, \ldots, \bar{k}_n, \bar{m}, 0)$, we cannot have $\bar{m} < u$. Since $\vdash (\forall y)(y < \bar{m} \Rightarrow \neg\mathscr{D}(\bar{k}_1, \ldots, \bar{k}_n, y, 0))$, we cannot have $u < \bar{m}$. Hence, $u = \bar{m}$. This shows the uniqueness.

Thus, we have proved that all recursive functions are representable in S.

COROLLARY 3.24 Every recursive relation is expressible in S.

Proof Let $\mathrm{R}(x_1, \ldots, x_n)$ be a recursive relation. Then its characteristic function C_{R} is recursive. By Proposition 3.23, C_{R} is representable in S and, therefore, by Proposition 3.12, R is expressible in S.

Exercises

3.30[A] (a) Show that, if a and b are relatively prime natural numbers, then there is a natural number c such that $ac \equiv 1 \pmod{b}$. [Two numbers a and b are said to be *relatively* *prime* if their greatest common divisor is 1. In general, $x \equiv y \pmod{z}$ means that x and y leave the same remainder upon division by z or, equivalently, that $x - y$ is divisible by z. This exercise amounts to showing that there exist integers u and v such that $1 = au + bv$.]

(b) Prove the Chinese Remainder Theorem: if $x_1, \ldots, x_k$ are relatively prime in pairs and $y_1, \ldots, y_k$ are any natural numbers, there is a natural number z such that $z \equiv y_1 \pmod{x_1}, \ldots, z \equiv y_k \pmod{x_k}$. Moreover, any two such z's differ by a multiple of $x_1 \cdot \ldots \cdot x_k$. [*Hint:* Let $x = x_1 \cdot \ldots \cdot x_k$ and let $x = w_1 x_1 = w_2 x_2 = \ldots = w_k x_k$. Then, for $1 \leqslant j \leqslant k$, w_j is relatively prime to x_j, and so, by part (a), there is some z_j such that $w_j z_j \equiv 1 \pmod{x_j}$. Now let $z = w_1 z_1 y_1 + w_2 z_2 y_2 + \cdots + w_k z_k y_k$. Then $z \equiv w_j z_j y_j \equiv y_j \pmod{x_j}$. In addition, the difference between any two such solutions is divisible by $x_1, \ldots, x_k$ and, hence, by $x_1 \cdot \ldots \cdot x_k$.]

3.31 Call a relation $R(x_1, \ldots, x_n)$ *arithmetical* if it is the interpretation of some wf $\mathscr{A}(x_1, \ldots, x_n)$ of S with respect to the standard model. Show that every recursive relation is arithmetical. [*Hint:* Use Corollary 3.24.]

3.32 For every theory with equality K that has the same symbols as S, prove that representability in K implies strong representability in K and, hence, that every recursive function is strongly representable in S. (V. H. Dyson)

3.33 This exercise will show the existence of a recursive, nonprimitive recursive function.

(a) Let $[\sqrt{n}]$ be the largest integer $\leqslant \sqrt{n}$. Show that $[\sqrt{n}]$ is defined by the recursion

$$\kappa(0) = 0$$

$$\kappa(n+1) = \kappa(n) + \overline{\mathrm{sg}}|(n+1) - (\kappa(n)+1)^2|$$

Hence, $[\sqrt{n}]$ is primitive recursive.

(b) The function $\mathrm{Quadrem}(n) = n \dot{-} [\sqrt{n}]^2$ is primitive recursive and represents the difference between n and the largest square $\leqslant n$.

(c) Let $\rho(x, y) = ((x+y)^2 + y)^2 + x$; $\rho_1(n) = \mathrm{Quadrem}(n)$ and $\rho_2(n) = \mathrm{Quadrem}([\sqrt{n}])$. These functions are primitive recursive. Prove the following.

(i) $\rho_1(\rho(x, y)) = x$ and $\rho_2(\rho(x, y)) = y$

(ii) $\rho(\rho_1(n), \rho_2(n)) = n$

(iii) ρ is a one–one function from ω^2 onto ω

(iv) $\rho_1(0) = \rho_2(0) = 0$ and

$$\left.\begin{aligned} \rho_1(n+1) &= \rho_1(n) + 1 \\ \rho_2(n+1) &= \rho_2(n) \end{aligned}\right\} \text{if } \rho_1(n+1) \neq 0$$

(v) Define for each $n \geqslant 3$, $\rho^n(x_1, \ldots, x_n) = \rho(\rho^{n-1}(x_1, \ldots, x_{n-1}), x_n)$. Let $\rho^2 = \rho$. Then each ρ^n is primitive recursive. Define $\rho_i^n(k) = \rho_i^{n-1}(\rho_1(k))$ for $1 \leqslant i \leqslant n-1$ and $\rho_n^n(k) = \rho_2(k)$. Then each $\rho_i^n (1 \leqslant i \leqslant n)$ is primitive recursive, $\rho_i^n(\rho^n(x_1, \ldots, x_n)) = x_i$ and $\rho^n(\rho_1^n(k), \rho_2^n(k), \ldots, \rho_n^n(k)) = k$. Hence, ρ^n is a one–one mapping of ω^n onto ω, and the ρ_i^n's are the corresponding "inverse" functions. The ρ^n's and ρ_i^n's are obtained from ρ, ρ_1, and ρ_2 by substitution.

(d) The Recursion Rule (V) can be limited to the form

$$\psi(x_1, \ldots, x_{n+1}, 0) = x_{n+1} \qquad (n \geqslant 0)$$

$$\psi(x_1, \ldots, x_{n+1}, y+1) = \varphi(x_1, \ldots, x_{n+1}, y, \psi(x_1, \ldots, x_{n+1}, y))$$

Suggestion: Given

$$\theta(x_1,\ldots,x_n,0) = \gamma(x_1,\ldots,x_n)$$

$$\theta(x_1,\ldots,x_n,y+1) = \delta(x_1,\ldots,x_n,y,\theta(x_1,\ldots,x_n,y))$$

Define ψ as above, letting $\varphi(x_1,\ldots,x_{n+1},y,z) = \delta(x_1,\ldots,x_n,y,z)$. Then $\theta(x_1,\ldots,x_n,y) = \psi(x_1,\ldots,x_n,\gamma(x_1,\ldots,x_n),y)$.

(e) Assuming ρ, ρ_1, and ρ_2 as additional initial functions, we can limit uses of the Recursion Rule (V) to the one-parameter form:

$$\psi(x,0) = \alpha(x)$$

$$\psi(x,y+1) = \beta(x,y,\psi(x,y))$$

Hint: Let $n \geqslant 2$. Given

$$\theta(x_1,\ldots,x_n,0) = \gamma(x_1,\ldots,x_n)$$

$$\theta(x_1,\ldots,x_n,y+1) = \delta(x_1,\ldots,x_n,y,\theta(x_1,\ldots,x_n,y))$$

Let $\eta(u,y) = \theta(\rho_1^n(u),\ldots,\rho_n^n(u),y)$. Define η by a permissible recursion.

(f) Assuming ρ, ρ_1, and ρ_2 as additional initial functions, we can use $\delta(y,\psi(x,y))$ instead of $\beta(x,y,\psi(x,y))$ in part (e). [*Hint*: Given

$$\psi(x,0) = \alpha(x)$$

$$\psi(x,y+1) = \beta(x,y,\psi(x,y))$$

let $\psi_1(x,y) = \rho(x,\psi(x,y))$. Then $x = \rho_1(\psi_1(x,y))$ and $\psi(x,y) = \rho_2(\psi_1(x,y))$. Define ψ_1 by an appropriate recursion.]

(g) Assuming ρ, ρ_1, and ρ_2 as additional initial functions, we can limit uses of the Recursion Rule (V) to the form

$$\psi(x,0) = x$$

$$\psi(x,y+1) = \beta(y,\psi(x,y))$$

Hint: Use part (f). Given

$$\varphi(x,0) = \alpha(x)$$

$$\varphi(x,y+1) = \beta(y,\varphi(x,y))$$

define ψ as above. Then $\varphi(x,y) = \psi(\alpha(x),y)$.

(h) Assuming $\rho, \rho_1, \rho_2, +, \cdot$, and $\overline{\text{sg}}$ as additional initial functions, we can limit all uses of the Recursion Rule (V) to those with one parameter of the form

$$f(0) = 0$$

$$f(y+1) = h(y,f(y))$$

Hint: Given, by part (g),

$$\psi(x,0) = x$$

$$\psi(x,y+1) = \beta(y,\psi(x,y))$$

let $f(n) = \psi(\rho_2(n), \rho_1(n))$. Then

$$f(0) = \psi(\rho_2(0), \rho_1(0)) = \psi(0,0) = 0$$

$$\begin{aligned}
f(n+1) &= \psi(\rho_2(n+1), \rho_1(n+1)) \\
&= \begin{cases} \rho_2(n+1) & \text{if } \rho_1(n+1) = 0 \\ \beta(\rho_1(n+1) \dot{-} 1, \psi(\rho_2(n+1), \rho_1(n+1) \dot{-} 1)) & \text{if } \rho_1(n+1) \neq 0 \end{cases} \\
&= \begin{cases} \rho_2(n+1) & \text{if } \rho_1(n+1) = 0 \\ \beta(\rho_1(n), \psi(\rho_1(n), \rho_2(n)) & \text{if } \rho_1(n+1) \neq 0 \end{cases} \\
&= \begin{cases} \rho_1(n+1) & \text{if } \rho_2(n+1) = 0 \\ \beta(\rho_1(n), f(n)) & \text{if } \rho_1(n+1) \neq 0 \end{cases} \\
&= \rho_2(n+1) \cdot \overline{\text{sg}}(\rho_1(n+1)) + \beta(\rho_1(n), f(n)) \cdot \text{sg}(\rho_1(n+1)) \\
&= h(n, f(n))
\end{aligned}$$

(Note that sg is obtainable by a recursion of the appropriate kind.) Then $\psi(x, y) = f(\rho(y, x))$.

(i) All primitive recursive functions are obtainable from the initial functions Z, N, U_i^n, ρ, ρ_1, ρ_2, $+$, $\cdot$, and $\overline{\text{sg}}$ by substitution and the Recursion Rule (V) in the form

$$f(0) = 0$$

$$f(y+1) = h(y, f(y))$$

[This is a restatement of part (h).]

(j) In part (i), $h(y, f(y))$ can be replaced by $h(f(y))$. *Hint:* Given

$$f(0) = 0$$

$$f(y+1) = h(y, f(y))$$

let $g(u) = \rho(u, f(u))$ and $\varphi(w) = \rho(\rho_1(w) + 1, h(\rho_1(w), \rho_2(w)))$. Then

$$g(0) = 0$$

$$g(y+1) = \varphi(g(y))$$

and

$$f(u) = \rho_2(g(u))$$

(k) Show that the equations

$$\psi(n, 0) = n + 1$$

$$\psi(0, m+1) = \psi(1, m)$$

$$\psi(n+1, m+1) = \psi(\psi(n, m+1), m)$$

define a number-theoretic function. In addition, prove the following.

(I) $\psi(n, m) > n$

(II) ψ is monotonic in each variable; that is, if $x < z$, then $\psi(x, y) < \psi(z, y)$ and $\psi(y, x) < \psi(y, z)$.

(III) $\psi(n, m+1) \geqslant \psi(n+1, m)$

(IV)^{D} $\psi(x, y)$ is a recursive function.

(V) For every primitive recursive function $f(x_1, \ldots, x_n)$, there is some fixed m such that $f(x_1, \ldots, x_n) < \psi(\max(x_1, \ldots, x_n), m)$ for all $x_1, \ldots, x_n$. [*Hint:* Prove this first for the initial functions Z, N, U_i^n, ρ, ρ_1, ρ_2, $+$, $\cdot$, and $\overline{sg}$, and then show that it is preserved by substitution and the recursion of part (j) above.] Hence, for every primitive recursive function $f(x)$ of one argument, there is some m such that $f(x) < \psi(x, m)$ for all x.

(VI) Prove that $\psi(x, x) + 1$ is recursive but not primitive recursive. [*Hint:* Use part (V).]

For other proofs of the existence of recursive functions that are not primitive recursive, see Ackermann (1928), Péter (1935, 1967), and R. M. Robinson (1948).

4. ARITHMETIZATION. GÖDEL NUMBERS

For an arbitrary first-order theory K, we correlate with each symbol u of K an odd positive integer $g(u)$, called the *Gödel number* of u, in the following manner:

$$g(()=3;\ g())=5;\ g(,)=7;\ g(\neg)=9;\ g(\Rightarrow)=11;\ g(\forall)=13$$

$$g(x_k) = 13 + 8k \qquad \text{for } k = 1, 2, \ldots$$

$$g(a_k) = 7 + 8k \qquad \text{for } k = 1, 2, \ldots$$

$$g(f_k^n) = 1 + 8(2^n 3^k) \qquad \text{for } k, n \geqslant 1$$

$$g(A_k^n) = 3 + 8(2^n 3^k) \qquad \text{for } k, n \geqslant 1$$

Clearly, every Gödel number is an odd positive integer. Moreover, when divided by 8, $g(u)$ leaves a remainder of 5 when u is a variable, a remainder of 7 when u is an individual constant, a remainder of 1 when u is a function letter, and a remainder of 3 when u is a predicate letter. Thus, different symbols have different Gödel numbers.

Examples $g(x_2) = 29$, $g(a_4) = 39$, $g(f_1^2) = 97$, $g(A_2^1) = 147$

Given an expression $u_0 u_1 \ldots u_r$, where each u_j is a symbol of K, we define its Gödel number $g(u_0 u_1 \ldots u_r)$ by the equation

$$g(u_0 u_1 \ldots u_r) = 2^{g(u_0)} 3^{g(u_1)} \ldots p_r^{g(u_r)}$$

where p_j denotes the jth prime and we assume that $p_0 = 2$. For example,

$$g(A_1^2(x_1, x_2)) = 2^{g(A_1^2)} 3^{g(()} 5^{g(x_1)} 7^{g(,)} 11^{g(x_2)} 13^{g())}$$

$$= 2^{99} 3^3 5^{21} 7^7 11^{29} 13^5$$

Observe that different expressions have different Gödel numbers, by virtue of the uniqueness of the factorization of integers into primes. In addition, expressions and symbols have different Gödel numbers, since the former have even Gödel numbers and the latter odd Gödel numbers. (Notice also that a single symbol, considered as an expression, has a different Gödel number from its Gödel number as a symbol.

For example, the symbol x_1 has Gödel number 21, whereas the expression that consists of only the symbol x_1 has Gödel number 2^{21}.)

If $e_0, e_1, \ldots, e_r$ is any finite sequence of expressions of K, we can assign a Gödel number to this sequence by setting

$$g(e_0, e_1, \ldots, e_r) = 2^{g(e_0)} 3^{g(e_1)} \ldots p_r^{g(e_r)}$$

Different sequences of expressions have different Gödel numbers. Since a Gödel number of a sequence of expressions is even and the exponent of 2 in its prime factorization is also even, it differs from Gödel numbers of symbols and expressions. Remember that a proof in K is a certain kind of finite sequence of expressions and, therefore, in particular, a proof has a Gödel number.

Thus, g is a one–one function from the set of symbols of K, expressions of K, and finite sequences of expressions of K, into the set of positive integers. The range of g is not the whole set of positive integers. For example, 10 is not a Gödel number.

Exercises

3.34 Determine the objects that have the following Gödel numbers:
(a) 1944 (b) 49 (c) 15 (d) 13,824 (e) $2^{51} 3^{11} 5^9$

3.35 Show that, if n is odd, $4n$ is not a Gödel number.

3.36 Find the Gödel numbers of the following expressions.
(a) $f_1^1(a_1)$ (b) $((\forall x_3)(\neg A_1^2(a_1, x_3)))$

This method of associating numbers with symbols, expressions, and sequences of expressions was originally devised by Gödel (1931) in order to *arithmetize* metamathematics*—that is, to replace assertions about a formal system by equivalent number-theoretic statements, and then to express these statements within the formal system itself. This idea turned out to be the key to many significant problems in mathematical logic.

The assignment of Gödel numbers given here is in no way unique. Other methods are found in Kleene (1952, chap. X) and in Smullyan (1961, chap. 1, § 6).

PROPOSITION 3.25 Let K be a theory about which we make the assumption that the following relations are primitive recursive (or recursive):

(a) IC(x): x is the Gödel number of an individual constant of K
(b) FL(x): x is the Gödel number of a function letter of K
(c) PL(x): x is the Gödel number of a predicate letter of K

Then the following relations and functions (1–16) are primitive recursive (or recursive). In each case, we give first the notation and intuitive definition for the relation or function, and then an equivalent formula from which its primitive recursiveness (or recursiveness) can be deduced.

* An *arithmetization* of a theory K is a one–one function g from the set of symbols of K, expressions of K, and finite sequences of expressions of K into the set of positive integers. The following conditions are to be satisfied by the function g: (1) g is effectively computable; (2) there is an effective procedure that determines whether any given positive integer m is in the range of g and, if m is in the range of g, the procedure finds the object x such that $g(x) = m$.

(1) EVbl(x): x is the Gödel number of an expression consisting of a variable, $(\exists z)_{z<x}(1 \leqslant z \wedge x = 2^{13+8z})$. By Proposition 3.17, this is primitive recursive.

EIC(x): x is the Gödel number of an expression consisting of an individual constant, $(\exists y)_{y<x}(\mathrm{IC}(y) \wedge x = 2^y)$ (Proposition 3.17).

EFL(x): x is the Gödel number of an expression consisting of a function letter, $(\exists y)_{y<x}(\mathrm{FL}(y) \wedge x = 2^y)$ (Proposition 3.17).

EPL(x): x is the Gödel number of an expression consisting of a predicate letter, $(\exists y)_{y<x}(\mathrm{PL}(y) \wedge x = 2^y)$ (Proposition 3.17).

(2) $\mathrm{Arg_T}(x) = (\mathrm{qt}(8, x \dot{-} 1))_0$: If x is the Gödel number of a function letter f_j^n, then $\mathrm{Arg_T}(x) = n$. $\mathrm{Arg_T}$ is primitive recursive.

$\mathrm{Arg_P}(x) = (\mathrm{qt}(8, x \dot{-} 3))_0$: If x is the Gödel number of a predicate letter A_j^n, then $\mathrm{Arg_P}(x) = n$. $\mathrm{Arg_p}$ is primitive recursive.

(3) Gd(x): x is the Gödel number of an expression of K, $\mathrm{EVbl}(x) \vee \mathrm{EIC}(x) \vee \mathrm{EFL}(x) \vee \mathrm{EPL}(x) \vee x = 2^3 \vee x = 2^5 \vee x = 2^7 \vee x = 2^9 \vee x = 2^{11} \vee x = 2^{13} \vee (\exists u)_{u<x}(\exists v)_{v<x}(x = u * v \wedge \mathrm{Gd}(u) \wedge \mathrm{Gd}(v))$. Use Corollary 3.20. Here, $*$ is the juxtaposition function defined in Example 4 on page 139.

(4) MP(x, y, z): The expression with Gödel number z is a direct consequence of the expressions with Gödel numbers x and y by modus ponens, $y = 2^3 * x * 2^{11} * z * 2^5 \wedge \mathrm{Gd}(x) \wedge \mathrm{Gd}(z)$.

(5) Gen(x, y): The expression with Gödel number y comes from the expression with Gödel number x by the Generalization Rule:

$$(\exists v)_{v<y}(\mathrm{EVbl}(v) \wedge y = 2^3 * 2^3 * 2^{13} * v * 2^5 * x * 2^5 \wedge \mathrm{Gd}(x))$$

(6) Trm(x): x is the Gödel number of a term of K. This holds when and only when either x is the Gödel number of an expression consisting of a variable or an individual constant, or there is a function letter f_k^n and terms $t_1, \ldots, t_n$ such that x is the Gödel number of $f_k^n(t_1, \ldots, t_n)$. The latter holds if and only if there is a sequence of $n + 1$ expressions

$$f_k^n(\quad f_k^n(t_1, \quad f_k^n(t_1, t_2, \quad \ldots \quad f_k^n(t_1, t_2, \ldots, t_{n-1}, \quad f_k^n(t_1, t_2, \ldots, t_n)$$

the last of which, $f_k^n(t_1, \ldots, t_n)$, has Gödel number x. This sequence can be represented by its Gödel number y. Clearly, $y < 2^x 3^x \cdots p_n^x = (2 \cdot 3 \cdots p_n)^x < (p_n!)^x < (p_x!)^x$. Note that $\ell h(y) = n + 1$ and also that $n = \mathrm{Arg_T}((x)_0)$, since $(x)_0$ is the Gödel number of f_k^n. Hence, Trm(x) is equivalent to the following relation:

$$\mathrm{EVbl}(x) \vee \mathrm{EIC}(x) \vee (\exists y)_{y<(p_x!)^x}[x = (y)_{\ell h(y) \dot{-} 1} \wedge \ell h(y) = \mathrm{Arg_T}((x)_0) + 1$$
$$\wedge\ \mathrm{FL}(((y)_0)_0) \wedge ((y)_0)_1 = 3 \wedge \ell h((y)_0) = 2 \wedge$$
$$(\forall u)_{u<\ell h(y) \dot{-} 2}(\exists v)_{v<x}((y)_{u+1} = (y)_u * v * 2^7 \wedge \mathrm{Trm}(v)) \wedge$$
$$(\exists v)_{v<x}((y)_{\ell h(y) \dot{-} 1} = (y)_{\ell h(y) \dot{-} 2} * v * 2^5 \wedge \mathrm{Trm}(v))]$$

Thus, Trm(x) is primitive recursive (or recursive) by Corollary 3.20, since the formula above involves Trm(v) for only $v < x$. In fact, if we replace both occurrences of Trm(v) in the formula by $(z)_v = 0$, then the new formula defines a primitive recursive (or recursive) relation H(x, z), and $\mathrm{Trm}(x) \Leftrightarrow \mathrm{H}(x, (C_{\mathrm{Trm}})^{\#}(x))$. Therefore, Corollary 3.20 is applicable.

(7) Atfml(x): x is the Gödel number of an atomic wf of K. This holds if and only if there are terms $t_1, \ldots, t_n$ and a predicate letter A_k^n such that x is the Gödel number of $A_k^n(t_1, \ldots, t_n)$. The latter holds if and only if there is a sequence of $n + 1$ expressions

$$A_k^n(\quad A_k^n(t_1, \quad A_k^n(t_1, t_2, \quad \ldots \quad A_k^n(t_1, \ldots, t_{n-1}, \quad A_k^n(t_1, \ldots, t_{n-1}, t_n)$$

the last of which, $A_k^n(t_1, \ldots, t_n)$, has Gödel number x. This sequence of expressions can be represented by its Gödel number y. Clearly, $y < (p_x!)^x$ [as in (6) above] and $n = \mathrm{Arg_P}((x)_0)$. Thus, Atfml($x$) is equivalent to the following relation:

$$(\exists y)_{y<(p_x!)^x}[x = (y)_{\ell h(y) \dot{-} 1} \wedge \ell h(y) = \mathrm{Arg_P}((x)_0) \wedge$$
$$\mathrm{PL}(((y)_0)_0) \wedge ((y)_0)_1 = 3 \wedge \ell h((y)_0) = 2 \wedge$$
$$(\forall u)_{u<\ell h(y) \dot{-} 2}(\exists v)_{v<x}((y)_{u+1} = (y)_u * v * 2^7 \wedge \mathrm{Trm}(v)) \wedge$$
$$(\exists v)_{v<x}((y)_{\ell h(y) \dot{-} 1} = (y)_{\ell h(y) \dot{-} 2} * v * 2^5 \wedge \mathrm{Trm}(v))]$$

Hence, by Proposition 3.17, Atfml(x) is primitive recursive (or recursive).

(8) Fml(y): y is the Gödel number of a wf of K:

$$\mathrm{Atfml}(y) \vee (\exists z)_{z<y}[(\mathrm{Fml}(z) \wedge y = 2^3 * 2^9 * z * 2^5) \vee$$
$$(\mathrm{Fml}((z)_0) \wedge \mathrm{Fml}((z)_1) \wedge y = 2^3 * (z)_0 * 2^{11} * (z_1) * 2^5) \vee$$
$$(\mathrm{Fml}((z)_0) \wedge \mathrm{EVbl}((z)_1) \wedge y = 2^3 * 2^3 * 2^{13} * (z)_1 * 2^5 * (z)_0 * 2^5)]$$

Now it is easy to verify that Corollary 3.20 is applicable.

(9) Subst(x, y, u, v): x is the Gödel number of the result of substituting in the expression with Gödel number y the term with Gödel number u for all free occurrences of the variable with Gödel number v:

$$\mathrm{Gd}(y) \wedge \mathrm{Trm}(u) \wedge \mathrm{EVbl}(2^v) \wedge [(y = 2^v \wedge x = u) \vee$$
$$(\exists w)_{w<y}(y = 2^w \wedge y \neq 2^v \wedge x = y) \vee$$
$$(\exists z)_{z<y}(\exists w)_{w<y}(\mathrm{Fml}(w) \wedge y = 2^3 * 2^{13} * 2^v * 2^5 * w * z \wedge$$
$$(\exists \alpha)_{\alpha<x}(x = 2^3 * 2^v * 2^5 * w * \alpha \wedge \mathrm{Subst}(\alpha, z, u, v))) \vee$$
$$((\neg(\exists z)_{z<y}(\exists w)_{w<y}(\mathrm{Fml}(w) \wedge y = 2^3 * 2^v * 2^5 * w * z)) \wedge$$
$$(\exists \alpha)_{\alpha<x}(\exists \beta)_{\beta<x}(\exists z)_{z<y}(1 < z \wedge y = 2^{(y)_0} * z \wedge x = \alpha * \beta \wedge$$
$$\mathrm{Subst}(\alpha, 2^{(y)_0}, u, v) \wedge \mathrm{Subst}(\beta, z, u, v)))]$$

Corollary 3.20 is applicable. The reader should verify that this formula actually captures the intuitive content of Subst(x, y, u, v).

(10) Sub(y, u, v): the Gödel number of the result of substituting the term with Gödel number u for all free occurrences in the expression with Gödel number y of the variable with Gödel number v:

$$\mathrm{Sub}(y, u, v) = \mu x_{x<(p_{uy}!)^{uy}} \mathrm{Subst}(x, y, u, v)$$

Therefore, Sub is primitive recursive (or recursive) by Proposition 3.17. [When the conditions on u, v, and y are not met, $\mathrm{Sub}(y, u, v)$ is defined, but its value is of no interest.]

(11) $\mathrm{Fr}(y, v)$: y is the Gödel number of a wf or a term of K that contains free occurrences of the variable with Gödel number v:

$$(\mathrm{Fml}(y) \vee \mathrm{Trm}(y)) \wedge \mathrm{EVbl}(2^v) \wedge \neg\mathrm{Subst}(y, y, 2^{13+8v}, v)$$

(That is, substitution in the wf or term with Gödel number y of a variable different from the variable with Gödel number v for all free occurrences of the variable with Gödel number v yields a different expression.)

(12) $\mathrm{Ff}(u, v, w)$: u is the Gödel number of a term that is free for the variable with Gödel number v in the wf with Gödel number w:

$$\begin{aligned}&\mathrm{Trm}(u) \wedge \mathrm{EVbl}(2^v) \wedge \mathrm{Fml}(w) \wedge\\ &[\mathrm{Atfml}(w) \vee (\exists y)_{y<w}(w = 2^3 * 2^9 * y * 2^5 \wedge \mathrm{Ff}(u, v, y)) \vee\\ &(\exists y)_{y<w}(\exists z)_{z<w}(w = 2^3 * y * 2^{11} * z * 2^5 \wedge \mathrm{Ff}(u, v, y) \wedge \mathrm{Ff}(u, v, z)) \vee\\ &(\exists y)_{y<w}(\exists z)_{z<w}(w = 2^3 * 2^3 * 2^{13} * 2^z * 2^5 * y * 2^5 \wedge \mathrm{EVbl}(2^z) \wedge\\ &(z \neq v \Rightarrow \mathrm{Ff}(u, v, y) \wedge (\mathrm{Fr}(u, z) \Rightarrow \neg\mathrm{Fr}(y, v))))]\end{aligned}$$

Use Corollary 3.20 again.

(13) (a) $\mathrm{Ax}_1(x)$: x is the Gödel number of an instance of axiom schema (A1):

$$\begin{aligned}&(\exists u)_{u<x}(\exists v)_{v<x}(\mathrm{Fml}(u) \wedge \mathrm{Fml}(v) \wedge\\ &x = 2^3 * u * 2^{11} * 2^3 * v * 2^{11} * u * 2^5 * 2^5)\end{aligned}$$

(b) $\mathrm{Ax}_2(x)$: x is the Gödel number of an instance of axiom schema (A2):

$$\begin{aligned}&(\exists u)_{u<x}(\exists v)_{v<x}(\mathrm{E}w)_{w<x}(\mathrm{Fml}(u) \wedge \mathrm{Fml}(v) \wedge \mathrm{Fml}(w) \wedge x = 2^3 *\\ &2^3 * u * 2^{11} * 2^3 * v * 2^{11} * w * 2^5 * 2^5 * 2^{11} * 2^3 * 2^3 * u * 2^{11} * v * 2^5 * 2^{11}\\ &* 2^3 * u * 2^{11} * w * 2^5 * 2^5 * 2^5)\end{aligned}$$

(c) $\mathrm{Ax}_3(x)$: x is the Gödel number of an instance of axiom schema (A3):

$$\begin{aligned}&(\exists u)_{u<x}(\exists v)_{v<x}(\mathrm{Fml}(u) \wedge \mathrm{Fml}(v) \wedge\\ &x = 2^3 * 2^3 * 2^3 * 2^9 * v * 2^5 * 2^{11} * 2^3 * 2^9 * v * 2^5 * 2^5 * 2^{11}\\ &* 2^3 * 2^3 * 2^3 * 2^9 * v * 2^5 * 2^{11} * u * 2^5 * 2^{11} * v * 2^5 * 2^5)\end{aligned}$$

(d) $\mathrm{Ax}_4(x)$: x is the Gödel number of an instance of axiom schema (A4):

$$\begin{aligned}&(\exists u)_{u<x}(\exists v)_{v<x}(\exists w)_{w<x}(\mathrm{Fml}(y) \wedge \mathrm{Trm}(u) \wedge \mathrm{EVbl}(2^v) \wedge \mathrm{Ff}(u, v, y) \wedge\\ &x = 2^3 * 2^3 * 2^3 * 2^{13} * 2^v * 2^5 * y * 2^5 * 2^{11} * \mathrm{Sub}(y, u, v) * 2^5)\end{aligned}$$

(e) $\mathrm{Ax}_5(x)$: x is the Gödel number of an instance of axiom schema (A5):

$$(\exists u)_{u<x}(\exists v)_{v<x}(\exists w)_{w<x}(\mathrm{Fml}(u) \wedge \mathrm{Fml}(w) \wedge \mathrm{EVbl}(2^v) \wedge \neg \mathrm{Fr}(u, v) \wedge x = 2^3 * 2^3 * 2^3 * 2^{13} * 2^v * 2^5 * 2^3 * u * 2^{11} * w * 2^5 * 2^5 * 2^{11} * 2^3 * u * 2^{11} * 2^3 * 2^3 * 2^{13} * 2^v * 2^5 * w * 2^5 * 2^5 * 2^5)$$

(f) LAX(y): y is the Gödel number of a logical axiom:

$$\mathrm{Ax}_1(y) \vee \mathrm{Ax}_2(y) \vee \mathrm{Ax}_3(y) \vee \mathrm{Ax}_4(y) \vee \mathrm{Ax}_5(y)$$

(14) The following *negation* function is primitive recursive.
Neg(x): the Gödel number of ($\neg \mathscr{A}$), if x is the Gödel number of $\mathscr{A}$:

$$\mathrm{Neg}(x) = 2^3 * 2^9 * x * 2^5$$

(15) The following *conditional* function is primitive recursive.
Cond(x, y): the Gödel number of ($\mathscr{A} \Rightarrow \mathscr{B}$) if x is the Gödel number of $\mathscr{A}$ and y is the Gödel number of $\mathscr{B}$:

$$\mathrm{Cond}(x, y) = 2^3 * x * 2^{11} * y * 2^5$$

(16) Clos(u): the Gödel number of the closure of $\mathscr{A}$ if u is the Gödel number of a wf $\mathscr{A}$.

First, let $V(u) = \mu v_{v \leqslant u}(\mathrm{EVbl}(2^v) \wedge \mathrm{Fr}(u, v))$. V is primitive recursive (or recursive). $V(u)$ is the least Gödel number of a free variable of u (if there are any). Let Sent(u) be $\mathrm{Fml}(u) \wedge \neg(\exists v)_{v \leqslant u} \mathrm{Fr}(u, v)$. Sent is primitive recursive (or recursive). Sent(u) holds when and only when u is the Gödel number of a sentence (i.e., a closed wf). Now let

$$G(u) = \begin{cases} 2^3 * 2^3 * 2^{13} * 2^{V(u)} * 2^5 * u * 2^5 & \text{if } \mathrm{Fml}(u) \wedge \neg \mathrm{Sent}(u) \\ u & \text{otherwise} \end{cases}$$

G is primitive recursive (or recursive). If u is the Gödel number of a wf $\mathscr{A}$ that is not a closed wf, then $G(u)$ is the Gödel number of $(\forall x)\mathscr{A}$, where x is the free variable of $\mathscr{A}$ that has the least Gödel number. Otherwise, $G(u) = u$. Now let

$$H(u, 0) = G(u)$$

$$H(u, y + 1) = G(H(u, y))$$

H is primitive recursive (or recursive). Finally,

$$\mathrm{Clos}(u) = H(u, \mu y_{y \leqslant u}(H(u, y) = H(u, y + 1)))$$

Thus, Clos is primitive recursive (or recursive).

REMARK The assumptions (a)–(c) of Proposition 3.25 hold for a first-order theory K that has only a finite number of individual constants, function letters, and predicate letters, since, in that case, IC(x), FL(x), and PL(x) are primitive recursive. For example, if the individual constants of K are a_{j_1}, a_{j_2}, ..., a_{j_n}, then IC(x) if and only if $x = 7 + 8j_1 \vee x = 7 + 8j_2 \vee \cdots \vee x = 7 + 8j_n$. In particular, the assumptions (a)–(c) hold for any theory that has the same symbols as S.

PROPOSITION 3.26 Let a theory K satisfy the assumptions (a)–(c) of Proposition 3.25 and contain the individual constant 0 and the function letter f_1^1 of S. (Thus, all the numerals are terms of K. In particular, K can be S itself.) Then the following functions and relation are primitive recursive (or recursive).

(17) Num(y): the Gödel number of the expression $\bar{y}$

$$\mathrm{Num}(0) = 2^{15}$$

$$\mathrm{Num}(y + 1) = 2^{49} * 2^3 * \mathrm{Num}(y) * 2^5$$

Num(y) is primitive recursive by virtue of the Recursion Rule (V).

(18) Nu(x): x is the Gödel number of a numeral:

$$(\exists y)_{y<x}(x = \mathrm{Num}(y))$$

Nu(x) is primitive recursive by Proposition 3.17.

(19) D(u): the Gödel number of $\mathscr{A}(\bar{u})$, if u is the Gödel number of a wf $\mathscr{A}(x_1)$:

$$\mathrm{D}(u) = \mathrm{Sub}(u, \mathrm{Num}(u), 21)$$

Thus, D(u) is primitive recursive (or recursive). D(u) is called the *diagonal function.*

PROPOSITION 3.27 Let a theory K satisfy the assumptions (a)–(c) of Proposition 3.25 as well as the following additional condition:

(d) The property

$$\mathrm{PrAx}(y)\text{: } y \text{ is the Gödel number of a proper axiom of K}$$

is primitive recursive (or recursive). Then the following relations are primitive recursive (or recursive).

(20) Ax(y): y is the Gödel number of an axiom of K:

$$\mathrm{LAx}(y) \vee \mathrm{PrAx}(y)$$

(21) Prf(y): y is the Gödel number of a proof in K:

$$(\exists u)_{u<y}(\exists v)_{v<y}(\exists z)_{z<y}(\exists w)_{w<y}([y = 2^w \wedge \mathrm{Ax}(w)] \vee$$
$$[\mathrm{Prf}(u) \wedge \mathrm{Fml}((u)_w) \wedge y = u * 2^v \wedge \mathrm{Gen}((u)_w, v)] \vee$$
$$[\mathrm{Prf}(u) \wedge \mathrm{Fml}((u)_z) \wedge \mathrm{Fml}((u)_w) \wedge y = u * 2^v \wedge \mathrm{MP}((u)_z, (u)_w, v)]$$
$$\vee [\mathrm{Prf}(u) \wedge y = u * 2^v \wedge \mathrm{Ax}(v)])$$

Apply Corollary 3.20.

(22) Pf(y, x): y is the Gödel number of a proof in K of the wf with Gödel number x:

$$\mathrm{Prf}(y) \wedge x = (y)_{\ell h(y) \dot{-} 1}$$

Notice that S satisfies assumption (d) of Proposition 3.27. Let $a_1, a_2, \ldots, a_8$ be the Gödel numbers of axioms (S1)–(S8). It is easy to see that a number u is the Gödel number of an instance of axiom schema (A9) if and only if

$$(\exists v)_{v<u}(\exists y)_{y<u}(\text{EVbl}(2^v) \wedge \text{Fml}(y) \wedge$$
$$u = 2^3 * \text{Sub}(y, 2^{15}, v) * 2^{11} * 2^3 * 2^3 * 2^3 * 2^{13} * 2^v * 2^5 * 2^3$$
$$* y * 2^{11} * \text{Sub}(y, 2^{49} * 2^3 * 2^v * 2^5, v) * 2^5 * 2^5 * 2^{11} * 2^3 * 2^3$$
$$* 2^{13} * 2^v * 2^5 * 2^5 * y * 2^5 * 2^5)$$

Denote the displayed formula by $A_9(u)$. Then y is the Gödel number of a proper axiom of S if and only if $y = a_1 \vee y = a_2 \vee \cdots \vee y = a_8 \vee A_9(y)$. Thus, PrAx($y$) is primitive recursive for S.

The relations and functions of Propositions 3.25–3.27 should have the subscript "K" attached to the corresponding signs to indicate the dependence on K. If we were considering a different theory, then, in general, we would obtain different relations and functions in Propositions 3.25–3.27.

PROPOSITION 3.28 Let K be a theory with equality that contains the individual constant 0 and the function letter f_1^1 of S and for which assumptions (a)–(d) of Propositions 3.25 and 3.27 hold. Also assume:

(*) For any natural numbers r and s, if $\vdash_K \bar{r} = \bar{s}$, then $r = s$.

Then any function $f(x_1, \ldots, x_n)$ that is representable in K is recursive.

Proof Let $\mathscr{A}(x_1, \ldots, x_n, x_{n+1})$ be a wf of K that represents f. Let $B_{\mathscr{A}}(u_1, \ldots, u_n, u_{n+1}, y)$ mean that y is the Gödel number of a proof in K of the wf $\mathscr{A}(\bar{u}_1, \ldots, \bar{u}_n, \bar{u}_{n+1})$. Note that, if $B_{\mathscr{A}}(u_1, \ldots, u_n, u_{n+1}, y)$, then $f(u_1, \ldots, u_n) = u_{n+1}$. [In fact, let $f(u_1, \ldots, u_n) = r$. Since $\mathscr{A}$ represents f in K, $\vdash_K \mathscr{A}(\bar{u}_1, \ldots, \bar{u}_n, \bar{r})$ and $\vdash_K (\exists_1 y)\mathscr{A}(\bar{u}_1, \ldots, \bar{u}_n, y)$. By hypothesis, $B_{\mathscr{A}}(u_1, \ldots, u_n, u_{n+1}, y)$. Hence, $\vdash_K \mathscr{A}(\bar{u}_1, \ldots, \bar{u}_n, \bar{u}_{n+1})$. But, K is a theory with equality. Therefore, $\vdash_K \bar{r} = \bar{u}_{n+1}$. By (*), $r = u_{n+1}$.] Now, let m be the Gödel number of $\mathscr{A}(x_1, \ldots, x_n, x_{n+1})$. Then, $B_{\mathscr{A}}(u_1, \ldots, u_n, u_{n+1}, y)$ is equivalent to:

$$\text{Pf}(y, \text{Sub}(\ldots \text{Sub}(\text{Sub}(m, \text{Num}(u_1), 21), \text{Num}(u_2), 29) \ldots, \text{Num}(u_{n+1}), 21 + 8n))$$

So, by Propositions 3.25–3.27, $B_{\mathscr{A}}(u_1, \ldots, u_n, u_{n+1}, y)$ is primitive recursive (or recursive). Now, consider any natural numbers $k_1, \ldots, k_n$. Let $f(k_1, \ldots, k_n) = r$. Then $\vdash_K \mathscr{A}(\bar{k}_1, \ldots, \bar{k}_n, \bar{r})$. Let j be the Gödel number of a proof in K of $\mathscr{A}(\bar{k}_1, \ldots, \bar{k}_n, \bar{r})$. Then $B_{\mathscr{A}}(k_1, \ldots, k_n, r, j)$. Thus, for any $x_1, \ldots, x_n$, there is some y such that $B_{\mathscr{A}}(x_1, \ldots, x_n, (y)_0, (y)_1)$. Then, by the μ-operator Rule (VI), $\mu y(B_{\mathscr{A}}(x_1, \ldots, x_n, (y)_0, (y)_1)$ is recursive. But, $f(x_1, \ldots, x_n) = (\mu y(B_{\mathscr{A}}(x_1, \ldots, x_n, (y)_0, (y)_1)))_0$ and, therefore, f is recursive.

Exercise

3.37 Let K be a theory containing the predicate letter =, the individual constant 0, and the function letter f_1^1.

(a) If K satisfies hypothesis (*) of Proposition 3.28, prove that K must be consistent.
(b) If K is inconsistent, prove that every number-theoretic function is representable in K.

(c) If K is consistent and the identity relation $x = y$ is expressible in K, show that K satisfies hypothesis (*) of Proposition 3.28.

A special case of Proposition 3.28 is that every function representable in S is recursive. [It is necessary here to take note of Exercise 3.37(b, c).] Together with Proposition 3.23, this shows that the class of recursive functions is identical with the class of functions representable in S. In Chapter 5, it will be made plausible that the notion of recursive function is a precise mathematical equivalent of the intuitive idea of *effectively computable function.*

COROLLARY 3.29 A number-theoretic relation $R(x_1, \ldots, x_n)$ is recursive if and only if $R(x_1, \ldots, x_n)$ is expressible in S.

Proof By definition, R is recursive if and only if C_R is recursive. By Proposition 3.28, C_R is recursive if and only if C_R is representable in S, and, by Proposition 3.12, C_R is representable in S if and only if R is expressible in S.

It will be helpful later to find weaker theories than S for which the representable functions are identical with the recursive functions. Analysis of the proof of Proposition 3.23 leads us to the following theory.

ROBINSON'S SYSTEM

Consider the theory that has the same symbols as S and the following finite list of proper axioms.

(1) $x_1 = x_1$
(2) $x_1 = x_2 \Rightarrow x_2 = x_1$
(3) $x_1 = x_2 \Rightarrow (x_2 = x_3 \Rightarrow x_1 = x_3)$
(4) $x_1 = x_2 \Rightarrow x_1' = x_2'$
(5) $x_1 = x_2 \Rightarrow (x_1 + x_3 = x_2 + x_3 \wedge x_3 + x_1 = x_3 + x_2)$
(6) $x_1 = x_2 \Rightarrow (x_1 \cdot x_3 = x_2 \cdot x_3 \wedge x_3 \cdot x_1 = x_3 \cdot x_2)$
(7) $x_1' = x_2' \Rightarrow x_1 = x_2$
(8) $0 \neq x_1'$
(9) $x_1 \neq 0 \Rightarrow (\exists x_2)(x_1 = x_2')$
(10) $x_1 + 0 = x_1$
(11) $x_1 + (x_2') = (x_1 + x_2)'$
(12) $x_1 \cdot 0 = 0$
(13) $x_1 \cdot (x_2') = (x_1 \cdot x_2) + x_1$
(14) $(x_2 = x_1 \cdot x_3 + x_4 \wedge x_4 < x_1 \wedge x_2 = x_1 \cdot x_5 + x_6 \wedge x_6 < x_1) \Rightarrow x_4 = x_6$ (uniqueness of remainder)

We shall call this theory RR. Clearly, RR is a subtheory of S, since all the axioms of RR are theorems of S. In addition, it follows from Proposition 2.25 and axioms (1)–(6) that RR is a theory with equality. [The system Q of axioms (1)–(13) is due to Raphael Robinson (1950). Axiom (14) has been added to make one of the proofs below easier.] Notice that RR has only a finite number of proper axioms.

LEMMA 3.30 In RR, the following are theorems.

(a) $\bar{n} + \bar{m} = \overline{n + m}$ for any natural numbers m and n
(b) $\bar{n} \cdot \bar{m} = \overline{n \cdot m}$ for any natural numbers m and n
(c) $\bar{n} \neq \bar{m}$ for any natural numbers m and n such that $n \neq m$
(d) $\bar{n} < \bar{m}$ for any natural numbers n and m such that $n < m$
(e) $x \not< 0$
(f) $x \leqslant \bar{n} \Rightarrow x = 0 \vee x = \bar{1} \vee \cdots \vee x = \bar{n}$ for any natural number n
(g) $x \leqslant \bar{n} \vee \bar{n} \leqslant x$ for any natural number n

Proof Parts (a)–(c) are proved the same way as Proposition 3.6(a). Parts (d)–(g) are left as exercises.

PROPOSITION 3.31 All recursive functions are representable in RR.

Proof The initial functions Z, N, and U_i^n are representable in RR by the same wfs as in Examples 1–3, pages 130–131. That the Substitution Rule doesn't lead out of the class of functions representable in RR is proved in the same way as in Example 4 on page 131. For the Recursion Rule, first notice that the proof of Proposition 3.21 is a demonstration that Gödel's beta function $\beta(x_1, x_2, x_3)$ is strongly representable in RR. [Axiom (14) is used for the uniqueness part.] Now, a careful examination of the treatment of the Recursion Rule in the proof of Proposition 3.23 reveals that all the required theorems are theorems of RR. The argument given for the μ-operator Rule also remains valid for RR.

By Proposition 3.31, all recursive functions are representable in any extension of RR. Hence, by Proposition 3.28 and Exercise 3.37(c), in any consistent extension of RR having the same symbols as S and satisfying the condition (d) that PrAx(y) is recursive, the class of representable functions is the same as the class of recursive functions. Likewise, by Proposition 3.12, the relations expressible in RR are the recursive relations.

Exercises

3.38 Prove parts (d)–(g) of Lemma 3.30.
3.39[D] Show that RR is a proper subtheory of S. (*Hint:* Find a model for RR that is not a model for S.) [*Remark:* Not only is S different from RR, but it is not finitely axiomatizable at all; that is, there is no theory K having only a finite number of proper axioms, whose theorems are the same as those of S. This was proved by Ryll-Nardzewski (1953) and Rabin (1961).]
3.40[D] Show that axiom (14) of RR is not provable from axioms (1)–(13) and, therefore, that Q is a proper subtheory of RR. [*Hint*: Find a model of (1)–(13) for which (14) is not true.]
3.41 Fill in all the details of the proof of Proposition 3.31.
3.42 Let K be a theory having the same symbols as S and with just one proper axiom: $(\forall x_1)(\forall x_2)x_1 = x_2$.

(a) Show that K is a consistent theory with equality.
(b) Prove that all number-theoretic functions are representable in K.
(c) Which number-theoretic relations are expressible in K? (*Hint:* Use elimination of quantifiers.)

(d) Show that the hypothesis $\vdash_K 0 \neq 1$ cannot be eliminated from Proposition 3.12.
(e) Show that, in Proposition 3.28, the hypothesis (*) cannot be replaced by the assumption that K is consistent.

3.43 Let R be the theory having the same symbols as S and as proper axioms, the equality axioms (1)–(6) of theory RR, as well as the following five axiom schemas, in which n and m are natural numbers:

(R1) $\bar{n} + \bar{m} = \overline{n+m}$
(R2) $\bar{n} \cdot \bar{m} = \overline{n \cdot m}$
(R3) $\bar{n} \neq \bar{m}$ if $n \neq m$
(R4) $x \leqslant \bar{n} \Rightarrow x = \bar{0} \vee \cdots \vee x = \bar{n}$
(R5) $x \leqslant \bar{n} \vee \bar{n} \leqslant x$

Prove:

(a) R is not finitely axiomatizable. (*Hint:* Show that every finite subset of the axioms of R has a model that is not a model of R.)
(b) R is a proper subtheory of Q.
(c)^D Every recursive function is representable in R. (See Monk, 1976, p. 248.)
(d) The functions representable in R are the recursive functions.
(e) The relations expressible in R are the recursive relations.

5. THE FIXED POINT THEOREM. GÖDEL'S INCOMPLETENESS THEOREM

If K is a theory that has the same symbols as S, recall that the diagonal function D has the property that, if u is the Gödel number of a wf $\mathscr{A}(x_1)$, then $D(u)$ is the Gödel number of the wf $\mathscr{A}(\bar{u})$.

NOTATION When $\mathscr{C}$ is an expression of a theory and the Gödel number of $\mathscr{C}$ is q, then we shall denote the numeral $\bar{q}$ by $\ulcorner \mathscr{C} \urcorner$. We can think of $\ulcorner \mathscr{C} \urcorner$ as being a "name" for $\mathscr{C}$ within the language of arithmetic.

PROPOSITION 3.32 (DIAGONALIZATION LEMMA) Let K be a theory with equality that has the same symbols as S and in which the diagonal function D is representable. Then, for any wf $\mathscr{B}(x_1)$ in which x_1 is the only free variable, there exists a closed wf $\mathscr{C}$ such that

$$\vdash_K \mathscr{C} \Leftrightarrow \mathscr{B}(\ulcorner \mathscr{C} \urcorner)$$

Proof Let $\mathscr{D}(x_1, x_2)$ be a wf representing D in K. Construct the wf

$$(\nabla) \quad (\forall x_2)(\mathscr{D}(x_1, x_2) \Rightarrow \mathscr{B}(x_2))$$

Let m be the Gödel number of (∇). Now substitute $\bar{m}$ for x_1 in (∇):

$$(\mathscr{C}) \quad (\forall x_2)(\mathscr{D}(\bar{m}, x_2) \Rightarrow \mathscr{B}(x_2))$$

Let q be the Gödel number of this wf $\mathscr{C}$. So, $\bar{q} = \ulcorner \mathscr{C} \urcorner$. Clearly, $D(m) = q$. [In fact, m is the Gödel number of a wf $\mathscr{A}(x_1)$—namely, (∇)—and q is the Gödel number of $\mathscr{A}(\bar{m})$.] Since $\mathscr{D}$ represents D in K,

$$(\partial) \quad \vdash_K \mathscr{D}(\bar{m}, \bar{q})$$

(a) Let us show $\vdash_K \mathscr{C} \Rightarrow \mathscr{B}(\bar{q})$.

1. $\mathscr{C}$	Hyp
2. $(\forall x_2)(\mathscr{D}(\bar{m}, x_2) \Rightarrow \mathscr{B}(x_2))$	Same as 1
3. $\mathscr{D}(\bar{m}, \bar{q}) \Rightarrow \mathscr{B}(\bar{q})$	2, Rule A4
4. $\mathscr{D}(\bar{m}, \bar{q})$	(∂)
5. $\mathscr{B}(\bar{q})$	3, 4, MP
6. $\mathscr{C} \vdash_K \mathscr{B}(\bar{q})$	1–5
7. $\vdash_K \mathscr{C} \Rightarrow \mathscr{B}(\bar{q})$	1–6, Corollary 2.5

(b) Let us prove $\vdash_K \mathscr{B}(\bar{q}) \Rightarrow \mathscr{C}$.

1. $\mathscr{B}(\bar{q})$	Hyp
2. $\mathscr{D}(\bar{m}, x_2)$	Hyp
3. $(\exists_1 x_2)\mathscr{D}(\bar{m}, x_2)$	$\mathscr{D}$ represents D
4. $\mathscr{D}(\bar{m}, \bar{q})$	(∂)
5. $x_2 = \bar{q}$	2–4, properties of $=$
6. $\mathscr{B}(x_2)$	1, 5, substitutivity of $=$
7. $\mathscr{B}(\bar{q}), \mathscr{D}(\bar{m}, x_2) \vdash_K \mathscr{B}(x_2)$	1–6
8. $\mathscr{B}(\bar{q}) \vdash_K \mathscr{D}(\bar{m}, x_2) \Rightarrow \mathscr{B}(x_2)$	1–7, Corollary 2.5
9. $\mathscr{B}(\bar{q}) \vdash_K (\forall x_2)\mathscr{D}(\bar{m}, x_2) \Rightarrow \mathscr{B}(x_2))$	8, Gen
10. $\vdash_K \mathscr{B}(\bar{q}) \Rightarrow (\forall x_2)(\mathscr{D}(\bar{m}, x_2) \Rightarrow \mathscr{B}(x_2))$	1–9, Corollary 2.6
11. $\vdash_K \mathscr{B}(\bar{q}) \Rightarrow \mathscr{C}$	Same as 10

From parts (a) and (b), by Biconditional Introduction, $\vdash_K \mathscr{C} \Leftrightarrow \mathscr{B}(\bar{q})$.

PROPOSITION 3.33 (FIXED POINT THEOREM)* Let K be a theory with equality that has the same symbols as S and in which all recursive functions are representable. Then, for any wf $\mathscr{B}(x_1)$ that contains x_1 as its only free variable, there is a closed wf $\mathscr{C}$ such that

$$\vdash_K \mathscr{C} \Leftrightarrow \mathscr{B}(\ulcorner \mathscr{C} \urcorner)$$

Proof By Proposition 3.26, D is primitive recursive. Hence, D is representable in K and Proposition 3.32 is applicable.

By Proposition 3.31, Proposition 3.33 holds when K is RR or any extension of RR (including S).

DEFINITION Let K be any theory containing the individual constant 0 and the function letter f_1^1. Then K is said to be *ω-consistent* if and only if, for every wf $\mathscr{A}(x)$ of K, if $\vdash_K \neg\mathscr{A}(\bar{n})$ for every natural number n, then it is not the case that $\vdash_K (\exists x)\mathscr{A}(x)$.

Any theory K that has the standard interpretation as a model must be ω-consistent. [In fact, if $\vdash_K \neg\mathscr{A}(\bar{n})$ for all n, then $\mathscr{A}(x)$ is false for every natural

*The terms "Fixed Point Theorem" and "Diagonalization Lemma" are often used interchangeably, but I have adopted the present terminology for convenience of reference. The central idea seems to have first received explicit mention by Carnap (1934), who pointed out that the result was implicit in the work of Gödel (1931). The use of indirect self-reference was the key idea in the explosion of progress in mathematical logic that began in the 1930s.

number, and, therefore, $(\exists x)\mathscr{A}(x)$ cannot be true for the standard model. Hence, $(\exists x)\mathscr{A}(x)$ cannot be a theorem of K.] In particular, RR and S are ω-consistent.

PROPOSITION 3.34 If K is ω-consistent, then K is consistent.

Proof Let $\mathscr{B}(x)$ be any wf of K containing x as a free variable. Let $\mathscr{A}(x)$ be $\mathscr{B}(x) \wedge \neg\mathscr{B}(x)$. Then, $\neg\mathscr{A}(\bar{n})$ is an instance of a tautology. Hence, $\vdash_K \neg\mathscr{A}(\bar{n})$ for every natural number n. By ω-consistency, not-$\vdash_K (\exists x)\mathscr{A}(x)$. Therefore, K is consistent. [Remember that *every* wf is provable in an inconsistent theory, by virtue of the tautology $\neg A \Rightarrow (A \Rightarrow B)$. Hence, if at least one wf is not provable, the theory must be consistent.]

It will turn out later that the converse of Proposition 3.34 does not hold.

DEFINITION An *undecidable sentence* of a theory K is a closed wf $\mathscr{A}$ of K such that neither $\mathscr{A}$ nor $\neg\mathscr{A}$ is a theorem of K—that is, such that not-$\vdash_K \mathscr{A}$ and not-$\vdash_K \neg\mathscr{A}$.

Let K be a theory with equality that has the same symbols as S and about which we shall make the following assumptions:

1. The property PrAx(y) is recursive.
2. $\vdash_K 0 \neq \bar{1}$.
3. Every recursive function is representable in K.

By assumption 1, Propositions 3.25–3.27 are applicable. By assumptions 2 and 3 and Proposition 3.12, every recursive relation is expressible in K. K can be taken to be RR, S, or, more generally, any extension of RR that satisfies assumption 1.

Recall that Pf(y, x) means that y is the Gödel number of a proof in K of a wf with Gödel number x. By Proposition 3.27, Pf is recursive. Hence, Pf is expressible in K by a wf $\mathscr{P}\!f(x_2, x_1)$.

Let $\mathscr{B}(x_1)$ be the wf $(\forall x_2)\neg\mathscr{P}\!f(x_2, x_1)$. By the Fixed Point Theorem, there must be a closed wf $\mathscr{G}$ such that

$$\vdash_K \quad \mathscr{G} \Leftrightarrow (\forall x_2)\neg\mathscr{P}\!f(x_2, \ulcorner\mathscr{G}\urcorner)$$

Observe that, in terms of the standard interpretation, $(\forall x_2)\neg\mathscr{P}\!f(x_2, \ulcorner\mathscr{G}\urcorner)$ says that there is no natural number x_2 that is the Gödel number of a proof in K of the wf $\mathscr{G}$, which is equivalent to asserting that there is no proof in K of $\mathscr{G}$. Hence, $\mathscr{G}$ is equivalent in K to an assertion that $\mathscr{G}$ is unprovable in K. In other words, $\mathscr{G}$ says "I am not provable in K". This is an analogue of the Liar Paradox: "I am lying" (that is, "I am not true"). However, although the Liar Paradox leads to a contradiction, Gödel (1931) showed that $\mathscr{G}$ is an undecidable sentence of K. We shall refer to $\mathscr{G}$ as a *Gödel sentence* for K.

PROPOSITION 3.35 (GÖDEL'S INCOMPLETENESS THEOREM)

(a) If K is consistent, not-$\vdash_K \mathscr{G}$.
(b) If K is ω-consistent, not-$\vdash_K \neg\mathscr{G}$.

Hence, if K is ω-consistent, $\mathscr{G}$ is an undecidable sentence of K.

Proof Let q be the Gödel number of $\mathscr{G}$.

(a) Assume $\vdash_K \mathscr{G}$. Let r be the Gödel number of a proof in K of $\mathscr{G}$. Then $\mathrm{Pf}(r, q)$. Hence, $\vdash_K \mathscr{Pf}(\overline{r}, \overline{q})$—that is, $\vdash_K \mathscr{Pf}(\overline{r}, \ulcorner\mathscr{G}\urcorner)$. We already have $\vdash_K \mathscr{G} \Leftrightarrow (\forall x_2) \neg \mathscr{Pf}(x_2, \ulcorner\mathscr{G}\urcorner)$. By Biconditional Elimination, $\vdash_K (\forall x_2) \neg \mathscr{Pf}(x_2, \ulcorner\mathscr{G}\urcorner)$. By Rule A4, $\vdash_K \neg \mathscr{Pf}(\overline{r}, \ulcorner\mathscr{G}\urcorner)$. Therefore, K is inconsistent.

(b) Assume K is ω-consistent and $\vdash_K \neg\mathscr{G}$. Since $\vdash_K \mathscr{G} \Leftrightarrow (\forall x_2) \neg \mathscr{Pf}(x_2, \ulcorner\mathscr{G}\urcorner)$, Biconditional Elimination yields $\vdash_K \neg(\forall x_2) \neg \mathscr{Pf}(x_2, \ulcorner\mathscr{G}\urcorner)$, which abbreviates to (∗) $\vdash_K (\exists x_2)\mathscr{Pf}(x_2, \ulcorner\mathscr{G}\urcorner)$. On the other hand, since K is ω-consistent, Proposition 3.34 implies that K is consistent. But, $\vdash_K \neg\mathscr{G}$. Hence, not-$\vdash_K \mathscr{G}$; that is, there is no proof in K of $\mathscr{G}$. So, $\mathrm{Pf}(n, q)$ is false for every natural number n and, therefore, $\vdash_K \neg \mathscr{Pf}(\overline{n}, \ulcorner\mathscr{G}\urcorner)$ for every natural number n. (Remember that $\ulcorner\mathscr{G}\urcorner$ is $\overline{q}$.) By ω-consistency, not-$\vdash_K (\exists x_2)\mathscr{Pf}(x_2, \ulcorner\mathscr{G}\urcorner)$, contradicting (∗).

REMARK Gödel's Incompleteness Theorem has been established for any theory with equality K having the same symbols as S and satisfying conditions 1–3 listed above. Assume that K also satisfies the following condition.

(+) The standard interpretation is a model of K.

(In particular, K can be S or a subtheory of S.) Now, part (a) of Proposition 3.35 shows that, if K is consistent, $\mathscr{G}$ is not provable in K. But, under the standard interpretation, $\mathscr{G}$ asserts its own unprovability in K. Therefore, $\mathscr{G}$ *is, in fact, true for the standard interpretation.* Moreover, when K satisfies (+), the following simple intuitive argument can be given for the undecidability of $\mathscr{G}$ in K.

(a) Assume $\vdash_K \mathscr{G}$. Since $\vdash_K \mathscr{G} \Leftrightarrow (\forall x_2) \neg \mathscr{Pf}(x_2, \ulcorner\mathscr{G}\urcorner)$, it follows that $\vdash_K (\forall x_2) \neg \mathscr{Pf}(x_2, \ulcorner\mathscr{G}\urcorner)$. By (+), $(\forall x_2) \neg \mathscr{Pf}(x_2, \ulcorner\mathscr{G}\urcorner)$ is true for the standard interpretation. But this wf says that $\mathscr{G}$ is not provable in K. So, $\mathscr{G}$ is not provable in K, contradicting our original assumption.

(b) Assume $\vdash_K \neg\mathscr{G}$. But, $\vdash_K \mathscr{G} \Leftrightarrow (\forall x_2) \neg \mathscr{Pf}(x_2, \ulcorner\mathscr{G}\urcorner)$. Hence, $\vdash_K \neg(\forall x_2) \neg \mathscr{Pf}(x_2, \ulcorner\mathscr{G}\urcorner)$. So, $\vdash_K (\exists x_2)\mathscr{Pf}(x_2, \ulcorner\mathscr{G}\urcorner)$. By (+), this wf is true for the standard interpretation; that is, $\mathscr{G}$ is provable in K. Thus, $\vdash_K \mathscr{G}$, contradicting part (a).

Exercises

3.44 Let $\mathscr{G}$ be a Gödel sentence for S. Let S_g be the extension of S obtained by adding $\neg\mathscr{G}$ as a new axiom. Prove that, if S is consistent, then S_g is consistent but not ω-consistent.

3.45 A theory K that has the individual constant 0 and function symbol f_1^1 is said to be *ω-incomplete* if there is a wf $\mathscr{A}(x)$ such that $\vdash_K \mathscr{A}(\overline{n})$ for every natural number n, but it is not the case that $\vdash_K (\forall x)\mathscr{A}(x)$. If K is a consistent theory with equality that has the same symbols as S and satisfies conditions 1–3 on page 161, show that K is ω-incomplete. (In particular, RR and S are ω-incomplete.)

3.46 Let K be a theory having the individual constant 0 and function letter f_1^1. Show that, if K is consistent and ω-inconsistent, then K is ω-incomplete.

The proof of undecidability of a Gödel sentence $\mathscr{G}$ required the assumption of ω-consistency. We will now prove a result of Rosser (1936b) showing that, at the cost of a slight increase in the complexity of the undecidable sentence, the assumption of ω-consistency can be replaced by consistency.

As before, let K be a theory with equality having the same symbols as S and satisfying conditions 1–3 on page 161. In addition, assume the following.

4. $\vdash_K x \leqslant \bar{n} \Rightarrow x = 0 \vee x = \bar{1} \vee \cdots \vee x = \bar{n}$ for every natural number n.
5. $\vdash_K x \leqslant \bar{n} \vee \bar{n} \leqslant x$ for every natural number n.

(Thus, K can be any extension of RR that satisfies condition 1. In particular, K can be RR or S.)

Recall that, by Proposition 3.25(14), Neg is a recursive function such that, if x is the Gödel number of a wf $\mathscr{A}$, then Neg(x) is the Gödel number of $\neg\mathscr{A}$. By condition 3, let $\mathscr{N}eg(x_1, x_2)$ be a wf that represents Neg in K. Now construct the following wf $\mathscr{E}(x_1)$:

$$(\forall x_2)(\mathscr{P}f(x_2, x_1) \Rightarrow (\forall x_3)(\mathscr{N}eg(x_1, x_3) \Rightarrow (\exists x_4)(x_4 \leqslant x_2 \wedge \mathscr{P}f(x_4, x_3))))$$

By the Fixed Point Theorem, there is a closed wf $\mathscr{R}$ such that

$$\vdash_K \mathscr{R} \Leftrightarrow \mathscr{E}(\ulcorner \mathscr{R} \urcorner)$$

$\mathscr{R}$ is called a *Rosser sentence* for K.

Notice what the intuitive meaning of $\mathscr{R}$ is under the standard interpretation. $\mathscr{R}$ asserts that, if $\mathscr{R}$ has a proof in K with Gödel number x_2, then $\neg\mathscr{R}$ has a proof in K with Gödel number smaller than x_2. This is a roundabout way for $\mathscr{R}$ to claim its own unprovability under the assumption of the consistency of K.

PROPOSITION 3.36 (GÖDEL-ROSSER THEOREM) Let K satisfy conditions 1–5. If K is consistent, $\mathscr{R}$ is an undecidable sentence in K.

Proof Let p be the Gödel number of $\mathscr{R}$. Thus, $\ulcorner \mathscr{R} \urcorner$ is $\bar{p}$. In addition, let j be the Gödel number of $\neg\mathscr{R}$.

(a) Assume $\vdash_K \mathscr{R}$. Since $\vdash_K \mathscr{R} \Leftrightarrow \mathscr{E}(\ulcorner \mathscr{R} \urcorner)$, Biconditional Elimination yields $\vdash_K \mathscr{E}(\ulcorner \mathscr{R} \urcorner)$; that is,

$$(\$) \quad \vdash_K (\forall x_2)(\mathscr{P}f(x_2, \bar{p}) \Rightarrow (\forall x_3)(\mathscr{N}eg(\bar{p}, x_3) \Rightarrow (\exists x_4)(x_4 \leqslant x_2 \wedge \mathscr{P}f(x_4, x_3))))$$

Let k be the Gödel number of a proof in K of $\mathscr{R}$. Then, Pf(k, p) and, therefore, $\vdash_K \mathscr{P}f(\bar{k}, \bar{p})$. Applying Rule A4 to (\$), we obtain

$$\vdash_K \mathscr{P}f(\bar{k}, \bar{p}) \Rightarrow (\forall x_3)(\mathscr{N}eg(\bar{p}, x_3) \Rightarrow (\exists x_4)(x_4 \leqslant \bar{k} \wedge \mathscr{P}f(x_4, x_3)))$$

So, by MP,

$$(\%) \quad \vdash_K (\forall x_3)(\mathscr{N}eg(\bar{p}, x_3) \Rightarrow (\exists x_4)(x_4 \leqslant \bar{k} \wedge \mathscr{P}f(x_4, x_3)))$$

Since j is the Gödel number of $\neg\mathscr{R}$, we have Neg(p,j) and, therefore, $\vdash_K \mathscr{N}eg(\bar{p}, \bar{j})$. Applying Rule A4 to (%), we obtain $\vdash_K \mathscr{N}eg(\bar{p}, \bar{j}) \Rightarrow (\exists x_4)(x_4 \leqslant \bar{k} \wedge \mathscr{P}f(x_4, \bar{j}))$. Hence, by MP, $\vdash_K (\exists x_4)(x_4 \leqslant \bar{k} \wedge \mathscr{P}f(x_4, \bar{j}))$, which is an abbreviation for

$$(\#) \quad \vdash_K \neg(\forall x_4)\neg(x_4 \leqslant \bar{k} \wedge \mathscr{P}f(x_4, \bar{j}))$$

Now, since $\vdash_K \mathscr{R}$, the consistency of K implies not-$\vdash_K \neg\mathscr{R}$. Hence, Pf(n,j) is false for all natural numbers n. Therefore, $\vdash_K \neg\mathscr{P}f(\bar{n}, \bar{j})$ for all natural numbers n. Since

K is a theory with equality, $\vdash_K x_4 = \bar{n} \Rightarrow \neg \mathscr{P}\!f(x_4, \bar{j})$ for all natural numbers n. By condition 4,

$$(\S) \quad \vdash_K x_4 \leqslant \bar{k} \Rightarrow x_4 = 0 \vee x_4 = \bar{1} \vee \cdots \vee x_4 = \bar{k}$$

But,

$$(\S\S) \quad \vdash_K x_4 = \bar{n} \Rightarrow \neg \mathscr{P}\!f(x_4, \bar{j}) \qquad \text{for } n = 0, 1, \ldots, k$$

So, by a suitable tautology, ($\S$) and ($\S\S$) yield $\vdash_K x_4 \leqslant \bar{k} \Rightarrow \neg \mathscr{P}\!f(x_4, \bar{j})$ and, then, by another tautology, $\vdash_K \neg(x_4 \leqslant \bar{k} \wedge \mathscr{P}\!f(x_4, \bar{j}))$. By Gen, $\vdash_K (\forall x_4)\neg(x_4 \leqslant \bar{k} \wedge \mathscr{P}\!f(x_4, \bar{j}))$. This, together with (#), contradicts the consistency of K.

(b) Assume $\vdash_K \neg\mathscr{R}$. Let t be the Gödel number of a proof of $\neg\mathscr{R}$ in K. So, Pf(t, j) is true and, therefore, $\vdash_K \mathscr{P}\!f(\bar{t}, \bar{j})$. Hence, by an application of Rule E4, $\vdash_K \bar{t} \leqslant x_2 \Rightarrow (\exists x_4)(x_4 \leqslant x_2 \wedge \mathscr{P}\!f(x_4, \bar{j}))$. By consistency of K, not-$\vdash_K \mathscr{R}$, and so, Pf(n, p) is false for all natural numbers n. Hence, $\vdash_K \neg\mathscr{P}\!f(\bar{n}, \bar{p})$ for all natural numbers n. By condition 4, $\vdash_K x_2 \leqslant \bar{t} \Rightarrow x_2 = 0 \vee x_2 = \bar{1} \vee \cdots \vee x_2 = \bar{t}$. Hence, $\vdash_K x_2 \leqslant \bar{t} \Rightarrow \neg\mathscr{P}\!f(x_2, \bar{p})$. Consider the following derivation.

1. $\mathscr{P}\!f(x_2, \bar{p})$	Hyp
2. $\mathscr{N}\!eg(\bar{p}, x_3)$	Hyp
3. $x_2 \leqslant \bar{t} \vee \bar{t} \leqslant x_2$	Condition 5
4. $\bar{t} \leqslant x_2 \Rightarrow (\exists x_4)(x_4 \leqslant x_2 \wedge \mathscr{P}\!f(x_4, \bar{j}))$	Proved above
5. $x_2 \leqslant \bar{t} \Rightarrow \neg\mathscr{P}\!f(x_2, \bar{p})$	Proved above
6. $\neg\mathscr{P}\!f(x_2, \bar{p}) \vee (\exists x_4)(x_4 \leqslant x_2 \wedge \mathscr{P}\!f(x_4, \bar{j}))$	3–5, Tautology
7. $(\exists x_4)(x_4 \leqslant x_2 \wedge \mathscr{P}\!f(x_4, \bar{j}))$	1, 6, Tautology
8. $\mathscr{N}\!eg(\bar{p}, \bar{j})$	Proved in part (a)
9. $(\exists_1 x_3)\mathscr{N}\!eg(\bar{p}, x_3)$	$\mathscr{N}\!eg$ represents Neg
10. $x_3 = \bar{j}$	2, 8, 9, properties of =
11. $(\exists x_4)(x_4 \leqslant x_2 \wedge \mathscr{P}\!f(x_4, x_3))$	7, 10, substitutivity of =
12. $\mathscr{P}\!f(x_2, \bar{p}), \mathscr{N}\!eg(\bar{p}, x_3) \vdash_K (\exists x_4)(x_4 \leqslant x_2 \wedge \mathscr{P}\!f(x_4, x_3))$	1–11
13. $\mathscr{P}\!f(x_2, \bar{p}) \vdash_K \mathscr{N}\!eg(\bar{p}, x_3) \Rightarrow (\exists x_4)(x_4 \leqslant x_2 \wedge \mathscr{P}\!f(x_4, x_3))$	1–12, Corollary 2.5
14. $\mathscr{P}\!f(x_2, \bar{p}) \vdash_K (\forall x_3)(\mathscr{N}\!eg(\bar{p}, x_3) \Rightarrow (\exists x_4)(x_4 \leqslant x_2 \wedge \mathscr{P}\!f(x_4, x_3)))$	13, Gen
15. $\vdash_K \mathscr{P}\!f(x_2, \bar{p}) \Rightarrow (\forall x_3)(\mathscr{N}\!eg(\bar{p}, x_3) \Rightarrow (\exists x_4)(x_4 \leqslant x_2 \wedge \mathscr{P}\!f(x_4, x_3)))$	1–14, Corollary 2.5
16. $\vdash_K (\forall x_2)(\mathscr{P}\!f(x_2, \bar{p}) \Rightarrow (\forall x_3)(\mathscr{N}\!eg(\bar{p}, x_3) \Rightarrow (\exists x_4)(x_4 \leqslant x_2 \wedge \mathscr{P}\!f(x_4, x_3))))$	15, Gen
17. $\vdash_K \mathscr{R}$	Same as 16

Thus, $\vdash_K \mathscr{R}$ and $\vdash_K \neg\mathscr{R}$, contradicting the consistency of K.

The Gödel and Rosser sentences for the theory S are undecidable sentences of S. They have a certain intuitive metamathematical meaning; for example, a Gödel sentence $\mathscr{G}$ asserts that $\mathscr{G}$ is unprovable in S. Until recently, no undecidable sentences of S were known that had intrinsic mathematical interest. However, in 1977, a mathematically significant sentence of combinatorics, related to the so-called

Finite Ramsey Theorem, was shown to be undecidable in S (see Kirby & Paris, 1977; Paris, 1978; and Paris & Harrington, 1977).

DEFINITION A theory K is said to be *recursively axiomatizable* if and only if there is a theory K′ having the same theorems as K such that the property $\mathrm{PrAx}_{K'}$ is recursive.

COROLLARY 3.37 Let K be a theory having the same symbols as S. If K is a consistent, recursively axiomatizable extension of RR, then K has an undecidable sentence.

Proof Let K′ be a theory having the same theorems as K and such that $\mathrm{PrAx}_{K'}$ is recursive. Conditions 1–5 of Proposition 3.36 hold for K′. Hence, a Rosser sentence for K′ is undecidable in K′ and, therefore, also undecidable in K.

An *effectively decidable* set of objects is a set for which there is a mechanical procedure that determines, for any given object, whether or not that object belongs to the set. By a *mechanical procedure* we mean a procedure that is carried out automatically without any need for originality or ingenuity in its application. On the other hand, a set A of natural numbers is said to be *recursive* if the property $x \in A$ is recursive.* It will appear plausible after Chapter 5 that *the precise notion of recursive set corresponds to the intuitive idea of an effectively decidable set of natural numbers.* This hypothesis is known as *Church's Thesis.*

Remember that a theory is said to be *axiomatic* if the set of its axioms is effectively decidable. Clearly, the set of axioms is effectively decidable if and only if the set of Gödel numbers of axioms is effectively decidable (since we can pass effectively from a wf to its Gödel number and, conversely, from the Gödel number to the wf). Hence, if we accept Church's Thesis, to say that PrAx_K is recursive is equivalent to saying that K is an axiomatic theory, and, therefore, Corollary 3.37 shows that RR is *essentially incomplete*—that is, that every consistent axiomatic extension of RR has an undecidable sentence. This result is very disturbing; it tells us that there is no complete axiomatization of arithmetic; that is, there is no way to set up an axiom system on the basis of which we can decide all problems of number theory.

Exercises

3.47 Church's Thesis is usually taken in the form that *a number-theoretic function is effectively computable if and only if the function is recursive.* Prove that this is equivalent to the form of Church's Thesis given above.

3.48 Let K be a theory that satisfies the hypotheses of the Gödel-Rosser Theorem (Proposition 3.36) and for which the standard interpretation is a model. Determine whether the Rosser sentence $\mathscr{R}$ for K is true for the standard interpretation.

3.49 (Church, 1936b) Let Tr be the set of Gödel numbers of all wfs of S that are true for the standard interpretation. Prove that Tr is not recursive. (Hence, under the assumption of

*To say that $x \in A$ is recursive means that the characteristic function C_A is a recursive function, where $C_A(x) = 0$ if $x \in A$ and $C_A(x) = 1$ if $x \notin A$ (see page 137).

Church's Thesis, there is no effective procedure for determining the truth or falsity of arbitrary sentences of arithmetic.)

3.50 Prove that there is no recursively axiomatizable theory that has Tr as the set of Gödel numbers of its theorems.

3.51 Let K be a theory with equality having the same symbols as S and satisfying conditions 4 and 5 on page 163. If every recursive relation is expressible in K, prove that every recursive function is representable in K.

GÖDEL'S SECOND THEOREM

Let K be an extension of S with the same symbols as S and for which $PrAx_K$ is recursive. Let $\mathcal{C}on_K$ be the following wf of K:

$$(\forall x_1)(\forall x_2)(\forall x_3)(\forall x_4)\neg(\mathcal{P}f(x_1, x_3) \wedge \mathcal{P}f(x_2, x_4) \wedge \mathcal{N}eg(x_3, x_4))$$

For the standard interpretation, $\mathcal{C}on_K$ asserts that there are no proofs in K of a wf and its negation—that is, that K is consistent. Consider the following sentence:

$$(\Xi) \quad \mathcal{C}on_K \Rightarrow \mathcal{G}$$

where $\mathcal{G}$ is a Gödel sentence for K. Remember that $\mathcal{G}$ asserts that $\mathcal{G}$ is unprovable in K. Hence, (Ξ) states that, if K is consistent, then $\mathcal{G}$ is not provable in K. But that is just the first half of Gödel's Incompleteness Theorem. The metamathematical reasoning used in the proof of that theorem can be expressed and carried through within K itself, so that one obtains a proof in K of (Ξ) (see Hilbert & Bernays, 1939, pp. 285–328; Feferman, 1960). Thus, $\vdash_K \mathcal{C}on_K \Rightarrow \mathcal{G}$. But, by Gödel's Incompleteness Theorem, if K is consistent, $\mathcal{G}$ is not provable in K. Hence, *if K is consistent, $\mathcal{C}on_K$ is not provable in K*. This is *Gödel's Second Theorem* (1931). One can paraphrase it by stating that, if K is consistent, then the consistency of K cannot be proved within K or, equivalently, a consistency proof of K must use ideas and methods that go beyond those available in K. Consistency proofs for S have been given by Gentzen (1936, 1938) and Schütte (1951), and these proofs do, in fact, employ notions and methods (for example, a portion of the theory of denumerable ordinal numbers) that apparently are not formalizable in S.

Gödel's Second Theorem is sometimes stated in the form that, if a "sufficiently strong" theory K is consistent, then the consistency of K cannot be proved within K. Aside from the vagueness of the phrase "sufficiently strong" (which can be made precise without much difficulty), the way in which the consistency of K is formulated is crucial. Feferman (1960, Cor. 5.10) has shown that there is a way of formalizing the consistency of S—say, $\mathcal{C}on_S^*$—such that $\vdash_S \mathcal{C}on_S^*$. A precise formulation of Gödel's Second Theorem may be found in Feferman (1960). [See Jeroslow (1971, 1972, 1973) for further clarification and development.]

In their proof of Gödel's Second Theorem, Hilbert and Bernays (1939) based their work on three so-called *derivability conditions*. For the sake of definiteness, we shall limit ourselves to the theory S, although everything we say also holds for recursively axiomatizable extensions of S. To formulate the Hilbert-Bernays results, let $\mathcal{B}ew(x_1)$ stand for $(\exists x_2)\mathcal{P}f(x_2, x_1)$. Thus, under the standard interpretation, $\mathcal{B}ew(x_1)$ means that there is a proof in S of the wf with Gödel number x_1; that is, the wf with Gödel number x_1 is provable in S.† Notice that a Gödel sentence satisfies the fixed point condition: $\vdash_K \mathscr{G} \Leftrightarrow \neg \mathscr{B}ew(\ulcorner \mathscr{G} \urcorner)$.

THE HILBERT-BERNAYS DERIVABILITY CONDITIONS

(HB1) If $\vdash_S \mathscr{A}$, then $\vdash_S \mathscr{B}ew(\ulcorner \mathscr{A} \urcorner)$
(HB2) $\vdash_S \mathscr{B}ew(\ulcorner \mathscr{A} \Rightarrow \mathscr{B} \urcorner) \Rightarrow (\mathscr{B}ew(\ulcorner \mathscr{A} \urcorner) \Rightarrow \mathscr{B}ew(\ulcorner \mathscr{B} \urcorner))$
(HB3) $\vdash_S \mathscr{B}ew(\ulcorner \mathscr{A} \urcorner) \Rightarrow \mathscr{B}ew(\ulcorner \mathscr{B}ew(\ulcorner \mathscr{A} \urcorner) \urcorner)$

Here, $\mathscr{A}$ and $\mathscr{B}$ are arbitrary closed wfs of S. (HB1) is straightforward and (HB2) is an easy consequence of properties of $\mathscr{P}f$. However, (HB3) requires a careful and difficult proof. [A clear treatment may also be found in Boolos (1979, chap. 2).]

A Gödel sentence $\mathscr{G}$ for S asserts its own unprovability in S: $\vdash_S \mathscr{G} \Leftrightarrow \neg \mathscr{B}ew(\ulcorner \mathscr{G} \urcorner)$. We also can apply the Fixed Point Theorem to obtain a sentence $\mathscr{H}$ such that $\vdash_S \mathscr{H} \Leftrightarrow \mathscr{B}ew(\ulcorner \mathscr{H} \urcorner)$. $\mathscr{H}$ is called a *Henkin sentence* for S. $\mathscr{H}$ asserts its own provability in S. On intuitive grounds it is not clear whether $\mathscr{H}$ is true for the standard interpretation, nor is it easy to determine whether $\mathscr{H}$ is provable, disprovable, or undecidable in S. The problem was solved by Löb (1955) on the basis of the following result.

PROPOSITION 3.38 (LÖB'S THEOREM) Let $\mathscr{C}$ be a sentence of S. If $\vdash_S \mathscr{B}ew(\ulcorner \mathscr{C} \urcorner) \Rightarrow \mathscr{C}$, then $\vdash_S \mathscr{C}$.

Proof Apply the Fixed Point Theorem to the wf $\mathscr{B}ew(x_1) \Rightarrow \mathscr{C}$ to obtain a sentence $\mathscr{L}$ such that:

1.	$\vdash_S \mathscr{L} \Leftrightarrow (\mathscr{B}ew(\ulcorner \mathscr{L} \urcorner) \Rightarrow \mathscr{C})$	
2.	$\vdash_S \mathscr{L} \Rightarrow (\mathscr{B}ew(\ulcorner \mathscr{L} \urcorner) \Rightarrow \mathscr{C})$	1, Biconditional Elimination
3.	$\vdash_S \mathscr{B}ew(\ulcorner \mathscr{L} \Rightarrow (\mathscr{B}ew(\ulcorner \mathscr{L} \urcorner) \Rightarrow \mathscr{C}) \urcorner)$	2, (HB1), MP
4.	$\vdash_S \mathscr{B}ew(\ulcorner \mathscr{L} \urcorner) \Rightarrow \mathscr{B}ew(\ulcorner \mathscr{B}ew(\ulcorner \mathscr{L} \urcorner) \Rightarrow \mathscr{C} \urcorner)$	3, (HB2), MP
5.	$\vdash_S \mathscr{B}ew(\ulcorner \mathscr{B}ew(\ulcorner \mathscr{L} \urcorner) \Rightarrow \mathscr{C} \urcorner)$ $\Rightarrow (\mathscr{B}ew(\ulcorner \mathscr{B}ew(\ulcorner \mathscr{L} \urcorner) \urcorner) \Rightarrow \mathscr{B}ew(\ulcorner \mathscr{C} \urcorner))$	(HB2)
6.	$\vdash_S \mathscr{B}ew(\ulcorner \mathscr{L} \urcorner) \Rightarrow (\mathscr{B}ew(\ulcorner \mathscr{B}ew(\ulcorner \mathscr{L} \urcorner) \urcorner)$ $\Rightarrow \mathscr{B}ew(\ulcorner \mathscr{C} \urcorner))$	4, 5, Tautology
7.	$\vdash_S \mathscr{B}ew(\ulcorner \mathscr{L} \urcorner) \Rightarrow \mathscr{B}ew(\ulcorner \mathscr{B}ew(\ulcorner \mathscr{L} \urcorner) \urcorner)$	(HB3)
8.	$\vdash_S \mathscr{B}ew(\ulcorner \mathscr{L} \urcorner) \Rightarrow \mathscr{B}ew(\ulcorner \mathscr{C} \urcorner)$	6, 7, Tautology
9.	$\vdash_S \mathscr{B}ew(\ulcorner \mathscr{C} \urcorner) \Rightarrow \mathscr{C}$	Hypothesis of the theorem
10.	$\vdash_S \mathscr{B}ew(\ulcorner \mathscr{L} \urcorner) \Rightarrow \mathscr{C}$	8, 9, Tautology
11.	$\vdash_S \mathscr{L}$	1, 10, Biconditional Elimination
12.	$\vdash_S \mathscr{B}ew(\ulcorner \mathscr{L} \urcorner)$	11, (HB1)
13.	$\vdash_S \mathscr{C}$	10, 12, MP

COROLLARY 3.39 Let $\mathscr{H}$ be a Henkin sentence for S. Then $\vdash_S \mathscr{H}$ and $\mathscr{H}$ is true for the standard model.

† "Bew" are the first three letters of the German word *beweisbar*, which means "provable".

Proof $\vdash_S \mathscr{H} \Leftrightarrow \mathscr{Bew}(\ulcorner \mathscr{H} \urcorner)$. By Biconditional Elimination, $\vdash_S \mathscr{Bew}(\ulcorner \mathscr{H} \urcorner) \Rightarrow \mathscr{H}$. So, by Löb's Theorem, $\vdash_S \mathscr{H}$. Since $\mathscr{H}$ asserts that $\mathscr{H}$ is provable in S, $\mathscr{H}$ is true.

Löb's Theorem also enables us to give a proof of Gödel's Second Theorem for S.

PROPOSITION 3.40 (GÖDEL'S SECOND THEOREM) If S is consistent, then not-$\vdash_S \mathscr{Con}_S$.

Proof Assume S is consistent. Since $\vdash_S 0 \neq \bar{1}$, the consistency of S implies not-$\vdash_S 0 = \bar{1}$. By Löb's Theorem, not-$\vdash_S (\mathscr{Bew}(\ulcorner 0 = \bar{1} \urcorner) \Rightarrow 0 = \bar{1})$. Hence, by the tautology $\neg A \Rightarrow (A \Rightarrow B)$, we have:

$$(*) \quad \text{not-}\vdash_S \neg \mathscr{Bew}(\ulcorner 0 = \bar{1} \urcorner)$$

But, since $\vdash_S 0 \neq \bar{1}$, (HB1) yields $\vdash_S \mathscr{Bew}(\ulcorner 0 \neq \bar{1} \urcorner)$. Now it is easy to show that $\vdash_S \mathscr{Con}_S \Rightarrow \neg \mathscr{Bew}(\ulcorner 0 = \bar{1} \urcorner)$. Then, by (*), not-$\vdash_S \mathscr{Con}_S$.

Boolos (1979) gives an elegant and extensive study of the Fixed Point Theorem and Löb's Theorem in the context of an axiomatic treatment of provability predicates. Such an axiomatic approach was first proposed and developed by Magari (1975).

Exercises

3.52 Prove (HB1).

3.53 Prove (HB2).

3.54 Give the details of the proof of $\vdash_S \mathscr{Con}_S \Rightarrow \neg \mathscr{Bew}(\ulcorner 0 = \bar{1} \urcorner)$, which was used in the proof of Proposition 3.40.

3.55[D] If $\mathscr{G}$ is a Gödel sentence for S, prove $\vdash_S \mathscr{G} \Leftrightarrow \neg \mathscr{Bew}(\ulcorner 0 = \bar{1} \urcorner)$. [Hence, any two Gödel sentences for S are provably equivalent. This is an instance of a more general phenomenon of equivalence of fixed point sentences, first noticed and verified independently by Bernardi (1975, 1976), De Jongh, and Sambin (1976). See Smorynski (1979, 1982).]

3.56 In each of the following cases, apply the Fixed Point Theorem for S to obtain a sentence of the indicated kind; determine whether that sentence is provable in S, disprovable in S, or undecidable in S; and determine the truth or falsity of the sentence for the standard model.

(a) A sentence $\mathscr{C}$ that asserts its own decidability in S (i.e., that $\vdash_S \mathscr{C}$ or $\vdash_S \neg \mathscr{C}$)
(b) A sentence that asserts its own undecidability in S
(c) A sentence $\mathscr{C}$ asserting that not-$\vdash_S \neg \mathscr{C}$
(d) A sentence $\mathscr{C}$ asserting that $\vdash_S \neg \mathscr{C}$

6. RECURSIVE UNDECIDABILITY. CHURCH'S THEOREM

If K is a theory, let T_K be the set of Gödel numbers of theorems of K. We shall say that K is *recursively decidable* if and only if T_K is a recursive set (that is, $x \in T_K$ is recursive). K is *recursively undecidable* if and only if T_K is not recursive. Moreover, K is said to be *essentially recursively undecidable* if and only if K and all consistent extensions of K are recursively undecidable.

If we accept Church's Thesis, then recursive undecidability is equivalent to effective undecidability—that is, nonexistence of a mechanical decision procedure for theoremhood. The nonexistence of such a mechanical procedure means that ingenuity is required for determining whether any given wf is a theorem.

PROPOSITION 3.41 Let K be a consistent theory with equality having the same symbols as S and in which the diagonal function D is representable. Then the property $x \in T_K$ is not expressible in K.

Proof Assume $x \in T_K$ is expressible by a wf $\mathscr{T}(x_1)$. Thus,

(a) If $n \in T_K$, $\vdash_K \mathscr{T}(\overline{n})$.
(b) If $n \notin T_K$, $\vdash_K \neg\mathscr{T}(\overline{n})$.

By the Diagonalization Lemma applied to $\neg\mathscr{T}(x_1)$, there is a sentence $\mathscr{C}$ such that $\vdash_K \mathscr{C} \Leftrightarrow \neg\mathscr{T}(\ulcorner\mathscr{C}\urcorner)$. Let q be the Gödel number of $\mathscr{C}$. Thus,

(c) $\vdash_K \mathscr{C} \Leftrightarrow \neg\mathscr{T}(\overline{q})$

Case 1. $\vdash_K \mathscr{C}$. Then $q \in T_K$. By (a), $\vdash_K \mathscr{T}(\overline{q})$. But, from $\vdash_K \mathscr{C}$ and (c), by Biconditional Elimination, $\vdash_K \neg\mathscr{T}(\overline{q})$. Hence, K is inconsistent, contradicting our hypothesis.

Case 2. Not-$\vdash_K \mathscr{C}$. So, $q \notin T_K$. By (b), $\vdash_K \neg\mathscr{T}(\overline{q})$. Hence, by (c) and Biconditional Elimination, $\vdash_K \mathscr{C}$. But, we saw in Case 1 that this is impossible.

Thus, in either case, a contradiction is reached.

DEFINITION A set B of natural numbers is said to be *arithmetical* if and only if there is a wf $\mathscr{B}(x)$ of S, with one free variable x, such that, for every natural number n, $n \in B$ if and only if $\mathscr{B}(\overline{n})$ is true for the standard model.

COROLLARY 3.42 [TARSKI'S THEOREM (1936)] Let Tr be the set of Gödel numbers of wfs of S that are true for the standard model. Then Tr is not arithmetical.

Proof Let $\mathscr{N}$ be the extension of S that has as proper axioms all those wfs that are true for the standard model. Since every theorem of $\mathscr{N}$ must be true for the standard model, the theorems of $\mathscr{N}$ are identical with the axioms of $\mathscr{N}$. Hence, $T_{\mathscr{N}} = \mathrm{Tr}$. Thus, for any closed wf $\mathscr{A}$, $\mathscr{A}$ holds in the standard model if and only if $\vdash_{\mathscr{N}} \mathscr{A}$. It follows that a set B is arithmetical if and only if the property $x \in B$ is expressible in $\mathscr{N}$. We may assume that $\mathscr{N}$ is consistent because it has the standard interpretation as a model. Since every recursive function is representable in RR, every recursive function is representable in $\mathscr{N}$ and, therefore, D is representable in $\mathscr{N}$. By Proposition 3.41, $x \in Tr$ is not expressible in $\mathscr{N}$. Hence, Tr is not arithmetical. (This result can be roughly paraphrased by saying that the notion of arithmetical truth is not arithmetically definable.)

PROPOSITION 3.43 Let K be a consistent theory with equality that has the same symbols as S and in which all recursive functions are representable. Moreover, assume $\vdash_K 0 \neq \overline{1}$. Then, K is recursively undecidable.

Proof D is primitive recursive and, therefore, representable in K. By Proposition 3.41, the property $x \in T_K$ is not expressible in K. By Proposition 3.12, the characteristic function C_{T_K} is not representable in K. Hence, C_{T_K} is not a recursive function. Therefore, T_K is not a recursive set, and so, by definition, K is recursively undecidable.

COROLLARY 3.44 RR is essentially recursively undecidable.

Proof RR and all consistent extensions of RR satisfy the conditions on K in Proposition 3.43 and, therefore, are recursively undecidable. [We take for granted that RR is consistent because it has the standard interpretation as a model. More constructive consistency proofs can be given along the same lines as the proofs by Beth (1959, § 84) or Kleene (1952, § 79).]

We shall now show how this result can be used to give another derivation of the Gödel-Rosser Theorem.

PROPOSITION 3.45 Let K be a theory for which conditions (a)–(c) of Proposition 3.25 hold; that is, IC(x), FL(x), and PL(x) are recursive. If K is recursively axiomatizable and recursively undecidable, then K is incomplete (i.e., K has an undecidable sentence).

Proof By the recursive axiomatizability of K, there is a theory J that has the same theorems as K and for which PrAx is recursive. Clearly, since K and J have the same theorems, $T_K = T_J$, and, therefore, J is recursively undecidable and K is incomplete if and only if J is incomplete. So, it suffices to prove J incomplete. Notice that, since J and K have the same theorems, J and K must have the same individual constants, function letters, and predicate letters (because all such symbols occur in logical axioms). Thus, the hypotheses of Propositions 3.25 and 3.27 hold for J. Moreover, J is consistent, since an inconsistent theory is recursively decidable (see Exercise 3.57).

Assume J is complete. Remember that, if x is the Gödel number of a wf, Clos(x) is the Gödel number of the closure of that wf. By Proposition 3.25(16), Clos is a recursive function. Define:

$$H(x) = \mu y[(\mathrm{Fml}(x) \wedge (\mathrm{Pf}(y, \mathrm{Clos}(x)) \vee \mathrm{Pf}(y, \mathrm{Neg}(\mathrm{Clos}(x))))) \vee \neg \mathrm{Fml}(x)]$$

Notice that, if x is not the Gödel number of a wf, $H(x) = 0$. If x is the Gödel number of a wf $\mathscr{A}$, the closure of $\mathscr{A}$ is a closed wf, and, by the completeness of J, there is a proof in J of either the closure of $\mathscr{A}$ or its negation. Hence, $H(x)$ is obtained by a legitimate application of the restricted μ-operator and, therefore, H is a recursive function. Remember that a wf is provable if and only if its closure is provable. So, $x \in T_J$ if and only if Pf($H(x)$, Clos(x)). But, Pf($H(x)$, Clos(x)) is recursive. Thus, T_J is recursive, contradicting the recursive undecidability of J.

The intuitive idea behind this proof is the following. Given any wf $\mathscr{A}$, we form its closure $\mathscr{B}$ and start listing all the theorems in J. (Since PrAx is recursive, Church's Thesis tells us that J is an axiomatic theory, and, therefore, by the argument on page

68, we have an effective procedure for generating all the theorems.) If J is complete, either $\mathscr{B}$ or $\neg\mathscr{B}$ will eventually appear in the list of theorems. If $\mathscr{B}$ appears, $\mathscr{A}$ is a theorem. If $\neg\mathscr{B}$ appears, then, by the consistency of J, $\mathscr{B}$ will not appear among the theorems and, therefore, $\mathscr{A}$ is not a theorem. Thus, we have a decision procedure for theoremhood, and, again by Church's Thesis, J would be recursively decidable.

COROLLARY 3.46 (GÖDEL-ROSSER THEOREM) Any consistent recursively axiomatizable extension of RR has undecidable sentences.

Proof This is an immediate consequence of Corollary 3.44 and Proposition 3.45.

Exercises

3.57 Prove that an inconsistent theory is recursively decidable.
3.58 Prove that a recursively decidable theory must be recursively axiomatizable.
3.59 Let K be any recursively axiomatizable theory with equality having the same symbols as S, such that all theorems of K are true for the standard model; that is, $T_K \subseteq \mathrm{Tr}$. Prove that K has an undecidable sentence. (*Hint:* Use Proposition 3.45 and Exercise 3.49.)
3.60 Two sets A and B of natural numbers are said to be *recursively inseparable* if there is no recursive set C such that $A \subseteq C$ and $B \subseteq \bar{C}$. ($\bar{C}$ is the complement $\omega - C$.) Let K be any consistent theory with equality having the same symbols as S, in which every recursive function is representable and such that $\vdash_K 0 \neq \bar{1}$. Let Ref_K be the set of Gödel numbers of refutable wfs of K—that is, $\{x \mid \mathrm{Neg}(x) \in T_K\}$. Prove that T_K and Ref_K are recursively inseparable.

DEFINITIONS Let K_1 and K_2 be two theories that have the same symbols. K_2 is called a *finite extension* of K_1 if and only if there is a set A of wfs and a finite set B of wfs such that: (1) the theorems of K_1 are precisely the wfs derivable from A, and (2) the theorems of K_2 are precisely the wfs derivable from $A \cup B$.

Let $K_1 \cup K_2$ denote the theory whose set of axioms is the union of the set of axioms of K_1 and the set of axioms of K_2. We say that K_1 and K_2 are *compatible* if and only if $K_1 \cup K_2$ is consistent.

PROPOSITION 3.47 Let K_1 and K_2 be two theories that have the same symbols. If K_2 is a finite extension of K_1 and if K_2 is recursively undecidable, then K_1 is recursively undecidable.

Proof Let A be a set of axioms of K_1 and $A \cup \{\mathscr{A}_1, \ldots, \mathscr{A}_n\}$, a set of axioms for K_2. We may assume that $\mathscr{A}_1, \ldots, \mathscr{A}_n$ are closed wfs. Then by the Deduction Theorem (Corollary 2.6), it is easy to see that a wf $\mathscr{B}$ is provable in K_2 if and only if $(\mathscr{A}_1 \wedge \cdots \wedge \mathscr{A}_n) \Rightarrow \mathscr{B}$ is provable in K_1. Let c be a Gödel number of $(\mathscr{A}_1 \wedge \cdots \wedge \mathscr{A}_n)$. Then b is a Gödel number of a theorem of K_2 when and only when $2^3 * c * 2^{11} * b * 2^5$ is a Gödel number of a theorem of K_1; that is, b is in T_{K_2} if and only if $2^3 * c * 2^{11} * b * 2^5$ is in T_{K_1}. Hence, $C_{T_{K_2}}(x) = C_{T_{K_1}}(2^3 * c * 2^{11} * x * 2^5)$. Hence, if T_{K_1} were recursive, T_{K_2} would also be recursive, contradicting the recursive undecidability of K_2.

PROPOSITION 3.48 Let K be a theory with the same symbols as S. If K is compatible with RR, then K is recursively undecidable.

Proof Since K is compatible with RR, the theory $K \cup RR$ is a consistent extension of RR. Therefore, by Corollary 3.44, $K \cup RR$ is recursively undecidable. Since RR has a finite number of axioms, $K \cup RR$ is a finite extension of K. Hence, by Proposition 3.47, K is recursively undecidable.

COROLLARY 3.49 Let K be a theory with the same symbols as S such that all the axioms of K are true for the standard model. Then K is recursively undecidable.

Proof $K \cup RR$ has the standard interpretation as a model and is, therefore, consistent; that is, K is compatible with RR. Now apply Proposition 3.48.

COROLLARY 3.50 Let P_S be the predicate calculus having the same symbols as S. Then, P_S is recursively undecidable.

Proof $P_S \cup RR = RR$. Hence, P_S is compatible with RR and, therefore, by Proposition 3.48, recursively undecidable.

By PF we mean the *full* first-order predicate calculus containing all predicate letters A_j^n, function letters f_j^n, and individual constants a_j. Let PP be the *pure* first-order predicate calculus, containing all predicate letters but no function letters or individual constants.

LEMMA 3.51 There is a recursive function h such that, for any wf $\mathscr{A}$ of PF having Gödel number u, there is a wf $\mathscr{A}'$ of PP having Gödel number $h(u)$ such that $\mathscr{A}$ is provable in PF if and only if $\mathscr{A}'$ is provable in PP.

Proof Let $\mathscr{A}$ be a wf of PF. With the distinct function letters f_j^n in $\mathscr{A}$, associate distinct predicate letters A_r^{n+1} not occurring in $\mathscr{A}$, and with the distinct individual constants a_j in $\mathscr{A}$, associate distinct predicate letters A_k^1 not occurring in $\mathscr{A}$. Find the first individual constant a_j in $\mathscr{A}$. Let z be the first variable not in $\mathscr{A}$, and let $\mathscr{A}^*$ result from $\mathscr{A}$ by replacing all occurrences of a_j in $\mathscr{A}$ by z. Form the wf $\mathscr{A}_1$: $(\exists z)A_k^1(z) \Rightarrow (\exists z)(A_k^1(z) \wedge \mathscr{A}^*)$, where A_k^1 is the predicate letter associated with a_j. It is easy to check (see the proof of Proposition 2.28) that $\mathscr{A}$ is logically valid if and only if $\mathscr{A}_1$ is logically valid. Keep on performing similar transformations until a wf $\mathscr{B}$ is reached without individual constants; then $\mathscr{B}$ is logically valid if and only if $\mathscr{A}$ is logically valid. Next, take the leftmost term $f_l^n(t_1, \ldots, t_n)$ in $\mathscr{B}$, where $t_1, \ldots, t_n$ do not contain function letters. Let w be the first variable not in $\mathscr{B}$, let $\mathscr{B}^\#$ result from $\mathscr{B}$ by replacing $f_l^n(t_1, \ldots, t_n)$ by w, and let $\mathscr{B}_1$ be the wf $(\exists w)A_r^{n+1}(w, t_1, \ldots, t_n) \Rightarrow (\exists w)(A_r^{n+1}(w, t_1, \ldots, t_n) \wedge \mathscr{B}^\#)$, where A_r^{n+1} is the predicate letter associated with f_l^n. It is easy to verify that $\mathscr{B}$ is logically valid if and only if $\mathscr{B}_1$ is logically valid. Repeat the same transformation on $\mathscr{B}_1$, and so on, until a wf $\mathscr{A}'$ is reached that contains no function letters. Then $\mathscr{A}'$ is a wf of PP, and $\mathscr{A}'$ is logically valid if and only if

$\mathscr{A}$ is logically valid. By Gödel's Completeness Theorem (Corollary 2.19), $\mathscr{A}$ is logically valid if and only if $\vdash_{PF} \mathscr{A}$, and $\mathscr{A}'$ is logically valid if and only if $\vdash_{PP} \mathscr{A}'$. Now, if u is the Gödel number of $\mathscr{A}$, let $h(u)$ be the Gödel number of $\mathscr{A}'$. When u is not the Gödel number of a wf of PF, define $h(u)$ to be 0. Clearly, h is effectively computable because we have described an effective procedure for obtaining $\mathscr{A}'$ from $\mathscr{A}$. Therefore, by Church's Thesis, h is recursive. Alternatively, an extremely diligent reader could avoid the use of Church's Thesis by "arithmetizing" all the steps described above in the computation of h.

PROPOSITION 3.52 [CHURCH'S THEOREM (1936a)] PF and PP are recursively undecidable.

Proof

(a) By Gödel's Completeness Theorem, a wf $\mathscr{A}$ of P_S is provable in P_S if and only if $\mathscr{A}$ is logically valid, and $\mathscr{A}$ is provable in PF if and only if $\mathscr{A}$ is logically valid. Hence, $\vdash_{P_S} \mathscr{A}$ if and only if $\vdash_{PF} \mathscr{A}$. However, the set Fml_{P_S} of Gödel numbers of wfs of P_S is recursive. Then, $T_{P_S} = T_{PF} \cap Fml_{P_S}$, where T_{P_S} and T_{PF} are, respectively, the sets of Gödel numbers of the theorems of P_S and PF. If T_{PF} were recursive, T_{P_S} would be recursive, contradicting Corollary 3.50. Therefore, PF is recursively undecidable.

(b) By Lemma 3.51, u is in T_{PF} if and only if $h(u)$ is in T_{PP}. Since h is recursive, the recursiveness of T_{PP} would imply the recursiveness of T_{PF}, contradicting part (a). Thus, T_{PP} is not recursive; that is, PP is recursively undecidable.

If we accept Church's Thesis, then "recursively undecidable" can be replaced everywhere by "effectively undecidable". In particular, Proposition 3.52 states that there is no decision procedure for recognizing theoremhood either for the pure predicate calculus PP or for the full predicate calculus PF. By Gödel's Completeness Theorem, this implies that *there is no effective method for determining whether any given wf is logically valid.*

Exercises

3.61[D] (a) By a wf of the *pure monadic predicate calculus (PMP)* we mean a wf of the pure predicate calculus that does not contain predicate letters of more than one argument. Show that, in contrast to Church's Theorem, there is an effective procedure for determining whether a wf of PMP is logically valid. [*Hint:* Let $B_1, \dots, B_k$ be the distinct predicate letters in a wf $\mathscr{A}$. Then $\mathscr{A}$ is logically valid if and only if $\mathscr{A}$ is true for every interpretation with at most 2^k elements. (In fact, assume $\mathscr{A}$ is true for every interpretation with at most 2^k elements, and let M be any interpretation. For any elements b and c of the domain D of M, call b and c *equivalent* if the truth values of $B_1(b), B_2(b), \dots, B_k(b)$ in M are, respectively, the same as those of $B_1(c), B_2(c), \dots, B_k(c)$. This defines an equivalence relation in D, and the corresponding set of equivalence classes has $\leqslant 2^k$ members and can be made the domain of an interpretation M′ by defining interpretations of $B_1, \dots, B_k$, in the obvious way, on the equivalence classes. By induction on the length of wfs $\mathscr{B}$ that contain no predicate letters other than $B_1, \dots, B_k$, one can show that $\mathscr{B}$ is true for M if and only if it is true for M′. Since $\mathscr{A}$ is true for M′, it is also true for M. Hence, $\mathscr{A}$ is true for every interpretation.) Note also that whether $\mathscr{A}$ is true for every interpretation that has at most 2^k elements can be effectively determined.†]

(b) Prove that a wf $\mathscr{A}$ of PMP is logically valid if and only if $\mathscr{A}$ is true for all finite interpretations. (This contrasts with the situation in the pure predicate calculus; see Exercise 2.56 on page 74.)

3.62 If a theory K* is consistent, if every theorem of an essentially recursively undecidable theory K_1 is a theorem of K*, and, if the property $\mathrm{Fml}_{K_1}(y)$ is recursive, prove that K* is essentially recursively undecidable.

3.63 (Tarski, Mostowski & Robinson, 1953, I)

(a) Let K be a theory with equality. If a predicate letter A_j^n, a function letter f_j^n, and an individual constant a_j are not symbols of K, then by *possible definitions* of A_j^n, f_j^n, and a_j in K we mean, respectively, expressions of the form

(i) $(\forall x_1)\ldots(\forall x_n)(A_j^n(x_1,\ldots,x_n)\Leftrightarrow\mathscr{A}(x_1,\ldots,x_n))$

(ii) $(\forall x_1)\ldots(\forall x_n)(\forall y)(f_j^n(x_1,\ldots,x_n)=y\Leftrightarrow\mathscr{B}(x_1,\ldots,x_n,y))$

(iii) $(\forall y)(a_j=y\Leftrightarrow\mathscr{C}(y))$

where $\mathscr{A}$, $\mathscr{B}$, and $\mathscr{C}$ are wfs of K; moreover, in case (ii), we must also have $\vdash_K(\forall x_1)\ldots(\forall x_n)(\exists_1 y)\mathscr{B}(x_1,\ldots,x_n,y)$, and, in case (iii), $\vdash_K(\exists_1 y)\mathscr{C}(y)$. If K is consistent, prove that the addition of any possible definitions to K as new axioms (using only one possible definition for each symbol) yields a consistent theory K′, and K′ is recursively undecidable if and only if K is.

(b) By a *nonlogical constant*, we mean a predicate letter, function letter, or individual constant. Let K_1 be a theory with equality that has a finite number of nonlogical constants. Then K_1 is said to be *interpretable* in a theory with equality K if we can associate with each nonlogical constant of K_1 that is not a nonlogical constant of K a possible definition in K such that, if K* is the theory obtained from K by adding these possible definitions as axioms, then every axiom (and hence every theorem) of K_1 is a theorem of K*. Notice that, if K_1 is interpretable in K, it is interpretable in every extension of K. Prove that, if K_1 is interpretable in K and K is consistent, and if K_1 is essentially recursively undecidable, then K is essentially recursively undecidable.

3.64 Let K be a theory with equality and A_j^1 a monadic predicate letter not in K. Given a closed wf $\mathscr{A}$, let $\mathscr{A}^{(A_j^1)}$ (called the *relativization* of $\mathscr{A}$ with respect to A_j^1) be the wf obtained from $\mathscr{A}$ by replacing every subformula (starting from the smallest subformulas) of the form $(\forall x)\mathscr{B}(x)$ by $(\forall x)(A_j^1(x)\Rightarrow\mathscr{B}(x))$. Let the proper axioms of a new theory with equality $K^{A_j^1}$ be: (i) all wfs $\mathscr{A}^{(A_j^1)}$, where $\mathscr{A}$ is the closure of any proper axiom of K; (ii) $(\exists x)A_j^1(x)$; (iii) $A_j^1(a_m)$ for each individual constant a_m of K; and (iv) $A_j^1(x_1)\wedge\cdots\wedge A_j^1(x_n)\Rightarrow A_j^1(f_k^n(x_1,\ldots,x_n))$ for any function letter f_k^n of K. Prove:

(a) As proper axioms of $K^{A_j^1}$ we could have taken all wfs $\mathscr{A}^{(A_j^1)}$, where $\mathscr{A}$ is the closure of any theorem of K.

(b) $K^{A_j^1}$ is interpretable in K.

(c) $K^{A_j^1}$ is consistent if and only if K is consistent.

(d) $K^{A_j^1}$ is essentially recursively undecidable if and only if K is (Tarski, Mostowski & Robinson, 1953, pp. 27–28).

† The result in this exercise is, in a sense, the best possible. By a theorem of Kalmár (1936), there is an effective procedure producing, for each wf $\mathscr{A}$ of the pure predicate calculus, another wf $\mathscr{A}_2$ of the pure predicate calculus such that $\mathscr{A}_2$ contains only one predicate letter, a binary one, and such that $\mathscr{A}$ is logically valid if and only if $\mathscr{A}_2$ is logically valid. [For another proof, see Church (1956, §47).] Hence, by Church's Theorem, there is no decision procedure for logical validity of wfs that contain only binary predicate letters. (For another proof, see Exercise 4.73 on page 209.)

3.65 K is said to be *relatively interpretable* in K′ if there is some predicate letter A_j^1 not in K such that $K^{A_j^1}$ is interpretable in K′. If K is relatively interpretable in K′ and K is essentially recursively undecidable, prove that K′ is essentially recursively undecidable.

3.66 Call a theory K in which RR is relatively interpretable *sufficiently strong*. Prove that any sufficiently strong consistent theory K is essentially recursively undecidable, and, if K is also recursively axiomatizable, prove that K is incomplete. Roughly speaking, we may say that K is sufficiently strong if the notions of natural number, 0, 1, addition, and multiplication are "definable" in K in such a way that the axioms of RR (relativized to the "natural numbers" of K) are provable in K. Clearly, any theory adequate for present-day mathematics will be sufficiently strong and so, if it is consistent, it will be recursively undecidable and, if it is recursively axiomatizable, it will be incomplete. If we accept Church's Thesis, this implies that any consistent sufficiently strong theory will be effectively undecidable and, if it is axiomatic, it will have undecidable sentences. [Similar results also hold for higher-order theories; for example, see Gödel (1931) and Hasenjaeger and Scholz (1961, §§ 237–38).] *This destroys all hope for a consistent and complete axiomatization of mathematics.*

CHAPTER FOUR

Axiomatic Set Theory

1. AN AXIOM SYSTEM

A prime reason for the increase in importance of mathematical logic in this century was the discovery of the paradoxes of set theory and the need for a revision of intuitive (and contradictory) set theory. Many different axiomatic theories have been proposed to serve as a foundation for set theory, but, no matter how they may differ at the fringes, they all have as a common core the fundamental theorems that mathematicians need in their daily work. A choice among the available theories is primarily a matter of taste, and we make no claim about the system we shall use *except* that it is an adequate basis for present-day mathematics.

We shall describe a first-order theory NBG, which is basically a system of the same type as one originally proposed by von Neumann (1925, 1928) and later thoroughly revised and simplified by R. Robinson (1937), Bernays (1937–1954), and Gödel (1940). (We shall follow Gödel's monograph to a great extent, although there will be some important differences.) NBG has a single predicate letter A_2^2 but no function letters or individual constants. In order to conform to the notation in Bernays (1937–1954) and Gödel (1940), we shall use capital italic letters X_1, X_2, $X_3, \ldots$ as variables, instead of x_1, x_2, $x_3, \ldots$. (As usual, we shall use $X, Y, Z, \ldots$ to represent arbitrary variables.) We shall abbreviate $A_2^2(X, Y)$ by $X \in Y$, and $\neg A_2^2(X, Y)$ by $X \notin Y$. Intuitively, $\in$ is to be thought of as the membership relation.

Let us define equality in the following way.

DEFINITION $X = Y$ for $(\forall Z)(Z \in X \Leftrightarrow Z \in Y)$

Thus, two objects are equal when and only when they have the same members.

DEFINITIONS $X \subseteq Y$ for $(\forall Z)(Z \in X \Rightarrow Z \in Y)$ (inclusion)
$X \subset Y$ for $X \subseteq Y \wedge X \neq Y$ (proper inclusion)

When $X \subseteq Y$, we say that X is a *subclass* of Y. When $X \subset Y$, we say that X is a *proper subclass* of Y.

As easy consequences of these definitions, we have the following.

PROPOSITION 4.1*

(a) $\vdash X = Y \Leftrightarrow (X \subseteq Y \wedge Y \subseteq X)$
(b) $\vdash X = X$
(c) $\vdash X = Y \Rightarrow Y = X$
(d) $\vdash X = Y \Rightarrow (Y = Z \Rightarrow X = Z)$

We shall now present the proper axioms of NBG, interspersing among the axioms some additional definitions and various consequences of the axioms. First, however, notice that in the "interpretation" we have in mind the variables take classes as values. Classes are the totalities corresponding to some, but not necessarily all, properties.† This "interpretation" is as imprecise as the notions of "totality" and "property".

We define a class to be a *set* if it is a member of some class. Those classes that are not sets are called *proper classes.*

DEFINITIONS $M(X)$ for $(\exists Y)(X \in Y)$ (X is a set)
$Pr(X)$ for $\neg M(X)$ (X is a proper class)

It will be seen later that the usual derivations of the paradoxes now no longer lead to contradictions but only yield the result that various classes are proper classes, not sets. The sets are intended to be those safe, comfortable classes that are used by mathematicians in their daily life and work, whereas proper classes are thought of as monstrously large collections that, if permitted to be sets (i.e., allowed to belong to other classes), would engender contradictions.

Exercise

4.1 Prove $\vdash Y \in X \Rightarrow M(Y)$.

The system NBG is designed to handle classes, not concrete individuals. The reason for this is that mathematics has no need for objects such as cows or molecules; all mathematical objects and relations can be formulated in terms of classes alone. If nonclasses are required for applications to other sciences, then the system NBG can be modified slightly so as to apply to both classes and nonclasses alike (see Mostowski, 1939).

Let us introduce lowercase letters $x_1, x_2, \ldots$ as special, restricted variables for sets. In other words, $(\forall x_j)\mathscr{A}(x_j)$ stands for $(\forall X)(M(X) \Rightarrow \mathscr{A}(X))$—that is, $\mathscr{A}$ holds for all sets; and $(\exists x_j)\mathscr{A}(x_j)$ stands for $(\exists X)(M(X) \wedge \mathscr{A}(X))$—that is, $\mathscr{A}$ holds for some set. Note that the variable X used in these definitions should be the first one that does not occur in $\mathscr{A}(x_j)$. As usual, we use $x, y, z, \ldots$ to stand for arbitrary set variables.

* The subscript NBG will be omitted from $\vdash_{NBG}$ in the rest of this chapter.

† Those wfs that actually do determine classes will be partially specified in the axioms. These axioms provide us with the classes we need in mathematics and appear (we hope) modest enough so that contradictions are not derivable from them.

Example $(\forall X_1)(\forall x)(\exists y)(\exists X_3)\mathscr{A}(X_1, x, y, X_3)$ stands for $(\forall X_1)(\forall X_2)(\mathrm{M}(X_2) \Rightarrow (\exists X_4)(\mathrm{M}(X_4) \wedge (\exists X_3)\mathscr{A}(X_1, X_2, X_4, X_3)))$.

Exercise

4.2 Prove that $\vdash X = Y \Leftrightarrow (\forall z)(z \in X \Leftrightarrow z \in Y)$. This is the so-called *Extensionality Principle*: two classes are equal when and only when they contain the same sets as members.

AXIOM T $X_1 = X_2 \Rightarrow (X_1 \in X_3 \Leftrightarrow X_2 \in X_3)$.

PROPOSITION 4.2 NBG is a first-order theory with equality.

Proof The proof is by Proposition 4.1, Axiom T, the definition of equality, and the discussion on page 78.

Exercise

4.3 Prove that $\vdash \mathrm{M}(Z) \wedge Z = Y \Rightarrow \mathrm{M}(Y)$.

AXIOM P (PAIRING AXIOM) $(\forall x)(\forall y)(\exists z)(\forall u)(u \in z \Leftrightarrow u = x \vee u = y)$

Thus, for any sets x and y, there is a set z that has x and y as its only members.*

Exercises

4.4 Prove $\vdash (\forall x)(\forall y)(\exists_1 z)(\forall u)(u \in z \Leftrightarrow u = x \vee u = y)$. This asserts that there is a unique set z, called the *unordered pair* of x and y, such that z has x and y as its only members. Use Axiom P and the definition of equality.

4.5 Prove $\vdash (\forall X)(\mathrm{M}(X) \Leftrightarrow (\exists y)(X \in y))$.

4.6 Prove $\vdash (\exists X)\mathrm{Pr}(X) \Rightarrow \neg(\forall X)(\forall Y)(\exists Z)(\forall U)(U \in Z \Leftrightarrow U = X \vee U = Y)$.

AXIOM N (NULL SET) $(\exists x)(\forall y)(y \notin x)$

Thus, there is a set that has no members.

Obviously, from Axiom N and the definition of equality, there is a unique set that has no members—that is, $\vdash (\exists_1 x)(\forall y)(y \notin x)$. Therefore, we can introduce a new individual constant 0 by means of the following condition.

DEFINITION $(\forall y)(y \notin 0)$

Since we have the uniqueness condition for the unordered pair, we can introduce a new function letter $g(x, y)$ to designate the unordered pair of x and y. We shall write $\{x, y\}$ instead of $g(x, y)$. Notice that we have to define a unique value for $\{X, Y\}$

* Strictly speaking, we should paraphrase Axiom P as follows: "For any sets x and y, there is a set z whose members are all sets equal to x or y". The reason for this is that a nonnormal model, where the interpretation of "=" is not the identity relation, may contain many objects that are "equal" to x or y. However, in what follows, I shall use the customary way of speaking used above, as if only normal models were being referred to.

for any classes X and Y, not only for sets x and y. We shall let $\{X, Y\} = 0$ whenever X is not a set or Y is not a set. One can prove $\vdash (\exists_1 Z)([(\neg M(X) \vee \neg M(Y)) \wedge Z = 0] \vee [M(X) \wedge M(Y) \wedge (\forall u)(u \in Z \Leftrightarrow u = X \vee u = Y)])$. This justifies the following definition of $\{X, Y\}$.

DEFINITION $[M(X) \wedge M(Y) \wedge (\forall u)(u \in \{X, Y\} \Leftrightarrow u = X \vee u = Y)] \vee [(\neg M(X) \vee \neg M(Y)) \wedge \{X, Y\} = 0]$

One can then prove $\vdash (\forall x)(\forall y)(\forall u)(u \in \{x, y\} \Leftrightarrow u = x \vee u = y)$ and $\vdash (\forall x)(\forall y)M(\{x, y\})$.

In connection with these definitions, the reader should review Section 9 of Chapter 2 and, in particular, Proposition 2.28, which assures us that the introduction of new individual constants and function letters, such as 0 and $\{X, Y\}$, adds nothing essentially new to the theory NBG.

Exercise

4.7 (a) Prove $\vdash \{X, Y\} = \{Y, X\}$.
(b) Define $\{X\}$ for $\{X, X\}$. Prove $\vdash (\forall x)(\forall y)(\{x\} = \{y\} \Rightarrow x = y)$.

DEFINITION $\langle X, Y\rangle$ for $\{\{X\}, \{X, Y\}\}$
$\langle X, Y\rangle$ is called the *ordered pair* of X and Y.

The definition of $\langle X, Y\rangle$ does not have any intrinsic intuitive meaning. It is just a convenient way (discovered by Kuratowski) to define ordered pairs so that one can prove the characteristic property of ordered pairs expressed in the following proposition.

PROPOSITION 4.3 $\vdash (\forall x)(\forall y)(\forall u)(\forall v)(\langle x, y\rangle = \langle u, v\rangle \Rightarrow x = u \wedge y = v)$

Proof Assume $\langle x, y\rangle = \langle u, v\rangle$. Then $\{\{x\}, \{x, y\}\} = \{\{u\}, \{u, v\}\}$. Since $\{x\} \in \{\{x\}, \{x, y\}\}$, $\{x\} \in \{\{u\}, \{u, v\}\}$. Hence, $\{x\} = \{u\}$ or $\{x\} = \{u, v\}$. In either case, $x = u$. Now, $\{u, v\} \in \{\{u\}, \{u, v\}\}$; so, $\{u, v\} \in \{\{x\}, \{x, y\}\}$. Then $\{u, v\} = \{x\}$ or $\{u, v\} = \{x, y\}$. Similarly, $\{x, y\} = \{u\}$ or $\{x, y\} = \{u, v\}$. If $\{u, v\} = \{x\}$ and $\{x, y\} = \{u\}$, then $x = y = u = v$; if not, $\{u, v\} = \{x, y\}$. Hence, $\{u, v\} = \{u, y\}$. So, if $v \neq u$, then $v = y$; if $v = u$, then $y = v$. Thus, in all cases, $y = v$.

Notice that the converse of Proposition 4.3 holds by virtue of the substitutivity of equality.

Exercise

4.8 (a) Show that, instead of the definition of ordered pair given in the text, we could have used $\langle X, Y\rangle = \{\{0, X\}, \{\{0\}, Y\}\}$; that is, Proposition 4.3 would still be provable with this new meaning of $\langle X, Y\rangle$.
(b) Show that the ordered pair also could be defined as $\{\{0, \{X\}\}, \{\{Y\}\}\}$. (This was the first such definition, discovered by Norbert Wiener in 1914.)

We now extend the definition of ordered pairs to ordered n-tuples.

DEFINITIONS $\langle X \rangle = X$

$$\langle X_1, \ldots, X_n, X_{n+1} \rangle = \langle \langle X_1, \ldots, X_n \rangle, X_{n+1} \rangle$$

Thus, $\langle X, Y, Z \rangle = \langle \langle X, Y \rangle, Z \rangle$ and $\langle X, Y, Z, U \rangle = \langle \langle \langle X, Y \rangle, Z \rangle, U \rangle$.

It is easy to establish the following generalization of Proposition 4.3:

$$\vdash (\forall x_1)\ldots(\forall x_n)(\forall y_1)\ldots(\forall y_n)(\langle x_1, \ldots, x_n \rangle = \langle y_1, \ldots, y_n \rangle \Rightarrow x_1 = y_1 \wedge \cdots \wedge x_n = y_n)$$

AXIOMS OF CLASS EXISTENCE These axioms state that, for certain properties expressed by wfs, there is a corresponding class of all those sets that satisfy the property.

(B1) $(\exists X)(\forall u)(\forall v)(\langle u, v \rangle \in X \Leftrightarrow u \in v)$ ($\in$-relation)
(B2) $(\forall X)(\forall Y)(\exists Z)(\forall u)(u \in Z \Leftrightarrow u \in X \wedge u \in Y)$ (intersection)
(B3) $(\forall X)(\exists Z)(\forall u)(u \in Z \Leftrightarrow u \notin X)$ (complement)
(B4) $(\forall X)(\exists Z)(\forall u)(u \in Z \Leftrightarrow (\exists v)(\langle u, v \rangle \in X))$ (domain)
(B5) $(\forall X)(\exists Z)(\forall u)(\forall v)(\langle u, v \rangle \in Z \Leftrightarrow u \in X)$
(B6) $(\forall X)(\exists Z)(\forall u)(\forall v)(\forall w)(\langle u, v, w \rangle \in Z \Leftrightarrow \langle v, w, u \rangle \in X)$
(B7) $(\forall X)(\exists Z)(\forall u)(\forall v)(\forall w)(\langle u, v, w \rangle \in Z \Leftrightarrow \langle u, w, v \rangle \in X)$

From axioms (B2)–(B4) and the definition of equality, we obtain:

$$\vdash (\forall X)(\forall Y)(\exists_1 Z)(\forall u)(u \in Z \Leftrightarrow u \in X \wedge u \in Y)$$

$$\vdash (\forall X)(\exists_1 Z)(\forall u)(u \in Z \Leftrightarrow u \notin X)$$

$$\vdash (\forall X)(\exists_1 Z)(\forall u)(u \in Z \Leftrightarrow (\exists v)(\langle u, v \rangle \in X))$$

These results justify the introduction of new function letters: $\cap$, $\bar{\ }$, and $\mathscr{D}$.

DEFINITIONS

$(\forall u)(u \in X \cap Y \Leftrightarrow u \in X \wedge u \in Y)$ (intersection of X and Y)
$(\forall u)(u \in \bar{X} \Leftrightarrow u \notin X)$ (complement of X)
$(\forall u)(u \in \mathscr{D}(X) \Leftrightarrow (\exists v)(\langle u, v \rangle \in X))$ (domain of X)
$X \cup Y = \overline{(\bar{X} \cap \bar{Y})}$ (union of X and Y)
$V = \bar{0}$ (universal class)
$X - Y = X \cap \bar{Y}$ (difference of X and Y)

Exercises

4.9 Prove the following wfs.
(a) $\vdash (\forall u)(u \in X \cup Y \Leftrightarrow u \in X \vee u \in Y)$
(b) $\vdash (\forall u)(u \in V)$
(c) $\vdash (\forall u)(u \in X - Y \Leftrightarrow u \in X \wedge u \notin Y)$

4.10 Prove the following.
(a) $\vdash X \cap Y = Y \cap X$
(b) $\vdash X \cup Y = Y \cup X$
(c) $\vdash (X \cap Y) \cap Z = X \cap (Y \cap Z)$
(d) $\vdash (X \cup Y) \cup Z = X \cup (Y \cup Z)$

(e) $\vdash X \cap X = X$
(f) $\vdash X \cup X = X$
(g) $\vdash X \cap 0 = 0$
(h) $\vdash X \cup 0 = X$
(i) $\vdash X \cap V = X$
(j) $\vdash X \cup V = V$
(k) $\vdash \overline{X \cup Y} = \bar{X} \cap \bar{Y}$
(l) $\vdash \overline{X \cap Y} = \bar{X} \cup \bar{Y}$
(m) $\vdash X - X = 0$
(n) $\vdash V - X = \bar{X}$
(o) $\vdash \bar{\bar{X}} = X$
(p) $\vdash \bar{V} = 0$
(q) $\vdash X \cap (Y \cup Z) = (X \cap Y) \cup (X \cap Z)$
(r) $\vdash X \cup (Y \cap Z) = (X \cup Y) \cap (X \cup Z)$

4.11 Prove the following wfs.

(a) $\vdash (\forall X)(\exists Z)(\forall u)(\forall v)(\langle u, v\rangle \in Z \Leftrightarrow \langle v, u\rangle \in X)$ [*Hint:* Apply axioms (B5), (B7), (B6), and (B4) successively.]

(b) $\vdash (\forall X)(\exists Z)(\forall u)(\forall v)(\forall w)(\langle u, v, w\rangle \in Z \Leftrightarrow \langle u, w\rangle \in X)$ [*Hint:* Use (B5) and (B7).]

(c) $\vdash (\forall X)(\exists Z)(\forall v)(\forall x_1)\dots(\forall x_n)(\langle x_1, \dots, x_n, v\rangle \in Z \Leftrightarrow \langle x_1, \dots, x_n\rangle \in X)$ [*Hint:* Use (B5).]

(d) $\vdash (\forall X)(\exists Z)(\forall v_1)\dots(\forall v_m)(\forall x_1)\dots(\forall x_n)(\langle x_1, \dots, x_n, v_1, \dots, v_m\rangle \in Z \Leftrightarrow \langle x_1, \dots, x_n\rangle \in X)$ [*Hint:* Iteration of part (c).]

(e) $\vdash (\forall X)(\exists Z)(\forall v_1)\dots(\forall v_m)(\forall x_1)\dots(\forall x_n)(\langle x_1, \dots, x_{n-1}, v_1, \dots, v_m, x_n\rangle \in Z \Leftrightarrow \langle x_1, \dots, x_n\rangle \in X)$ [*Hint:* For $m = 1$, use part (b), substituting $\langle x_1, \dots, x_{n-1}\rangle$ for u and x_n for w; the general case then follows by iteration.]

(f) $\vdash (\forall X)(\exists Z)(\forall x)(\forall v_1)\dots(\forall v_m)(\langle v_1, \dots, v_m, x\rangle \in Z \Leftrightarrow x \in X)$ [*Hint:* Use axiom (B5) and part (a).]

(g) $\vdash (\forall X)(\exists Z)(\forall x_1)\dots(\forall x_n)(\langle x_1, \dots, x_n\rangle \in Z \Leftrightarrow (\exists y)(\langle x_1, \dots, x_n, y\rangle \in X))$ [*Hint:* In axiom (B4), substitute $\langle x_1, \dots, x_n\rangle$ for u and y for v.]

(h) $\vdash (\forall X)(\exists Z)(\forall u)(\forall v)(\forall w)(\langle v, u, w\rangle \in Z \Leftrightarrow \langle u, w\rangle \in X)$ [*Hint:* Substitute $\langle u, w\rangle$ for u in axiom (B5) and apply (B6).]

(i) $\vdash (\forall X)(\exists Z)(\forall v_1)\dots(\forall v_k)(\forall u)(\forall w)(\langle v_1, \dots, v_k, u, w\rangle \in Z \Leftrightarrow \langle u, w\rangle \in X)$ [*Hint:* Substitute $\langle v_1, \dots, v_k\rangle$ for v in part (h).]

Now we can derive a general Class Existence Theorem.

PROPOSITION 4.4 Let $\phi(X_1, \dots, X_n, Y_1, \dots, Y_m)$ be a wf, the variables of which occur among $X_1, \dots, X_n, Y_1, \dots, Y_m$, and in which only set variables are quantified (i.e., ϕ can be abbreviated in such a way that only set variables are quantified). We call such a wf *predicative*. Then

$$\vdash (\exists Z)(\forall x_1)\dots(\forall x_n)(\langle x_1, \dots, x_n\rangle \in Z \Leftrightarrow \phi(x_1, \dots, x_n, Y_1, \dots, Y_m))$$

Proof We shall consider only wfs ϕ in which no wf of the form $Y_i \in W$ occurs, since $Y_i \in W$ can be replaced by $(\exists x)(x = Y_i \wedge x \in W)$, which is equivalent to $(\exists x)[(\forall z)(z \in x \Leftrightarrow z \in Y_i) \wedge x \in W]$. Moreover, we may assume that ϕ contains no wf of the form $X \in X$, since this may be replaced by $(\exists u)(u = X \wedge u \in X)$, which is equivalent to $(\exists u)[(\forall z)(z \in u \Leftrightarrow z \in X) \wedge u \in X]$. We shall proceed now by induction on the number k of connectives and quantifiers in ϕ (written with restricted set variables).

Case 1. When $k = 0$, ϕ has the form $x_i \in x_j$ or $x_j \in x_i$ or $x_i \in Y_l$, where $1 \leqslant i < j \leqslant n$. For $x_i \in x_j$, axiom (B1) guarantees that there is some W_1 such that $(\forall x_i)(\forall x_j)(\langle x_i, x_j\rangle \in W_1 \Leftrightarrow x_i \in x_j)$. For $x_j \in x_i$, axiom (B1) implies that there is some W_2 such that $(\forall x_i)(\forall x_j)(\langle x_j, x_i\rangle \in W_2 \Leftrightarrow x_j \in x_i)$, and then, by Exercise 4.11(a), there is some W_3 such that $(\forall x_i)(\forall x_j)(\langle x_i, x_j\rangle \in W_3 \Leftrightarrow x_j \in x_i)$. So, in both cases, there is some W such that $(\forall x_i)(\forall x_j)(\langle x_i, x_j\rangle \in W \Leftrightarrow \phi(x_1, \dots, x_n, Y_1, \dots, Y_m))$. Then, by

Exercise 4.11(i) with $W = X$, there is some Z_1 such that

$$(\forall x_1)\dots(\forall x_{i-1})(\forall x_i)(\forall x_j)(\langle x_1,\dots,x_{i-1},x_i,x_j\rangle \in Z_1 \Leftrightarrow \phi(x_1,\dots,x_n, Y_1,\dots, Y_m))$$

Hence, by Exercise 4.11(e) with $Z_1 = X$, there is some Z_2 such that

$$(\forall x_1)\dots(\forall x_i)(\forall x_{i+1})\dots(\forall x_j)(\langle x_1,\dots,x_j\rangle \in Z_2 \Leftrightarrow \phi(x_1,\dots,x_n, Y_1,\dots, Y_m))$$

Then, by Exercise 4.11(d) with $Z_2 = X$, there is some Z such that

$$(\forall x_1)\dots(\forall x_n)(\langle x_1,\dots,x_n\rangle \in Z \Leftrightarrow \phi(x_1,\dots,x_n, Y_1,\dots, Y_m))$$

In the remaining case, $x_i \in Y_l$, the theorem follows by application of Exercise 4.11(f, d).

Case 2. Let the theorem be provable for all $k < r$, and assume that ϕ has r connectives and quantifiers.

(a) ϕ is $\neg\psi$. By inductive hypothesis, there is some W such that $(\forall x_1)\dots(\forall x_n)(\langle x_1,\dots,x_n\rangle \in W \Leftrightarrow \psi(x_1,\dots,x_n, Y_1,\dots, Y_m))$. Let $Z = \overline{W}$.

(b) ϕ is $\psi \Rightarrow \theta$. By inductive hypothesis, there are classes Z_1 and Z_2 such that $(\forall x_1)\dots(\forall x_n)(\langle x_1,\dots,x_n\rangle \in Z_1 \Leftrightarrow \psi(x_1,\dots,x_n, Y_1,\dots, Y_m))$ and $(\forall x_1)\dots(\forall x_n)(\langle x_1,\dots,x_n\rangle \in Z_2 \Leftrightarrow \theta(x_1,\dots,x_n, Y_1,\dots, Y_m))$. Let $Z = \overline{(Z_1 \cap \overline{Z_2})}$.

(c) ϕ is $(\forall x)\psi$. By inductive hypothesis, there is some W such that

$$(\forall x_1)\dots(\forall x_n)(\forall x)(\langle x_1,\dots,x_n,x\rangle \in W \Leftrightarrow \psi(x_1,\dots,x_n,x, Y_1,\dots, Y_m))$$

Apply Exercise 4.11(g) with $X = \overline{W}$ to obtain a class Z_1 such that

$$(\forall x_1)\dots(\forall x_n)(\langle x_1,\dots,x_n\rangle \in Z_1 \Leftrightarrow (\exists x)\neg\psi(x_1,\dots,x_n,x, Y_1,\dots, Y_m))$$

Now let $Z = \bar{Z}_1$, noting that $(\forall x)\psi$ is equivalent to $\neg(\exists x)\neg\psi$.

Examples

1. Let $\phi(X, Y_1, Y_2)$ be $(\exists u)(\exists v)(X = \langle u,v\rangle \wedge u \in Y_1 \wedge v \in Y_2)$. The only quantifiers in ϕ involve set variables. Hence, by the Class Existence Theorem, $\vdash (\exists Z)(\forall x)(x \in Z \Leftrightarrow (\exists u)(\exists v)(x = \langle u,v\rangle \wedge u \in Y_1 \wedge v \in Y_2))$. By the definition of equality,

$$\vdash (\exists_1 Z)((\forall x)(x \in Z \Leftrightarrow (\exists u)(\exists v)(x = \langle u,v\rangle \wedge u \in Y_1 \wedge v \in Y_2))$$

So, we can introduce a new function letter $\times$.

DEFINITION $(\forall x)(x \in Y_1 \times Y_2 \Leftrightarrow (\exists u)(\exists v)(x = \langle u,v\rangle \wedge u \in Y_1 \wedge v \in Y_2))$

(*Cartesian Product* of Y_1 and Y_2)

DEFINITIONS

Y^2	for	$Y \times Y$	
Y^n	for	$Y^{n-1} \times Y$	when $n > 2$
$\text{Rel}(X)$	for	$X \subseteq V^2$	(X is a *relation*.)*

* More precisely, Rel(X) means that X is a *binary relation*.

V^2 is the class of all ordered pairs, and V^n is the class of all ordered n-tuples. In ordinary language, the word "relation" indicates some kind of connection between objects. For example, the *parenthood relation* holds between parents and their children. For our purposes, we interpret the parenthood relation to be the class of all ordered pairs $\langle u, v\rangle$ such that u is a parent of v. In this way, we have replaced a hazy abstraction by a specific mathematical construction.

2. Let $\phi(X, Y)$ be $X \subseteq Y$. By the Class Existence Theorem and the definition of equality, $\vdash (\exists_1 Z)(\forall x)(x \in Z \Leftrightarrow x \subseteq Y)$. Thus, there is a unique class Z that has as its members all *subsets* of Y.

DEFINITION $(\forall x)(x \in \mathscr{P}(Y) \Leftrightarrow x \subseteq Y)$ [$\mathscr{P}(Y)$: the *power class* of Y].

3. Let $\phi(X, Y)$ be $(\exists v)(X \in v \wedge v \in Y)$. By the Class Existence Theorem and the definition of equality, $\vdash (\exists_1 Z)(\forall x)(x \in Z \Leftrightarrow (\exists v)(x \in v \wedge v \in Y))$. Thus, there is a unique class Z that contains all members of members of Y.

DEFINITION $(\forall x)(x \in \bigcup(Y) \Leftrightarrow (\exists v)(x \in v \wedge v \in Y))$ [$\bigcup(Y)$ is called the *sum class* of Y.]

4. Let $\phi(X)$ be $(\exists u)(X = \langle u, u\rangle)$. By the Class Existence Theorem and the definition of equality, there is a unique class Z such that $(\forall x)(x \in Z \Leftrightarrow (\exists u)(x = \langle u, u\rangle))$.

DEFINITION $(\forall x)(x \in I \Leftrightarrow (\exists u)(x = \langle u, u\rangle))$ [I is called the *identity relation.*]

COROLLARY 4.5 If $\phi(X_1, \ldots, X_n, Y_1, \ldots, Y_m)$ is a predicative wf, then $\vdash (\exists_1 W)(W \subseteq V^n \wedge (\forall x_1) \ldots (\forall x_n)(\langle x_1, \ldots, x_n\rangle \in W \Leftrightarrow \phi(x_1, \ldots, x_n, Y_1, \ldots, Y_m)))$.

Proof By Proposition 4.4, there is some Z such that $(\forall x_1) \ldots (\forall x_n)(\langle x_1, \ldots, x_n\rangle \in Z \Leftrightarrow \phi(x_1, \ldots, x_n, Y_1, \ldots, Y_m))$. Clearly, $W = Z \cap V^n$ satisfies the corollary, and the uniqueness follows from the definition of equality.

DEFINITION Given any predicative wf $\phi(X_1, \ldots, X_n, Y_1, \ldots, Y_m)$, we shall use $\{\langle x_1, \ldots, x_n\rangle \mid \phi(x_1, \ldots, x_n, Y_1, \ldots, Y_m)\}$ to denote the class of all n-tuples $\langle x_1, \ldots, x_n\rangle$ that satisfy $\phi(x_1, \ldots, x_n, Y_1, \ldots, Y_m)$; that is,

$$(\forall u)(u \in \{\langle x_1, \ldots, x_n\rangle \mid \phi(x_1, \ldots, x_n, Y_1, \ldots, Y_m)\} \Leftrightarrow (\exists x_1) \ldots (\exists x_n)(u = \langle x_1, \ldots, x_n\rangle \wedge \phi(x_1, \ldots, x_n, Y_1, \ldots, Y_m)))$$

This definition is justified by Corollary 4.5. In particular, when $n = 1$, $\vdash (\forall u)(u \in \{x \mid \phi(x, Y_1, \ldots, Y_m)\} \Leftrightarrow \phi(u, Y_1, \ldots, Y_m))$.

Examples

1. Take ϕ to be $\langle x_2, x_1\rangle \in Y$. Let $\breve{Y}$ be an abbreviation for $\{\langle x_1, x_2\rangle \mid \langle x_2, x_1\rangle \in Y\}$. Hence, $\breve{Y} \subseteq V^2 \wedge (\forall x_1)(\forall x_2)(\langle x_1, x_2\rangle \in \breve{Y} \Leftrightarrow \langle x_2, x_1\rangle \in Y)$. Call $\breve{Y}$ the *inverse relation* of Y.

2. Take ϕ to be $(\exists v)(\langle v, x \rangle \in Y)$. Let $\mathscr{R}(Y)$ stand for $\{x | (\exists v)(\langle v, x \rangle \in Y)\}$. Then $\vdash (\forall u)(u \in \mathscr{R}(Y) \Leftrightarrow (\exists v)(\langle v, u \rangle \in Y))$. $\mathscr{R}(Y)$ is called the *range* of Y. Clearly, $\vdash \mathscr{R}(Y) = \mathscr{D}(\breve{Y})$.

Notice that axioms (B1)–(B7) are special cases of the Class Existence Theorem, Proposition 4.4. Thus, instead of having to assume Proposition 4.4 as an axiom schema, we need take only a finite number of instances of that schema.

Until now, although we can prove, using Proposition 4.4, the existence of a great many classes, the existence of only a few sets, such as 0, $\{0\}$, $\{0, \{0\}\}$, and $\{\{0\}\}$, is known to us. To guarantee the existence of sets of greater complexity, we require more axioms.

AXIOM U (SUM SET) $(\forall x)(\exists y)(\forall u)(u \in y \Leftrightarrow (\exists v)(u \in v \wedge v \in x))$

This axiom asserts that the sum class $\bigcup(x)$ of a set x (see Example 3 on page 183) is also a set, which we shall call the *sum set* of x—that is, $\vdash (\forall x)\mathrm{M}(\bigcup(x))$. The sum set $\bigcup(x)$ is usually referred to as the *union* of all the sets in the set x and is often denoted $\bigcup_{v \in x} v$.

Exercises

4.12 Show that $\vdash (\forall x)(\forall y)(\bigcup(\{x, y\}) = x \cup y)$. Hence, $\vdash (\forall x)(\forall y)\mathrm{M}(x \cup y)$.

4.13 Prove the following.

(a) $\vdash \bigcup(0) = 0$

(b) $\vdash \bigcup(\{0\}) = 0$

(c) $\vdash (\forall x)(\bigcup(\{x\}) = x)$

(d) $\vdash (\forall x)(\forall y)(\bigcup(\langle x, y \rangle) = \{x, y\})$

4.14 We can define, by induction, $\{x_1, \ldots, x_n\}$ to be $\{x_1, \ldots, x_{n-1}\} \cup \{x_n\}$. Prove $\vdash (\forall x_1)\ldots(\forall x_n)(\forall u)(u \in \{x_1, \ldots, x_n\} \Leftrightarrow u = x_1 \vee \ldots \vee u = x_n)$. Thus, for any given sets $x_1, \ldots, x_n$, there is a set that has $x_1, \ldots, x_n$ as its only members.

Another means of generating new sets from old is the formation of the set of subsets of a given set.

AXIOM W (POWER SET) $(\forall x)(\exists y)(\forall u)(u \in y \Leftrightarrow u \subseteq x)$

This axiom asserts that the power class $\mathscr{P}(x)$ of a set x (see Example 2 on page 183) is itself a set—that is, $\vdash (\forall x)\mathrm{M}(\mathscr{P}(x))$.

Examples

$$\vdash \mathscr{P}(0) = \{0\}$$
$$\vdash \mathscr{P}(\{0\}) = \{0, \{0\}\}$$
$$\vdash \mathscr{P}(\{0, \{0\}\}) = \{0, \{0\}, \{0, \{0\}\}, \{\{0\}\}\}$$

A much more general way to produce sets is the following *Axiom of Subsets.*

AXIOM S (SUBSETS) $(\forall x)(\forall Y)(\exists z)(\forall u)(u \in z \Leftrightarrow u \in x \wedge u \in Y)$

Thus, for any set x and class Y, there is a set consisting of the common elements of x and Y. Hence, $\vdash (\forall x)(\forall Y)\mathrm{M}(x \cap Y)$—that is, the intersection of a set and a class is a set.

PROPOSITION 4.6 $\vdash (\forall x)(\forall Y)(Y \subseteq x \Rightarrow \mathrm{M}(Y))$ (that is, any subclass of a set is a set).

Proof $\vdash (\forall x)(Y \subseteq x \Rightarrow Y \cap x = Y)$ and $\vdash (\forall x)\mathrm{M}(Y \cap x)$.

Since any predicative wf $\mathscr{A}(y)$ generates a corresponding class (by the Class Existence Theorem), Axiom S implies that, given any set x, the class of all elements y of x that satisfy $\mathscr{A}(y)$ is a set.

A stronger axiom than the Axiom of Subsets (S) will be necessary for the full development of set theory. First, we introduce a few definitions.

DEFINITIONS

$\mathrm{Un}(X)$	for	$(\forall x)(\forall y)(\forall z)(\langle x, y\rangle \in X \wedge \langle x, z\rangle \in X \Rightarrow y = z)$ (X is *univocal.*)
$\mathrm{Fnc}(X)$	for	$X \subseteq V^2 \wedge \mathrm{Un}(X)$ (X is a *function.*)
$X: Y \to Z$	for	$\mathrm{Fnc}(X) \wedge \mathscr{D}(X) = Y \wedge \mathscr{R}(X) \subseteq Z$ (X is a *function from Y into Z.*)
$Y \,\rceil\, X$	for	$X \cap (Y \times V)$ (*restriction* of X to the domain Y)
$\mathrm{Un}_1(X)$	for	$\mathrm{Un}(X) \wedge \mathrm{Un}(\check{X})$ (X is *one–one.*)

$$X`Y = \begin{cases} z & \text{if } (\forall u)(\langle Y, u\rangle \in X \Leftrightarrow u = z) \\ 0 & \text{otherwise} \end{cases}$$

$$X``Y = \mathscr{R}(Y \,\rceil\, X)$$

If there is a unique z such that $\langle y, z\rangle \in X$, then $z = X`y$; otherwise, $X`y = 0$. If X is a function and y is a set in its domain, $X`y$ is the value of the function applied to y.[†] If X is a function, $X``Y$ is the range of X restricted to Y.

Exercise

4.15 Prove $\vdash \mathrm{Fnc}(X) \Rightarrow (\forall v)(v \in X``Y \Leftrightarrow (\exists u)(u \in Y \cap \mathscr{D}(X) \wedge v = X`u))$.

AXIOM R (REPLACEMENT)

$$(\forall x)(\mathrm{Un}(Y) \Rightarrow (\exists y)(\forall u)(u \in y \Leftrightarrow (\exists v)(\langle v, u\rangle \in Y \wedge v \in x)))$$

Axiom R asserts that, if Y is univocal, then the class of second components of ordered pairs in Y whose first components are in x is a set [or, equivalently, $\mathrm{M}(\mathscr{R}(x \,\rceil\, Y))$]. When Y is a function, this implies that the range of the restriction of the function Y to a domain that is a set is also a set.

[†] From here on, we shall introduce new function letters or individual constants wherever it is made clear that the definition is based upon a uniqueness theorem. In this case, we have introduced a new function letter $h(X, Y)$, abbreviated $X`Y$.

Exercises

4.16 Show that the Axiom of Replacement (R) implies the Axiom of Subsets (S).

4.17 Prove $\vdash (\forall x)(\mathrm{M}(\mathscr{D}(x)) \wedge \mathrm{M}(\mathscr{R}(x)))$.

4.18 (a) Prove $\vdash x \times y \subseteq \mathscr{P}(\mathscr{P}(x \cup y))$.

(b) Prove $\vdash (\forall x)(\forall y)\mathrm{M}(x \times y)$.

4.19 (a) Prove $\vdash \mathrm{M}(\mathscr{D}(Y)) \wedge \mathrm{M}(\mathscr{R}(Y)) \wedge \mathrm{Rel}(Y) \Rightarrow \mathrm{M}(Y)$.

(b) Prove $\vdash (\forall x)(\mathrm{Fnc}(Y) \Rightarrow \mathrm{M}(Y``x))$.

To ensure the existence of an infinite set, we add the following axiom.

AXIOM I (AXIOM OF INFINITY) $(\exists x)(0 \in x \wedge (\forall u)(u \in x \Rightarrow u \cup \{u\} \in x))$

Axiom I states that there is a set x that contains 0 and such that, whenever a set u belongs to x, then $u \cup \{u\}$ also belongs to x. Clearly, for such a set x, $\{0\} \in x$, $\{0, \{0\}\} \in x$, $\{0, \{0\}, \{0, \{0\}\}\} \in x$, and so on. Intuitively, if we let 1 stand for $\{0\}$, 2 for $\{0, 1\}$, 3 for $\{0, 1, 2\}, \ldots, n$ for $\{0, 1, 2, \ldots, n - 1\}, \ldots$, then, for all integers $n \geqslant 0$, $n \in x$; and $0 \neq 1, 0 \neq 2, 1 \neq 2, 0 \neq 3, 1 \neq 3, 2 \neq 3, \ldots$.

Exercise

4.20 (a) Prove that any wf that implies $(\exists X)\mathrm{M}(X)$ would, together with Axiom S, imply Axiom N.

(b) Show that Axiom I is equivalent to the following sentence (I*):

$$(\exists x)((\exists y)(y \in x \wedge (\forall u)(u \notin y)) \wedge (\forall u)(u \in x \Rightarrow u \cup \{u\} \in x))$$

Then prove that (I*) implies Axiom N. [Hence, if we assumed (I*) instead of (I), Axiom N would become superfluous.]

This completes the list of axioms of NBG, and we see that NBG has only a finite number of axioms—namely, Axiom T, Axiom P (Pairing), Axiom N (Null Set), Axiom S (Subsets), Axiom U (Sum Set), Axiom W (Power Set), Axiom R (Replacement), Axiom I (Infinity), and the seven class existence axioms (B1)–(B7). We have also seen that Axiom S is provable from the other axioms; it has been included here because it is of interest in the study of certain weaker subtheories of NBG.

Let us verify now that Russell's Paradox is not derivable in NBG. Let $Y = \{x | x \notin x\}$. Hence, $(\forall x)(x \in Y \Leftrightarrow x \notin x)$. (Such a class Y exists by the Class Existence Theorem, Proposition 4.4, since $x \notin x$ is a predicative wf.) In unabbreviated notation, this becomes $(\forall X)(\mathrm{M}(X) \Rightarrow (X \in Y \Leftrightarrow X \notin X))$. Assume $\mathrm{M}(Y)$. Then, $Y \in Y \Leftrightarrow Y \notin Y$, which, by the tautology $(A \Leftrightarrow \neg A) \Rightarrow (A \wedge \neg A)$, implies $Y \in Y \wedge Y \notin Y$. Hence, by the Deduction Theorem, $\vdash \mathrm{M}(Y) \Rightarrow (Y \in Y \wedge Y \notin Y)$, and so, by the tautology $(B \Rightarrow (A \wedge \neg A)) \Rightarrow \neg B$, we obtain $\vdash \neg \mathrm{M}(Y)$. Thus, in NBG, the argument for Russell's Paradox merely shows that Russell's Class Y is a proper class, not a set. NBG avoids the paradoxes of Cantor and Burali-Forti in a similar way.

Exercise

4.21 Prove $\vdash \neg \mathrm{M}(V)$. (*The universal class is not a set.*)

2. ORDINAL NUMBERS

Let us first define some familiar notions concerning relations.

DEFINITIONS

$X \text{ Irr } Y$	for	$\text{Rel}(X) \wedge (\forall y)(y \in Y \Rightarrow \langle y, y \rangle \notin X)$ (X is an *irreflexive* relation on Y.)
$X \text{ Tr } Y$	for	$\text{Rel}(X) \wedge (\forall u)(\forall v)(\forall w)([u \in Y \wedge v \in Y \wedge w \in Y \wedge \langle u, v \rangle \in X \wedge \langle v, w \rangle \in X] \Rightarrow \langle u, w \rangle \in X)$ (X is a *transitive* relation on Y.)
$X \text{ Part } Y$	for	$(X \text{ Irr } Y) \wedge (X \text{ Tr } Y)$ (X *partially orders* Y.)
$X \text{ Con } Y$	for	$\text{Rel}(X) \wedge (\forall u)(\forall v)([u \in Y \wedge v \in Y \wedge u \neq v] \Rightarrow \langle u, v \rangle \in X \vee \langle v, u \rangle \in X)$ (X is a *connected* relation on Y.)
$X \text{ Tot } Y$	for	$(X \text{ Irr } Y) \wedge (X \text{ Tr } Y) \wedge (X \text{ Con } Y)$ (X *totally orders* Y.)
$X \text{ We } Y$	for	$(X \text{ Irr } Y) \wedge (\forall Z)([Z \subseteq Y \wedge Z \neq 0] \Rightarrow (\exists y)(y \in Z \wedge (\forall v)(v \in Z \wedge v \neq y \Rightarrow \langle y, v \rangle \in X \wedge \langle v, y \rangle \notin X)))$ (X *well-orders* Y; that is, the relation X is irreflexive on Y and every nonempty subclass of Y has a least element with respect to X.)

Exercises

4.22 Prove $\vdash (X \text{ We } Y) \Rightarrow (X \text{ Tot } Y)$. [*Hint:* To show $X \text{ Con } Y$, let $x, y \in Y$ with $x \neq y$. Then $\{x, y\}$ has a least element, say x. Then $\langle x, y \rangle \in X$. To show $X \text{ Tr } Y$, let x, y, $z \in Y$ with $\langle x, y \rangle \in X$ and $\langle y, z \rangle \in X$. Then $\{x, y, z\}$ has a least element, which must be x.]

4.23 Prove $\vdash (X \text{ We } Y) \wedge (Z \subseteq Y) \Rightarrow (X \text{ We } Z)$.

Examples (from intuitive set theory)

1. The relation $<$ on the set P of positive integers well-orders P.
2. The relation $<$ on the set of all integers totally orders, but does not well-order, this set. The set has no least element.
3. The relation $\subset$ on the set W of all subsets of the set of integers partially orders W, but does not totally order W. For example, $\{1\} \not\subset \{2\}$ and $\{2\} \not\subset \{1\}$.

DEFINITION $\text{Sim}(Z, W_1, W_2)$ for $(\exists x_1)(\exists x_2)(\exists r_1)(\exists r_2)(\text{Rel}(r_1) \wedge \text{Rel}(r_2) \wedge W_1 = \langle r_1, x_1 \rangle \wedge W_2 = \langle r_2, x_2 \rangle \wedge \text{Fnc}(Z) \wedge \text{Un}_1(Z) \wedge \mathscr{D}(Z) = x_1 \wedge \mathscr{R}(Z) = x_2 \wedge (\forall u)(\forall v)(u \in x_1 \wedge v \in x_1 \Rightarrow (\langle u, v \rangle \in r_1 \Leftrightarrow \langle Z`u, Z`v \rangle \in r_2)))$ (Z *is a similarity mapping of the relation* r_1 *on* x_1 *onto the relation* r_2 *on* x_2.)

DEFINITION $\text{Sim}(W_1, W_2)$ for $(\exists z)\text{Sim}(z, W_1, W_2)$ (W_1 and W_2 are *similar ordered structures.*)

Example Let r_1 be the relation $<$ on the set A of nonnegative integers, and let r_2 be the relation $<$ on the set B of positive integers. Let z be the set of all ordered pairs $\langle x, x+1 \rangle$ for $x \in A$. Then z is a similarity mapping of $\langle r_1, A \rangle$ onto $\langle r_2, B \rangle$.

Exercises

4.24 Prove $\vdash \text{Sim}(Z, X, Y) \Rightarrow \text{Sim}(\breve{Z}, Y, X)$.

4.25 Prove $\vdash \text{Sim}(Z, X, Y) \Rightarrow \text{M}(Z) \wedge \text{M}(X) \wedge \text{M}(Y)$.

DEFINITIONS

Fld(X)	for	$\mathscr{D}(X) \cup \mathscr{R}(X)$	(the *field* of X)
TOR(X)	for	$\text{Rel}(X) \wedge (X \text{ Tot}(\text{Fld}(X)))$	(X is a *total order*.)
WOR(X)	for	$\text{Rel}(X) \wedge (X \text{ We}(\text{Fld}(X)))$	(X is a *well-ordering relation*.)

Exercises

4.26 Prove $\vdash (\text{Sim}(X, Y) \Rightarrow \text{Sim}(Y, X)) \wedge (\text{Sim}(X, Y) \wedge \text{Sim}(Y, U) \Rightarrow \text{Sim}(X, U))$.
4.27 Prove $\vdash \text{Sim}(\langle X, \text{Fld}(X)\rangle, \langle Y, \text{Fld}(Y)\rangle) \Rightarrow (\text{TOR}(X) \Leftrightarrow \text{TOR}(Y)) \wedge (\text{WOR}(X) \Leftrightarrow \text{WOR}(Y))$.

If x is a total order, then the class of all total orders similar to x is called the *order type* of x. We are especially interested in the order types of well-ordering relations, but, since it turns out that all order types in NBG are proper classes (except the order type $\{0\}$ of 0), it will be convenient to find a class W of well-ordered structures such that every well-ordering is similar to a unique member of W. This leads us to the study of *ordinal numbers*.

DEFINITIONS

E	for	$\{\langle x, y\rangle \mid x \in y\}$ (the *membership relation*)
Trans(X)	for	$(\forall u)(u \in X \Rightarrow u \subseteq X)$ (X is *transitive*.)
$\text{Sect}_Y(X, Z)$	for	$Z \subseteq X \wedge (\forall u)(\forall v)([u \in X \wedge v \in Z \wedge \langle u, v\rangle \in Y] \Rightarrow u \in Z)$ (Z is a Y-*section* of X; that is, Z is a subclass of X and every member of X that Y-precedes a member of Z is also a member of Z.)
$\text{Seg}_Y(X, U)$	for	$\{x \mid x \in X \wedge \langle x, U\rangle \in Y\}$ (the Y-*segment of* X *determined by* U—that is, the class of all members of X that Y-precede U)

Exercises

Prove the following.
4.28 $\vdash \text{Trans}(X) \Leftrightarrow (\forall u)(\forall v)(v \in u \wedge u \in X \Rightarrow v \in X)$
4.29 $\vdash \text{Trans}(X) \Leftrightarrow \bigcup(X) \subseteq X$
4.30 $\vdash \text{Trans}(X) \wedge \text{Trans}(Y) \Rightarrow \text{Trans}(X \cup Y) \wedge \text{Trans}(X \cap Y)$
4.31 $\vdash \text{Seg}_E(X, u) = X \cap u \wedge M(\text{Seg}_E(X, u))$
4.32 $\vdash \text{Trans}(X) \Leftrightarrow (\forall u)(u \in X \Rightarrow \text{Seg}_E(X, u) = u)$
4.33 $\vdash \text{E We } X \wedge \text{Sect}_E(X, Z) \wedge Z \neq X \Rightarrow (\exists u)(u \in X \wedge Z = \text{Seg}_E(X, u))$

DEFINITIONS

Ord(X)	for	$\text{E We } X \wedge \text{Trans}(X)$ (X is an *ordinal class* if and only if the $\in$-relation well-orders X and any member of X is a subset of X.)
On	for	$\{x \mid \text{Ord}(x)\}$

Thus, $\vdash (\forall x)(x \in On \Leftrightarrow \text{Ord}(x))$. An ordinal class that is a set is called an *ordinal number*. On is the class of all ordinal numbers. Notice that a wf $x \in On$ is equivalent to a predicative wf—namely, the conjunction of the following wfs.

(a) $(\forall u)(u \in x \Rightarrow u \notin u)$
(b) $(\forall u)(u \subseteq x \wedge u \neq 0 \Rightarrow (\exists v)(v \in u \wedge (\forall w)(w \in u \wedge w \neq v \Rightarrow v \in w \wedge w \notin v)))$
(c) $(\forall u)(u \in x \Rightarrow u \subseteq x)$

[The conjunction of (a) and (b) is equivalent to $E\,We\,x$ and (c) is $Trans(x)$.] In addition, any wf $On \in Y$ can be replaced by the wf $(\exists y)(y \in Y \wedge (\forall z)(z \in y \Leftrightarrow z \in On))$. Hence, any wf that is predicative except for the presence of "On" is equivalent to a predicative wf and, therefore, can be used in connection with the Class Existence Theorem.

Exercises

4.34 Prove $\vdash 0 \in On$.
4.35 Let 1 stand for $\{0\}$. Prove $\vdash 1 \in On$.

We shall use lowercase Greek letters $\alpha, \beta, \gamma, \delta, \tau, \ldots$ as restricted variables for ordinal numbers. Thus, $(\forall \alpha)\mathscr{A}(\alpha)$ stands for $(\forall x)(x \in On \Rightarrow \mathscr{A}(x))$, and $(\exists \alpha)\mathscr{A}(\alpha)$ stands for $(\exists x)(x \in On \wedge \mathscr{A}(x))$.

PROPOSITION 4.7

(a) $\vdash \mathrm{Ord}(X) \Rightarrow (X \notin X \wedge (\forall u)(u \in X \Rightarrow u \notin u))$
(b) $\vdash \mathrm{Ord}(X) \wedge Y \subset X \wedge \mathrm{Trans}(Y) \Rightarrow Y \in X$
(c) $\vdash (\mathrm{Ord}(X) \wedge \mathrm{Ord}(Y)) \Rightarrow (Y \subset X \Leftrightarrow Y \in X)$
(d) $\vdash \mathrm{Ord}(X) \wedge \mathrm{Ord}(Y) \Rightarrow (X \in Y \vee X = Y \vee Y \in X) \wedge \neg(X \in Y \wedge Y \in X) \wedge \neg(X \in Y \wedge X = Y)$
(e) $\vdash \mathrm{Ord}(X) \wedge Y \in X \Rightarrow Y \in On$
(f) $\vdash \mathrm{E\,We}\,On$
(g) $\vdash \mathrm{Ord}(On)$
(h) $\vdash \neg \mathrm{M}(On)$
(i) $\vdash \mathrm{Ord}(X) \Rightarrow X = On \vee X \in On$
(j) $\vdash y \subseteq On \wedge \mathrm{Trans}(y) \Rightarrow y \in On$

Proof

(a) If $\mathrm{Ord}(X)$, then E is irreflexive on X; so, $(\forall u)(u \in X \Rightarrow u \notin u)$; and, if $X \in X$, $X \notin X$. Hence, $X \notin X$.
(b) Assume $\mathrm{Ord}(X) \wedge Y \subset X \wedge \mathrm{Trans}(Y)$. It is easy to see that Y is a proper E-section of X. Hence, by Exercises 4.32 and 4.33, $Y \in X$.
(c) Assume $\mathrm{Ord}(X) \wedge \mathrm{Ord}(Y)$. If $Y \in X$, then $Y \subseteq X$, since X is transitive; but $Y \neq X$ by (a); so, $Y \subset X$. Conversely, if $Y \subset X$, then, since Y is transitive, we have $Y \in X$, by (b).
(d) Assume $\mathrm{Ord}(X) \wedge \mathrm{Ord}(Y) \wedge X \neq Y$. Now, $X \cap Y \subseteq X$ and $X \cap Y \subseteq Y$. Since X and Y are transitive, so is $X \cap Y$. If $X \cap Y \subset X$ and $X \cap Y \subset Y$, then, by (b), $X \cap Y \in X$ and $X \cap Y \in Y$; hence, $X \cap Y \in X \cap Y$, contradicting the irreflexivity of E on X. Hence, either $X \cap Y = X$ or $X \cap Y = Y$; that

is, $X \subseteq Y$ or $Y \subseteq X$. But $X \neq Y$. Hence, by (c), $X \in Y$ or $Y \in X$. Also, if $X \in Y$ and $Y \in X$, then, by (c), $X \subset Y$ and $Y \subset X$, which is impossible. Clearly, $X \in Y \wedge X = Y$ is impossible, by (a).

(e) Assume $\mathrm{Ord}(X) \wedge Y \in X$. We must show E We Y and $\mathrm{Trans}(Y)$. Since $Y \in X$ and $\mathrm{Trans}(X)$, $Y \subset X$. Hence, since E We X, E We Y. Moreover, if $u \in Y$ and $v \in u$, then, by $\mathrm{Trans}(X)$, $v \in X$. Since E Con X and $Y \in X \wedge v \in X$, then $v \in Y$ or $v = Y$ or $Y \in v$. If either $v = Y$ or $Y \in v$, then, since E Tr X and $u \in Y \wedge v \in u$, we would have $u \in u$, contradicting (a). Hence, $v \in Y$. So, if $u \in Y$, then $u \subseteq Y$—that is, $\mathrm{Trans}(Y)$.

(f) By (a), E Irr On. Now, assume $X \subseteq On \wedge X \neq 0$. Let $\alpha \in X$. If α is the least element of X, we are done. [By *least element* of X, we mean an element $v \in X$ such that $(\forall u)(u \in X \wedge u \neq v \Rightarrow v \in u)$.] If not, then E We α and $X \cap \alpha \neq 0$; let β be the least element of $X \cap \alpha$. It is obvious, using (d), that β is the least element of X.

(g) We must show E We On and $\mathrm{Trans}(On)$. The first part is (f). For the second, if $u \in On$ and $v \in u$, then, by (e), $v \in On$. Hence, $\mathrm{Trans}(On)$.

(h) If $\mathrm{M}(On)$, then, by (g), $On \in On$, contradicting (a).

(i) Assume $\mathrm{Ord}(X)$. Then $X \subseteq On$. If $X \neq On$, then, by (c), $X \in On$.

(j) Substitute On for X and y for Y in (b). By (h), $y \subset On$.

We see, from Proposition 4.7(i), that the only ordinal class that is not an ordinal number is the class On itself.

DEFINITION $x <_0 y$ for $x \in On \wedge y \in On \wedge x \in y$
$x \leqslant_0 y$ for $y \in On \wedge (x = y \vee x <_0 y)$

Thus, for ordinals, $<_0$ is the same as $\in$; so, $<_0$ well-orders On. In particular, from Proposition 4.7(e), we see that any ordinal x is equal to the set of smaller ordinals.

PROPOSITION 4.8 (TRANSFINITE INDUCTION)

$$\vdash (\forall \beta)[(\forall \alpha)(\alpha \in \beta \Rightarrow \alpha \in X) \Rightarrow \beta \in X] \Rightarrow On \subseteq X$$

(If, for any β, whenever all ordinals $<_0 \beta$ are in X, β must also be in X, then all ordinals are in X.)

Proof Assume that $(\forall \beta)[(\forall \alpha)(\alpha \in \beta \Rightarrow \alpha \in X) \Rightarrow \beta \in X]$. Assume there is an ordinal in $On - X$. Then, since On is well-ordered by E, there is a least ordinal β in $On - X$. Hence, all ordinals $<_0 \beta$ are in X. So, by hypothesis, β is in X, which is a contradiction.

Proposition 4.8 is used to prove that all ordinals have a given property $\mathscr{A}(\alpha)$. We let $X = \{x \mid \mathscr{A}(x) \wedge x \in On\}$ and show that $(\forall \beta)[(\forall \alpha)(\alpha \in \beta \Rightarrow \mathscr{A}(\alpha)) \Rightarrow \mathscr{A}(\beta)]$.

DEFINITION x' for $x \cup \{x\}$

PROPOSITION 4.9

(a) $\vdash (\forall x)(x \in On \Leftrightarrow x' \in On)$
(b) $\vdash (\forall \alpha) \neg (\exists \beta)(\alpha <_0 \beta <_0 \alpha')$
(c) $\vdash (\forall \alpha)(\forall \beta)(\alpha' = \beta' \Rightarrow \alpha = \beta)$

Proof

(a) $x \in x'$. Hence, if $x' \in On$, then $x \in On$ by Proposition 4.7(e). Conversely, assume $x \in On$. We must prove $\mathrm{E\,We}(x \cup \{x\})$ and $\mathrm{Trans}(x \cup \{x\})$. Since $\mathrm{E\,We}\, x$ and $x \notin x$, $\mathrm{E\,Irr}(x \cup \{x\})$. Also, if $y \neq 0 \wedge y \subseteq (x \cup \{x\})$, then either $y = \{x\}$, in which case the least element of y is x, or $y \cap x \neq 0$ and the least element of $y \cap x$ is then the least element of y. Hence, $\mathrm{E\,We}(x \cup \{x\})$. In addition, if $y \in x \cup \{x\}$ and $u \in y$, then $u \in x$. Thus, $\mathrm{Trans}(x \cup \{x\})$.
(b) Assume $\alpha <_0 \beta <_0 \alpha'$. Then $\alpha \in \beta \wedge \beta \in \alpha'$. Since $\alpha \in \beta$, $\beta \notin \alpha$, and $\beta \neq \alpha$, by Proposition 4.7(d), contradicting $\beta \in \alpha'$.
(c) Assume $\alpha' = \beta'$. Then $\beta <_0 \alpha'$ and, by part (b), $\beta \leqslant_0 \alpha$. Similarly, $\alpha \leqslant_0 \beta$. Hence, $\alpha = \beta$.

DEFINITIONS

$\mathrm{Suc}(X)$	for	$X \in On \wedge (\exists \alpha)(X = \alpha')$ (X is a *successor ordinal*.)
K_I	for	$\{x \mid x = 0 \vee \mathrm{Suc}(x)\}$ (the class of *ordinals of the first kind*)
ω	for	$\{x \mid x \in K_\mathrm{I} \wedge (\forall u)(u \in x \Rightarrow u \in K_\mathrm{I})\}$ (ω is the class of all ordinals α of the first kind such that all ordinals $<_0 \alpha$ are also of the first kind.)

Example $\vdash 0 \in \omega \wedge 1 \in \omega$ (Remember that $1 = \{0\}$.)

PROPOSITION 4.10

(a) $\vdash (\forall \alpha)(\alpha \in \omega \Leftrightarrow \alpha' \in \omega)$
(b) $\vdash \mathrm{M}(\omega)$
(c) $\vdash 0 \in X \wedge (\forall u)(u \in X \Rightarrow u' \in X) \Rightarrow \omega \subseteq X$
(d) $\vdash (\forall \alpha)(\alpha \in \omega \wedge \beta <_0 \alpha \Rightarrow \beta \in \omega)$

Proof

(a) Assume $\alpha \in \omega$. Now, $\mathrm{Suc}(\alpha')$. Hence, $\alpha' \in K_\mathrm{I}$. Also, if $\beta \in \alpha'$, then $\beta \in \alpha$ or $\beta = \alpha$. Hence, $\beta \in K_\mathrm{I}$. Thus, $\alpha' \in \omega$. Conversely, if $\alpha' \in \omega$, then, since $\alpha \in \alpha'$ and $(\forall \beta)(\beta \in \alpha \Rightarrow \beta \in \alpha')$, it follows that $\alpha \in \omega$.
(b) By the Axiom of Infinity (I), there is a set x such that $0 \in x$ and $(\forall u)(u \in x \Rightarrow u' \in x)$. We shall prove $\omega \subseteq x$. Assume not. Let α be the least ordinal in $\omega - x$. Clearly, $\alpha \neq 0$, since $0 \in x$. Hence, $\mathrm{Suc}(\alpha)$. So, $(\exists \beta)(\alpha = \beta')$. Let δ be an ordinal such that $\alpha = \delta'$. Then, $\delta <_0 \alpha$ and, by part (a), $\delta \in \omega$. Therefore, $\delta \in x$. Hence, $\delta' \in x$. But $\alpha = \delta'$. Therefore, $\alpha \in x$, which yields a contradiction. Thus, $\omega \subseteq x$. So, $\mathrm{M}(\omega)$, by Proposition 4.6.
(c) This is proved by a procedure similar to that used for part (b).
(d) This is left as an exercise.

The elements of ω are called *finite ordinals*. We shall use the standard notation: 1 for $0'$, 2 for $1'$, 3 for $2'$, and so on. Thus, $0 \in \omega$, $1 \in \omega$, $2 \in \omega$, $3 \in \omega, \ldots$.

The nonzero ordinals that are not successor ordinals are called *limit ordinals*, or *ordinals of the second kind*.

DEFINITION $\operatorname{Lim}(x)$ for $x \in On \wedge x \notin K_{\mathrm{I}}$

Exercise

4.36 Prove $\vdash \operatorname{Lim}(\omega)$.

PROPOSITION 4.11

(a)
$$\vdash (\forall x)(x \subseteq On \Rightarrow (\bigcup(x) \in On \wedge (\forall \alpha)(\alpha \in x \Rightarrow \alpha \leqslant_0 \bigcup(x)) \wedge (\forall \beta)((\forall \alpha)(\alpha \in x \Rightarrow \alpha \leqslant_0 \beta) \Rightarrow \bigcup(x) \leqslant_0 \beta)))$$

[If x is a set of ordinals, then $\bigcup(x)$ is an ordinal that is the least upper bound of x.]

(b)
$$\vdash (\forall x)([x \subseteq On \wedge x \neq 0 \wedge (\forall \alpha)(\alpha \in x \Rightarrow (\exists \beta)(\beta \in x \wedge \alpha <_0 \beta))] \Rightarrow \operatorname{Lim}(\bigcup(x)))$$

[If x is a nonempty set of ordinals without a maximum, then $\bigcup(x)$ is a limit ordinal.]

Proof

(a) Assume $x \subseteq On$. $\bigcup(x)$, as a set of ordinals, is well-ordered by E. Also, if $\alpha \in \bigcup(x) \wedge \beta \in \alpha$, then there is some γ with $\gamma \in x$ and $\alpha \in \gamma$. Then $\beta \in \alpha$ and $\alpha \in \gamma$; since every ordinal is transitive, $\beta \in \gamma$. So, $\beta \in \bigcup(x)$. Hence, $\bigcup(x)$ is transitive and, therefore, $\bigcup(x) \in On$. In addition, if $\alpha \in x$, then $\alpha \subseteq \bigcup(x)$; so, $\alpha \leqslant_0 \bigcup(x)$, by Proposition 4.7(c). Assume now that $(\forall \alpha)(\alpha \in x \Rightarrow \alpha \leqslant_0 \beta)$. Clearly, if $\delta \in \bigcup(x)$, then there is some γ such that $\delta \in \gamma \wedge \gamma \in x$. Hence, $\gamma \leqslant_0 \beta$ and so, $\delta <_0 \beta$. Therefore, $\bigcup(x) \subseteq \beta$ and, by Proposition 4.7(c), $\bigcup(x) \leqslant_0 \beta$.

(b) Assume $x \neq 0 \wedge x \subseteq On \wedge (\forall \alpha)(\alpha \in x \Rightarrow (\exists \beta)(\beta \in x \wedge \alpha <_0 \beta))$. If $\bigcup(x) = 0$, then $\alpha \in x$ implies $\alpha = 0$. So, $x = 0$ or $x = 1$, which contradicts our assumption. Hence, $\bigcup(x) \neq 0$. Assume $\operatorname{Suc}(\bigcup(x))$. Then $\bigcup(x) = \gamma'$ for some γ. By part (a), $\bigcup(x)$ is a least upper bound of x. Therefore, γ is not an upper bound of x; there is some $\delta \in x$ with $\gamma <_0 \delta$. But then $\delta = \bigcup(x)$, since $\bigcup(x)$ is an upper bound of x. Thus, $\bigcup(x)$ is a maximum element of x, contradicting our hypothesis. Hence, $\neg \operatorname{Suc}(\bigcup(x))$, and $\operatorname{Lim}(x)$ is the only possibility left.

Exercise

4.37 Prove $\vdash (\forall \alpha)([\operatorname{Suc}(\alpha) \Rightarrow (\bigcup(\alpha))' = \alpha] \wedge [\operatorname{Lim}(\alpha) \Rightarrow \bigcup(\alpha) = \alpha])$.

We can now state and prove another form of transfinite induction.

PROPOSITION 4.12 (TRANSFINITE INDUCTION: SECOND FORM)

(a) $$\vdash [0 \in X \wedge (\forall\alpha)(\alpha \in X \Rightarrow \alpha' \in X) \wedge (\forall\alpha)(\text{Lim}(\alpha) \wedge (\forall\beta)(\beta <_0 \alpha \Rightarrow \beta \in X) \Rightarrow \alpha \in X)] \Rightarrow On \subseteq X$$

(b) *Induction up to δ.*

$$\vdash [0 \in X \wedge (\forall\alpha)(\alpha <_0 \delta \wedge \alpha \in X \Rightarrow \alpha' \in X) \wedge (\forall\alpha)(\alpha <_0 \delta \wedge \text{Lim}(\alpha) \wedge (\forall\beta)(\beta <_0 \alpha \Rightarrow \beta \in X) \Rightarrow \alpha \in X)] \Rightarrow \delta \subseteq X$$

Proof

(a) Assume the antecedent. Let $Y = \{x | x \in On \wedge (\forall\alpha)(\alpha \leqslant_0 x \Rightarrow \alpha \in X)\}$. It is then easy to prove that $(\forall\alpha)(\alpha <_0 \gamma \Rightarrow \alpha \in Y) \Rightarrow \gamma \in Y$. Hence, by Proposition 4.8, $On \subseteq Y$. But $Y \subseteq X$. Hence, $On \subseteq X$.

(b) The proof is left as an exercise.

Set theory depends heavily upon definitions by transfinite induction, which are justified by the following theorems.

PROPOSITION 4.13

(a) $$\vdash (\forall X)(\exists_1 Y)(\text{Fnc}(Y) \wedge \mathscr{D}(Y) = On \wedge (\forall\alpha)(Y`\alpha = X`(\alpha \upharpoonright Y)))$$

(Given X, there is a unique function Y defined on all ordinals such that the value of Y at α is the value of X applied to the restriction of Y to the set of ordinals less than α.)

(b) $$\vdash (\forall x)(\forall X_1)(\forall X_2)(\exists_1 Y)(\text{Fnc}(Y) \wedge \mathscr{D}(Y) = On \wedge Y`0 = x \wedge (\forall\alpha)(Y`(\alpha') = X_1`(Y`\alpha)) \wedge (\forall\alpha)(\text{Lim}(\alpha) \Rightarrow Y`\alpha = X_2`(\alpha \upharpoonright Y)))$$

(c) *Induction up to δ.*

$$\vdash (\forall x)(\forall X_1)(\forall X_2)(\exists_1 Y)(\text{Fnc}(Y) \wedge \mathscr{D}(Y) = \delta \wedge Y`0 = x \wedge (\forall\alpha)(\alpha' <_0 \delta \Rightarrow Y`(\alpha') = X_1`(Y`\alpha)) \wedge (\forall\alpha)(\text{Lim}(\alpha) \wedge \alpha <_0 \delta \Rightarrow Y`\alpha = X_2`(\alpha \upharpoonright Y)))$$

Proof (a) Let $Y_1 = \{u | \text{Fnc}(u) \wedge \mathscr{D}(u) \in On \wedge (\forall\alpha)(\alpha \in \mathscr{D}(u) \Rightarrow u`\alpha = X`(\alpha \upharpoonright u))\}$. Now, if $u_1 \in Y_1$ and $u_2 \in Y_1$, then $u_1 \subseteq u_2$ or $u_2 \subseteq u_1$. In fact, let $\gamma_1 = \mathscr{D}(u_1)$ and $\gamma_2 = \mathscr{D}(u_2)$. Either $\gamma_1 \leqslant_0 \gamma_2$ or $\gamma_2 \leqslant_0 \gamma_1$—say, $\gamma_1 \leqslant_0 \gamma_2$. Let w be the set of all ordinals $\alpha <_0 \gamma_1$ such that $u_1`\alpha \neq u_2`\alpha$; assume $w \neq 0$ and let η be the least ordinal in w. Then for all $\beta <_0 \eta$, $u_1`\beta = u_2`\beta$. Hence, $\eta \upharpoonright u_1 = \eta \upharpoonright u_2$. But, $u_1`\eta = X`(\eta \upharpoonright u_1)$ and $u_2`\eta = X`(\eta \upharpoonright u_2)$; and so, $u_1`\eta = u_2`\eta$, contradicting our assumption. Therefore, $w = 0$; that is, for all $\alpha \leqslant_0 \gamma_1$, $u_1`\alpha = u_2`\alpha$. Hence, $u_1 = \gamma_1 \upharpoonright u_1 = \gamma_1 \upharpoonright u_2 \subseteq u_2$. Thus, any two functions in Y_1 agree in their common domain. Let $Y = \bigcup(Y_1)$. We leave it to the reader to prove that Y is a function, the domain of which is either an ordinal or the class On, and $(\forall\alpha)(\alpha \in \mathscr{D}(Y) \Rightarrow Y`\alpha = X`(\alpha \upharpoonright Y))$. That $\mathscr{D}(Y) = On$ follows easily from the observation that, if $\mathscr{D}(Y) = \delta$ and if we let $W = Y \cup \{\langle \delta, X`Y \rangle\}$, then $W \in Y_1$; so, $W \subseteq Y$ and $\delta \in \mathscr{D}(Y) = \delta$, which contradicts the fact that $\delta \notin \delta$. The uniqueness of Y follows by a simple transfinite induction (Proposition 4.12).

The proof of part (b) is similar to that of (a), and part (c) follows from (b).

Using Proposition 4.13, one can introduce new function letters by transfinite induction.

Examples

1. *Ordinal addition.* In Proposition 4.13(b), take

$$x = \beta \qquad X_1 = \{\langle u, v\rangle | v = u'\} \qquad X_2 = \{\langle u, v\rangle | v = \bigcup(\mathscr{R}(u))\}$$

Hence, for each ordinal β, there is a unique function Y_β such that

$$Y_\beta\text{‘}0 = \beta \wedge (\forall\alpha)(Y_\beta\text{‘}(\alpha') = (Y_\beta\text{‘}\alpha)') \wedge (\forall\alpha)(\text{Lim}(\alpha) \Rightarrow Y_\beta\text{‘}\alpha = \bigcup(Y_\beta\text{“}\alpha))$$

Hence, there is a unique binary function $+_0$ with domain $(On)^2$ such that, for any ordinals β and γ, $+_0(\beta, \gamma) = Y_\beta\text{‘}\gamma$. As usual, we write $\beta +_0 \gamma$ instead of $+_0(\beta, \gamma)$. Notice that:

$$\beta +_0 0 = \beta$$

$$\beta +_0 (\gamma') = (\beta +_0 \gamma)'$$

$$\text{Lim}(\alpha) \Rightarrow \beta +_0 \alpha = \bigcup_{\tau <_0 \alpha} (\beta +_0 \tau)$$

In particular,

$$\beta +_0 1 = \beta +_0 (0') = (\beta +_0 0)' = \beta'$$

2. *Ordinal multiplication.* In Proposition 4.13(b), take $x = 0$, $X_1 = \{\langle u, v\rangle | v = u +_0 \beta\}$, and $X_2 = \{\langle u, v\rangle | v = \bigcup(\mathscr{R}(u))\}$. Then, as in Example 1, one obtains a function $\beta \times_0 \gamma$ with the properties

$$\beta \times_0 0 = 0$$

$$\beta \times_0 (\gamma') = (\beta \times_0 \gamma) +_0 \beta$$

$$\text{Lim}(\alpha) \Rightarrow \beta \times_0 \alpha = \bigcup_{\tau <_0 \alpha} (\beta \times_0 \tau)$$

Exercise

4.38 Justify the following definition of ordinal exponentiation.*

$$\exp(\beta, 0) = 1$$

$$\exp(\beta, \gamma') = \exp(\beta, \gamma) \times_0 \beta$$

$$\text{Lim}(\alpha) \Rightarrow \exp(\beta, \alpha) = \bigcup_{0 <_0 \tau <_0 \alpha} \exp(\beta, \tau)$$

* We use the notation $\exp(\beta, \alpha)$ instead of β^α in order to avoid confusion with the notation X^Y to be introduced later (see page 197).

For any set X, let E_X be the membership relation restricted to X; that is, $E_X = \{\langle x, y\rangle | x \in y \wedge x \in X \wedge y \in X\}$.

PROPOSITION 4.14* Let R be a well-ordering relation on a class Y; that is, R We Y. Let F be a function from Y into Y such that, for any u, v in Y, if $\langle u, v\rangle \in R$, then $\langle F\text{`}u, F\text{`}v\rangle \in R$. Then, for all u in Y, $u = F\text{`}u$ or $\langle u, F\text{`}u\rangle \in R$.

Proof Let $X = \{u | \langle F\text{`}u, u\rangle \in R\}$. We wish to show that $X = 0$. Assume $X \neq 0$. Since $X \subseteq Y$ and $R$ well-orders $Y$, there is an $R$-least element $u_0$ of $X$. Hence, $\langle F\text{`}u_0, u_0\rangle \in R$. Therefore, $\langle F\text{`}(F\text{`}u_0), F\text{`}u_0\rangle \in R$. Thus, $F\text{`}u_0 \in X$, but $F\text{`}u_0$ is R-smaller than u_0, contradicting the definition of u_0.

As a special case of Proposition 4.14, if Y is a class of ordinals, $F\colon Y \to Y$ and F is increasing on Y (that is, $\alpha \in Y \wedge \beta \in Y \wedge \alpha <_0 \beta \Rightarrow F\text{`}\alpha <_0 F\text{`}\beta$), then $\alpha \leqslant_0 F\text{`}\alpha$ for all α in Y.

COROLLARY 4.15 Let $\alpha <_0 \beta$ and $y \subseteq \alpha$; that is, let y be a subset of a segment of β. Then $\langle E_\beta, \beta\rangle$ is not similar to $\langle E_y, y\rangle$.

Proof Assume f is a function from β onto y such that, for u, v in β, if $u <_0 v$, then $f\text{`}u <_0 f\text{`}v$. Since the range of f is y, $f\text{`}\alpha \in y$. But $y \subseteq \alpha$. Hence, $f\text{`}\alpha <_0 \alpha$. But, by the special case of Proposition 4.14 mentioned above (with $y = \beta$ and $R = E_\beta$), $\alpha \leqslant_0 f\text{`}\alpha$, which yields a contradiction.

COROLLARY 4.16

(a) For $\alpha \neq \beta$, $\langle E_\alpha, \alpha\rangle$ and $\langle E_\beta, \beta\rangle$ are not similar.
(b) For any α, if f is a similarity mapping of $\langle E_\alpha, \alpha\rangle$ with $\langle E_\alpha, \alpha\rangle$, then f is the identity mapping; that is, $f\text{`}\beta = \beta$ for all $\beta <_0 \alpha$.

Proof

(a) The proof is by Corollary 4.15.
(b) By Proposition 4.14, $f\text{`}\beta \geqslant_0 \beta$ for all $\beta <_0 \alpha$. But, again by Proposition 4.14, $(\check{f})\text{`}\beta \geqslant_0 \beta$ for all $\beta <_0 \alpha$. Hence, $\beta = (\check{f})\text{`}(f\text{`}\beta) \geqslant_0 f\text{`}\beta \geqslant_0 \beta$ and, therefore, $f\text{`}\beta = \beta$.

PROPOSITION 4.17 Assume that R is a well-ordering of a nonempty set u; that is, $R \operatorname{We} u \wedge u = \operatorname{Fld}(R) \wedge u \neq 0$. Then there is a unique ordinal γ and a unique similarity mapping of $\langle E_\gamma, \gamma\rangle$ with $\langle R, u\rangle$; that is, every nonempty well-ordered set is similar to a unique ordinal.

Proof Let $Z = \{\langle v, w\rangle | w \in u - v \wedge (\forall z)(z \in u - v \Rightarrow \langle z, w\rangle \notin R)\}$. Z is a function such that, if v is a subset of u and $u - v \neq 0$, then $Z\text{`}v$ is the R-least element

*From this point on we shall express many theorems of NBG in English by using the corresponding informal English translations. This is done to avoid writing mile-long wfs that are difficult to decipher and only in cases where the reader can easily produce from the English version the precise wf of NBG.

of $u - v$. Let $X = \{\langle v, w\rangle | \langle \mathscr{R}(v), w\rangle \in Z\}$. Now we use a definition by transfinite induction (Proposition 4.13) to obtain a function Y with On as its domain such that $(\forall\alpha)(Y`\alpha = X`(\alpha \upharpoonright Y))$. Let $W = \{\alpha | Y``\alpha \subseteq u \wedge u - Y``\alpha \neq 0\}$. Clearly, if $\alpha \in W$ and $\beta \in \alpha$, then $\beta \in W$. Hence, either $W = On$ or W is some ordinal γ. (If $W \neq On$, let γ be the least ordinal in $On - W$.) If $\alpha \in W$, then $Y`\alpha = X`(\alpha \upharpoonright Y)$ is the R-least element of $u - Y``\alpha$; so, $Y`\alpha \in u$ and, if $\beta \in \alpha$, $Y`\alpha \neq Y`\beta$. Thus, Y is a one–one function on W and the range of Y restricted to W is a subset of u. Now, let $h = (W \upharpoonright Y)$ and $f = \breve{h}$; that is, let f be the inverse of Y restricted to W. So, by the Replacement Axiom (R), W is a set. Hence, W is some ordinal γ. Let $g = \gamma \upharpoonright Y$. Then g is a one–one function with domain γ and range a subset u_1 of u. We must show that $u_1 = u$ and that, if α and β are in γ and $\beta <_0 \alpha$, $\langle g`\beta, g`\alpha\rangle \in R$. Assume α and β are in γ and $\beta <_0 \alpha$. Then $g``\beta \subseteq g``\alpha$ and, since $g`\alpha \in u - g``\alpha$, $g`\alpha \in u - g``\beta$. But $g`\beta$ is the R-least element of $u - g``\beta$. Hence, $\langle g`\beta, g`\alpha\rangle \in R$. It remains to prove that $u_1 = u$. Now, $u_1 = Y``\gamma$. Assume $u - u_1 \neq 0$. Then $\gamma \in W$. But $W = \gamma$, which yields a contradiction. Hence, $u = u_1$. That γ is unique follows from Corollary 4.16.

PROPOSITION 4.18 Let R be a well-ordering of a proper class X such that, for each $y \in X$, the class of all R-predecessors of y in X (i.e., the R-segment in X determined by y) is a set. Then R is similar to E_{On}; that is, there is a one–one mapping h of On onto X such that $\alpha \in \beta$ implies $\langle h`\alpha, h`\beta\rangle \in R$.

Proof Proceed as in the proof of Proposition 4.17. Here, however, $W = On$; also, one proves that $\mathscr{R}(Y) = X$ by using the hypothesis that every R-segment of X is a set. [If $X - \mathscr{R}(Y) \neq 0$, then, if w is the R-least element of $X - \mathscr{R}(Y)$, the proper class On is the range of $\breve{Y}$, while the domain of $\breve{Y}$ is the R-segment of X determined by w, contradicting the Replacement Axiom.]

3. EQUINUMEROSITY. FINITE AND DENUMERABLE SETS

We say that two classes X and Y are *equinumerous* if and only if there is a one–one function F with domain X and range Y. We shall denote this by $X \cong Y$.

DEFINITIONS

$$X \underset{F}{\cong} Y \quad \text{for} \quad (\mathrm{Fnc}(F) \wedge \mathrm{Un}_1(F) \wedge \mathscr{D}(F) = X \wedge \mathscr{R}(F) = Y)$$

$$X \cong Y \quad \text{for} \quad (\exists F)(X \underset{F}{\cong} Y)$$

Notice that $\vdash (\forall x)(\forall y)(x \cong y \Leftrightarrow (\exists z)(x \underset{z}{\cong} y))$. Hence, a wf $x \cong y$ is predicative (i.e., is equivalent to a wf using only set quantifiers).

Clearly, if $X \underset{F}{\cong} Y$, then $Y \underset{\breve{F}}{\cong} X$; and if $X \underset{F}{\cong} Y$ and $Y \underset{G}{\cong} Z$, then $X \underset{H}{\cong} Z$, where H is the composition $G \circ F$ of F and G; that is, $G \circ F = \{\langle x, y\rangle | (\exists z)(\langle x, z\rangle \in F \wedge \langle z, y\rangle \in G)\}$. Hence, we have the following theorem.

PROPOSITION 4.19

(a) $X \cong X$
(b) $X \cong Y \Rightarrow Y \cong X$
(c) $X \cong Y \wedge Y \cong Z \Rightarrow X \cong Z$

PROPOSITION 4.20

(a) $(X \cong Y \wedge X_1 \cong Y_1 \wedge X \cap X_1 = 0 \wedge Y \cap Y_1 = 0) \Rightarrow X \cup X_1 \cong Y \cup Y_1$
(b) $(X \cong Y \wedge X_1 \cong Y_1) \Rightarrow X \times X_1 \cong Y \times Y_1$
(c) $X \times \{y\} \cong X$
(d) $X \times Y \cong Y \times X$
(e) $(X \times Y) \times Z \cong X \times (Y \times Z)$

Proof

(a) Let $X \underset{F}{\cong} Y$ and $X_1 \underset{G}{\cong} Y_1$. Then $X \cup X_1 \underset{F \cup G}{\cong} Y \cup Y_1$.

(b) Let $X \underset{F}{\cong} Y$ and $X_1 \underset{G}{\cong} Y_1$. Let

$$W = \{\langle u, v\rangle \mid (\exists x)(\exists y)(x \in X \wedge y \in X_1 \wedge u = \langle x, y\rangle \wedge v = \langle F`x, G`y\rangle)\}$$

Then $X \times X_1 \underset{W}{\cong} Y \times Y_1$.

(c) Let $F = \{\langle u, v\rangle \mid u \in X \wedge v = \langle u, y\rangle\}$. Then $X \underset{F}{\cong} X \times \{y\}$.

(d) Let

$$F = \{\langle u, v\rangle \mid (\exists x)(\exists y)(x \in X \wedge y \in Y \wedge u = \langle x, y\rangle \wedge v = \langle y, x\rangle)\}.$$

Then $X \times Y \underset{F}{\cong} Y \times X$.

(e) Let

$$F = \{\langle u, v\rangle \mid (\exists x)(\exists y)(\exists z)(x \in X \wedge y \in Y \wedge z \in Z \wedge u = \langle\langle x, y\rangle, z\rangle \wedge v = \langle x, \langle y, z\rangle\rangle)\}$$

Then $(X \times Y) \times Z \underset{F}{\cong} X \times (Y \times Z)$.

DEFINITION $X^Y = \{u \mid u \colon Y \to X\}$

X^Y is the class of all sets that are functions from Y into X.

Exercises

Prove the following.

4.39 $\vdash (\forall X)(\forall Y)(\exists X_i)(\exists Y_1)(X \cong X_i \wedge Y \cong Y_1 \wedge X_1 \cap Y_1 = 0)$
4.40 $\vdash \mathscr{P}(x) \cong 2^x$. (Remember that $2 = \{0, 1\}$.)
4.41 $\vdash \neg \mathrm{M}(Y) \Rightarrow X^Y = 0$
4.42 $\vdash \mathrm{M}(x^y)$
4.43 (a) $\vdash X^0 = \{0\} = 1$
(b) $\vdash X \cong X^{\{u\}}$

4.44 $\vdash Y \neq 0 \Rightarrow 0^Y = 0$
4.45 $\vdash X \cong Y \wedge Z \cong Z_1 \Rightarrow X^Z \cong Y^Z$
4.46 $\vdash X \cap Y = 0 \Rightarrow Z^{X \cup Y} \cong Z^x \times Z^Y$
4.47 $\vdash (X^Y)^Z \cong X^{Y \times Z}$, except when $Y = 0 \wedge \neg \mathrm{M}(Z)$
4.48 $\vdash (X \times Y)^Z \cong X^Z \times Y^Z$
4.49 $\vdash (\forall x)(\forall R)(R \text{ We } x \Rightarrow (\exists \alpha)(x \cong \alpha))$ (Every well-ordered set is equinumerous with some ordinal.)

One can define a partial order $\preccurlyeq$ on classes such that, intuitively, $X \preccurlyeq Y$ if X has the same number or fewer elements than Y.

DEFINITION $X \preccurlyeq Y$ for $(\exists Z)(Z \subseteq Y \wedge X \cong Z)$ (X is equinumerous with a subclass of Y.)

DEFINITION $X \prec Y$ for $X \preccurlyeq Y \wedge \neg(X \cong Y)$

Hence, $\vdash X \preccurlyeq Y \Leftrightarrow (X \prec Y \vee X \simeq Y)$.

Exercises

4.50 Prove $\vdash X \preccurlyeq Y \wedge \neg \mathrm{M}(X) \Rightarrow \neg \mathrm{M}(Y)$.
4.51 Prove $\vdash X \preccurlyeq Y \wedge (\exists Z)(Z \text{ We } Y) \Rightarrow (\exists Z)(Z \text{ We } X)$.

PROPOSITION 4.21

(a) $\vdash X \preccurlyeq X \wedge \neg(X \prec X)$
(b) $\vdash X \subseteq Y \Rightarrow X \preccurlyeq Y$
(c) $\vdash X \preccurlyeq Y \wedge Y \preccurlyeq Z \Rightarrow X \preccurlyeq Z$
(d) (Schröder-Bernstein) $\vdash X \preccurlyeq Y \wedge Y \preccurlyeq X \Rightarrow X \cong Y$

Proof

(a) and (b). These proofs are obvious.
(c) Assume $X \underset{F}{\cong} Y_1 \wedge Y_1 \subseteq Y \wedge Y \underset{G}{\cong} Z_1 \wedge Z_1 \subseteq Z$. Let H be the composition of F and G. Then $\mathscr{R}(H) \subseteq Z \wedge X \underset{H}{\cong} \mathscr{R}(H)$.
(d) There are many proofs of this nontrivial theorem. The following one was devised by Hellman (1961). First we derive a lemma.

Lemma. Assume $X \cap Y = 0$, $X \cap Z = 0$, and $Y \cap Z = 0$, and let $X \underset{F}{\cong} X \cup Y \cup Z$. Then there is a G such that $X \underset{G}{\cong} X \cup Y$. [*Proof:* Define a function H on a subclass of $X \times \omega$ as follows: $\langle (u, k), v \rangle \in H$ if and only if $u \in X$ and $k \in \omega$ and there is a function f with domain k' such that $f`0 = F`u$ and, if $j \in k$, then $f`j \in X$ and $f`(j') = F`(f`j)$ and $f`k = v$. Thus, $H`(\langle u, 0 \rangle) = F`u$, $H`(\langle u, 1 \rangle) = F`(F`u)$ if $F`u \in X$, and $H`(\langle u, 2 \rangle) = F`(F`(F`u))$ if $F`u$ and $F`(F`u)$ are in X, and so on. Let X^* be the class of all $u \in X$ such that $(\exists y)(y \in \omega \wedge \langle u, y \rangle \in \mathscr{D}(H) \wedge H`(\langle u, y \rangle) \in Z)$. Let $Y^*$ be the class of all $u \in X$ such that $(\forall y)(y \in \omega \wedge \langle u, y \rangle \in \mathscr{D}(H) \Rightarrow H`(\langle u, y \rangle) \notin Z)$. Then $X = X^* \cup Y^*$.

Now define G as follows: $\mathscr{D}(G) = X$ and, if $u \in X^*$, then $G`u = u$, whereas if $u \in Y^*$, then $G`u = F`u$. Then $X \underset{G}{\cong} X \cup Y$. (This is left as an exercise.)]

Now, to prove the Schröder-Bernstein Theorem: let $X \underset{F}{\cong} Y_1 \wedge Y_1 \subseteq Y \wedge Y \underset{G}{\cong} X_1 \wedge X_1 \subseteq X$. Let $A = G``Y_1 \subseteq X_1 \subseteq X$. But $A \cap (X_1 - A) = 0$, $A \cap (X - X_1) = 0$, and $(X - X_1) \cap (X_1 - A) = 0$. Also, $X = (X - X_1) \cup (X_1 - A) \cup A$, and the composition H of F and G is a one–one functon with domain X and range A. Hence, $A \underset{H}{\cong} X$. So, by the lemma, there is a one–one function D such that $A \underset{D}{\cong} X_1$ [since $(X_1 - A) \cup A = X_1$]. Let T be the composition of the functions H, D, and $\breve{G}$; that is, let $T`u = (\breve{G})`(D`(H`u))$. Then $X \underset{T}{\cong} Y$, since $X \underset{H}{\cong} A$ and $A \underset{D}{\cong} X_1$ and $X_1 \underset{\breve{G}}{\cong} Y$.

Exercises

4.52 Carry out the details of the following proof (due to J. Whitaker) of the Schröder-Bernstein Theorem in the case where X and Y are sets. Let $X \underset{F}{\cong} Y_1 \wedge Y_1 \subseteq Y \wedge Y \underset{G}{\cong} X_1 \wedge X_1 \subseteq X$. We wish to find a set $Z \subseteq X$ such that G, restricted to $Y - F``Z$, is a one–one function of $Y - F``Z$ onto $X - Z$. [If we have such a set Z, let $H = (Z \rceil F) \cup ((X - Z) \rceil G)$; that is, $H`x = F`x$ for $x \in Z$, and $H`x = \breve{G}`x$ for $x \in X - Z$. Then $X \underset{H}{\cong} Y$.] Let $Z = \{x | (\exists u)(u \subseteq X \wedge x \in u \wedge G``(Y - F``u) \subseteq X - u)\}$. Notice that this proof does not presuppose the definition of ω nor any other part of the theory of ordinals. For still another proof, see Kleene (1952, § 4).

PROPOSITION 4.22 Assume $X \preccurlyeq Y$ and $A \preccurlyeq B$. Then,

(a) $Y \cap B = 0 \Rightarrow X \cup A \preccurlyeq Y \cup B$
(b) $X \times A \preccurlyeq Y \times B$
(c) $X^A \preccurlyeq Y^B$ if B is a set and $\neg(X = A = Y = 0 \wedge B \neq 0)$

Proof

(a) Assume $X \underset{F}{\cong} Y_1 \subseteq Y$ and $A \underset{G}{\cong} B_1 \subseteq B$. Let H be a function with domain $X \cup A$ such that $H`x = F`x$ if $x \in X$ and $H`x = G`x$ if $x \in A - X$. Then $X \cup A \underset{H}{\cong} H``(X \cup A) \subseteq Y \cup B$.

(b) and (c) are left as exercises.

Exercises

4.53 Prove $\vdash X \preccurlyeq X \cup Y$.
4.54 Prove $\vdash X \prec Y \Rightarrow \neg(Y \prec X)$.
4.55 Prove $\vdash X \prec Y \wedge Y \preccurlyeq Z \Rightarrow X \prec Z$.

PROPOSITION 4.23 (CANTOR'S THEOREM)

(a) $\vdash \neg(\exists f)(\mathrm{Fnc}(f) \wedge \mathscr{D}(f) = x \wedge \mathscr{R}(f) = \mathscr{P}(x))$ [There is no function from x onto $\mathscr{P}(x)$.]
(b) $\vdash x \prec \mathscr{P}(x)$

Proof

(a) Assume $\text{Fnc}(f) \wedge \mathscr{D}(f) = x \wedge \mathscr{R}(f) = \mathscr{P}(x)$. Let $y = \{u | u \in x \wedge u \notin f`u\}$. Then $y \in \mathscr{P}(x)$. Hence, there is some $z$ in $x$ such that $f`z = y$. But, $(\forall u)(u \in y \Leftrightarrow u \in x \wedge u \notin f`u)$. Hence, $(\forall u)(u \in f`z \Leftrightarrow u \in x \wedge u \notin f`u)$. By Rule A4, $z \in f`z \Leftrightarrow z \in x \wedge z \notin f`z$. Since $z \in x$, we obtain $z \in f`z \Leftrightarrow z \notin f`z$, which yields a contradiction.

(b) Let f be the function with domain x such that $f`u = \{u\}$ for each u in x. Then $f``x \subseteq \mathscr{P}(x)$ and f is one–one. Hence, $x \preccurlyeq \mathscr{P}(x)$. By part (a), $x \cong \mathscr{P}(x)$ is impossible. Hence, $x \prec \mathscr{P}(x)$.

In naive set theory, Proposition 4.23(b) gave rise to Cantor's Paradox. If we let $x = V$, then $V \prec \mathscr{P}(V)$. But, $\mathscr{P}(V) \subseteq V$ and, therefore, $\mathscr{P}(V) \preccurlyeq V$. In NBG, this argument is just another proof that V is not a set.

Observe that we have not proved $\vdash (\forall x)(\forall y)(x \preccurlyeq y \vee y \preccurlyeq x)$. This proposition is, in fact, not yet provable, since it turns out to be equivalent to the Axiom of Choice (which will be discussed in Section 5).

The equinumerosity relation $\cong$ has all the properties of an equivalence relation. We are inclined, therefore, to partition the class of all sets into equivalence classes under this relation. The equivalence class of a set x would be the class of all sets equinumerous with x. The equivalence classes are called *cardinal numbers*. For example, if u is a set and $x = \{u\}$, then the equivalence class of x is the class of all unit sets $\{v\}$ and is called the cardinal number 1_c. Likewise, if $u \neq v$ and $y = \{u, v\}$, then the equivalence class of y is the class of all sets that contain exactly two elements, and is the cardinal number 2_c; that is, $2_c = \{z | (\exists x)(\exists y)(x \neq y \wedge z = \{x, y\})\}$. Notice that all the cardinal numbers, except the cardinal number of 0 (which is $\{0\}$), are proper classes. For example, $V \cong 1_c$, where V is the universal class. (Let $F`x = \{x\}$ for every x in V. Then $V \underset{F}{\cong} 1_c$.) But, $\neg M(V)$; hence, by the Replacement Axiom, $\neg M(1_c)$.

Exercise

4.56 Prove $\vdash \neg M(2_c)$.

Because the cardinal numbers are proper classes, we cannot talk about classes of cardinal numbers, and it is difficult or impossible to say and prove many interesting things about them. Most assertions one would like to make about cardinal numbers can be paraphrased by the suitable use of $\cong$ and $\preccurlyeq$. In addition, we shall see later that, given certain additional plausible axioms, there are other ways of defining a notion that does the same job as that of cardinal number.

To see how everything we want to say about cardinal numbers can be said without explicit mention of cardinal numbers, consider the following treatment of the "sum" of cardinal numbers.

DEFINITION $X +_c Y$ for $(X \times \{0\}) \cup (Y \times \{1\})$.

Since $X \times \{0\}$ and $Y \times \{1\}$ are disjoint, their union is a class whose "size" is the "sum" of the "sizes" of X and Y.

Exercise

Prove the following.

4.57 (a) $\vdash X \preccurlyeq X +_c Y \wedge Y \preccurlyeq X +_c Y$

(b) $\vdash X \cong A \wedge Y \cong B \Rightarrow X +_c Y \cong A +_c B$

(c) $\vdash X +_c Y \cong Y +_c X$

(d) $\vdash \mathrm{M}(X +_c Y) \Leftrightarrow \mathrm{M}(X) \wedge \mathrm{M}(Y)$

(e) $\vdash X +_c (Y +_c Z) \cong (X +_c Y) +_c Z$

(f) $\vdash X \preccurlyeq Y \Rightarrow X +_c Z \preccurlyeq Y +_c Z$

(g) $\vdash X +_c X = X \times 2$

(h) $\vdash X^{Y +_c Z} \cong X^Y \times X^Z$

(i) $\vdash x \cong x +_c 1 \Rightarrow 2^x +_c x \cong 2^x$

FINITE SETS

Remember that ω is the set of all ordinals α such that α and all smaller ordinals are successor ordinals or 0. The elements of ω will be called *finite ordinals*. A set will be called *finite* if and only if it is equinumerous with a finite ordinal.

DEFINITION $\quad \mathrm{Fin}(X) \quad$ for $\quad (\exists \alpha)(\alpha \in \omega \wedge X \cong \alpha)$

Clearly, by the Replacement Axiom, $\vdash \mathrm{Fin}(X) \Rightarrow \mathrm{M}(X)$. Trivially, all finite ordinals are finite sets, and $\vdash \mathrm{Fin}(X) \wedge X \cong Y \Rightarrow \mathrm{Fin}(Y)$.

PROPOSITION 4.24

(a) $\vdash (\forall \alpha)(\alpha \in On - \omega \Rightarrow \alpha \cong \alpha')$

(b) $\vdash (\forall \alpha)(\forall \beta)(\alpha \in \omega \wedge \alpha \neq \beta \Rightarrow \neg \alpha \cong \beta)$ (No finite ordinal is equinumerous with any other ordinal. Hence, a finite set is equinumerous with exactly one finite ordinal; moreover, a nonfinite ordinal, that is, a member of $On - \omega$, is not finite.)

(c) $\vdash (\forall \alpha)(\forall x)(\alpha \in \omega \wedge x \subset \alpha \Rightarrow \neg \alpha \cong x)$ (No finite ordinal is equinumerous with a proper subset of itself.)

Proof

(a) Assume $\alpha \in On - \omega$. Define a function f with domain α' as follows: $f`\delta = \delta'$ if $\delta \in \omega$; $f`\delta = \delta$ if $\delta \in \alpha' \wedge \delta \notin \omega \cup \{\alpha\}$; and $f`\alpha = 0$. Then $\alpha' \underset{f}{\cong} \alpha$.

(b) Assume this is false, and let α be the least ordinal such that $\alpha \in \omega$ and there is a $\beta \neq \alpha$ such that $\alpha \cong \beta$. Hence, $\alpha <_0 \beta$. (Otherwise, β would be a smaller ordinal than α, and β would be equinumerous with some ordinal $\neq \beta$.) Let $\alpha \underset{f}{\cong} \beta$. If $\alpha = 0$, then $f = 0$ and $\beta = 0$, contradicting $\alpha \neq \beta$. So, $\alpha \neq 0$. Since $\alpha \in \omega$, $\alpha \overset{f}{=} \delta'$ for some $\delta \in \omega$. We may assume that $\beta = \gamma'$ for some γ. [If $\beta \in \omega$, then $\beta \neq 0$; and if $\beta \notin \omega$, then, by part (a), $\beta \cong \beta'$, and we can take β' instead of β.] Thus, $\delta' = \alpha \underset{f}{\cong} \gamma'$. Also, $\delta \neq \gamma$, since $\alpha \neq \beta$.

Case 1. $f`\delta = \gamma$. Then $\delta \underset{\delta \upharpoonright f}{\cong} \gamma$.

Case 2. $f`\delta \neq \gamma$. Then there is some $\mu \in \delta$ such that $f`\mu = \gamma$. Let $h = ((\delta \upharpoonright f) - \{\langle \mu, \gamma \rangle\}) \cup \{\langle \mu, f`\delta \rangle\}$; that is, let $h`\tau = f`\tau$ if $\tau \notin \{\delta, \mu\}$; $h`\mu = f`\delta$. Then $\delta \underset{h}{\cong} \gamma$. In both cases, δ is a finite ordinal smaller than α that is equinumerous with a different ordinal γ, contradicting the minimality of α.

(c) Assume $\beta \in \omega \wedge x \subset \beta \wedge \beta \cong x$ holds for some β, and let α be the least such β. Clearly, $\alpha \neq 0$; hence, $\alpha = \gamma'$ for some γ; but, as in the proof of part (b), one can then show that γ is also equinumerous with a proper subset of itself, contradicting the minimality of α.

Exercise

4.58 Prove that the Axiom of Infinity (I) is equivalent to the following sentence:

$$(*) \quad (\exists x)((\exists u)(u \in x) \wedge (\forall y)(y \in x \Rightarrow (\exists z)(z \in x \wedge y \subset z)))$$

PROPOSITION 4.25

(a) $\vdash \mathrm{Fin}(X) \wedge Y \subseteq X \Rightarrow \mathrm{Fin}(Y)$

(b) $\vdash \mathrm{Fin}(X) \Rightarrow \mathrm{Fin}(X \cup \{y\})$

(c) $\vdash \mathrm{Fin}(X) \wedge \mathrm{Fin}(Y) \Rightarrow \mathrm{Fin}(X \cup Y)$

(d) A set is said to be *Dedekind-finite* if and only if it is not equinumerous with any proper subset of itself. Then every finite set is Dedekind-finite. (The converse is not provable without additional assumptions, such as the Axiom of Choice.)

Proof

(a) Assume $\mathrm{Fin}(X) \wedge Y \subseteq X$. Then $X \cong \alpha$, where $\alpha \in \omega$. Let $g = Y \upharpoonright f$ and $W = g``Y \subseteq \alpha$. W is a set of ordinals, and so, E_W is a well-ordering of W. By Proposition 4.17, $\langle \mathrm{E}_W, W \rangle$ is similar to $\langle \mathrm{E}_\beta, \beta \rangle$ for some ordinal β. Hence, $W \cong \beta$. In addition, $\beta \leqslant_0 \alpha$. (If $\beta >_0 \alpha$, then the similarity of $\langle \mathrm{E}_\beta, \beta \rangle$ to $\langle \mathrm{E}_W, W \rangle$ contradicts Corollary 4.15.) Since $\alpha \in \omega$, $\beta \in \omega$. From $Y \underset{g}{\cong} W \wedge W \cong \beta$, it follows that $\mathrm{Fin}(Y)$.

(b) If $y \in X$, then $X \cup \{y\} = X$ and the result is trivial. So, assume $y \notin X$. From $\mathrm{Fin}(X)$ it follows that there is a finite ordinal α and a function f such that $\alpha \underset{f}{\cong} X$. Let $g = f \cup \{\langle \alpha, y \rangle\}$. Then $\alpha' \underset{g}{\cong} X \cup \{y\}$. Hence, $\mathrm{Fin}(X \cup \{y\})$.

(c) Let $Z = \{u | u \in \omega \wedge (\forall x)(\forall y)(\forall f)(x \underset{f}{\cong} u \wedge \mathrm{Fin}(y) \Rightarrow \mathrm{Fin}(x \cup y))\}$. We must show that $Z = \omega$. Clearly, $0 \in Z$, for, if $x \cong 0$, then $x = 0$ and $x \cup y = y$. Assume that $\alpha \in Z$. Let $x \underset{f}{\cong} \alpha'$ and $\mathrm{Fin}(y)$. Let $f`w = \alpha$ and $x_1 = x - \{w\}$. Then $x_1 \cong \alpha$. Since $\alpha \in Z$, $\mathrm{Fin}(x_1 \cup y)$. But $x \cup y = (x_1 \cup y) \cup \{w\}$. Hence, by part (b), $\mathrm{Fin}(x \cup y)$. Thus, $\alpha' \in Z$. Hence, by Proposition 4.10(c), $Z = \omega$.

(d) This follows from Proposition 4.24(c).

DEFINITIONS $\mathrm{Inf}(X)$ for $\neg\mathrm{Fin}(X)$ (X is *infinite.*)
$\mathrm{Den}(X)$ for $X \cong \omega$ (X is *denumerable.*)

Clearly, $\vdash \mathrm{Inf}(X) \wedge X \cong Y \Rightarrow \mathrm{Inf}(Y)$ and $\vdash \mathrm{Den}(X) \wedge X \cong Y \Rightarrow \mathrm{Den}(Y)$. By the Replacement Axiom and the fact that $\mathrm{M}(\omega)$, it follows that $\vdash \mathrm{Den}(X) \Rightarrow \mathrm{M}(X)$.

PROPOSITION 4.26

(a) $\vdash \text{Inf}(X) \wedge X \subseteq Y \Rightarrow \text{Inf}(Y)$
(b) $\vdash \text{Inf}(X) \Leftrightarrow \text{Inf}(X \cup \{y\})$
(c) A class is called *Dedekind-infinite* if and only if it is equinumerous with a proper subset of itself. Then every Dedekind-infinite class is infinite.
(d) $\vdash \text{Inf}(\omega)$

Proof

(a) This follows from Proposition 4.25(a).
(b) $\vdash \text{Inf}(X) \Rightarrow \text{Inf}(X \cup \{y\})$ by part (a). By Proposition 4.25(c), $\vdash \text{Inf}(X \cup \{y\}) \Rightarrow \text{Inf}(X)$.
(c) Use Proposition 4.25(d).
(d) $\vdash \omega \notin \omega$ and Proposition 4.24(b).

PROPOSITION 4.27 $\vdash \text{Den}(v) \wedge z \subseteq v \Rightarrow (\text{Den}(z) \vee \text{Fin}(z))$

Proof It suffices to prove that $z \subseteq \omega \Rightarrow (\text{Den}(z) \vee \text{Fin}(z))$. Assume $z \subseteq \omega \wedge \neg\text{Fin}(z)$. Since $\neg\text{Fin}(z)$, for any $\alpha \in z$, there is some $\beta \in z$ with $\alpha <_0 \beta$. [Otherwise, $z \subseteq \alpha'$ and, since $\text{Fin}(\alpha')$, $\text{Fin}(z)$.] Let X be a function such that, for any $\alpha \in \omega$, $X`\alpha$ is the least ordinal $\beta$ in $z$ with $\alpha <_0 \beta$. Then, by Proposition 4.13(c) (with $\delta = \omega$), there is a function $Y$ with domain $\omega$ such that $Y`0$ is the least ordinal in z, and for any γ in ω, $Y`(\gamma')$ is the least ordinal $\beta$ in $z$ with $\beta >_0 (Y`\gamma)$. Clearly, Y is one–one, $\mathscr{D}(Y) = \omega$, and $Y``\omega \subseteq z$. Also, $Y``\omega = z$; for, if $z - Y``\omega \neq 0$, $\delta$ is the least ordinal in $z - Y``\omega$, and τ is the least ordinal in $Y``\omega$ with $\tau >_0 \delta$, then $\tau = Y`\sigma$ for some $\sigma$ in $\omega$. Since $\delta <_0 \tau$, $\sigma \neq 0$. So, $\sigma = \mu'$ for some $\mu$ in $\omega$. Then $\tau = Y`\sigma =$ the least ordinal in z that is greater than $Y`\mu$. But $\delta >_0 Y`\mu$, since τ is the least ordinal in $Y``\omega$ that is greater than δ. Hence $\tau \leqslant_0 \delta$, which contradicts $\delta <_0 \tau$.

Exercises

Prove Exercises 4.59–4.69.

4.59 $\vdash \text{Fin}(x) \Rightarrow \text{Fin}(\mathscr{P}(x))$
4.60 $\vdash \text{Fin}(x) \wedge (\forall y)(y \in x \Rightarrow \text{Fin}(y)) \Rightarrow \text{Fin}(\bigcup(x))$
4.61 $\vdash x \preccurlyeq y \wedge \text{Fin}(y) \Rightarrow \text{Fin}(x)$
4.62 $\vdash \text{Fin}(\mathscr{P}(x)) \Rightarrow \text{Fin}(x)$
4.63 $\vdash \text{Fin}(\bigcup(x)) \Rightarrow (\text{Fin}(x) \wedge (\forall y)(y \in x \Rightarrow \text{Fin}(y)))$
4.64 $\vdash \text{Fin}(x) \Rightarrow (x \preccurlyeq y \vee y \preccurlyeq x)$
4.65 $\vdash \text{Fin}(x) \wedge \text{Inf}(Y) \Rightarrow x \prec Y$
4.66 $\vdash \text{Fin}(x) \wedge y \subset x \Rightarrow y \prec x$
4.67 $\vdash \text{Fin}(x) \wedge \text{Fin}(y) \Rightarrow \text{Fin}(x \times y)$
4.68 $\vdash \text{Fin}(x) \wedge \text{Fin}(y) \Rightarrow \text{Fin}(x^y)$
4.69 $\vdash \text{Fin}(x) \wedge y \notin x \Rightarrow x \prec (x \cup \{y\})$
4.70 Define x to be a *minimal* (respectively, *maximal*) element of Y if and only if $x \in Y$ and $(\forall y)(y \in Y \Rightarrow \neg y \subset x)$ [respectively, $(\forall y)(y \in Y \Rightarrow \neg x \subset y)$]. Prove that a set Z is finite if and only if every nonempty set of subsets of Z has a minimal (respectively, maximal) element (Tarski, 1925).

4.71 Prove the following results.

(a) $\vdash \mathrm{Fin}(x) \wedge \mathrm{Den}(y) \Rightarrow \mathrm{Den}(x \cup y)$

(b) $\vdash \mathrm{Fin}(x) \wedge \mathrm{Den}(y) \wedge x \neq 0 \Rightarrow \mathrm{Den}(x \times y)$

(c) A set x contains a denumerable subset if and only if x is Dedekind-infinite.

(d) If $y \notin x$, then x is Dedekind-infinite if and only if $x \cong x \cup \{y\}$.

(e) $\vdash \omega \preccurlyeq x \Rightarrow x +_c 1 \cong x$

4.72 Notice that, if NBG is consistent, then it has a denumerable model (by Proposition 2.17). Explain why this does not contradict Cantor's Theorem, which implies that there exist nondenumerable infinite sets (for example, 2^{ω}). This apparent, but not real, contradiction is sometimes called *Skolem's Paradox*.

4. HARTOGS' THEOREM. INITIAL ORDINALS. ORDINAL ARITHMETIC

An unjustly neglected proposition with manifold uses in set theory is Hartogs' Theorem.

PROPOSITION 4.28 (Hartogs, 1915) For any set x, there is an ordinal that is not equinumerous with any subset of x (and hence, there is a least such ordinal).

Proof Assume that every ordinal α is equinumerous with some subset y of x. Hence, $y \underset{f}{\cong} \alpha$ for some f. Define a relation R on y by stipulating that $\langle u, v\rangle \in R$ if and only if $(f`u) \in (f`v)$. Then R is a well-ordering of y such that $\langle R, y\rangle$ is similar to $\langle E_{\alpha}, \alpha\rangle$. Now define a function F with domain On such that, for any α, $F`\alpha$ is the set w of all pairs $\langle z, y\rangle$ such that $y \subseteq x$, z is a well-ordering of y, and $\langle E_{\alpha}, \alpha\rangle$ is similar to $\langle z, y\rangle$. [w is a set, since $w \subseteq \mathscr{P}(x \times x) \times \mathscr{P}(x)$.] Hence, $F``(On)$ is a set. $F$ is one–one; hence, $On = \breve{F}``(F``(On))$ is a set by the Replacement Axiom, contradicting Proposition 4.7(h).

Let $\mathscr{H}$ be the function that assigns to each set x the least ordinal α that is not equinumerous with any subset of x. Notice that, to each $\beta <_0 \mathscr{H}`x$, we can associate the set of relations $r$ such that $r \subseteq x \times x$, $r$ is a well-ordering of its field $y$, and $\langle r, y\rangle$ is similar to $\langle E_{\beta}, \beta\rangle$. This defines a one–one function from $\mathscr{H}`x$ into $\mathscr{P}\mathscr{P}(x \times x)$. Hence, $\mathscr{H}`x \preccurlyeq \mathscr{P}\mathscr{P}(x \times x)$ and, since $x \times x \subseteq \mathscr{P}\mathscr{P}(x)$ by Exercise 4.18(a), we obtain $\mathscr{H}`x \preccurlyeq \mathscr{P}\mathscr{P}\mathscr{P}\mathscr{P}(x)$.

By an *initial ordinal* we mean an ordinal that is not equinumerous with any smaller ordinal. By Proposition 4.24(b), every finite ordinal is an initial ordinal and ω is the smallest infinite initial ordinal. It is obvious that, for any x, $\mathscr{H}`x$ is an initial ordinal. Moreover, for any ordinal $\alpha$, $\mathscr{H}`\alpha$ is the least initial ordinal greater than α.

By transfinite induction [Proposition 4.13(b)], there is a function G with domain On such that

$$G`0 = \omega$$

$$G`(\alpha') = \mathscr{H}`(G`\alpha)$$

$$G`\lambda = \bigcup (G``\lambda) \qquad \text{if } \lambda \text{ is a limit ordinal}$$

G is an increasing function; that is, $\alpha \in \beta \Rightarrow G`\alpha \in G`\beta$; therefore, if λ is a limit ordinal and each $G`\alpha$, for $\alpha <_0 \lambda$, is an initial ordinal, then $\bigcup (G``\lambda)$ is also an initial ordinal. {In fact, $\delta = \bigcup (G``\lambda)$ is the least upper bound of $G``\lambda$. Assume $\delta \cong \gamma$ with $\gamma <_0 \delta$. Hence, there is some $\alpha <_0 \lambda$ such that $\gamma <_0 G`\alpha$. But $G`(\alpha') <_0 \delta$. So, by the Schröder-Bernstein Theorem [Proposition 4.21(d)], using $G`\alpha \preccurlyeq G`(\alpha')$ and $G`(\alpha') \preccurlyeq \delta \cong \gamma \preccurlyeq G`\alpha$, we have $G`\alpha \cong G`(\alpha') = \mathscr{H}`(G`\alpha)$, contradicting the definition of $\mathscr{H}$.} Hence, $G`\alpha$ is an initial ordinal for all $\alpha$. In addition, every infinite initial ordinal is equal to $G`\alpha$ for some α. [Assume not. Let σ be the least infinite initial ordinal not in $G``On$. By the Replacement Axiom R, $G``On$ is not a set; hence, there is some ordinal greater than σ in $G``On$. Let μ be the least such ordinal, and let $\mu = G`\beta$. Clearly, $\beta \neq 0$; if $\beta = \gamma'$ for some $\gamma$, then $G`\gamma <_0 \sigma <_0 G`(\gamma') = \mathscr{H}`(G`\gamma)$, contradicting the definition of $\mathscr{H}$. If $\beta$ is a limit ordinal, then there is some $\alpha <_0 \beta$ such that $\sigma <_0 G`\alpha <_0 G`\beta$, contradicting the definition of μ.] Thus, G is an $\in$-preserving "isomorphism" of On with the class of infinite initial ordinals.

We denote $G`\alpha$ by ω_α. Thus, $\omega_0 = \omega$; $\omega_{\alpha'}$ is the least initial ordinal greater than ω_α; and, for limit ordinals λ, ω_λ is the initial ordinal that is the least upper bound of the set of all ω_α with $\alpha <_0 \lambda$. It follows from Proposition 4.14 that $\omega_\alpha \geqslant_0 \alpha$ for all α. Also, any infinite ordinal α is equinumerous with a unique initial ordinal $\omega_\beta \leqslant_0 \alpha$—namely, with the least ordinal equinumerous with α.

Let us turn now to ordinal arithmetic. We have already defined (see page 194) addition, multiplication, and exponentiation:

1.
$$\beta +_0 0 = \beta$$
$$\beta +_0 \gamma' = (\beta +_0 \gamma)'$$
$$\mathrm{Lim}(\alpha) \Rightarrow \beta +_0 \alpha = \bigcup_{\tau <_0 \alpha} (\beta +_0 \tau)$$

2.
$$\beta \times_0 0 = 0$$
$$\beta \times_0 (\gamma') = (\beta \times_0 \gamma) +_0 \beta$$
$$\mathrm{Lim}(\alpha) \Rightarrow \beta \times_0 \alpha = \bigcup_{\tau <_0 \alpha} (\beta \times_0 \tau)$$

3.
$$\exp(\beta, 0) = 1$$
$$\exp(\beta, \gamma') = \exp(\beta, \gamma) \times_0 \beta$$
$$\mathrm{Lim}(\alpha) \Rightarrow \exp(\beta, \alpha) = \bigcup_{0 <_0 \tau <_0 \alpha} \exp(\beta, \tau)$$

PROPOSITION 4.29 The following wfs are theorems.

(a) $\beta +_0 1 = \beta'$
(b) $0 +_0 \beta = \beta$
(c) $0 <_0 \beta \Rightarrow \alpha <_0 \alpha +_0 \beta \wedge \beta \leqslant_0 \alpha +_0 \beta$
(d) $\beta <_0 \gamma \Rightarrow \alpha +_0 \beta <_0 \alpha +_0 \gamma$
(e) $\alpha +_0 \beta = \alpha +_0 \delta \Rightarrow \beta = \delta$
(f) $\alpha <_0 \beta \Rightarrow (\exists_1 \delta)(\alpha +_0 \delta = \beta)$
(g) $0 \neq x \subseteq On \Rightarrow \alpha +_0 \bigcup_{\beta \in x} \beta = \bigcup_{\beta \in x} (\alpha +_0 \beta)$

(h) $0 <_0 \alpha \wedge 1 <_0 \beta \Rightarrow \alpha <_0 \alpha \times_0 \beta$
(i) $0 <_0 \alpha \wedge 0 <_0 \beta \Rightarrow \alpha \leqslant_0 \alpha \times_0 \beta$
(j) $\gamma <_0 \beta \wedge 0 <_0 \alpha \Rightarrow \alpha \times_0 \gamma <_0 \alpha \times_0 \beta$
(k) $x \subseteq On \Rightarrow \alpha \times_0 \bigcup_{\beta \in x} \beta = \bigcup_{\beta \in x} (\alpha \times_0 \beta)$

Proof

(a) $\beta +_0 1 = \beta +_0 (0') = (\beta +_0 0)' = \beta'$

(b) Prove $0 +_0 \beta = \beta$ by transfinite induction (Proposition 4.12). Let $X = \{\beta | 0 +_0 \beta = \beta\}$. First, $0 \in X$, since $0 +_0 0 = 0$. If $0 +_0 \gamma = \gamma$, then $0 +_0 \gamma' = (0 +_0 \gamma)' = \gamma'$. If Lim($\alpha$) and $0 +_0 \tau = \tau$ for all $\tau <_0 \alpha$, then $0 +_0 \alpha = \bigcup_{\tau <_0 \alpha} (0 +_0 \tau) = \bigcup_{\tau <_0 \alpha} \tau = \alpha$, since $\bigcup_{\tau <_0 \alpha} \tau$ is the least upper bound of the set of all $\tau <_0 \alpha$.

(c) Let $X = \{\beta | 0 <_0 \beta \Rightarrow \alpha <_0 \alpha +_0 \beta\}$. Prove $X = On$ by transfinite induction. Clearly, $0 \in X$. If $\gamma \in X$, then $\alpha \leqslant_0 \alpha +_0 \gamma$; hence, $\alpha \leqslant_0 \alpha +_0 \gamma <_0 (\alpha +_0 \gamma)' = \alpha +_0 \gamma'$. If Lim($\lambda$) and $\tau \in X$ for all $\tau <_0 \lambda$, then $\alpha <_0 \alpha' = \alpha +_0 1 \leqslant_0 \bigcup_{\tau <_0 \lambda} (\alpha +_0 \tau) = \alpha +_0 \lambda$. The second part is left as an exercise.

(d) Use transfinite induction. Let

$$X = \{\gamma | (\forall \alpha)(\forall \beta)(\beta <_0 \gamma \Rightarrow \alpha +_0 \beta <_0 \alpha +_0 \gamma)\}$$

Clearly, $0 \in X$. Assume $\gamma \in X$. Assume $\beta <_0 \gamma'$. Then $\beta <_0 \gamma$ or $\beta = \gamma$. If $\beta <_0 \gamma$, then, since $\gamma \in X$, $\alpha +_0 \beta <_0 \alpha +_0 \gamma <_0 (\alpha +_0 \gamma)' = \alpha +_0 \gamma'$. If $\beta = \gamma$, then $\alpha +_0 \beta = \alpha +_0 \gamma <_0 (\alpha +_0 \gamma)' = \alpha +_0 \gamma'$. Hence, $\gamma' \in X$. Assume Lim(λ) and $\tau \in X$ for all $\tau <_0 \lambda$. Assume $\beta <_0 \lambda$. Then $\beta <_0 \tau$ for some $\tau <_0 \lambda$, since Lim(λ). Hence, since $\tau \in X$, $\alpha +_0 \beta <_0 \alpha +_0 \tau \leqslant_0 \bigcup_{\tau <_0 \lambda} (\alpha +_0 \tau) = \alpha +_0 \lambda$. Hence, $\lambda \in X$.

(e) Assume $\alpha +_0 \beta = \alpha +_0 \delta$. Now, either $\beta <_0 \delta$ or $\delta <_0 \beta$ or $\delta = \beta$. If $\beta <_0 \delta$, then $\alpha +_0 \beta <_0 \alpha +_0 \delta$, and if $\delta <_0 \beta$, then $\alpha +_0 \delta <_0 \alpha +_0 \beta$, by part (d), contradicting $\alpha +_0 \beta = \alpha +_0 \delta$. Hence, $\delta = \beta$.

(f) The uniqueness follows from part (e). Prove the existence by induction on β. Let $X = \{\beta | \alpha <_0 \beta \Rightarrow (\exists_1 \delta)(\alpha +_0 \delta = \beta)\}$. Clearly, $0 \in X$. Assume $\gamma \in X$ and $\alpha <_0 \gamma'$. Hence, $\alpha = \gamma$ or $\alpha <_0 \gamma$. If $\alpha = \gamma$, then $(\exists \delta)(\alpha +_0 \delta = \gamma')$—namely, $\delta = 1$. If $\alpha <_0 \gamma$, then $(\exists_1 \delta)(\alpha +_0 \delta = \gamma)$. Take an ordinal σ such that $\alpha +_0 \sigma = \gamma$. Then $\alpha +_0 \sigma' = (\alpha +_0 \sigma)' = \gamma'$; thus, $(\exists \delta)(\alpha +_0 \delta = \gamma')$; that is, $\gamma' \in X$. Assume now that Lim(λ) and $\tau \in X$ for all $\tau <_0 \lambda$. Assume $\alpha <_0 \lambda$. Now define a function f such that, for $\alpha <_0 \mu <_0 \lambda$, $f`\mu$ is the unique ordinal $\delta$ such that $\alpha +_0 \delta = \mu$. But $\lambda = \bigcup_{\alpha <_0 \mu <_0 \lambda} \mu = \bigcup_{\alpha <_0 \mu <_0 \lambda} (\alpha +_0 f`\mu)$. Let $\rho = \bigcup_{\alpha <_0 \mu <_0 \lambda} (f`\mu)$. Notice that, if $\alpha <_0 \mu <_0 \lambda$, then $f`\mu <_0 f`(\mu')$; hence, $\rho$ is a limit ordinal. Then $\lambda = \bigcup_{\alpha <_0 \mu <_0 \lambda} (\alpha +_0 f`\mu) = \bigcup_{\sigma <_0 \rho} (\alpha +_0 \sigma) = \alpha +_0 \rho$.

(g) Assume $0 \neq x \subseteq On$. By part (f), there is some δ such that $\alpha +_0 \delta = \bigcup_{\beta \in x} (\alpha +_0 \beta)$. We must show that $\delta = \bigcup_{\beta \in x} \beta$. If $\beta \in x$, $\alpha +_0 \beta \leqslant_0 \alpha +_0 \delta$. Hence, $\beta \leqslant_0 \delta$, by part (d). Therefore, δ is an upper bound of the set of all $\beta \in x$. So, $\bigcup_{\beta \in x} \beta \leqslant_0 \delta$. On the other hand, if $\beta \in x$, then $\alpha +_0 \beta \leqslant_0 \alpha +_0 \bigcup_{\beta \in x} \beta$. Hence, $\alpha +_0 \delta = \bigcup_{\beta \in x} (\alpha +_0 \beta) \leqslant_0 \alpha +_0 \bigcup_{\beta \in x} \beta$, and so, by part (d), $\delta \leqslant_0 \bigcup_{\beta \in x} \beta$. Therefore, $\delta = \bigcup_{\beta \in x} \beta$.

(h)–(k) are left as exercises.

PROPOSITION 4.30 The following wfs are theorems.

(a) $\beta \times_0 1 = \beta \wedge 1 \times_0 \beta = \beta$
(b) $0 \times_0 \beta = 0$
(c) $(\alpha +_0 \beta) +_0 \gamma = \alpha +_0 (\beta +_0 \gamma)$
(d) $(\alpha \times_0 \beta) \times_0 \gamma = \alpha \times_0 (\beta \times_0 \gamma)$
(e) $\alpha \times_0 (\beta +_0 \gamma) = (\alpha \times_0 \beta) +_0 (\alpha \times_0 \gamma)$
(f) $\exp(\beta, 1) = \beta \wedge \exp(1, \beta) = 1$
(g) $\exp(\exp(\beta, \gamma), \delta) = \exp(\beta, \gamma \times_0 \delta)$
(h) $\exp(\beta, \gamma +_0 \delta) = \exp(\beta, \gamma) \times_0 \exp(\beta, \delta)$
(i) $\alpha >_0 1 \wedge \beta <_0 \gamma \Rightarrow \exp(\alpha, \beta) <_0 \exp(\alpha, \gamma)$

Proof

(a) $\beta \times_0 1 = \beta \times_0 0' = (\beta \times_0 0) +_0 \beta = 0 +_0 \beta = \beta$, by Proposition 4.29(b). Prove $1 \times_0 \beta = \beta$ by transfinite induction on β.
(b) Prove $0 \times_0 \beta = 0$ by transfinite induction on β.
(c) Let $X = \{\gamma | (\forall\alpha)(\forall\beta)((\alpha +_0 \beta) +_0 \gamma = \alpha +_0 (\beta +_0 \gamma))\}$. $0 \in X$, since $(\alpha +_0 \beta) +_0 0 = \alpha +_0 \beta = \alpha +_0 (\beta +_0 0)$. Now, assume $\gamma \in X$. Then $(\alpha +_0 \beta) +_0 \gamma' = ((\alpha +_0 \beta) +_0 \gamma)' = (\alpha +_0 (\beta +_0 \gamma))' = \alpha +_0 (\beta +_0 \gamma)' = \alpha +_0 (\beta +_0 \gamma')$. Hence, $\gamma' \in X$. Assume now that $\mathrm{Lim}(\lambda)$ and $\tau \in X$ for all $\tau <_0 \lambda$. Then $(\alpha +_0 \beta) +_0 \lambda = \bigcup_{\tau <_0 \lambda}((\alpha +_0 \beta) +_0 \tau) = \bigcup_{\tau <_0 \lambda}(\alpha +_0 (\beta +_0 \tau)) = \alpha +_0 \bigcup_{\tau <_0 \lambda}(\beta +_0 \tau)$ [by Proposition 4.29(g)] and this is equal to $\alpha +_0 (\beta +_0 \lambda)$.
(d)–(i) are left as exercises.

We would like to consider for a moment the properties of ordinal addition and multiplication when restricted to ω.

PROPOSITION 4.31 Assume α, β, γ are in ω. Then,

(a) $\alpha +_0 \beta \in \omega$
(b) $\alpha \times_0 \beta \in \omega$
(c) $\exp(\alpha, \beta) \in \omega$
(d) $\alpha +_0 \beta = \beta +_0 \alpha$
(e) $\alpha \times_0 \beta = \beta \times_0 \alpha$
(f) $(\alpha +_0 \beta) \times_0 \gamma = (\alpha \times_0 \gamma) +_0 (\beta \times_0 \gamma)$
(g) $\exp(\alpha \times_0 \beta, \gamma) = \exp(\alpha, \gamma) \times_0 \exp(\beta, \gamma)$

Proof

(a) This proof is by induction on β. Let $X = \{\beta | (\forall\alpha)(\alpha \in \omega \Rightarrow \alpha +_0 \beta \in \omega)\}$. Clearly, $0 \in X$. Assume $\beta \in X$ and $\alpha \in \omega$. Then $\alpha +_0 \beta \in \omega$. Hence, $\alpha +_0 (\beta') = (\alpha +_0 \beta)' \in \omega$ by Proposition 4.10(a). So, by Proposition 4.10(c), $\omega \subseteq X$.
(b) and (c) are left as exercises.
(d) *Lemma.* $\vdash \alpha \in \omega \wedge \beta \in \omega \Rightarrow \alpha' +_0 \beta = \alpha +_0 \beta'$. Let $Y = \{\beta | \beta \in \omega \wedge (\forall\alpha)(\alpha \in \omega \Rightarrow \alpha' +_0 \beta = \alpha +_0 \beta')\}$. Observe that $0 \in Y$. Assume that $\beta \in Y$ and let $\alpha \in \omega$. So, $\alpha' +_0 \beta = \alpha +_0 \beta'$. Then $\alpha' +_0 \beta' = (\alpha' +_0 \beta)' = (\alpha +_0 \beta')' = \alpha +_0 (\beta')'$. Hence, $\beta' \in Y$.

Now, to prove part (d), let

$$X = \{\beta | \beta \in \omega \wedge (\forall\alpha)(\alpha \in \omega \Rightarrow \alpha +_0 \beta = \beta +_0 \alpha)\}$$

Then $0 \in X$, and it is easy to prove, using the lemma, that $\beta \in X \Rightarrow \beta' \in X$. (e)–(g) are left as exercises.

The reader will have noticed that we have not stated for ordinals certain well-known laws that hold for other familiar number systems—for example, the commutative laws for addition and multiplication. In fact, these laws fail for ordinals, as the following examples show.

Examples

1. $(\exists\alpha)(\exists\beta)(\alpha +_0 \beta \neq \beta +_0 \alpha)$

$$1 +_0 \omega = \bigcup_{\alpha <_0 \omega} (1 +_0 \alpha) = \omega$$

$$\omega +_0 1 = \omega' >_0 \omega$$

2. $(\exists\alpha)(\exists\beta)(\alpha \times_0 \beta \neq \beta \times_0 \alpha)$

$$2 \times_0 \omega = \bigcup_{\alpha <_0 \omega} (2 \times_0 \alpha) = \omega$$

$$\omega \times_0 2 = \omega \times_0 (1 +_0 1) = (\omega \times_0 1) +_0 (\omega \times_0 1) = \omega +_0 \omega >_0 \omega$$

3. $(\exists\gamma)(\exists\alpha)(\exists\beta)((\alpha +_0 \beta) \times_0 \gamma \neq (\alpha \times_0 \gamma) +_0 (\beta \times_0 \gamma))$

$$(1 +_0 1) \times_0 \omega = 2 \times_0 \omega = \omega$$

$$(1 \times_0 \omega) +_0 (1 \times_0 \omega) = \omega +_0 \omega >_0 \omega$$

4. $(\exists\alpha)(\exists\beta)(\exists\gamma)(\exp(\alpha \times_0 \beta, \gamma) \neq \exp(\alpha, \gamma) \times_0 \exp(\beta, \gamma))$

$$\exp(2 \times_0 2, \omega) = \exp(4, \omega) = \bigcup_{\alpha <_0 \omega} \exp(4, \alpha) = \omega$$

$$\exp(2, \omega) = \bigcup_{\alpha <_0 \omega} \exp(2, \alpha) = \omega$$

So, $\exp(2, \omega) \times_0 \exp(2, \omega) = \omega \times_0 \omega >_0 \omega$.

Given any wf $\mathscr{A}$ of formal number theory S (see Chapter 3), we can associate with $\mathscr{A}$ a wf $\mathscr{A}^*$ of NBG as follows: first, replace every "$+$" by "$+_0$" and every "$\cdot$" by "$\times_0$"; then, if $\mathscr{A}$ is $\mathscr{B} \Rightarrow \mathscr{C}$ or $\neg\mathscr{B}$, respectively, and we already have found $\mathscr{B}^*$ and $\mathscr{C}^*$, let $\mathscr{A}^*$ be $\mathscr{B}^* \Rightarrow \mathscr{C}^*$ or $\neg(\mathscr{B}^*)$, respectively; if $\mathscr{A}$ is $(\forall x)\mathscr{B}(x)$, replace it by $(\forall x)(x \in \omega \Rightarrow \mathscr{B}^*(x))$. This completes the definition of $\mathscr{A}^*$. Now, if $x_1, \ldots, x_n$ are the free variables (if any) of $\mathscr{A}$, prefix $(x_1 \in \omega \wedge x_2 \in \omega \wedge \cdots \wedge x_n \in \omega) \Rightarrow$ to $\mathscr{A}^*$, obtaining a wf $\mathscr{A}\#$. This amounts to restricting all variables to ω and interpreting addition, multiplication, and the successor function on natural numbers as the corresponding operations on ordinals. Then every axiom $\mathscr{A}$ of S is transformed into a theorem $\mathscr{A}\#$ of NBG. [Axioms (S1)–(S3) are obviously transformed into theorems. (S4)# is a theorem, by Proposition 4.9(c), and (S5)#–(S8)# are properties of ordinal addition and multiplication (see page 205).] Now, for any wf

$\mathscr{A}$ of S, $\mathscr{A}\#$ is predicative. Hence, by Proposition 4.4, all instances of (S9)# are provable by transfinite induction [Proposition 4.12(b)]. [In fact, assume $\mathscr{A}\#(0) \wedge (\forall x)(x \in \omega \Rightarrow (\mathscr{A}\#(x) \Rightarrow \mathscr{A}\#(x')))$. Let $X = \{y | y \in \omega \wedge \mathscr{A}\#(y)\}$. Then, by Proposition 4.12(b), $(\forall x)(x \in \omega \Rightarrow \mathscr{A}\#(x))$.] Applications of modus ponens are easily seen to be preserved under the transformation of $\mathscr{A}$ into $\mathscr{A}\#$. As for the Generalization Rule, consider a wf $\mathscr{A}(x)$ and assume that $\mathscr{A}\#(x)$ is provable in NBG. But $\mathscr{A}\#(x)$ is of the form $x \in \omega \wedge y_1 \in \omega \wedge \cdots \wedge y_m \in \omega \Rightarrow \mathscr{A}^*(x)$. Hence, $y_1 \in \omega \wedge \cdots \wedge y_m \in \omega \Rightarrow (\forall x)(x \in \omega \Rightarrow \mathscr{A}^*(x))$ is provable in NBG. But this wf is just $((\forall x)\mathscr{A}(x))\#$. Hence, application of Gen leads from theorems to theorems. Therefore, for every theorem $\mathscr{A}$ of S, $\mathscr{A}\#$ is a theorem of NBG, and we can translate into NBG all the theorems of S proved in Chapter 3.

One can check that the number-theoretic function h such that, if x is the Gödel number of a wf $\mathscr{A}$ of S, then $h(x)$ is the Gödel number of $\mathscr{A}\#$ in NBG, and if x is not the Gödel number of a wf of S, then $h(x) = 0$, is recursive (in fact, primitive recursive). Let K be any consistent extension of NBG. As we saw above, if x is the Gödel number of a theorem of S, $h(x)$ is the Gödel number of a theorem of NBG and, hence, also a theorem of K. Let S′ be the extension of S obtained by taking as axioms all wfs $\mathscr{A}$ of S such that $\mathscr{A}\#$ is a theorem of K. Since K is consistent, S′ must be consistent. Therefore, since S is essentially recursively undecidable (by Corollary 3.44), S′ is recursively undecidable; that is, the set $T_{S'}$ of Gödel numbers of theorems of S′ is not recursive. Now, assume K is recursively decidable; that is, the set T_K of Gödel numbers of theorems of K is recursive. But, $C_{T_{S'}}(x) = C_{T_K}(h(x))$ for any x, where $C_{T_{S'}}$ and C_{T_K} are the characteristic functions of $T_{S'}$ and T_K. Hence $T_{S'}$ would be recursive, contradicting the recursive undecidability of S′. Therefore K is recursively undecidable, and, thus, if NBG is consistent, NBG is essentially recursively undecidable. Recursive undecidability of a recursively axiomatizable theory implies incompleteness (see Proposition 3.45). Hence NBG is also essentially incomplete. Thus, we have the following result: *if NBG is consistent, then NBG is essentially recursively undecidable and essentially incomplete.* (It is possible to prove this result directly in the same way that the corresponding result was proved for S in Chapter 3. Also see Exercise 3.66 on page 175.) Since NBG apparently can serve as a foundation for all of present-day mathematics (i.e., it is clear to every mathematician that every mathematical theorem can be translated and proved within NBG, or within extensions of NBG obtained by adding various extra axioms such as the Axiom of Choice), the essential incompleteness of NBG seems to indicate that the "axiomatic approach to mathematics" is inadequate. This conclusion does not depend upon the peculiarities of the theory NBG. Any other consistent theory (including "higher-order theories" as well as first-order theories) in which the theory of natural numbers can be developed far enough so as to include all the theorems of S (or even of RR) must also be essentially recursively undecidable and essentially incomplete, as the proof given above for NBG shows.

Exercise

4.73 Prove that a predicate calculus with a single binary predicate letter is recursively undecidable. (*Hint:* Use Proposition 3.47.)

There are a few facts about the "cardinal arithmetic" of ordinals that we would like to deal with now. By "cardinal arithmetic", we mean properties connected with the operations of union ($\bigcup$) and Cartesian product ($\times$) and X^Y, as opposed to the properties of $+_0$ and $\times_0$ and ordinal exponentiation. Observe that $\times$ is distinct from $\times_0$; also notice that ordinal exponentiation $\exp(\alpha, \beta)$, in spite of the ambiguous notation, has nothing to do with the operation of forming X^Y, the class of all functions from Y into X. [From Example 4 on page 208, we see that $\exp(2, \omega)$, in the sense of ordinal exponentiation, is ω; while, from Cantor's Theorem, $\omega \prec 2^\omega$, where, in the latter formula, we mean by 2^ω the set of functions from ω into 2.]

PROPOSITION 4.32

(a) $\vdash \omega \times \omega \cong \omega$

(b) If each of X and Y contains at least two elements, then $X \cup Y \preccurlyeq X \times Y$.

(c) $\text{Den}(x) \wedge \text{Den}(y) \Rightarrow \text{Den}(x \cup y)$

Proof

(a) Let f be a function with domain ω such that, if $\alpha \in \omega$, then $f`\alpha = \langle \alpha, 0 \rangle$. Then f is a one–one function from ω into a subset of $\omega \times \omega$. Hence, $\omega \preccurlyeq \omega \times \omega$. Conversely, let g be a function with domain $\omega \times \omega$ such that, for any $\langle \alpha, \beta \rangle \in \omega \times \omega$, $g`\langle \alpha, \beta \rangle = 2^\alpha \times_0 3^\beta$. We leave it as an exercise to show that g is one–one. Hence, $\omega \times \omega \preccurlyeq \omega$. So, by the Schröder-Bernstein Theorem, $\omega \times \omega \cong \omega$.

(b) Assume $a_1 \in X$, $a_2 \in X$, $a_1 \neq a_2$ and $b_1 \in Y$, $b_2 \in Y$, and $b_1 \neq b_2$. Define:

$$f`x = \begin{cases} \langle x, b_1 \rangle & \text{if } x \in X \\ \langle a_1, x \rangle & \text{if } x \in Y - X \text{ and } x \neq b_1 \\ \langle a_2, b_2 \rangle & \text{if } x = b_1 \text{ and } x \in Y - X \end{cases}$$

Then f is a one–one function with domain $X \cup Y$ and range a subset of $X \times Y$. Hence, $X \cup Y \preccurlyeq X \times Y$.

(c) Assume $\text{Den}(A)$ and $\text{Den}(B)$. Hence, each of A and B contains at least two elements. Then, by part (b), $A \cup B \preccurlyeq A \times B$. But $A \cong \omega$ and $B \cong \omega$. Hence, $A \times B \cong \omega \times \omega$. Therefore, $A \cup B \preccurlyeq \omega \times \omega \cong \omega$. By Proposition 4.27, either $\text{Den}(A \cup B)$ or $\text{Fin}(A \cup B)$. But $A \subseteq A \cup B$ and $\text{Den}(A)$; hence, $\neg \text{Fin}(A \cup B)$.

For the further study of ordinal addition and multiplication, it is quite useful to obtain concrete interpretations of these operations.

PROPOSITION 4.33 (ADDITION) Assume that $\langle R, A \rangle$ is similar to $\langle E_\alpha, \alpha \rangle$, that $\langle S, B \rangle$ is similar to $\langle E_\beta, \beta \rangle$, and that $A \cap B = 0$. Define the relation T on $A \cup B$ by: $\langle x, y \rangle \in T \Leftrightarrow (x \in A \wedge y \in B) \vee (x \in A \wedge y \in A \wedge \langle x, y \rangle \in R) \vee (x \in B \wedge y \in B \wedge \langle x, y \rangle \in S)$ (i.e., T is the same as R in the set A, the same as S in the set B, and every element of A T-precedes every element of B). Then T is a well-ordering of $A \cup B$, and $\langle T, A \cup B \rangle$ is similar to $\langle E_{\alpha +_0 \beta}, \alpha +_0 \beta \rangle$.

Proof First, it is simple to verify that T is a well-ordering of $A \cup B$, since R is a well-ordering of A and S is a well-ordering of B. To show that $\langle T, A \cup B \rangle$ is similar to $\langle E_{\alpha +_0 \beta}, \alpha +_0 \beta \rangle$, perform transfinite induction on β. For $\beta = 0$, $B = 0$. Hence, $T = R$, $A \cup B = A$, and $\alpha +_0 \beta = \alpha$. So, $\langle T, A \cup B \rangle$ is similar to

$\langle E_{\alpha+_0\beta}, \alpha +_0 \beta\rangle$. Assume the proposition for γ and let $\beta = \gamma'$. Since $\langle S, B\rangle$ is similar to $\langle E_\beta, \beta\rangle$, we have a function f with domain B and range β such that, for any x, y in B, $\langle x, y\rangle \in S$ if and only if $f\text{`}x \in f\text{`}y$. Let $b = (\check{f})\text{`}\gamma$, let $B_1 = B - \{b\}$, and let $S_1 = S \cap (B_1 \times B_1)$. Since b is the S-maximum of B, it follows easily that S_1 well-orders B_1. Also, $B_1 \upharpoonright f$ is a similarity mapping of B_1 onto γ. Let $T_1 = T \cap ((A \cup B_1) \times (A \cup B_1))$. By inductive hypothesis, $\langle T_1, A \cup B_1\rangle$ is similar to $\langle E_{\alpha+_0\gamma}, \alpha +_0 \gamma\rangle$, by means of some similarity mapping g with domain $A \cup B_1$ and range $\alpha +_0 \gamma$. Extend g to $g_1 = g \cup \{\langle b, \alpha +_0 \gamma\rangle\}$, which is a similarity mapping of $A \cup B$ onto $(\alpha +_0 \gamma)' = \alpha +_0 \gamma' = \alpha +_0 \beta$. Finally, if $\text{Lim}(\beta)$ and our proposition holds for all $\tau <_0 \beta$, assume that f is a similarity mapping of B onto β. Now, for each $\tau <_0 \beta$, let $B_\tau = (\check{f})\text{``}\tau$, $S_\tau = S \cap (B_\tau \times B_\tau)$, and $T_\tau = T \cap ((A \cup B_\tau) \times (A \cup B_\tau))$. By inductive hypothesis and Corollary 4.16(b), there is a unique similarity mapping g_τ of $\langle T_\tau, A \cup B_\tau\rangle$ with $\langle E_{\alpha+_0\tau}, \alpha +_0 \tau\rangle$; also, if $\tau_1 <_0 \tau_2 <_0 \beta$, then, since $A \cup B_{\tau_1} \upharpoonright g_{\tau_2}$ is a similarity mapping of $\langle T_{\tau_1}, A \cup B_{\tau_1}\rangle$ with $\langle E_{\alpha+_0\tau_1}, \alpha +_0 \tau_1\rangle$ and, by the uniqueness of g_{τ_1}, $A \cup B_{\tau_1} \upharpoonright g_{\tau_2} = g_{\tau_1}$; that is, g_{τ_2} is an extension of g_{τ_1}. Hence, if $g = \bigcup_{\tau<_0\beta}(g_\tau)$, then g is a similarity mapping of $\langle T, \bigcup_{\tau<_0\beta}(A \cup B_\tau)\rangle$ with $\langle E_{\bigcup_{\tau<_0\beta}(\alpha+_0\tau)}, \bigcup_{\tau<_0\beta}(\alpha +_0 \tau)\rangle$. But, $\bigcup_{\tau<_0\beta}(A \cup B_\tau) = A \cup B$ and $\bigcup_{\tau<_0\beta}(\alpha +_0 \tau) = \alpha +_0 \beta$. This completes the transfinite induction.

PROPOSITION 4.34 (MULTIPLICATION) Assume that $\langle R, A\rangle$ is similar to $\langle E_\alpha, \alpha\rangle$ and that $\langle S, B\rangle$ is similar to $\langle E_\beta, \beta\rangle$. Define the relation W on $A \times B$ as follows: $\langle\langle x, y\rangle, \langle u, v\rangle\rangle \in W \Leftrightarrow (x \in A \wedge u \in A \wedge y \in B \wedge v \in B) \wedge ((\langle y, v\rangle \in S) \vee (y = v \wedge \langle x, u\rangle \in R))$. Then W is a well-ordering of $A \times B$ and $\langle W, A \times B\rangle$ is similar to $\langle E_{\alpha\times_0\beta}, \alpha \times_0 \beta\rangle$.†

Proof This is left as an exercise. (Proceed as in the proof of Proposition 4.33.)

Examples

1. $2 \times_0 \omega = \omega$. Let $\langle R, A\rangle = \langle E_2, 2\rangle$ and $\langle S, B\rangle = \langle E_\omega, \omega\rangle$. Then the pairs in $2 \times \omega$ can be well-ordered as follows: $\langle 0,0\rangle, \langle 1,0\rangle, \langle 0,1\rangle, \langle 1,1\rangle, \langle 0,2\rangle, \langle 1,2\rangle, \ldots, \langle 0,n\rangle, \langle 1,n\rangle, \langle 0,n+1\rangle, \langle 1,n+1\rangle, \ldots$.
2. By Proposition 4.30(e), $\omega \times_0 2 = \omega +_0 \omega$. Let $\langle R, A\rangle = \langle E_\omega, \omega\rangle$ and $\langle S, B\rangle = \langle E_2, 2\rangle$. Then $\omega \times 2$ can be well-ordered (see Proposition 4.34) as follows: $\langle 0,0\rangle, \langle 1,0\rangle, \langle 2,0\rangle, \ldots, \langle 0,1\rangle, \langle 1,1\rangle, \langle 2,1\rangle, \ldots$.

PROPOSITION 4.35 For all α, $\omega_\alpha \times \omega_\alpha \cong \omega_\alpha$.

Proof (Sierpinski, 1958) Assume this is false, and let α be the least ordinal such that $\neg(\omega_\alpha \times \omega_\alpha \cong \omega_\alpha)$. Then $\omega_\beta \times \omega_\beta \cong \omega_\beta$ for all $\beta <_0 \alpha$. By Proposition 4.32(a), $\alpha >_0 0$. Now, let $P = \omega_\alpha \times \omega_\alpha$ and, for $\beta <_0 \omega_\alpha$, let $P_\beta = \{\langle\gamma, \delta\rangle | \gamma +_0 \delta = \beta\}$. First, we wish to show that $P = \bigcup_{\beta<_0\omega_\alpha} P_\beta$. Now, if $\gamma +_0 \delta = \beta <_0 \omega_\alpha$, then $\gamma \leqslant_0 \beta <_0 \omega_\alpha$ and $\delta \leqslant_0 \beta <_0 \omega_\alpha$; hence, $\langle\gamma, \delta\rangle \in \omega_\alpha \times \omega_\alpha = P$. Thus, $\bigcup_{\beta<_0\omega_\alpha} P_\beta \subseteq P$. To show that $P \subseteq \bigcup_{\beta<_0\omega_\alpha} P_\beta$, it suffices to show that, if $\gamma <_0 \omega_\alpha$ and $\delta <_0 \omega_\alpha$, then $\gamma +_0 \delta <_0 \omega_\alpha$.

†The ordering W is called an *inverse lexicographical ordering* because it orders pairs as follows: first, according to the size of their second components and then, if their second components are equal, according to the size of the first components.

This is clear when γ or δ is finite. Hence we may assume that γ and δ are equinumerous with initial ordinals $\omega_\sigma \leqslant_0 \gamma$ and $\omega_\rho \leqslant_0 \delta$, respectively. Let ζ be the larger of σ and ρ. Since $\gamma <_0 \omega_\alpha$ and $\delta <_0 \omega_\alpha$, then $\omega_\zeta <_0 \omega_\alpha$. Hence, by the minimality of α, $\omega_\zeta \times \omega_\zeta \cong \omega_\zeta$. Let $A = \gamma \times \{0\}$, $B = \delta \times \{1\}$. Then, by Proposition 4.33, $A \cup B \cong \gamma +_0 \delta$. Since $\gamma \cong \omega_\sigma$ and $\delta \cong \omega_\rho$, $A \cong \omega_\sigma \times \{0\}$ and $B \cong \omega_\rho \times \{1\}$. Hence, since $A \cap B = 0$, $A \cup B \cong (\omega_\sigma \times \{0\}) \cup (\omega_\rho \times \{1\})$. But, by Proposition 4.32(b), $(\omega_\sigma \times \{0\}) \cup (\omega_\rho \times \{1\}) \preccurlyeq (\omega_\sigma \times \{0\}) \times (\omega_\rho \times \{1\}) \cong \omega_\sigma \times \omega_\rho \preccurlyeq \omega_\zeta \times \omega_\zeta \cong \omega_\zeta$. Hence, $\gamma +_0 \delta \preccurlyeq \omega_\zeta <_0 \omega_\alpha$. Since ω_α is an initial ordinal, $\gamma +_0 \delta <_0 \omega_\alpha$. (If $\omega_\alpha \leqslant_0 \gamma +_0 \delta$, then $\omega_\alpha \preccurlyeq \omega_\zeta$ and $\omega_\zeta \preccurlyeq \omega_\alpha$; so, by the Schröder-Bernstein Theorem, $\omega_\alpha = \omega_\zeta$, contradicting $\omega_\zeta <_0 \omega_\alpha$.) Thus, $P = \bigcup_{\beta <_0 \omega_\alpha} P_\beta$. Consider P_β for any $\beta <_0 \omega_\alpha$. By Proposition 4.29(f), for each $\gamma \leqslant_0 \beta$, there is exactly one ordinal δ such that $\gamma +_0 \delta = \beta$. Hence there is a similarity mapping from β' onto P_β, where P_β is ordered according to the size of the first component γ of the pairs $\langle \gamma, \delta \rangle$. Define the following relation R on P. For any $\gamma <_0 \omega_\alpha$, $\delta <_0 \omega_\alpha$, $\mu <_0 \omega_\alpha$, $\nu <_0 \omega_\alpha$, $\langle\langle \gamma, \delta\rangle, \langle \mu, \nu \rangle\rangle \in R$ if and only if either $\gamma +_0 \delta <_0 \mu +_0 \nu$ or $(\gamma +_0 \delta = \mu +_0 \nu \wedge \gamma <_0 \mu)$. Thus, if $\beta_1 <_0 \beta_2 <_0 \omega_\alpha$, the pairs in P_{β_1} R-precede the pairs in P_{β_2}, and, within each P_β, the pairs are R-ordered according to the size of their first components. One easily verifies that R well-orders P. Since $P = \omega_\alpha \times \omega_\alpha$, it suffices now to show that $\langle R, P \rangle$ is similar to $\langle E_{\omega_\alpha}, \omega_\alpha \rangle$. By Proposition 4.17, $\langle R, P \rangle$ is similar to some $\langle E_\xi, \xi \rangle$, where ξ is an ordinal. Hence, $P \cong \xi$. Assume that $\xi >_0 \omega_\alpha$. There is a similarity mapping f between $\langle E_\xi, \xi \rangle$ and $\langle R, P \rangle$. Let $b = f`\omega_\alpha$; then b is an ordered pair $\langle \gamma, \delta \rangle$ with $\gamma <_0 \omega_\alpha, \delta <_0 \omega_\alpha$, and $\omega_\alpha \rceil f$ is a similarity mapping between $\langle E_{\omega_\alpha}, \omega_\alpha \rangle$ and the R-segment $Y = \mathrm{Seg}_R(P, \langle \gamma, \delta \rangle)$ of P determined by $\langle \gamma, \delta \rangle$. Then $Y \cong \omega_\alpha$. Also, letting $\beta = \gamma +_0 \delta$, if $\langle \sigma, \rho \rangle \in Y$, we have $\sigma +_0 \rho \leqslant_0 \gamma +_0 \delta = \beta$; hence, $\sigma \leqslant_0 \beta$ and $\rho \leqslant_0 \beta$. Therefore, $Y \subseteq \beta' \times \beta'$. But $\beta' <_0 \omega_\alpha$. Since β is obviously not finite, $\beta' \cong \omega_\mu$ with $\mu <_0 \alpha$. By the minimality of α, $\omega_\mu \times \omega_\mu \cong \omega_\mu$. So, $\omega_\alpha \cong Y \preccurlyeq \omega_\mu$, contradicting $\omega_\mu \prec \omega_\alpha$. Thus, $\xi \leqslant_0 \omega_\alpha$ and, therefore, $P \preccurlyeq \omega_\alpha$. Let h be the function with domain ω_α such that $h`\beta = \langle \beta, 0 \rangle$ for every $\beta <_0 \omega_\alpha$. Then h is a one–one correspondence between ω_α and the subset $\omega_\alpha \times \{0\}$ and, therefore, $\omega_\alpha \preccurlyeq P$. By the Schröder-Bernstein Theorem, $\omega_\alpha \cong P$, contradicting the definition of α. Hence, $\omega_\beta \times \omega_\beta \cong \omega_\beta$ for all β.

COROLLARY 4.36 If $A \cong \omega_\alpha$ and $B \cong \omega_\beta$, and if γ is the maximum of α and β, then $A \times B \cong \omega_\gamma$ and $A \cup B \cong \omega_\gamma$. In particular, $\omega_\alpha \times \omega_\beta \cong \omega_\gamma$.

Proof By Propositions 4.35 and 4.32(b), $\omega_\gamma \preccurlyeq A \cup B \preccurlyeq A \times B \cong \omega_\alpha \times \omega_\beta \preccurlyeq \omega_\gamma \times \omega_\gamma \cong \omega_\gamma$. Hence, by the Schröder-Bernstein Theorem, $A \times B \cong \omega_\gamma$ and $A \cup B \cong \omega_\gamma$.

This is really only the beginning of ordinal arithmetic. For further study, see Sierpinski (1958) and Bachmann (1955).

Exercises

Prove in Exercises 4.74–4.77 that the following are theorems of NBG.

4.74 (a) $x \preccurlyeq \omega_\alpha \Rightarrow x \cup \omega_\alpha \cong \omega_\alpha$

(b) $\omega_\alpha +_c \omega_\alpha \cong \omega_\alpha$

4.75 $0 \neq x \preccurlyeq \omega_\alpha \Rightarrow x \times \omega_\alpha \cong \omega_\alpha$

4.76 $0 \neq x \prec \omega \Rightarrow \omega_\alpha^x \cong \omega_\alpha$

4.77 (a) $\mathscr{P}(\omega_\alpha) \times \mathscr{P}(\omega_\alpha) \cong \mathscr{P}(\omega_\alpha)$
(b) $x \preccurlyeq \mathscr{P}(\omega_\alpha) \Rightarrow x \cup \mathscr{P}(\omega_\alpha) \cong \mathscr{P}(\omega_\alpha)$
(c) $0 \neq x \preccurlyeq \mathscr{P}(\omega_\alpha) \Rightarrow x \times \mathscr{P}(\omega_\alpha) \cong \mathscr{P}(\omega_\alpha)$
(d) $0 \neq x \preccurlyeq \omega_\alpha \Rightarrow (\mathscr{P}(\omega_\alpha))^x \cong \mathscr{P}(\omega_\alpha)$
(e) $1 \prec x \preccurlyeq \omega_\alpha \Rightarrow x^{\omega_\alpha} \cong \omega_\alpha^{\omega_\alpha} \cong (\mathscr{P}(\omega_\alpha))^{\omega_\alpha} \cong \mathscr{P}(\omega_\alpha)$

4.78 Assume $y \neq 0 \wedge y \cong y +_c y$. Remember that $y +_c y = (y \times \{0\}) \cup (y \times \{1\})$. [This assumption holds for $y = \omega_\alpha$ by Corollary 4.36 and for $y = \mathscr{P}(\omega_\alpha)$ by Exercise 4.77(b). It will turn out to hold for all infinite sets y if the Axiom of Choice holds.] Prove that the following are theorems of NBG.
(a) $\mathrm{Inf}(y)$
(b) $y \cong 1 +_c y$
(c) $(\exists u)(\exists v)(y = u \cup v \wedge u \cap v = 0 \wedge u \cong y \wedge v \cong y)$
(d) $\{z | z \subseteq y \wedge z \cong y\} \cong \mathscr{P}(y)$
(e) $\{z | z \subseteq y \wedge \mathrm{Inf}(z)\} \cong \mathscr{P}(y)$
(f) $(\exists f)(y \underset{f}{\cong} y \wedge (\forall u)(u \in y \Rightarrow f`u \neq u))$

4.79 Assume $y \cong y \times y \wedge 1 \prec y$. [This holds when $y = \omega_\alpha$ by Proposition 4.35 and for $y = \mathscr{P}(\omega_\alpha)$ by Exercise 4.77(a). It is true for all infinite sets y if the Axiom of Choice holds.] Prove that the following are theorems of NBG.
(a) $y \cong y +_c y$
(b)[D] Let $\mathrm{Perm}(y) = \{f | y \underset{f}{\cong} y\}$. Then $\mathrm{Perm}(y) \cong \mathscr{P}(y)$.

5. THE AXIOM OF CHOICE. THE AXIOM OF REGULARITY

The Axiom of Choice is one of the most celebrated and contested statements of the theory of sets. We shall state it in the next proposition and show its equivalence to several other important assertions.

PROPOSITION 4.37 The following wfs are equivalent.

(a) *Axiom of Choice (AC).* For any set x, there is a function f such that, for any nonempty subset y of x, $f`y \in y$. (f is called a *choice function* for x.)
(b) *Multiplicative Axiom (Mult).* If x is a set of pairwise disjoint nonempty sets, then there is a set y (called a *choice set* for x) such that y contains exactly one element of each set in x: $(\forall u)(u \in x \Rightarrow u \neq 0 \wedge (\forall v)(v \in x \wedge v \neq u \Rightarrow v \cap u = 0)) \Rightarrow (\exists y)(\forall u)(u \in x \Rightarrow (\exists_1 w)(w \in u \cap y))$.
(c) *Well-Ordering Principle (WO).* Every set can be well-ordered: $(\forall x)(\exists y)(y \,\mathrm{We}\, x)$.
(d) *Trichotomy (Trich).* $(\forall x)(\forall y)(x \preccurlyeq y \vee y \preccurlyeq x)$
(e) *Zorn's Lemma (Zorn).* Any nonempty partially ordered set x, in which every chain (i.e., every totally ordered subset) has an upper bound, has a maximal element:

$$(\forall x)(\forall y)((y \,\mathrm{Part}\, x) \wedge (\forall u)(u \subseteq x \wedge y \,\mathrm{Tot}\, u \Rightarrow (\exists v)(v \in x \wedge (\forall w)(w \in u \Rightarrow$$
$$w = v \vee \langle w, v\rangle \in y))) \Rightarrow (\exists v)(v \in x \wedge (\forall w)(w \in x \Rightarrow \langle v, w\rangle \notin y)))$$

Proof

1. $\vdash$ (WO) $\Rightarrow$ Trich. Given sets x, y, then, by (WO), x and y can be well-ordered; hence, by Proposition 4.17, $x \cong \alpha$ and $y \cong \beta$ for some ordinals α, β. But $\alpha \preccurlyeq \beta$ or $\beta \preccurlyeq \alpha$. Hence, $x \preccurlyeq y$ or $y \preccurlyeq x$.

2. $\vdash$ Trich $\Rightarrow$ (WO). Given a set x, then, by Hartogs' Theorem, there is an ordinal α such that α is not equinumerous with any subset of x. By Trich, x is equinumerous with some subset y of α. Hence, by translating the well-ordering E_y of y to x, x can be well-ordered.

3. $\vdash$ (WO) $\Rightarrow$ Mult. Let x be a set of nonempty disjoint sets. By (WO), there is a well-ordering R of $\bigcup(x)$. Hence, there is a function f with domain x such that, for any u in x, $f`u$ is the R-least element of u. [Notice that $u \subseteq \bigcup(x)$.]

4. $\vdash$ Mult $\Rightarrow$ AC. For any set x, we may define a one–one function g such that, for each nonempty subset u of x, $g`u = u \times \{u\}$. Let x_1 be the range of g. Then x_1 is a set of nonempty disjoint sets. Hence, by Mult, there is a choice set y for x_1. Therefore, if $0 \neq u$ and $u \subseteq x$, then $u \times \{u\}$ is in x_1, and so, y contains exactly one element $\langle v, u \rangle$ in $u \times \{u\}$. Then the function f such that $f`u = v$ is a choice function for x.

5. $\vdash$ AC $\Rightarrow$ Zorn. Let y partial order a nonempty set x such that every y-chain in x has an upper bound in x. By AC, there is a choice function f for x. Let b be any element of x. By transfinite induction (Proposition 4.13), we define a function F such that $F`0 = b$ and, for any $\alpha > 0$, $F`\alpha$ is $f`u$, where u is the set of y-upper bounds v in x of $F``\alpha$ such that $v \notin F``\alpha$. Let β be the least ordinal such that the set of y-upper bounds in x of $F``\beta$ that are not in $F``\beta$ is empty. (There must be such an ordinal; otherwise, F is a one–one function with domain On and range a subset of x, which, by the Replacement Axiom R, implies that On is a set.) Let $g = \beta \upharpoonright F$. Then it is an easy exercise to check that g is one–one and, if $\alpha <_0 \gamma <_0 \beta$, $\langle g`\alpha, g`\gamma \rangle \in y$. Hence, $g``\beta$ is a $y$-chain in $x$; by hypothesis, there is an upper bound $w$ of $g``\beta$. Since the set of upper bounds of $F``\beta (= g``\beta)$ that are not in $g``\beta$ is empty, $w \in g``\beta$ and w is the only upper bound of $g``\beta$ (because a set can contain at most one upper bound). Hence, $w$ is a $y$-maximal element. (If $\langle w, z \rangle \in y$ and $z \in x$, then $z$ is a $y$-upper bound of $g``\beta$, which is impossible.)

6. $\vdash$ Zorn $\Rightarrow$ (WO). Given a set z, let X be the class of all one–one functions with domain an ordinal and range a subset of z. By Hartogs' Theorem, X is a set. Clearly, $0 \in X$. X is partially ordered by the proper inclusion relation $\subset$. Given any chain of functions in X, of any two, one is an extension of the other. Hence, the union of all the functions in the chain is also a one–one function from an ordinal into z, which is an $\subset$-upper bound of the chain. Hence, by Zorn, X has a maximal element g, which is a one–one function from an ordinal α into z. Assume $z - g``\alpha \neq 0$ and let $b \in z - g``\alpha$. Let $f = g \cup \{\langle \alpha, b \rangle\}$. Then $f \in X$ and $g \subset f$, contradicting the maximality of g. So, $g``\alpha = z$. Thus, $\alpha \underset{g}{\cong} z$. We can transfer by means of g the well-ordering E_α of α to a well-ordering of z.

Exercises

4.80 Show that the following are equivalent to the Axiom of Choice.

(a) Any set x is equinumerous with some ordinal.

(b) *Special case of Zorn's Lemma.* If x is a nonempty set and if the union of each nonempty $\subset$-chain in x is also in x, then x has a $\subset$-maximal element.

(c) *Hausdorff Maximal Principle.* If x is a set, then every $\subset$-chain in x is a subset of some maximal $\subset$-chain in x.

(d) *Teichmüller-Tukey Lemma.* Any set of finite character has a $\subset$-maximal element. [A nonempty set x is said to be of *finite character* if and only if (i) every finite subset of an element of x is also an element of x and (ii) if every finite subset of a set y is a member of x, then $y \in x$.]

(e) $(\forall x)(\text{Rel}(x) \Rightarrow (\exists y)(\text{Fnc}(y) \wedge \mathscr{D}(x) = \mathscr{D}(y) \wedge y \subseteq x))$

(f) For any nonempty sets x and y, either there is a function with domain x and range y or there is a function with domain y and range x.

4.81 Show that the following Finite Axiom of Choice is provable in NBG: if x is a finite set of nonempty disjoint sets, then there is a choice set y for x. (*Hint:* Assume $x \cong \alpha$ where $\alpha \in \omega$. Use induction on α.)

PROPOSITION 4.38 The following are consequences of the Axiom of Choice

(a) Any infinite set has a denumerable subset.

(b) Any infinite set is Dedekind-infinite.

(c) If x is a denumerable set whose elements are denumerable sets, then $\bigcup(x)$ is denumerable.

Proof

(a) Assume AC. Let x be an infinite set. By Exercise 4.80(a), x is equinumerous with some ordinal α. Since x is infinite, so is α. Hence, $\omega \leqslant_0 \alpha$; therefore, ω is equinumerous with some subset of x.

(b) The proof is by part (a) and Exercise 4.71(c).

(c) Assume x is a denumerable set of denumerable sets. Let f be a function assigning to each $u \in x$ the set of all one–one correspondences between u and ω. Let z be the union of the range of f. Then, by AC applied to z, there is a function g such that $g\text{‘}v \in v$ for each nonempty $v \subseteq z$. In particular, if $u \in x$, then $g\text{‘}(f\text{‘}u)$ is a one–one correspondence between u and ω. Let h be a one–one correspondence between ω and x. Define a function F on $\bigcup(x)$ as follows: let $y \in \bigcup(x)$ and let n be the smallest element of ω such that $y \in h\text{‘}n$. Now, $h\text{‘}n \in x$; so, $g\text{‘}(f\text{‘}(h\text{‘}n))$ is a one–one correspondence between $h\text{‘}n$ and ω. Define $F\text{‘}y = \langle n, (g\text{‘}(f\text{‘}(h\text{‘}n)))\text{‘}y\rangle$. Then F is a one–one function with domain $\bigcup(x)$ and range a subset of $\omega \times \omega$. Hence, $\bigcup(x) \preccurlyeq \omega \times \omega$. But $\omega \times \omega \cong \omega$ and, therefore, $\bigcup(x) \preccurlyeq \omega$. If $v \in x$, then $v \subseteq \bigcup(x)$ and $v \cong \omega$. Hence, $\omega \preccurlyeq \bigcup(x)$. By the Schröder-Bernstein Theorem, $\bigcup(x) \cong \omega$.

Exercises

4.82 If x is a set, the Cartesian product $\prod_{u \in x} u$ is the set of functions f with domain x such that $f\text{‘}u \in u$ for all $u \in x$. Show that AC is equivalent to the proposition that the Cartesian product of any set x of nonempty sets is also nonempty.

4.83 Show that AC implies that any partial ordering of a set x is included in a total ordering of x.

4.84 Prove that the following assertion is a consequence of AC: for any ordinal α, if x is a set such that $x \preccurlyeq \omega_\alpha$ and such that $(\forall u)(u \in x \Rightarrow u \preccurlyeq \omega_\alpha)$, then $\bigcup(x) \preccurlyeq \omega_\alpha$. [*Hint:* The proof is analogous to that of Proposition 4.38(c).]

4.85 (a) Prove $y \preccurlyeq x \Rightarrow (\exists f)(\text{Fnc}(f) \wedge \mathscr{D}(f) = x \wedge \mathscr{R}(f) = y)$.

(b) Prove that AC implies the converse of part (a):

$$(\exists f)(\text{Fnc}(f) \wedge \mathscr{D}(f) = x \wedge \mathscr{R}(f) = y) \Rightarrow y \preccurlyeq x$$

4.86^{D} (a) Prove $(u +_c v)^2 \cong u^2 +_c (2 \times (u \times v)) +_c v^2$.

(b) Assume y is a well-ordered set such that $x \times y \cong x +_c y$ and $\neg(y \preccurlyeq x)$. Prove that $x \preccurlyeq y$.

(c) Assume $y \cong y \times y$ for all infinite sets y. Prove that, if Inf(x) and $z = \mathscr{H}\text{`}x$, then $x \times z \cong x +_c z$.

(d) Prove that AC is equivalent to $(\forall y)(\text{Inf}(y) \Rightarrow y \cong y \times y)$ (Tarski, 1923).

A stronger form of the Axiom of Choice is the following sentence (UCF): $(\exists X)(\text{Fnc}(X) \wedge (\forall u)(u \neq 0 \Rightarrow X\text{`}u \in u))$. (There is a *universal choice function*—that is, a function that assigns to every nonempty set u an element of u.) UCF obviously implies AC, but it was proved by W. B. Easton in 1964 that UCF is not provable from AC if NBG is consistent. However, Felgner (1971b) proved that, for any sentence $\mathscr{A}$ in which all quantifiers are restricted to sets, if $\mathscr{A}$ is provable from NBG + (UCF), then $\mathscr{A}$ is also provable in NBG + (AC). [See Felgner (1976) for a thorough treatment of the relations between UCF and AC.]

The theory of cardinal numbers is simplified if we assume AC; for AC implies that every set is equinumerous with some ordinal and, therefore, that every set x is equinumerous with a unique initial ordinal, which we shall call the *cardinal number* of x. Thus, the cardinal numbers are identified with the initial ordinals. To conform with the standard notation for ordinals, we let $\aleph_\alpha$ stand for ω_α. Proposition 4.35 and Corollary 4.36 establish some of the basic properties of addition and multiplication of cardinal numbers.

The status of the Axiom of Choice has become less controversial in recent years. To most mathematicians it seems quite plausible, and it has so many important applications in practically all branches of mathematics that not to accept it would seem to be a willful hobbling of the practicing mathematician. We shall discuss its consistency and independence later in this section.

Another hypothesis that has been proposed as a basic principle of set theory is the so-called *Axiom of Regularity* (Axiom D):

$$(\forall X)(X \neq 0 \Rightarrow (\exists y)(y \in X \wedge y \cap X = 0))$$

(Every nonempty class X contains a member that is disjoint from X.)

PROPOSITION 4.39

(a) The Axiom of Regularity implies the *Fundierungsaxiom*:

$$\neg(\exists x)(\text{Fnc}(x) \wedge \mathscr{D}(x) = \omega \wedge (\forall u)(u \in \omega \Rightarrow x\text{`}(u') \in x\text{`}u))$$

that is, there is no infinitely descending $\in$-sequence $x_1 \ni x_2 \ni x_3 \ni \cdots$.

(b) If we assume the Axiom of Choice, then the Fundierungsaxiom implies the Axiom of Regularity.

(c) The Axiom of Regularity implies the nonexistence of finite $\in$-cycles—that is, of functions f on a nonzero finite ordinal α' such that $f\text{`}0 \in f\text{`}1 \in \cdots \in f\text{`}\alpha \in f\text{`}0$. In particular, it implies that there is no set y such that $y \in y$.

Proof

(a) Assume $\mathrm{Fnc}(x) \wedge \mathscr{D}(x) = \omega \wedge (\forall u)(u \in \omega \Rightarrow x`(u') \in x`u)$. Let $z = x``\omega$. By the Axiom of Regularity, there is some element y in z such that $y \cap z = 0$. Since $y \in z$, there is some finite ordinal α such that $y = x`\alpha$. Then $x`(\alpha') \in y \cap z$, contradicting $y \cap z = 0$.

(b) First, we define the *transitive closure* of a set u. Define by induction a function g on ω such that $g`0 = \{u\}$, and $g`(\alpha') = \bigcup (g`\alpha)$ for each $\alpha \in \omega$. Thus, $g`1 = u, g`2 = \bigcup (u)$, and so on. Let $\mathrm{TC}(u) = \bigcup (g``\omega)$ be called the *transitive closure* of u. For any u, $\mathrm{TC}(u)$ is transitive; that is, $(\forall v)(v \in \mathrm{TC}(u) \Rightarrow v \subseteq \mathrm{TC}(u))$. Now, assume AC and the Fundierungsaxiom; also, assume $X \neq 0$ but there is no $y \in X$ such that $y \cap X = 0$. Let b be some element of X; hence, $b \cap X \neq 0$. Let $c = \mathrm{TC}(b) \cap X$. By AC, let h be a choice function for c. Define a function f on ω such that $f`0 = b$ and, for any $\alpha \in \omega$, $f`(\alpha') = h`((f`\alpha) \cap X)$. It follows easily that, for each $\alpha \in \omega$, $f`(\alpha') \in f`\alpha$, contradicting the Fundierungsaxiom. (The proof can be summarized as follows: we start with an element b of X; then, using h, we pick an element $f`1$ in $b \cap X$; since, by assumption, $f`1$ and X cannot be disjoint, we pick an element $f`2$ in $f`1 \cap X$, and so on.)

(c) Assume given a finite $\in$-cycle: $f`0 \in f`1 \in \cdots \in f`n \in f`0$. Let X be the range of f: $\{f`0, f`1, \ldots, f`n\}$. By the Axiom of Regularity, there is some $f`i \in X$ such that $f`i \cap X = 0$. But each element of X has an element in common with X.

Remark: The use of the Axiom of Choice in deriving the Axiom of Regularity from the Fundierungsaxiom is necessary. It can be shown (see Mendelson, 1958) that, if NBG is consistent and if we add the Fundierungsaxiom as an axiom, then the Axiom of Regularity is not provable in this enlarged theory.

Exercises

4.87 If v is a transitive set such that $u \in v$, prove that $\mathrm{TC}(u) \subseteq v$.

4.88 By the *Principle of Dependent Choices* (PDC) we mean the following assertion: if r is a nonempty relation whose range is a subset of its domain, then there is a function $f\colon \omega \to \mathscr{D}(r)$ such that $(\forall u)(u \in \omega \Rightarrow \langle f`u, f`(u')\rangle \in r)$ (Mostowski, 1948).

(a) Prove $\vdash \mathrm{AC} \Rightarrow \mathrm{PDC}$.

(b) Show that PDC implies the following *Denumerable Axiom of Choice* (DAC):

$$\mathrm{Den}(x) \wedge (\forall u)(u \in x \Rightarrow u \neq 0) \Rightarrow (\exists f)(f\colon x \to \bigcup (x) \wedge (\forall u)(u \in x \Rightarrow f`u \in u))$$

(c) Prove $\vdash \mathrm{PDC} \Rightarrow (\forall x)(\mathrm{Inf}(x) \Rightarrow \omega \preccurlyeq x)$. [Hence, by Exercise 4.71(c), PDC implies that a set is infinite if and only if it is Dedekind-infinite.]

(d) Prove that the conjunction of PDC and the Fundierungsaxiom implies the Axiom of Regularity.

Let us define by transfinite induction a function Ψ, which was originally devised by von Neumann:

$$\Psi`0 = 0$$

$$\Psi`(\alpha') = \mathscr{P}(\Psi`\alpha)$$

$$\mathrm{Lim}(\lambda) \Rightarrow \Psi`\lambda = \bigcup_{\beta <_0 \lambda} (\Psi`\beta)$$

Let $H = \bigcup(\Psi``On)$ and let H_β stand for $\Psi`(\beta')$. Define a function $\rho$ on $H$ such that, for any $x$ in $H$, $\rho`x$ is the least ordinal α such that $x \in \Psi`\alpha$. $\rho`x$ is called the *rank* of x. Observe that $\rho`x$ must be a successor ordinal.

Exercises

Prove Exercises 4.89–4.96.

4.89 $\vdash (\forall\alpha)\,\mathrm{Trans}(\Psi`\alpha)$

4.90 $\vdash \mathrm{Trans}(H)$

4.91 $\vdash \Psi`\alpha \subseteq \Psi`(\alpha')$

4.92 $\vdash \alpha <_0 \beta \Rightarrow \Psi`\alpha \subseteq \Psi`\beta$

4.93 $\vdash On \subseteq H$

4.94 $\vdash \rho`\alpha = \alpha'$

4.95 $\vdash u \in H \land v \in H \land u \in v \Rightarrow \rho`u <_0 \rho`v$

4.96 $\vdash u \subseteq H \Rightarrow u \in H$

PROPOSITION 4.40 The Axiom of Regularity is equivalent to the assertion that $V = H$—that is, that every set is a member of H.

Proof

(a) Assume $V = H$ and let $X \neq 0$. Let α be the least of the ranks of all the elements of X, and let b be an element of X such that $\rho`b = \alpha$. Then $b \cap X = 0$; for, if $u \in b \cap X$, then, by Exercise 4.95, $\rho`u \in p`b = \alpha$, contradicting the minimality of α.

(b) Assume the Axiom of Regularity and assume that $V - H \neq 0$. By the Axiom of Regularity, there is some $y \in V - H$ such that $y \cap (V - H) = 0$. Hence, $y \subseteq H$, and so, by Exercise 4.96, $y \in H$, contradicting $y \in V - H$.

Exercises

4.97 Show that the Axiom of Regularity is equivalent to the special case: $x \neq 0 \Rightarrow (\exists y)(y \in x \land y \cap x = 0)$.

4.98 Show that, if we assume the Axiom of Regularity, then $\mathrm{Ord}(X)$ is equivalent to: $\mathrm{Trans}(X) \land E\,\mathrm{Con}\,X$—that is, to the wf

$$(\forall u)(u \in X \Rightarrow u \subseteq X) \land (\forall u)(\forall v)(u \in X \land v \in X \land u \neq v \Rightarrow u \in v \lor v \in u)$$

Thus, with the Axiom of Regularity, a much simpler definition of the notion of ordinal class is available, a definition in which all quantifiers are restricted to sets.

4.99 Show that the Axiom of Regularity implies that every nonempty transitive class contains 0.

Proposition 4.40 certainly increases the attractiveness of adding the Axiom of Regularity as a new axiom to NBG. The proposition $V = H$ asserts that every set can be obtained by starting with 0 and applying the power set and union operations any transfinite number of times, and the assumption that this is so would clarify our rather hazy ideas about sets. By Exercise 4.98, the Axiom of Regularity would also simplify the definition of ordinal numbers. In addition, we can develop the theory of cardinal numbers on the basis of the Axiom of Regularity; namely, just define the cardinal number of a set x to be the set of all those y of lowest rank such that $y \cong x$. (The basic requirement of the theory of cardinal numbers is that there be a function Card whose domain is V such that $\text{Card}`x = \text{Card}`y \Leftrightarrow x \cong y$.) There is no unanimity among mathematicians about whether we have sufficient grounds for adding the Axiom of Regularity as a new axiom, for, although it has great simplifying power, it does not have the immediate plausibility that even the Axiom of Choice has, nor has it had any mathematical applications.

The class H defined above determines an *inner model* of NBG in the following sense: for any wf $\mathscr{A}$ (written in unabbreviated notation) that contains the free variables $Y_1, \ldots, Y_n$, let $\text{Rel}_H(\mathscr{A})$ be the wf obtained from $\mathscr{A}$ by replacing every subformula $(\forall X)\mathscr{B}(X)$ by $(\forall X)(X \subseteq H \Rightarrow \mathscr{B}(X))$ (in making the replacements, we start with the innermost subformulas) and then prefixing $(Y_1 \subseteq H \wedge Y_2 \subseteq H \wedge \cdots \wedge Y_n \subseteq H) \Rightarrow$. In other words, in forming $\text{Rel}_H(\mathscr{A})$, we interpret "class" as "subclass of H". Then, for any theorem $\mathscr{A}$ of NBG, $\text{Rel}_H(\mathscr{A})$ is also a theorem of NBG.

Exercises

4.100 Verify that, for each axiom $\mathscr{A}$ of NBG, $\text{Rel}_H(\mathscr{A})$ is a theorem of NBG. Notice that $\text{Rel}_H((\forall x)\mathscr{B})$ is equivalent to $(\forall x)(x \in H \Rightarrow \mathscr{B}^{\#})$, where $\mathscr{B}^{\#}$ is $\text{Rel}_H(\mathscr{B})$. In particular, $\text{Rel}_H(M(X))$ is $(\exists Y)(Y \subseteq H \wedge X \in Y)$, which is equivalent to $X \in H$; thus, the "sets" of the model are the elements of H. If we adopt a semantic approach, then one need only observe that, if N is a model for NBG (in the usual sense of "model"), then the objects X of N that satisfy the wf $X \subseteq H$ also form a model for NBG. In addition, one can verify that the Axiom of Regularity holds in this model; this is just part (a) of Proposition 4.40. A direct consequence of this fact is the consistency of the Axiom of Regularity; that is, if NBG is consistent, so is the theory obtained by adding the Axiom of Regularity as a new axiom. That the Axiom of Regularity is independent of NBG can also be proved (see Bernays, 1937–1954, part VII) by means of a suitable model, though the model is more complex than that given above for the consistency proof. Thus, the Axiom of Regularity is both consistent and independent with respect to NBG; we can consistently add either it or its negation as an axiom to NBG, if NBG is consistent. [Practically the same proofs also show the independence and consistency of the Axiom of Regularity with respect to NBG + (AC).]

4.101 Consider the model whose domain is H_α and whose interpretation of $\in$ is E_{H_α}, the membership relation restricted to H_α. Notice that the "sets" of this model are the sets of rank $\leqslant_0 \alpha$ and the "proper classes" are the sets of rank α'. Show that the model H_α satisfies all axioms of NBG (except possibly the Axioms of Infinity and Replacement) if and only if $\text{Lim}(\alpha)$. Prove also that H satisfies the Axiom of Infinity if and only if $\alpha >_0 \omega$.

4.102 Show that the Axiom of Infinity is not provable from the other axioms of NBG, if the latter are consistent.

4.103 Show that the Axiom of Replacement R is not provable from the other axioms [T, P, N, (B1)–(B7), U, W, S] if these latter are consistent.

4.104^{D} An ordinal α such that H_α is a model for NBG is called *inaccessible*. Since NBG has only a finite number of proper axioms, the assertion that α is inaccessible can be expressed by the conjunction of the relativization to H_α of the proper axioms of NBG. Show that the existence of inaccessible ordinals is not provable in NBG if the latter is consistent, and the same is true even if the Axiom of Choice and the Generalized Continuum Hypothesis are added as axioms. [Compare Shepherdson (1951–53), Montague and Vaught (1959), and, for related results, Bernays (1961) and Levy (1960).] Inaccessible ordinals have been shown to have connections with problems in measure theory and algebra (see Ulam, 1930; Zeeman, 1955; and Erdös & Tarski, 1961).* The consistency of the theory obtained from NBG by adding an axiom asserting the existence of an inaccessible ordinal is still an open question.

The Axiom of Choice turns out to be consistent and independent with respect to the theory NBG + (Axiom of Regularity). More precisely, if NBG is consistent, AC is an undecidable sentence of the theory NBG + (Axiom of Regularity). In fact, Gödel (1938, 1939, 1940) showed that, if NBG is consistent, then the theory NBG + (AC) + (Axiom of Regularity) + (GCH) is also consistent, where (GCH) stands for the *Generalized Continuum Hypothesis*:

$$(\forall x)(\mathrm{Inf}(x) \Rightarrow \neg(\exists y)(x \prec y \wedge y \prec \mathscr{P}(x)))$$

[Our statement of Gödel's result is a bit redundant, since $\vdash$ (GCH) $\Rightarrow$ (AC) has been proved by Sierpinski (1947) and Specker (1954). This result will be proved below.] The unprovability of AC from NBG + (Axiom of Regularity), if NBG is consistent, has been proved by P. J. Cohen (1963), who also has shown the independence of the special *Continuum Hypothesis*, $2^{\omega} \cong \omega_1$, in the theory NBG + (AC) + (Axiom of Regularity). Expositions of the work of Cohen and its further development can be found in Cohen (1966) and Shoenfield (1971b), as well as in Rosser (1969), Felgner (1971a), Takeuti and Zaring (1973), and Bell (1977). For a thorough treatment of these results and other independence proofs in set theory, Jech (1978) and Kunen (1980) should be consulted.

We shall present here a modified form of the proof in Cohen (1966) of Sierpinski's theorem that GCH implies AC.

DEFINITION For any set v, let $\mathscr{P}^0(v) = v$, $\mathscr{P}^1(v) = \mathscr{P}(v)$, $\mathscr{P}^2(v) = \mathscr{P}(\mathscr{P}(v)), \ldots, \mathscr{P}^{k+1}(v) = \mathscr{P}(\mathscr{P}^k(v))$ for all k in ω.

LEMMA 4.41 If $\omega \preccurlyeq v$, then $\mathscr{P}^k(v) +_c \mathscr{P}^k(v) \cong \mathscr{P}^k(v)$ for all $k \geqslant_0 1$.

Proof Remember that $\mathscr{P}(x) \cong 2^x$ (see Exercise 4.40). From $\omega \preccurlyeq v$, we obtain $\omega \preccurlyeq \mathscr{P}^k(v)$ for all k in ω. Hence, $\mathscr{P}^k(v) +_c 1 \cong \mathscr{P}^k(v)$ for all k in ω, by Exercise 4.71(e). Now, for any $k \geqslant_0 1$,

$$\mathscr{P}^k(v) +_c \mathscr{P}^k(v) = \mathscr{P}^k(v) \times 2 = \mathscr{P}(\mathscr{P}^{k-1}(v)) \times 2 \cong 2^{\mathscr{P}^{k-1}(v)} \times 2$$
$$\cong 2^{\mathscr{P}^{k-1}(v)} \times 2^1 \cong 2^{\mathscr{P}^{k-1}(v) +_c 1} \cong 2^{\mathscr{P}^{k-1}(v)} \cong \mathscr{P}(\mathscr{P}^{k-1}(v)) = \mathscr{P}^k(v)$$

LEMMA 4.42 If $y +_c x \cong \mathscr{P}(x +_c x)$, then $\mathscr{P}(x) \preccurlyeq y$.

Proof Notice that $\mathscr{P}(x +_c x) \cong 2^{x +_c x} \cong 2^x \times 2^x \cong \mathscr{P}(x) \times \mathscr{P}(x)$. Let $y^* = y \times \{0\}$ and $x^* = x \times \{1\}$. Since $y +_c x \cong \mathscr{P}(x +_c x) \cong \mathscr{P}(x) \times \mathscr{P}(x)$, there is a function f such that $y^* \cup x^* \underset{f}{\cong} \mathscr{P}(x) \times \mathscr{P}(x)$. Let h be the function that takes each u in x^* into the first component of the pair $f`u$. Thus, $h: x^* \rightarrow \mathscr{P}(x)$. By Proposition

* Inaccessible ordinals are involved also with attempts to provide a suitable set-theoretic foundation for category theory (see Maclane, 1971; Gabriel, 1962; Sonner, 1962; Kruse, 1966; and Isbell, 1966).

4.23(a), there must exist $c \in \mathscr{P}(x) - h``x^*$. Then, for all z in $\mathscr{P}(x)$, there exists a unique v in y^* such that $f`v = \langle c, z \rangle$. This determines a one–one function from $\mathscr{P}(x)$ into y. Hence, $\mathscr{P}(x) \preccurlyeq y$.

PROPOSITION 4.43 Assume GCH.

(a) For any ordinal β, if u cannot be well-ordered, $u +_c u \cong u$ and $\beta \preccurlyeq 2^u$, then $\beta \preccurlyeq u$.
(b) The Axiom of Choice holds.

Proof

(a) Notice that $u +_c u \cong u$ implies $1 +_c u \cong u$, by Exercise 4.78(b); therefore, by Exercise 4.57(i), $2^u +_c u \cong 2^u$. Now, $u \preccurlyeq \beta +_c u \preccurlyeq 2^u +_c u \cong 2^u$. By GCH, either (i) $u \cong \beta +_c u$ or (ii) $\beta +_c u \cong 2^u$. If (ii) holds, $\beta +_c u \cong 2^u +_c u \cong \mathscr{P}(u +_c u)$. Hence, by Lemma 4.42, $\mathscr{P}(u) \preccurlyeq \beta$ and, therefore, $u \preccurlyeq \beta$. Then, since u would be equinumerous with a subset of an ordinal, u could be well-ordered, contradicting our assumption. Hence, (i) must hold. But then, $\beta \preccurlyeq \beta +_c u \cong u$.

(b) We shall prove AC by proving the equivalent sentence (WO) asserting that every set can be well-ordered. To that end, consider any set x and assume, for the sake of contradiction, that x cannot be well-ordered. Let $v = 2^{x \cup \omega}$. Then $\omega \preccurlyeq x \cup \omega \preccurlyeq v$. Hence, by Lemma 4.41, $\mathscr{P}^k(v) +_c \mathscr{P}^k(v) \cong \mathscr{P}^k(v)$ for all $k \geqslant_0 1$. Also, since $x \preccurlyeq x \cup \omega \preccurlyeq v \prec \mathscr{P}(v) \prec \mathscr{P}\mathscr{P}(v) \prec \cdots$, and x cannot be well-ordered, each $\mathscr{P}^k(v)$ cannot be well-ordered, for $k \geqslant_0 0$. Let $\beta = \mathscr{H}`v$. We know that $\beta \preccurlyeq \mathscr{P}^4(v)$ (see page 204). Hence, by part (a), with $u = \mathscr{P}^3(v)$, we obtain $\beta \preccurlyeq \mathscr{P}^3(v)$. Using part (a) twice more [successively with $u = \mathscr{P}^2(v)$ and $u = \mathscr{P}(v)$], we obtain $\mathscr{H}`v = \beta \preccurlyeq v$. But this contradicts the definition of $\mathscr{H}`v$ as the least ordinal not equinumerous with a subset of v.

Exercise

4.105 An *α-sequence* is a function f whose domain is α. If the range of f consists of ordinals, f is called an *ordinal α-sequence* and, if, in addition, $\beta <_0 \gamma <_0 \alpha$ implies $f(\beta) <_0 f(\gamma)$, then f is called an *increasing* ordinal α-sequence. By Proposition 4.11, if f is an increasing ordinal α-sequence, then $\bigcup(f``\alpha)$ is the least upper bound of the range of $f$. An ordinal $\delta$ is said to be *regular* if, for any increasing ordinal α-sequence such that $\alpha <_0 \delta$ and the ordinals in the range of $f$ are all $<_0 \delta$, $\bigcup(f``\alpha) +_0 1 <_0 \delta$. Nonregular ordinals are called *singular* ordinals.

(a) Which finite ordinals are regular?
(b) Show that ω_0 is regular and ω_ω is singular.
(c) Prove that every regular ordinal is an initial ordinal.
(d) Assuming the Axiom of Choice (AC), prove that every ordinal of the form $\omega_{\gamma +_0 1}$ is regular.
(e) If ω_α is regular and Lim(α), prove that $\omega_\alpha = \alpha$. [A regular ordinal ω_α such that Lim(α) is called a *weakly inaccessible ordinal*.]
(f) Show that, if ω_α has the property that $\gamma <_0 \omega_\alpha$ implies $\mathscr{P}(\gamma) \prec \omega_\alpha$, then Lim($\alpha$). The converse is implied by the Generalized Continuum Hypothesis. A regular ordinal ω_α, such that $\alpha >_0 0$ and such that $\gamma <_0 \omega_\alpha$ implies $\mathscr{P}(\gamma) \prec \omega_\alpha$, is called *strongly inaccessible*. Thus, every strongly inaccessible ordinal is weakly inaccessible and, if the (GCH) holds, the strongly inaccessible ordinals coincide with the weakly inaccessible ordinals.

(g) (Shepherdson, 1951–53; Montague & Vaught, 1959) (i) if γ is inaccessible (i.e., if H_γ is a model of NBG), prove that γ is weakly inaccessible. (ii)[D] In the theory NBG + (AC), show that γ is inaccessible if and only if γ is strongly inaccessible.

(h) If NBG is consistent, then, in the theory NBG + (AC) + (GCH), show that it is impossible to prove the existence of weakly inaccessible ordinals.

6. OTHER AXIOMATIZATIONS OF SET THEORY

We have chosen to develop set theory on the basis of NBG because it is relatively simple and convenient for the practicing mathematician. There are, of course, many other varieties of axiomatic set theory, of which we will now make a brief survey.

I. MORSE-KELLEY (MK)

Strengthening NBG, we can replace axioms (B1)–(B7) by the axiom schema:

$$(\square) \quad (\exists Y)(\forall x)(x \in Y \Leftrightarrow \mathscr{A}(x))$$

where $\mathscr{A}(x)$ is *any* wf (not necessarily predicative) of NBG and Y is not free in $\mathscr{A}(x)$. This new theory MK, called *Morse-Kelley set theory*, became well known through its appearance as an appendix in a book on general topology by Kelley (1955). The basic idea was proposed independently by Mostowski, Quine, and A. Morse [whose rather unorthodox system may be found in Morse (1965)]. Axioms (B1)–(B7) follow easily from ($\square$), and, therefore, NBG is a subtheory of MK. Mostowski (1951) showed that, if NBG is consistent, then MK is really stronger than NBG. He did this by constructing a "truth definition" in MK on the basis of which he proved $\vdash_{MK} \mathscr{C}on_{NBG}$, where $\mathscr{C}on_{NBG}$ is a standard arithmetic sentence asserting the consistency of NBG. On the other hand, by Gödel's Second Theorem (see page 166), $\mathscr{C}on_{NBG}$ is not provable in NBG if the latter is consistent.

The simplicity and power of schema ($\square$) make MK very suitable for use by mathematicians who are not interested in the subtleties of axiomatic set theory. But this very strength makes the consistency of MK a riskier gamble. However, if we add to NBG + (AC) the axiom (In) asserting the existence of a strongly inaccessible ordinal θ, then H_θ is a model of MK. Hence, MK involves no more risk than NBG + (AC) + (In).

There are several textbooks that develop axiomatic set theory on the basis of MK (Rubin, 1967; Monk, 1980; and Chuquai, 1981). Some of Cohen's independence results have been extended to MK by Chuquai (1972).

Exercises

4.106 Prove that axioms (B1)–(B7) are theorems of MK.

4.107 Verify that, if θ is a strongly inaccessible ordinal, H_θ is a model of MK.

II. ZERMELO-FRAENKEL (ZF)

The earliest axiom system for set theory was devised by Zermelo (1908). The objects of the theory are thought of intuitively as *sets*, not the *classes* of NBG or MK. Zermelo's theory Z was formulated in a language that contains only one predicate letter $\in$. Equality is defined extensionally: $x = y$ stands for $(\forall z)(z \in x \Leftrightarrow z \in y)$. The proper axioms are:

T: $x = y \Rightarrow (x \in z \Leftrightarrow y \in z)$
P: $(\exists z)(\forall u)(u \in z \Leftrightarrow u = x \vee u = y)$
N: $(\exists x)(\forall y)(y \notin x)$
U: $(\exists y)(\forall u)(u \in y \Leftrightarrow (\exists v)(u \in v \wedge v \in x))$
W: $(\exists y)(\forall u)(u \in y \Leftrightarrow u \subseteq x)$
S*: $(\exists y)(\forall u)(u \in y \Leftrightarrow (u \in x \wedge \mathscr{A}(u)))$
I: $(\exists x)(0 \in x \wedge (\forall z)(z \in x \Rightarrow z \cup \{z\} \in x))$

Here we have assumed the same definitions of $\subseteq$, 0, $\cup$, and $\{u\}$ as in NBG.

Zermelo's intention was to build up mathematics by starting with a few simple sets (0 and ω) and then constructing further sets by various well-defined operations (such as formation of pairs, unions, and power sets). In fact, a good deal of mathematics can be built up within Z. However, Fraenkel (1922) observed that Z was too weak for a full development of mathematics. For example, for each finite ordinal n, the ordinal $\omega +_0 n$ can be shown to exist, but the set A of all such ordinals cannot be proved to exist, and, therefore, $\omega +_0 \omega$, the least upper bound of A, cannot be shown to exist. Fraenkel proposed a way of overcoming such difficulties, but his idea could not be clearly expressed in the language of Z. However, Skolem (1923) succeeded in recasting Fraenkel's idea in the following way: for any wf $\mathscr{A}(x, y)$, let Fun($\mathscr{A}$) stand for $(\forall x)(\forall u)(\forall v)(\mathscr{A}(x, u) \wedge \mathscr{A}(x, v) \Rightarrow u = v)$. Thus, Fun($\mathscr{A}$) asserts that *$\mathscr{A}$ determines a function.* Skolem's *Axiom Schema of Replacement* can then be formulated as follows:

($R^{\#}$) For any wf $\mathscr{A}(x, y)$,

$$\text{Fun}(\mathscr{A}) \Rightarrow (\forall w)(\exists z)(\forall v)(v \in z \Leftrightarrow (\exists u)(u \in w \wedge \mathscr{A}(u, v)))$$

This is the best approximation that can be found for the Axiom of Replacement R of NBG.

The system Z + ($R^{\#}$) is denoted ZF and is called *Zermelo-Fraenkel set theory.* In recent years, ZF is often assumed to contain a set-theoretic Axiom of Regularity (Reg): $x \neq 0 \Rightarrow (\exists y)(y \in x \wedge y \cap x = 0)$. The reader should always check to see whether (Reg) is included within ZF. ZF is now the most popular form of axiomatic set theory; most of the modern research in set theory on independence and consistency proofs has been carried out with respect to ZF. For expositions of ZF, see Krivine (1971), Suppes (1960), Zuckerman (1974), Lévy (1978), and Hrbacek and Jech (1978).

ZF and NBG yield essentially equivalent developments of set theory. Every sentence of ZF is an abbreviation of a sentence of NBG, since, in NBG, lowercase variables $x, y, z, \ldots$ serve as restricted set variables. For example, Axiom N is an abbreviation of $(\exists x)(M(x) \wedge (\forall y)(M(y) \Rightarrow y \notin x))$ in NBG. It is a simple matter to verify that all axioms of ZF are theorems in NBG. Indeed, NBG was originally constructed so that this would be the case. We can conclude that, if NBG is consistent, then so is ZF. In fact, if a contradiction could be derived in ZF, the same proof would yield a contradiction in NBG.

The presence of class variables in NBG seems to make it much more powerful than ZF. At any rate, it is possible to express propositions in NBG that either are impossible to formulate in ZF (such as the universal choice axiom UCF) or are much more unwieldy in ZF (such as transfinite induction theorems). Nevertheless, it is a surprising fact that NBG is no riskier than ZF. An even stronger result can be proved: NBG is a *conservative extension* of ZF in the sense that, for any sentence $\mathscr{A}$ of ZF, if $\vdash_{\text{NBG}} \mathscr{A}$, then $\vdash_{\text{ZF}} \mathscr{A}$ (see Novak (Gal) 1951; Rosser & Wang, 1950; and Shoenfield, 1954). This implies that, if ZF is consistent, NBG is consistent. Thus, NBG is consistent if and only if ZF is consistent, and NBG seems to be no stronger than ZF. However, NBG and ZF do differ with respect to the existence of certain kinds of models (see Montague & Vaught, 1959). Moreover, another important difference is that NBG is finitely axiomatizable, whereas Montague (1961a) showed that ZF (as well as Z) is not finitely axiomatizable. Montague (1961b) proved the stronger result that ZF cannot be obtained by adding a finite number of axioms to Z.

Exercise

4.108 Let $H_\alpha^* = \bigcup H_\alpha$ (see page 218).

(a) Verify that H_α^* consists of all sets of rank less than α.

(b) If α is a limit ordinal $>_0 \omega$, show that H_α^* is a model for Z.

(c)[D] Find an instance of the Axiom Schema of Replacement ($\text{R}^\#$) that is false in $H^*_{\omega+_0\omega}$. [*Hint:* Let $\mathscr{A}(x, y)$ be $x \in \omega \wedge y = \omega +_0 x$. Observe that $\omega +_0 \omega \notin H^*_{\omega+_0\omega}$ and $\omega +_0 \omega = \bigcup \{v | (\exists u)(u \in \omega \wedge \mathscr{A}(u, v))\}$.]

(d) Show that, if ZF is consistent, then ZF is a proper extension of Z.

III. *THE THEORY OF TYPES (ST)*

Russell's Paradox is based on the set K of all those sets that are not members of themselves: $K = \{x | x \notin x\}$. Clearly, $K \in K$ if and only if $K \notin K$. In NBG this argument simply shows that K is a proper class, not a set. In ZF the conclusion is just that there is no such set K.

Russell himself chose to find the source of his paradox elsewhere. He maintained that $x \in x$ and $x \notin x$ should be considered "illegitimate" and "ungrammatical" formulas and, therefore, that the definition of K makes no sense. However, this alone is not adequate because paradoxes analogous to Russell's can be obtained from slightly more complicated circular properties, like $x \in y \wedge y \in x$.

Exercise

4.109

(a) Derive a Russell-style paradox by using $x \in y \wedge y \in x$.

(b) Use $x \in y_1 \wedge y_1 \in y_2 \wedge \cdots \wedge y_{n-1} \in y_n \wedge y_n \in x$ to obtain a paradox, where $n \geqslant 1$.

Thus, to avoid paradoxes, one must forbid any kind of indirect circularity. For this purpose, we can think of the universe divided up into types in the following way. Start with a collection W of *nonsets* or *individuals*. The elements of W are said to have *type* 0. Sets whose members are of type 0 are the objects of type 1. Sets whose members are of type 1 will be the objects of type 2, and so on.

Our language will have variables of different types. The superscript of a variable will indicate its type. Thus, x^0 is a variable of type 0, y^1 is a variable of type 1, and so on. There are no variables other than type variables. The atomic wfs are of the form $x^n \in y^{n+1}$, where n is one of the natural numbers 0, 1, 2, ... The rest of the wfs are built up from the atomic wfs by means of logical connectives and quantifiers. Observe that $\neg(x \in x)$ and $\neg(x \in y \wedge y \in x)$ are not wfs.

The equality relation must be defined piecemeal: one definition for each type.

DEFINITION $x^n = y^n$ for $(\forall z^{n+1})(x^n \in z^{n+1} \Leftrightarrow y^n \in z^{n+1})$

Notice that two objects are defined to be equal if they belong to the same sets of the next higher type. The basic property of equality is provided by the following axiom scheme.

ST1 (EXTENSIONALITY AXIOM)

$$(\forall x^n)(x^n \in y^{n+1} \Leftrightarrow x^n \in z^{n+1}) \Rightarrow y^{n+1} = z^{n+1}$$

This asserts that two sets that have the same members must be equal. On the other hand, observe that the property of having the same members could not be taken as a general definition of equality because it is not suitable for objects of type 0.

Given any wf $\mathscr{A}(x^n)$, we wish to be able to define a set $\{x^n | \mathscr{A}(x^n)\}$.

ST2 (COMPREHENSION AXIOM SCHEME) For any wf $\mathscr{A}(x^n)$, the following wf is an axiom:

$$(\exists y^{n+1})(\forall x^n)(x^n \in y^{n+1} \Leftrightarrow \mathscr{A}(x^n))$$

Here, y^{n+1} is any variable not free in $\mathscr{A}(x^n)$. If we use the Extensionality Axiom, then the set y^{n+1} is unique and can be denoted by $\{x^n | \mathscr{A}(x^n)\}$.

We can begin to develop mathematics within this system, defining the usual set-theoretic notions and operations, as well as the natural numbers, ordinal numbers, cardinal numbers, and so on. However, these concepts are not unique but are repeated for each type (or, in some cases, for all but the first few types). For example, the Comprehension Scheme provides a null set $\Lambda^{n+1} = \{x^n | x^n \neq x^n\}$ for each nonzero type. But there is no null set per se. The same thing happens for natural numbers. In type theory the natural numbers are not defined as they are in NBG. Here they are the finite cardinal numbers. For example, the set of natural numbers of type 2 is the intersection of all sets containing $\{\Lambda^1\}$ and closed under the following successor operation: the *successor* $S(y^2)$ of a set y^2 is the set $\{v^1 | (\exists u^1)(\exists z^0)(u^1 \in y^2 \wedge z^0 \notin u^1 \wedge v^1 = u^1 \cup \{z^0\})\}$. Then $0 = \{\Lambda^1\}$, $1 = S(0)$, $2 = S(1)$, and so on. Here, the numerals 0, 1, 2, ... should really have a superscript ^{2} to indicate their type, but the superscripts were omitted for the sake of legibility. Note that 0 is the set of all sets (of type 1) that contain no elements, 1 is the set of all sets (of type 1) that contain one element, 2 is the set of all sets (of type 1) that contain two elements, and so on.

This repetition of the same notion in different types makes it very inconvenient for mathematicians to work within a type theory. Moreover, it is easy to show that the existence of an infinite set cannot be proved from the Extensionality and Comprehension Schema.* To see this, consider the "model" in which each variable of type n ranges over the sets of rank $\leqslant n +_0 1$. (There is nothing wrong about assigning overlapping ranges to variables of different types.)

We shall assume an axiom that guarantees the existence of an infinite set. As a preliminary, we shall adopt the usual definition of ordered pair: $\langle x^n, y^n \rangle = \{\{x^n\}, \{x^n, y^n\}\}$, where $\{x^n, y^n\} = \{u^n | u^n = x^n \vee u^n = y^n\}$. Notice that $\langle x^n, y^n \rangle$ is of type $n + 2$. Hence, a binary relation on a set A, being a set of ordered pairs of elements of A, will have type 2 greater than the type of A. In particular, a binary relation on the universe $V^1 = \{x^0 | x^0 = x^0\}$ of all objects of type 0 will be a set of type 3.

ST3 (AXIOM OF INFINITY)

$$(\exists x^3)([(\exists u^0)(\exists v^0)(\langle u^0, v^0 \rangle \in x^3)] \wedge (\forall u^0)(\forall v^0)(\forall w^0)(\langle u^0, u^0 \rangle \notin x^3$$
$$\wedge [\langle u^0, v^0 \rangle \in x^3 \wedge \langle v^0, w^0 \rangle \in x^3 \Rightarrow \langle u^0, w^0 \rangle \in x^3]$$
$$\wedge [\langle u^0, v^0 \rangle \in x^3 \Rightarrow (\exists z^0)(\langle v^0, z^0 \rangle \in x^3)]))$$

This asserts that there is a nonempty irreflexive, transitive binary relation x^3 on V^1 such that every member of the range of x^3 also belongs to the domain of x^3. Since no such relation exists on a finite set, V^1 must be infinite.

The system based on ST1–ST3 is called the *simple theory of types* and is denoted ST. Because of its somewhat complex notation and the repetition of concepts at all (or, in some cases, almost all) type levels, ST is not generally used as a foundation of mathematics and is not the subject of much contemporary research. Turing (1948) proposed ways of making type theory more acceptable to practicing mathematicians, but his work probably had an effect opposite to that intended.

With ST we can associate a first-order theory ST′. The nonlogical constants of ST′ are $\in$ and monadic predicates T_n for each natural number n. We then translate any wf $\mathscr{A}$ of ST into ST′ by replacing subformulas $(\forall x^n)\mathscr{B}(x^n)$ by $(\forall x)(T_n(x) \Rightarrow \mathscr{B}(x))$ and, if $y^{j_1}, \ldots, y^{j_k}$ are the free variables of $\mathscr{A}$, prefixing to the result $T_{j_1}(y_1) \wedge \cdots \wedge T_{j_k}(y_k) \Rightarrow$ and changing each y^{j_i} into y_i. In a rigorous presentation, we would have to specify clearly that the replacements are made by proceeding from smaller to larger subformulas and that the variables $x, y_1, \ldots, y_k$ are new variables. The axioms of ST′ are the translations of the axioms of ST. Any theorem of ST translates into a theorem of ST′.

Exercise

4.110 Exhibit a model of ST′ within NBG.

* This fact seemed to undermine Russell's doctrine of *logicism*, according to which all of mathematics could be reduced to basic axioms that were of an essentially *logical* character; an axiom of infinity could not be thought of as a logical truth.

By virtue of Exercise 4.110, NBG (or ZF) is stronger than ST: (1) any theorem of ST can be translated into a corresponding theorem of NBG, and (2) if NBG is consistent, so is ST.*

To provide a type theory that is easier to work with, one can add axioms that impose additional structure on the set V^1 of objects of type 0. For example, Peano's axioms for the natural numbers were adopted at level 0 in Gödel's system P, for which he originally proved his famous incompleteness theorem (see Gödel, 1931).

In *Principia Mathematica* (1910–13), the three-volume work that he co-authored with A. N. Whitehead, Russell worked with a theory of types that was further complicated by an additional hierarchy of *orders*. This hierarchy was introduced so that the Comprehension Scheme could be suitably restricted in order not to generate an *impredicatively defined set*—that is, a set A defined by a formula in which some quantified variable ranges over a set that turns out to contain the set A itself. Along with the mathematician Henri Poincaré, Russell believed impredicatively defined concepts to be the root of all evil. Unfortunately, such concepts are required in analysis (for example, in the proof that any nonempty set of real numbers that is bounded above has a least upper bound). *Principia Mathematica* had to add the so-called *Axiom of Reducibility* to overcome the order restrictions imposed on the Comprehension Axiom Schema. The Russell-Whitehead system without the Axiom of Reducibility is called *ramified type theory*; it is mathematically weak but is still of some interest to those who wish an extreme constructivist approach to mathematics. The Axiom of Reducibility vitiates the effect of the order hierarchy; therefore, it is much simpler to drop the notion of order and the Axiom of Reducibility. The result is the simple theory of types ST, which we have described above.

In ST, the types are the natural numbers. For a smoother presentation, some logicians allow a larger set of types, including types for relations and/or functions defined on objects taken from previously defined types. Such a system may be found in Church (1940).

Principia Mathematica is the only source for a thorough treatment of type theory. However, it must be read critically; for example, it often overlooks the distinction between a formal theory and its metalanguage. The idea of a simple theory of types goes back to Ramsey (1925) and, independently, to Chwistek (1924–25). Discussions of type theory are found in Hatcher (1982) and Quine (1963).

IV. QUINE'S THEORIES NF AND ML

Quine (1937) invented a theory that was designed to do away with some of the unpleasant aspects of type theory while keeping the essential idea of the Comprehension Axiom ST2. Quine's theory NF (New Foundations) uses only one kind of variable $x, y, z, \ldots$ and one binary predicate letter $\in$. Equality is defined as in type theory: $x = y$ stands for $(\forall z)(x \in z \Leftrightarrow y \in z)$. The first axiom is familiar:

NF1 (EXTENSIONALITY) $(\forall z)(z \in x \Leftrightarrow z \in y) \Rightarrow x = y$

*A stronger result was proved by J. Kemeny (1949) by means of a truth definition within Z: if Z is consistent, so is ST.

In order to formulate the Comprehension Axiom, we introduce the notion of *stratification*. A wf $\mathscr{A}$ is said to be *stratified* if one can assign integers to the variables of $\mathscr{A}$ so that: (1) all occurrences of the same free variable are assigned the same integer; (2) all bound occurrences of a variable that are bound by the same quantifier must be assigned the same integer; and (3) for every subformula $x \in y$ of $\mathscr{A}$, the integer assigned to y is 1 greater than the integer assigned to x.

Examples

1. $(\exists y)(x \in y \wedge y \in z) \vee u \in x$ is stratified by virtue of the assignment indicated below by superscripts:

$$(\exists y^2)(x^1 \in y^2 \wedge y^2 \in z^3) \vee u^0 \in x^1$$

2. $(\exists y)(x \in y) \wedge (\exists y)(y \in x)$ is stratified as follows: $(\exists y^2)(x^1 \in y^2) \wedge (\exists y^0)(y^o \in x^1)$. Notice that the y's in the second conjunct do not have to have the same integers assigned to them as the y's in the first conjunct.
3. $x \in y \vee y \in x$ is not stratified; if x is assigned an integer n, then the first y must be assigned $n+1$ and the second y must be assigned $n-1$, contradicting condition (1).

NF2 (COMPREHENSION) For any stratified wf $\mathscr{A}(x)$,

$$(\exists y)(\forall x)(x \in y \Leftrightarrow \mathscr{A}(x))$$

is an axiom. [Here, y is assumed to be the first variable not free in $\mathscr{A}(x)$.]

Although NF2 is an axiom scheme, it turns out that NF is finitely axiomatizable (Hailperin, 1944).

Exercise

4.111 Prove that equality could have been defined as follows: $x = y$ for $(\forall z)(x \in z \Rightarrow y \in z)$. (More precisely, in the presence of NF2, this definition is equivalent to the original one.)

The theory of natural numbers, ordinal numbers, and cardinal numbers is developed in much the same way as in type theory, except that there is no longer a multiplicity of similar concepts. There is a unique empty set $\Lambda = \{x \mid x \neq x\}$ and a unique universal set $V = \{x \mid x = x\}$. We can easily prove $V \in V$, which immediately distinguishes NF from type theory (and from NBG, MK, and ZF).

Russell's Paradox is not derivable in NF, since $x \notin x$ is not stratified. Almost all of standard set theory and mathematics is derivable in NF; this is done in full detail in Rosser (1953). However, NF has some very strange properties. First of all, the usual proof of Cantor's Theorem, $A \prec \mathscr{P}(A)$, does not go through in NF; at a key step in the proof, a set that is needed is not available because its defining condition is not stratified. The underivability of Cantor's Theorem has the desirable effect of preventing the proof of Cantor's Paradox. If we could prove $A \prec \mathscr{P}(A)$, then, since $\mathscr{P}(V) = V$, we could obtain a contradiction from $V \prec \mathscr{P}(V)$. In NF, the standard proof of Cantor's Theorem does yield $\mathrm{USC}(A) \prec \mathscr{P}(A)$, where $\mathrm{USC}(A) =$

$\{x|(\exists u)(u \in A \wedge x = \{u\})\}$. If we let $A = V$, we conclude that $USC(V) \prec V$. Thus, V has the peculiar property that it is not equinumerous with the set of all unit sets of its elements. In NBG, the function f, defined by $f(u) = \{u\}$ for all u in A, establishes a correspondence between A and USC(A) for any set A. However, the defining condition for f is not stratified, so that f may not exist in NF. If f does exist, A is said to be *strongly Cantorian*.

Other surprising properties of NF are the following.

1. The Axiom of Choice is disprovable in NF (Specker, 1953).
2. Any model for NF must be *nonstandard* in the sense that the usual ordering of the finite cardinals or of the ordinals is not actually a well-ordering in the metalanguage (Rosser & Wang, 1950).
3. The Axiom of Infinity is provable in NF (Specker, 1953).

Although property 3 would ordinarily be thought of as a great advantage, the fact of the provability of an Axiom of Infinity appeared to many logicians to be too strong a result. If *that* can be proved, then probably anything can be proved; that is, NF is likely to be inconsistent. The disprovability of the Axiom of Choice seems to make NF a poor choice for practicing mathematicians. However, if we restrict attention to so-called *Cantorian* sets, sets A for which A and USC(A) are equinumerous, then it might be consistent to assume the Axiom of Choice for Cantorian sets and to do mathematics within the universe of Cantorian sets.

NF has another attractive feature. Category theory (see Maclane, 1971) can be developed in a straightforward way in NF, whereas this is not possible in ZF, NBG, or MK. Since category theory has become an important branch of mathematics, this is a distinct advantage for NF.

If the system obtained from NF by assuming the existence of an inaccessible ordinal is consistent, then ZF is consistent (see Orey, 1956a; Collins, 1955). If we add to NF the assumption of the existence of a strongly Cantorian set, then Zermelo's set theory Z is consistent (see Rosser, 1954). It is still an open question whether the consistency of ZF implies the consistency of NF.

Let ST^- be the simple theory of types ST without the Axiom of Infinity. Given any closed wf $\mathscr{A}$ of ST, let $\mathscr{A}^+$ denote the result of adding 1 to the types of all variables in $\mathscr{A}$. Let SP denote the theory obtained from ST^- by adding as axioms the wfs $\mathscr{A} \Leftrightarrow \mathscr{A}^+$ for all closed wfs $\mathscr{A}$. Specker (1958, 1962) proved that NF is consistent if and only if SP is consistent.

Let NFU denote the theory obtained from NF by restricting the Extensionality Axiom to nonempty sets:

$$\text{NF1*} \quad (\exists u)(u \in x) \wedge (\forall z)(z \in x \Leftrightarrow z \in y) \Rightarrow x = y$$

Jensen (1968–69) proved that NFU is consistent if and only if ST^- is consistent, and the equiconsistency continues to hold when both theories are supplemented by an Axiom of Infinity or by Axioms of Infinity and Choice.

Discussions of NF may be found in Hatcher (1982) and Quine (1963). Forster (1983) gives a survey of more recent results.

Quine also proposed a system ML that is formally related to NF in much the same way that MK is related to ZF. The variables are capital italic letters X, Y,

$Z, \ldots$; these variables are called *class variables*. We define $M(X)$, *X is a set*,* by $(\exists Y)(X \in Y)$, and we introduce lowercase italic letters $x, y, z, \ldots$ as variables restricted to sets. Equality is defined as in NBG: $X = Y$ for $(\forall Z)(Z \in X \Leftrightarrow Z \in Y)$. Then we introduce an axiom of equality:

ML1: $X = Y \wedge X \in Z \Rightarrow Y \in Z$

There is an unrestricted Comprehension Axiom:

ML2: $(\exists Y)(\forall x)(x \in Y \Leftrightarrow \mathscr{A}(x))$

where $\mathscr{A}(x)$ is any wf of ML. Finally, we wish to introduce an axiom that has the same effect as the Comprehension Axiom Scheme NF2:

ML3: For any wf $\mathscr{A}(x)$ whose free variables are $x, y_1, \ldots, y_n$ $(n \geqslant 0)$,

$$(\forall y_1) \ldots (\forall y_n)(\exists z)(\forall x)(x \in z \Leftrightarrow \mathscr{A}(x))$$

is an axiom, provided that $\mathscr{A}(x)$ is stratified and all its quantifiers are set quantifiers.

All theorems of NF are provable in ML. Hence, if ML is consistent, so is NF. Surprisingly, the converse has been proved by Wang (1950). In fact, any closed wf of NF provable in ML is already provable in NF.

ML has the same advantages over NF that MK and NBG have over ZF: a greater ease and power of expression. Moreover, the natural numbers of ML behave much better than those of NF; the principle of mathematical induction can be proved in full generality in ML.

The prime source for ML is Quine (1951).† For further discussion, consult Quine (1963) and Fraenkel, Bar-Hillel, and Lévy (1973).

* Quine uses the word "element" instead of "set".

† An earlier version of ML, published in 1950, was proved inconsistent by Rosser (1942). The present version is due to Wang (1950).

CHAPTER FIVE

Effective Computability

1. ALGORITHMS. TURING MACHINES

An *algorithm* is a computational method for solving each and every problem from a large class of problems. The computation has to be precisely specified so that it requires no ingenuity for its performance. The familiar technique for adding integers is an algorithm, as are the techniques for computing the other arithmetic operations of subtraction, multiplication, and division. The truth table procedure to determine whether a statement form is a tautology is an algorithm within logic itself.

It is often easy to see that a specified procedure yields a desired algorithm. In recent years, however, many classes of problems have been proved not to have an algorithmic solution. Examples are: (1) Is a given wf of quantification theory logically valid? (2) Is a given wf of formal number theory S true (in the standard interpretation)? (3) Is a given wf of S provable in S? (4) Does a given polynomial $f(x_1, x_2, \ldots, x_n)$ with integral coefficients have integral roots (Hilbert's Tenth Problem)? In order to rigorously prove that there does *not* exist an algorithm for answering such questions, it is necessary to supply a precise definition of the notion of algorithm.

Various proposals for such a definition were independently offered in 1936 by Church (1936b), Turing (1936–37), and Post (1936). All of these definitions, as well as others proposed later, have been shown to be equivalent. Moreover, it is intuitively clear that every procedure given by these definitions is an algorithm. Conversely, every known algorithm falls under these definitions. Our exposition will use Turing's ideas.

First of all, the objects with which an algorithm deals may be assumed to be the symbols of a finite alphabet $\mathrm{A} = \{\mathrm{a}_0, \mathrm{a}_1, \ldots, \mathrm{a}_n\}$. Nonsymbolic objects can be represented by symbols, and languages actually used for computation require only finitely many symbols.

A finite sequence of symbols of a language A is called a *word* of A. It is convenient to admit an empty word Λ consisting of no symbols at all. If P and Q are words, then PQ denotes the word obtained by writing Q to the right of P. For any positive integer k, P^k stands for the word made up of k consecutive occurrences of P.

The work space of an algorithm often consists of a piece of paper or a blackboard. However, we shall make the simplifying assumption that all calculations take place on a tape that is divided into squares (see Figure 5.1). The tape is potentially infinite in both directions, in the sense that, although at any moment it is finite, additional squares always can be added to the right- and left-hand ends of the tape. Each square may contain one symbol of the alphabet A. However, at any one time, only a finite number of squares contain symbols, while the rest are blank. The symbol a_0 will be reserved for the content of a blank square. (In ordinary languages, a space is used for the same purpose.) Thus, the condition of the tape at a given moment can be represented by a word of A; the tape in Figure 5.1 is $a_2a_0a_5a_1$. It turns out that our use of a one-dimensional tape does not limit the algorithms that can be handled; the information in a two-dimensional array can be encoded as a finite sequence.

Our computing device, which we shall refer to as a *Turing machine*, works in the following way. The machine operates at discrete moments of time, not continuously. It has a *reading head*, which, at each moment, will be scanning one square of the tape. (We assume that the observation of a larger domain can be reduced to consecutive observations of individual squares.) The device then reacts in any one of four different ways:

1. It prints a symbol in the square, erasing the previous symbol.
2. It moves to the next square on the right.
3. It moves to the next square on the left.
4. It stops.

What the machine does depends not only on the observed symbol but also on the *internal state* of the machine at that moment (which, in turn, depends on the previous steps of the computation and on the structure of the machine). We shall follow Turing in making the plausible assumption that a machine has only a finite number of internal states $\{q_0, q_1, \ldots, q_m\}$. The machine will always begin its operation in the *initial state* q_0.

A step in a computation corresponds to a quadruple of one of the following three forms: (1) $q_ja_ia_kq_r$, (2) $q_ja_iRq_r$, or (3) $q_ja_iLq_r$. In each case, q_j is the present internal state, a_i is the symbol being observed, and q_r is the internal state after the step. In form (1), the machine erases a_i and prints a_k. In form (2), it moves one square to the right, and, in form (3), it moves one square to the left. We shall indicate later how the machine is told to stop.

Now we can give a precise definition. A *Turing machine* with an alphabet A of *tape symbols* $\{a_0, a_1, \ldots, a_n\}$ and with *internal states* $\{q_0, q_1, \ldots, q_m\}$ is a finite set $\mathcal{T}$ of quadruples of the forms (1) $q_ja_ia_kq_r$, (2) $q_ja_iRq_r$, and (3) $q_ja_iLq_r$ such that no two quadruples of $\mathcal{T}$ have the same first two symbols. Thus, for fixed q_ja_i, no two

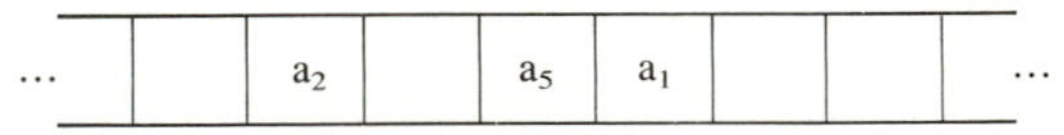

Figure 5.1

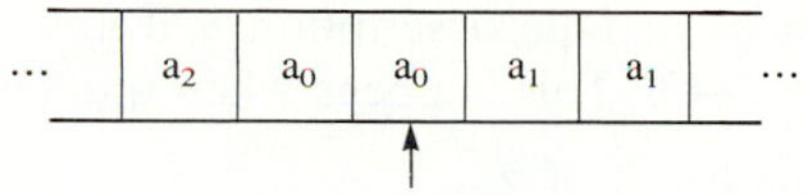

Figure 5.2

quadruples of types (1), (2), and (3) are in $\mathscr{T}$.† q_0 is called the *initial state* of $\mathscr{T}$.

The Turing machine $\mathscr{T}$ operates in accordance with its list of quadruples. This can be made precise in the following manner: A *tape description* (TD) of $\mathscr{T}$ is a word such that (1) all symbols in the word but one are tape symbols, (2) the only symbol that is not a tape symbol is some internal state q_j, and (3) q_j is not the last symbol of the word.

A tape description describes the condition of the machine and the tape at a given moment. When read from left to right, the tape symbols in the description represent the symbols on the tape at that moment. The internal state q_j in the description is the internal state of the machine at that moment, and the tape symbol that occurs immediately to the right of q_j in the tape description represents the symbol being scanned by the reading head at that moment. If the internal state q_j is q_0, then the tape description is called an *initial* tape description.

Example The tape description $a_2a_0q_1a_0a_1a_1$ indicates that the machine is in the internal state q_1, the tape is as shown in Figure 5.2, and the reading head is scanning the square indicated by the arrow.

We say that $\mathscr{T}$ *moves* one tape description α into another one β (abbreviated $\alpha \underset{\mathscr{T}}{\mapsto} \beta$) if and only if one of the following is true.

1. α is of the form Pq_ja_iQ, β is of the form Pq_ra_kQ, and $q_ja_ia_kq_r$ is one of the quadruples of $\mathscr{T}$.‡
2. α is of the form $Pa_sq_ja_iQ$, β is $Pq_ra_sa_iQ$, and $q_ja_iLq_r$ is one of the quadruples of $\mathscr{T}$.
3. α is of the form q_ja_iQ, β is $q_ra_0a_iQ$, and $q_ja_iLq_r$ is one of the quadruples of $\mathscr{T}$.
4. α is of the form $Pq_ja_ia_kQ$, β is $Pa_iq_ra_kQ$, and $q_ja_iRq_r$ is one of the quadruples of $\mathscr{T}$.
5. α is of the form Pq_ja_i, β is $Pa_iq_ra_0$, and $q_ja_iRq_r$ is one of the quadruples of $\mathscr{T}$.

According to our intuitive picture, "$\mathscr{T}$ moves α into β" means that, if the condition at time t of the Turing machine and tape is described by α, then the condition at time $t + 1$ is described by β. Notice that, by clause 3, whenever the machine reaches the left-hand end of the tape and is ordered to move left, a blank square is attached to the tape on the left; similarly, by clause 5, a blank square is added on the right when the machine reaches the right-hand end and has to move right.

† This condition ensures that there is never a situation in which the machine is instructed to perform two contradictory operations.

‡ Here and below, P and Q are arbitrary (possibly empty) words of the alphabet of $\mathscr{T}$.

We say that $\mathscr{T}$ *stops* at tape description α if and only if there is no tape description β such that $\alpha \underset{\mathscr{T}}{\mapsto} \beta$. This happens when $q_j a_i$ occurs in α but $q_j a_i$ is not the beginning of any quadruple of $\mathscr{T}$.

A *computation* of $\mathscr{T}$ is a finite sequence of tape descriptions $\alpha_0, \ldots, \alpha_k$ $(k \geqslant 0)$ such that the following conditions hold.

1. α_0 is an initial tape description; that is, the internal state occurring in α is q_0.
2. $\alpha_i \underset{\mathscr{T}}{\mapsto} \alpha_{i+1}$ for $0 \leqslant i < k$.
3. $\mathscr{T}$ stops at α_k.

This computation is said to *begin* at α_0 and *end* at α_k. If there is a computation beginning with α_0, we say that $\mathscr{T}$ is *applicable* to α_0.

The algorithm $Alg_{\mathscr{T}}$ determined by $\mathscr{T}$ is defined as follows:

> For any words P and Q of A, $Alg_{\mathscr{T}}(P) = Q$ if and only if there is a computation of $\mathscr{T}$ that begins with the tape description $q_0 P$ and ends with a tape description of the form $R_1 q_j R_2$, where $Q = R_1 R_2$.

This means that, when $\mathscr{T}$ begins at the left-hand end of P and there is nothing else on the tape, $\mathscr{T}$ stops with Q as the entire content of the tape. Notice that $Alg_{\mathscr{T}}$ need not be defined for certain words P. An algorithm $Alg_{\mathscr{T}}$ determined by a Turing machine $\mathscr{T}$ is said to be a *Turing algorithm.*

Example In any computation of the Turing machine $\mathscr{T}$ given by

$$q_0 a_0 R q_0,\ q_0 a_1 a_0 q_1,\ q_0 a_2 a_0 q_1, \ldots, q_0 a_n a_0 q_1$$

$\mathscr{T}$ locates the first nonblank symbol (if any) at or to the right of the square scanned at the beginning of the computation, erases that symbol, and then stops. If there are only blank squares at or to the right of the initial square, $\mathscr{T}$ keeps on moving right forever.

Let us now consider computations of number-theoretic functions. For convenience, we sometimes will write | instead of a_1 and B instead of a_0. (Think of B as standing for "blank".) For any natural number k, its *tape representation* $\overline{k}$ will stand for the word $|^{k+1}$—that is, the word consisting of $k + 1$ occurrences of |. Thus, $\overline{0} = |$, $\overline{1} = ||$, $\overline{2} = |||$, and so on. We represent k by $k + 1$ occurrences of | instead of k occurrences because we wish $\overline{0}$ to be a nonempty word, so that we will be aware of its presence. The *tape representation* $\overline{(k_1, k_2, \ldots, k_n)}$ of an n-tuple of natural numbers $(k_1, k_2, \ldots, k_n)$ is defined to be the word $\overline{k}_1 B \overline{k}_2 B \ldots B \overline{k}_n$. For example, $\overline{(3, 1, 0, 5)}$ is ||||B||B|B||||||.

A Turing machine $\mathscr{T}$ will be thought of as computing the following partial function $f_{\mathscr{T},1}(x)$ of one variable.*

> $f_{\mathscr{T},1}(k) = m$ if and only if the following condition holds: $Alg_{\mathscr{T}}(\overline{k})$ is defined and $Alg_{\mathscr{T}}(\overline{k}) = E_1 \overline{m} E_2$, where E_1 and E_2 are certain (possibly empty) words consisting of only B's (blanks).

* Remember that a partial function *may* fail to be defined for some values of its arguments. A total function is also considered to be a partial function.

The function $f_{\mathcal{T},1}$ is said to be *Turing-computable*. Thus, a one-place partial function f is Turing-computable if and only if there is a Turing machine $\mathcal{T}$ such that $f = f_{\mathcal{T},1}$.

For each $n > 1$, a Turing machine $\mathcal{T}$ also computes a function $f_{\mathcal{T},n}$ of n variables. For any natural numbers $k_1, \ldots, k_n$:

> $f_{\mathcal{T},n}(k_1, \ldots, k_n) = m$ if and only if the following condition holds: $Alg_{\mathcal{T}}(\overline{(k_1, \ldots, k_n)})$ is defined and $Alg_{\mathcal{T}}(\overline{(k_1, \ldots, k_n)}) = \mathrm{E}_1\,\overline{m}\,\mathrm{E}_2$, where E_1 and E_2 are (possibly empty) words consisting of B's.

The partial function $f_{\mathcal{T},n}$ is said to be *Turing-computable*. Thus, an n-place partial function f is Turing-computable if and only if there is a Turing machine $\mathcal{T}$ such that $f = f_{\mathcal{T},n}$.

Notice that, at the end of a computation of a value of a Turing-computable function, only the value appears on the tape, aside from blank squares, and the location of the reading head does not matter. Also observe that, whenever the function is not defined, either the Turing machine will never stop or, if it does stop, the resulting tape is not of the appropriate form $\mathrm{E}_1\,\overline{m}\,\mathrm{E}_2$.

Examples

1. Consider the Turing machine $\mathcal{T}$, with alphabet $\{\mathrm{B}, |\}$, defined by $q_0|Lq_1$, $q_1 \mathrm{B}|q_2$. $\mathcal{T}$ computes the successor function $N(x)$, since $q_0\,\overline{k} \underset{\mathcal{T}}{\mapsto} q_1\,\mathrm{B}\,\overline{k} \underset{\mathcal{T}}{\mapsto} q_2\,\overline{k+1}$. Hence, $N(x)$ is Turing-computable.
2. The Turing machine $\mathcal{T}$ defined by

$$q_0|\mathrm{B}q_1,\ q_1\mathrm{BR}q_0,\ q_0\mathrm{B}|q_2$$

computes the zero function $Z(x)$. Given $\overline{k}$ on the tape, $\mathcal{T}$ moves right, erasing all |'s until it reaches a blank, which it changes to a |. So, $\overline{0}$ is the final result. Thus, $Z(x)$ is Turing-computable.
3. The addition function is computed by the Turing machine defined by the following seven quadruples:

$$q_0|\mathrm{B}q_0,\ q_0\mathrm{BR}q_1,\ q_1|\mathrm{R}q_1,\ q_1\mathrm{B}|q_2,\ q_2|\mathrm{R}q_2,\ q_2\mathrm{BL}q_3,\ q_3|\mathrm{B}q_3$$

In fact, for any natural numbers m and n,

$$\begin{aligned} q_0\,\overline{(m,n)} &= q_0\,|^{m+1}\,\mathrm{B}\,|^{n+1} \underset{\mathcal{T}}{\mapsto} q_0\,\mathrm{B}\,|^m\,\mathrm{B}\,|^{n+1} \underset{\mathcal{T}}{\mapsto} \mathrm{B}\,q_1\,|^m\,\mathrm{B}\,|^{n+1} \underset{\mathcal{T}}{\mapsto} \cdots \\ &\underset{\mathcal{T}}{\mapsto} \mathrm{B}\,|^m\,q_1\,\mathrm{B}\,|^{n+1} \underset{\mathcal{T}}{\mapsto} \mathrm{B}\,|^m\,q_2\,||^{n+1} \underset{\mathcal{T}}{\mapsto} \cdots \underset{\mathcal{T}}{\mapsto} \mathrm{B}\,|^m\,|^{n+2}\,q_2\,\mathrm{B} \\ &\underset{\mathcal{T}}{\mapsto} \mathrm{B}\,|^{m+n+1}\,q_3\,|\,\mathrm{B} \underset{\mathcal{T}}{\mapsto} \mathrm{B}\,|^{m+n+1}\,q_3\,\mathrm{B}\,\mathrm{B} = \mathrm{B}\,\overline{m+n}\,q_3\,\mathrm{B}\,\mathrm{B} \end{aligned}$$

Exercises

5.1 Show that the function U_2^2 such that $U_2^2(x_1, x_2) = x_2$ is Turing-computable.

5.2 (a) What function $f(x_1, x_2, x_3)$ is computed by the following Turing machine?

$q_0\,||\,q_1$	$q_2\,|\,\mathrm{R}\,q_2$
$q_1\,|\,\mathrm{B}\,q_0$	$q_2\,\mathrm{B}\,\mathrm{R}\,q_3$

$$q_0 B R q_1 \qquad q_3 | B q_4$$

$$q_1 B R q_2 \qquad q_4 B R q_3$$

(b) What function $f(x)$ is computed by the following Turing machine?

$$q_0 | B q_1, \ q_1 B R q_2, \ q_2 B | q_2$$

5.3 (a) State in plain language the operation of the Turing machine, described in Example 3, for computing the addition function.
(b) Starting with the tape description $q_0|||B||||$, write the sequence of tape descriptions that make up the computation by the addition machine of Example 3.

5.4 What function $f(x)$ is computed by the following Turing machine?

$q_0 \| R q_1$	$q_4 \| R q_4$	$q_6 B \| q_0$
$q_1 \| B q_2$	$q_4 B \| q_5$	$q_1 B \| q_7$
$q_2 B R q_3$	$q_5 \| L q_5$	$q_7 \| L q_7$
$q_3 \| R q_3$	$q_5 B L q_6$	$q_7 B R q_8$
$q_3 B R q_4$	$q_6 \| L q_6$	$q_8 \| B q_8$

5.5 Find a Turing machine that computes the function sg(x). [Remember that sg(0) = 0 and sg(x) = 1 for $x > 0$.]

5.6 Find a Turing machine that computes the function $x \dot{-} y$. [Remember that $x \dot{-} y = x - y$ if $x \geqslant y$, and $x \dot{-} y = 0$ if $x < y$.]

5.7 If a function is Turing-computable, show that it is computable by infinitely many different Turing machines.

2. DIAGRAMS

Many Turing machines that compute even relatively simple functions (like multiplication) require a large number of quadruples. It is difficult and tedious to construct such machines, and even more difficult to check that they do the desired job. We shall introduce a pictorial technique for constructing Turing machines so that their operation is easier to comprehend. The basic ideas and notation are due to Hermes (1965).

1. Let $\mathcal{T}_1, \ldots, \mathcal{T}_r$ be any Turing machines with alphabet $A = \{a_0, a_1, \ldots, a_k\}$.
2. Select a finite set of points in a plane. These points will be called *vertices.*
3. To each vertex attach the name of one of the machines $\mathcal{T}_1, \ldots, \mathcal{T}_r$. Copies of the same machine may be assigned to more than one vertex.
4. Connect some vertices to others by arrows. An arrow may go from a vertex to itself. Each arrow is labeled with one of the numbers 0, 1, ..., k. No two arrows that emanate from the same vertex are allowed to have the same label.
5. One vertex is enclosed in a circle and is called the *initial vertex.*

The resulting graph is called a *diagram.*

Example See Figure 5.3.

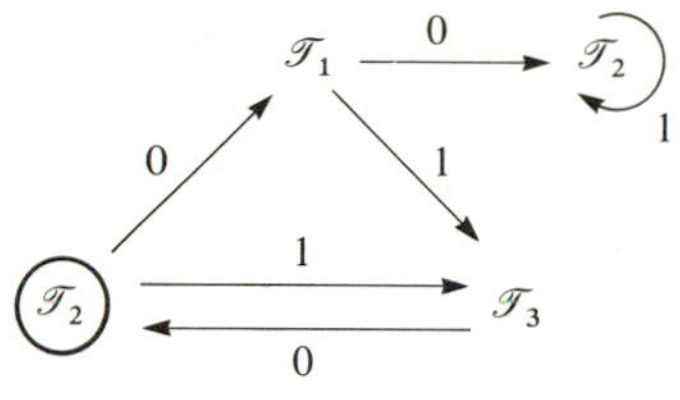

Figure 5.3

We shall show that every diagram determines a Turing machine whose operation can be described in the following manner. Given a tape and a specific square on the tape, the Turing machine of the initial vertex V of the diagram begins to operate, with its reading head scanning the specified square of the tape. If this machine finally stops and the square being scanned at the end of the computation contains the symbol a_i, then we look for an arrow with label i emanating from the vertex V. If there is no such arrow, the computation stops. If there is such an arrow, it leads to a vertex to which another Turing machine has been assigned. Start that machine on the tape produced by the previous computation, at the square that was being scanned at the end of the computation. Repeat the same procedure that was just performed, and keep on doing this until the machine stops. The resulting tape is the output of the machine determined by the diagram. If the machine never stops, then it is not applicable to the initial tape description.

The quadruples for this Turing machine can be specified in the following way.

1. For each occurrence in the diagram of a machine $\mathcal{T}_j$, write its quadruples, changing internal states so that no two machine occurrences have an internal state in common. The initial vertex machine is not to be changed. This retains q_0 as the initial internal state of the machine assigned to the initial vertex. For every other machine occurrence, the original initial state q_0 has been changed to a new internal state.

2. If an occurrence of some $\mathcal{T}_i$ is connected by an arrow $\overset{u}{\to}$ to some $\mathcal{T}_j$, then, for every (new) internal state q_s of that occurrence of $\mathcal{T}_i$ such that no (new) quadruple of $\mathcal{T}_i$ begins with $q_s a_u$, add the quadruple $q_s a_u a_u q_t$, where q_t is the (new) initial state for $\mathcal{T}_j$. (Step 2 ensures that, whenever $\mathcal{T}_i$ stops while scanning a_u, $\mathcal{T}_j$ will begin operating.)

ABBREVIATIONS IN DIAGRAMS

1. If one vertex is connected to another vertex by all arrows $\overset{0}{\to}, \overset{1}{\to}, \ldots, \overset{k}{\to}$, we replace the arrows by one unlabeled arrow.
2. If one vertex is connected to another by all arrows except $\overset{u}{\to}$, we replace all the arrows by the one arrow $\overset{\neq u}{\to}$.
3. Let $\mathcal{T}_1\mathcal{T}_2$ stand for $\mathcal{T}_1 \to \mathcal{T}_2$, $\mathcal{T}_1\mathcal{T}_2\mathcal{T}_3$ for $\mathcal{T}_1 \to \mathcal{T}_2 \to \mathcal{T}_3$, and so on. Let $\mathcal{T}^2$ be $\mathcal{T}\mathcal{T}$, let $\mathcal{T}^3$ be $\mathcal{T}\mathcal{T}\mathcal{T}$, and so forth.
4. If no vertex is circled, then the leftmost vertex is to be initial.

To construct diagrams, we need a few simple Turing machines as building blocks.

1. **r** (right machine). Let $\{a_0, a_1, \ldots, a_k\}$ be the alphabet. **r** consists of the quadruples $q_0 a_i R q_1$ for all a_i. This machine, which has $k + 1$ quadruples, moves one square to the right and stops.
2. **l** (left machine) for the alphabet $\{a_0, a_1, \ldots, a_k\}$. **l** consists of the quadruples $q_0 a_i L q_1$ for all a_i. This machine moves one square to the left and stops.
3. $\mathbf{a_j}$ (constant machine) for the alphabet $\{a_0, a_1, \ldots, a_k\}$. $\mathbf{a_j}$ consists of the quadruples $q_0 a_i a_j q_1$ for all a_i. This machine replaces the initial scanned symbol by a_j and then stops. (In particular, $\mathbf{a_0}$ erases the scanned symbol, and $\mathbf{a_1}$ prints |.)

Examples of Machines Defined by Diagrams

1. **P**

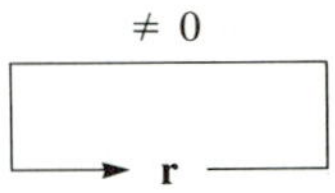

Figure 5.4

P (Figure 5.4) finds the first blank to the right of the initially scanned square. In an alphabet $\{a_0, a_1, \ldots, a_k\}$, the quadruples for the machine **P** are:

$$\begin{cases} q_0 a_i R q_1 & \text{for all } a_i \\ q_1 a_i a_i q_0 & \text{for all } a_i \neq a_0 \end{cases}$$

2. **Λ**

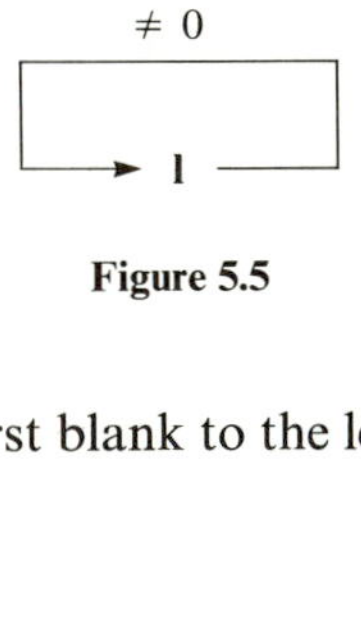

Figure 5.5

Λ (Figure 5.5) finds the first blank to the left of the initially scanned square.

Exercises

5.8 Describe the operation of the Turing machines determined by the diagrams in Figures 5.6 and 5.7 and write the list of quadruples for each machine.

(a) **ρ**

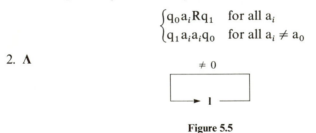

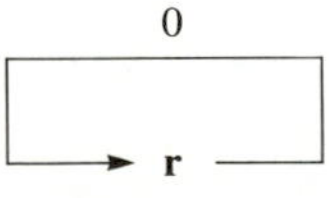

Figure 5.6

(b) λ

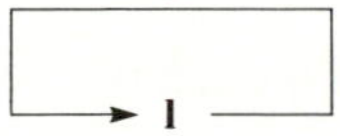

Figure 5.7

5.9 Show that the machine **S** in Figure 5.8 searches the tape for a nonblank square. If there are such squares, **S** finds one and stops. Otherwise, **S** never stops.

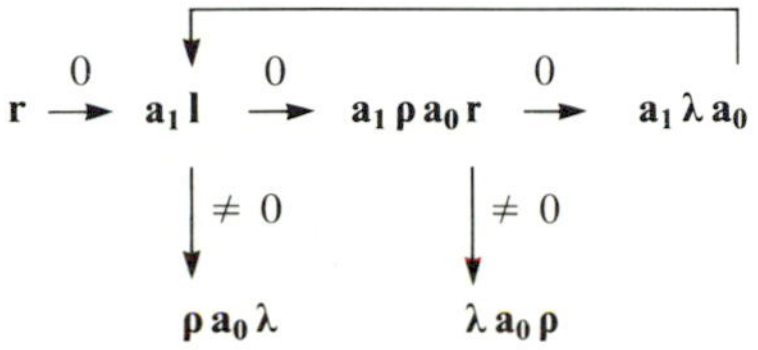

Figure 5.8

To describe some aspects of the operation of a Turing machine on part of a tape, we introduce the following notation of Hermes:

$\sim$	arbitrary symbol
B.... B	sequence of blanks
B....	everything blank to the right
.... B	everything blank to the left
W	nonempty word consisting of nonblanks
X	$W_1 B W_2 B \ldots W_n$ $(n \geqslant 1)$, a sequence of nonempty words of nonblanks, separated by blanks

Underlining will indicate the scanned symbol.

Some More Turing Machines Defined by Diagrams

3. $\mathcal{R}$ (right-end machine). See Figure 5.9.

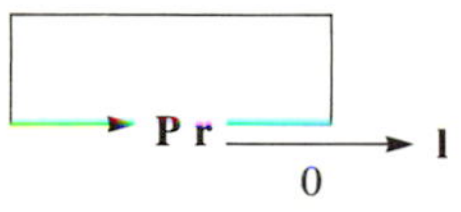

Figure 5.9

$$\underline{\sim} X B B \Rightarrow \sim X \underline{B} B$$

Squares on the rest of the tape are not affected. The same assumption is made in similar places below. When the machine $\mathcal{R}$ begins on a square preceding a sequence of one or more nonempty words, followed by at least two blank squares, it moves right to the first of those blank squares, and then it stops.

4. $\mathscr{L}$ (left-end machine). See Figure 5.10.

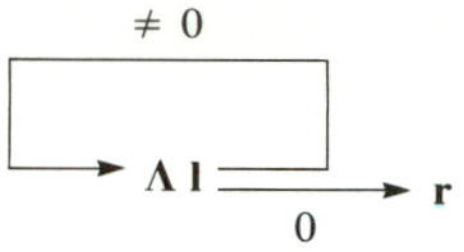

Figure 5.10

$$B\,BX\underline{\sim} \Rightarrow B\,\underline{B}\,X\sim$$

5. **T** (left-translation machine). See Figure 5.11.*

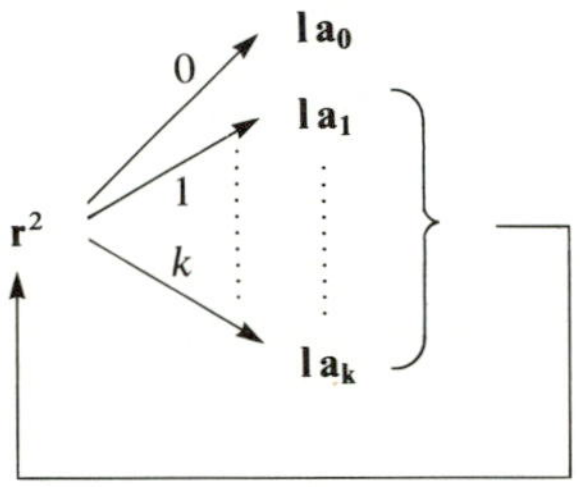

Figure 5.11

$$\underline{\sim}B\,W\,B \Rightarrow \sim W\,\underline{B}\,B$$

This machine shifts the whole word W one square to the left.

6. **σ** (shift machine). See Figure 5.12.

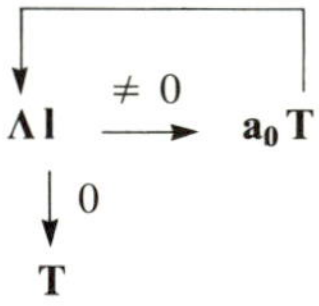

Figure 5.12

$$B\,W_1\,B\,W_2\,\underline{B} \Rightarrow B\,W_2\,\underline{B}\ldots B$$

In the indicated situation, W_1 is erased and W_2 is shifted leftward so that it begins where W_1 originally began.

7. **C** (clean-up machine). See Figure 5.13.

*There is a separate arrow from $\mathbf{r}^2$ to each of the groups on the right and a separate arrow from each of these, except $\mathbf{l\,a_0}$, back to $\mathbf{r}^2$.

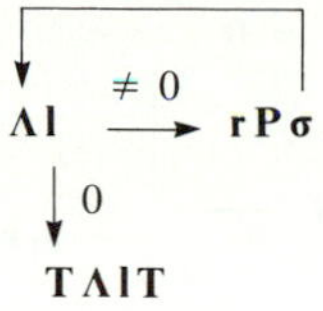

Figure 5.13

$$\sim B\,B\,X\,B\,W\,\underline{B} \Rightarrow \sim W\,\underline{B} \ldots B$$

8. **K** (word-copier). See Figure 5.14.

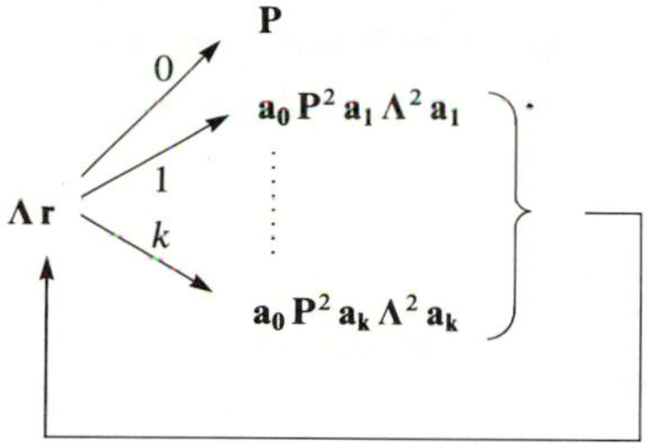

Figure 5.14

$$B\,W\,\underline{B} \ldots \Rightarrow B\,W\,B\,W\,\underline{B} \ldots$$

9. $\mathbf{K}_n$ (*n*-shift copier). See Figure 5.15.

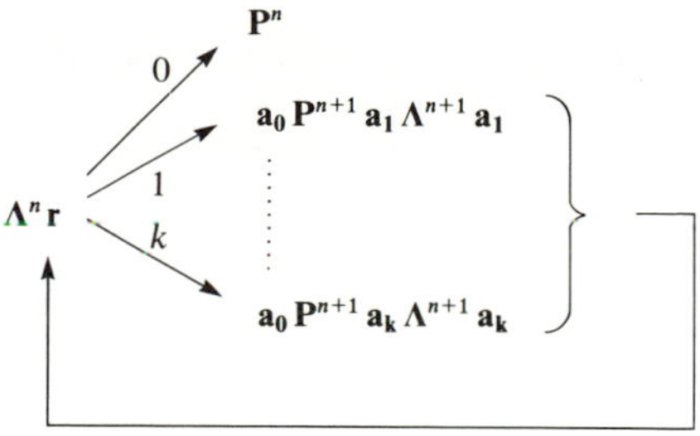

Figure 5.15

$$B\,W_n\,B\,W_{n-1}\,B \ldots W_1\,\underline{B} \ldots \Rightarrow B\,W_n\,B\,W_{n-1}\,B \ldots W_1\,B\,W_n\,\underline{B} \ldots$$

Exercises

5.10 Find the number-theoretic function $f(x)$ computed by each of the following Turing machines.

(a) $\mathbf{l}\,\mathbf{a}_1$

(b) Figure 5.16

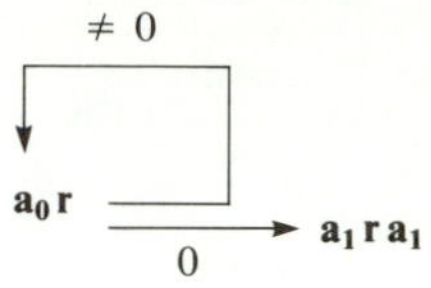

Figure 5.16

(c) $\mathbf{P K \Lambda a_1 \Lambda (r a_0)^2}$

5.11 Verify that the given functions are computed by the indicated Turing machines.

(a) $|x - y|$ (Figure 5.17)

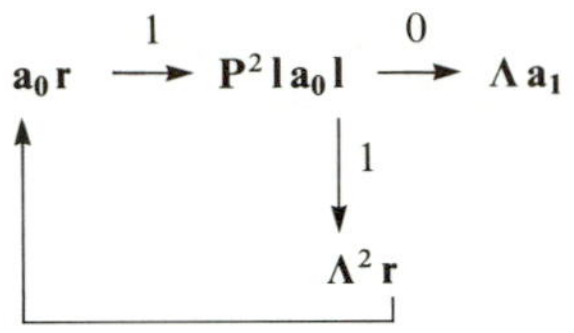

Figure 5.17

(b) $x + y \qquad \mathbf{P a_1 \Lambda (r a_0)^2}$

(c) $x \cdot y$ (Figure 5.18)

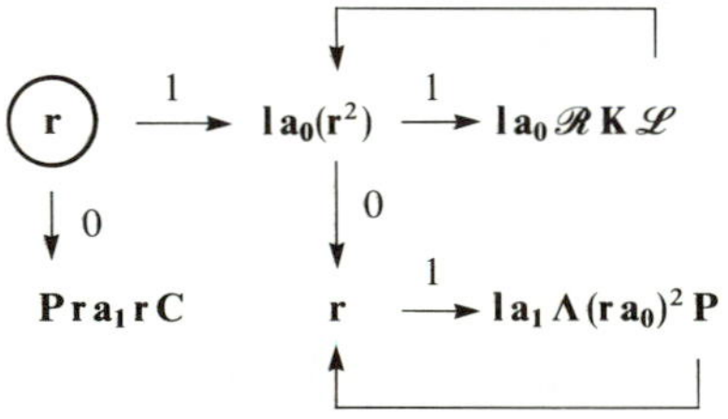

Figure 5.18

5.12 Draw diagrams for Turing machines that will compute the following functions: (a) $\max(x, y)$ (b) $\min(x, y)$ (c) $x \dot{-} y$ (d) $[x/2]$ = the greatest integer $\leqslant x/2$

5.13 Prove that, for any Turing machine $\mathscr{T}$ with alphabet $\{a_0, \ldots, a_k\}$, there is a diagram using the Turing machines **r**, **l**, $\mathbf{a_0}, \ldots, \mathbf{a_k}$ that defines a Turing machine $\mathscr{S}$ such that $\mathscr{T}$ and $\mathscr{S}$ have the same effect on all tapes. (In fact, $\mathscr{S}$ can be defined so that, except for two additional trivial initial moves left and right, it carries out the same computations as $\mathscr{T}$.)

3. PARTIAL RECURSIVE FUNCTIONS. UNSOLVABLE PROBLEMS

Recall, from Chapter 3, Section 3, that the recursive functions are obtained from the initial functions [the zero function $Z(x)$, the successor function $N(x)$, and the projection functions $U_i^n(x_1, \ldots, x_n)$] by means of substitution, recursion, and the

μ-operator. Instead of the μ-operator, let us introduce the *unrestricted μ-operator*:

If $f(x_1,\ldots,x_n) = \mu y(g(x_1,\ldots,x_n,y) = 0)$
$= \text{the least } y \text{ such that } g(x_1,\ldots,x_n,y) = 0$
then f is said to arise from g by means of the unrestricted μ-operator.

Notice that, for some $x_1, \ldots, x_n$, the value $f(x_1,\ldots,x_n)$ need not be defined; this happens when there is no y such that $g(x_1,\ldots,x_n,y) = 0$.

If we replace the μ-operator by the unrestricted μ-operator in the definition of the recursive functions, we obtain a definition of the partial recursive functions. In other words, the *partial recursive functions* are those functions obtained from the initial functions by means of substitution, recursion, and the unrestricted μ-operator. Thus, whereas all recursive functions are total functions, some partial recursive functions will not be total functions. For example, $\mu y(x + y = 0)$ is defined only when $x = 0$.

Since partial recursive functions may not be defined for certain arguments, the definition of the unrestricted μ-operator should be made more precise:

$\mu y(g(x_1,\ldots,x_n,y) = 0) = k$ means that, for $0 \leqslant u < k$, $g(x_1,\ldots,x_n,u)$ is defined and $g(x_1,\ldots,x_n,u) \neq 0$, and $g(x_1,\ldots,x_n,k) = 0$.

Observe that, if $R(x_1,\ldots,x_n,y)$ is a recursive relation, then $\mu y(R(x_1,\ldots,x_n,y))$ can be considered an admissible application of the unrestricted μ-operator: $\mu y(R(x_1,\ldots,x_n,y)) = \mu y(C_R(x_1,\ldots,x_n,y) = 0)$, where C_R is the characteristic function of R. Since R is a recursive relation, C_R is, by definition, a recursive function.

Exercises

5.14 Describe the following partial recursive functions.
(a) $\mu y(x + y + 1 = 0)$ (b) $\mu y(y > x)$ (c) $\mu y(y + x = x)$

5.15 Show that all recursive functions are partial recursive.

5.16 Show that every partial function whose domain is a finite set of natural numbers is a partial recursive function.

It is easy to convince ourselves that every partial recursive function $f(x_1,\ldots,x_n)$ is effectively computable in the sense that there is an algorithm that computes $f(x_1,\ldots,x_n)$ when $f(x_1,\ldots,x_n)$ is defined and gives no result when $f(x_1,\ldots,x_n)$ is undefined. This property is clear for the initial functions and is inherited under the operations of substitution, recursion, and the unrestricted μ-operator.

It turns out that the partial recursive functions are identical with the Turing-computable functions. To show this, it is convenient to introduce a different kind of Turing-computability.

A partial number-theoretic function $f(x_1,\ldots,x_n)$ is said to be *standard Turing-computable* if there is a Turing machine $\mathscr{T}$ such that, for any natural numbers $k_1, \ldots, k_n$, the following holds.

Let $B\,\overline{k}_1\,B\,\overline{k}_2\,B\ldots B\,\overline{k}_n$ be called the *argument strip*.* Notice that the argument

* For a function of one variable, the argument strip is taken to be $B\,\overline{k}_1$.

strip is $\mathrm{B}\overline{(k_1,\ldots,k_n)}$. Take any tape containing the argument strip but without any symbols to the right of it. (It may contain symbols to the left.) The machine $\mathscr{T}$ is begun on this tape with its reading head scanning the first | of $\overline{k}_1$. Then:

1. $\mathscr{T}$ stops if and only if $f(k_1,\ldots,k_n)$ is defined.
2. If $\mathscr{T}$ stops, the tape contains the same argument strip as before, followed by $\mathrm{B}\,\overline{f(k_1,\ldots,k_n)}$. Thus, the final tape contains:

$$\mathrm{B}\,\overline{k}_1\,\mathrm{B}\,\overline{k}_2\,\mathrm{B}\ldots\mathrm{B}\,\overline{k}_n\,\mathrm{B}\,\overline{f(k_1,\ldots,k_n)}$$

Moreover,

3. The reading head is scanning the first | of $\overline{f(k_1,\ldots,k_n)}$.
4. There is no nonblank symbol on the tape to the right of $\overline{f(k_1,\ldots,k_n)}$.
5. During the entire computation, the reading head never scans any square to the left of the argument strip.

For the sake of brevity, we shall say that the machine $\mathscr{T}$ described above *ST-computes* the function $f(x_1,\ldots,x_n)$.

Thus, the additional requirement of standard Turing-computability is that the original arguments are preserved, the machine stops if and only if the function is defined for the given arguments, and the machine operates on or to the right of the argument strip. So, anything to the left of the argument strip remains unchanged.

PROPOSITION 5.1 Every standard Turing-computable function is Turing-computable.

Proof Let $\mathscr{T}$ be a Turing machine that ST-computes a partial function $f(x_1,\ldots,x_n)$. Then f is Turing-computable by the Turing machine $\mathscr{T}\,\mathbf{P}\,\mathbf{C}$. In fact, after $\mathscr{T}$ operates, we obtain $\mathrm{B}\,\overline{x}_1\,\mathrm{B}\ldots\mathrm{B}\,\overline{x}_n\,\mathrm{B}\,\overline{f(x_1,\ldots,x_n)}$, with the reading head at the leftmost | of $\overline{f(x_1,\ldots,x_n)}$. **P** then moves the reading head to the right of $\overline{f(x_1,\ldots,x_n)}$, and then **C** removes the original argument strip.

PROPOSITION 5.2 Every partial recursive function is standard Turing-computable.

Proof

(a) $\mathbf{P}\,\mathbf{r}\,\mathbf{a}_1$ ST-computes the zero function $Z(x)$.
(b) The successor function $N(x)$ is ST-computed by $\mathbf{P}\,\mathbf{K}\,\mathbf{a}_1\,\mathbf{\Lambda}\,\mathbf{r}$.
(c) The projection function $U_i^n(x_1,\ldots,x_n)=x_i$ is ST-computed by $\mathscr{R}\,\mathbf{K}_{n-i+1}\,\mathbf{\Lambda}\,\mathbf{r}$.
(d) *Substitution*. Let $f(x_1,\ldots,x_n) = g(h_1(x_1,\ldots,x_n),\ldots,h_m(x_1,\ldots,x_n))$, and assume that $\mathscr{T}$ ST-computes g and $\mathscr{T}_j$ ST-computes h_j for $1 \leqslant j \leqslant m$. Now, let $\mathscr{S}_j$ be the machine $\mathscr{T}_j\,\mathbf{P}\,\boldsymbol{\sigma}^n\,(\mathbf{K}_{n+j})^n\,\mathbf{\Lambda}^n\,\mathbf{r}$. The reader should verify that f is ST-computed by

$$\mathscr{T}_1\,\mathbf{P}\,(\mathbf{K}_{n+1})^n\,\mathbf{\Lambda}^n\,\mathbf{r}\,\mathscr{S}_2\,\mathscr{S}_3\ldots\mathscr{S}_{m-1}\,\mathscr{T}_m\,\mathbf{P}\,\boldsymbol{\sigma}^n\,\mathbf{\Lambda}^m\,\mathbf{r}\,\mathscr{T}\,\boldsymbol{\sigma}^m\,\mathbf{\Lambda}\,\mathbf{r}$$

We take advantage of the ST-computability when, storing $\overline{x}_1,\ldots,\overline{x}_n$, $\overline{h_1(x_1,\ldots,x_n)},\ldots,\overline{h_i(x_1,\ldots,x_n)}$ on the tape, we place $\overline{(x_1,\ldots,x_n)}$ on the tape to the right and compute $\overline{h_{i+1}(x_1, \ldots, x_n)}$ without disturbing what we have stored on the left.

(e) *Recursion.* Let

$$f(x_1, \ldots, x_n, 0) = g(x_1, \ldots, x_n)$$
$$f(x_1, \ldots, x_n, y + 1) = h(x_1, \ldots, x_n, y, f(x_1, \ldots, x_n, y))$$

Assume that $\mathscr{S}$ ST-computes g and $\mathscr{T}$ ST-computes h. Then the reader should verify that the machine in Figure 5.19 ST-computes f.

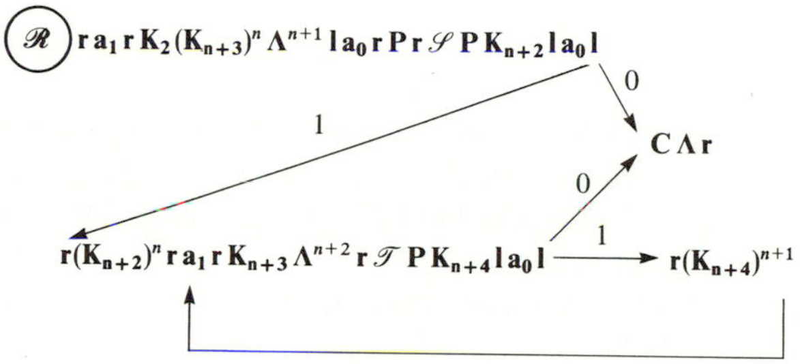

Figure 5.19

(f) *Unrestricted μ-operator.* Let $f(x_1, \ldots, x_n) = \mu y(g(x_1, \ldots, x_n, y) = 0)$ and assume that $\mathscr{T}$ ST-computes g. Then the machine in Figure 5.20 ST-computes f.

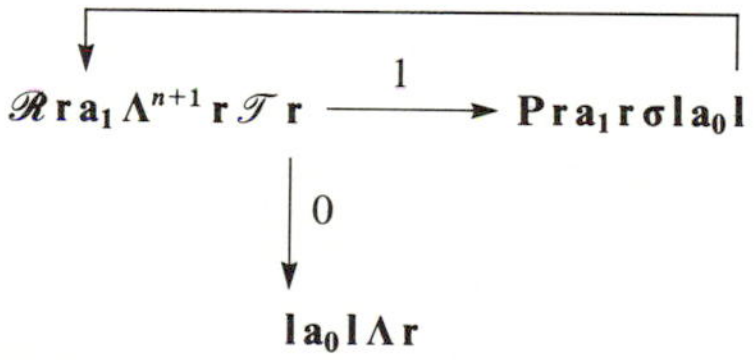

Figure 5.20

Exercise

5.17 For a recursion of the form

$$f(0) = k$$
$$f(y + 1) = h(y, f(y))$$

show how the diagram in Figure 5.19 must be modified.

COROLLARY 5.3 Every partial recursive function is Turing-computable.

Exercise

5.18 Verify that every partial recursive function is Turing-computable by a Turing machine with alphabet $\{a_0, a_1\}$.

In order to prove the converse of Corollary 5.3, we must arithmetize the language of Turing-computability by assigning numbers, called *Gödel numbers*, to the expressions arising in our study of Turing machines. "R" and "L" are assigned the Gödel numbers 3 and 5, respectively. The tape symbols a_i are assigned the numbers $7 + 4i$, while the internal state symbols q_i are given the numbers $9 + 4i$.

Symbol	*Gödel Number*
R	3
L	5
a_i	$7 + 4i$
q_i	$9 + 4i$

For example, the blank B, which is a_0, receives the number 7; the stroke |, which is a_1, has the number 11; and the initial internal state symbol q_0 has the number 9. Notice that all symbols have odd Gödel numbers, and different symbols have different numbers assigned to them.

As in Chapter 3, Section 4, a finite sequence $u_0, u_1, \ldots, u_k$ of symbols is assigned the Gödel number $p_0^{g(u_0)} p_1^{g(u_1)} \ldots p_k^{g(u_k)}$, where $p_0, p_1, p_2, \ldots$ are the prime numbers 2, 3, 5, ... in ascending order, and $g(u_i)$ is the Gödel number assigned to u_i. For example, the quadruple $q_0 a_0 a_1 q_0$ receives the Gödel number $2^9 3^7 5^{11} 7^9$.

By an *expression* we mean a finite sequence of symbols. We have just shown how to assign Gödel numbers to expressions. In a similar manner, to any finite sequence $E_0, E_1, \ldots, E_m$ of expressions we assign the number $p_0^{g(E_0)} p_1^{g(E_1)} \ldots p_m^{g(E_m)}$. For example, this assigns Gödel numbers to finite sequences of Turing machine quadruples and to finite sequences of tape descriptions. (Observe that the Gödel number of an expression is even and, therefore, different from the Gödel number of a symbol, which is odd; moreover, the Gödel number of a sequence of expressions has an even number as an exponent of p_0 and is, therefore, different from the Gödel number of an expression, which has an odd number as exponent of p_0.)

The reader should review Chapter 3, Sections 3 and 4, especially the functions $\ell h(x)$, $(x)_i$, and $x * y$. Assume that x is the Gödel number of a finite sequence $w_0, w_1, \ldots, w_k$; that is, $x = p_0^{g(w_0)} p_1^{g(w_1)} \ldots p_k^{g(w_k)}$, where $g(w_j)$ is the Gödel number of w_j. Recall that $\ell h(x) = k + 1$, the length of the sequence, and $(x)_j = g(w_j)$, the Gödel number of the jth term of the sequence. If, in addition, y is the Gödel number of a finite sequence $v_0, v_1, \ldots, v_m$, then $x * y$ is the Gödel number of the juxtaposition $w_0, w_1, \ldots, w_k, v_0, v_1, \ldots, v_m$ of the two sequences.

PROPOSITION 5.4 The following number-theoretic relations and functions are primitive recursive. In each case, we write first the notation for the relation or function, then the intuitive interpretation in terms of Turing machines, and, finally, the exact definition. (For the proofs of primitive recursiveness, use Proposition 3.17 and various primitive recursive relations and functions defined in Chapter 3, Section 3. At a first reading, it may be advisable to concentrate on just the intuitive meanings and postpone the technical verifications until later.)

IS(x): x is the Gödel number of an internal state symbol q_u:

$$(\exists u)_{u<x}(x = 9 + 4u)$$

Sym(x): x is the Gödel number of an alphabet symbol a_u:

$$(\exists u)_{u<x}(x = 7 + 4u)$$

Quad(x): x is the Gödel number of a Turing machine quadruple:

$$\ell h(x) = 4 \wedge \mathrm{IS}((x)_0) \wedge \mathrm{Sym}((x)_1) \wedge \mathrm{IS}((x)_3) \wedge$$
$$[\mathrm{Sym}((x)_2) \vee (x)_2 = 3 \vee (x)_2 = 5]$$

TM(x): x is the Gödel number of a Turing machine (in the form of a finite sequence of appropriate quadruples):

$$(\forall u)_{u<\ell h(x)}\mathrm{Quad}((x)_u) \wedge x > 1 \wedge (\forall u)_{u<\ell h(x)}(\forall v)_{v<\ell h(x)}(u \neq v$$
$$\Rightarrow [((x)_u)_0 \neq ((x)_v)_0 \vee ((x)_u)_1 \neq ((x)_v)_1]$$

TD(x): x is the Gödel number of a tape description:

$$x > 1 \wedge (\forall u)_{u<\ell h(x)}[\mathrm{IS}((x)_u) \vee \mathrm{Sym}((x)_u)] \wedge (\exists_1 u)_{u<\ell h(x)}\mathrm{IS}((x)_u) \wedge$$
$$(\forall u)_{u<\ell h(x)}(\mathrm{IS}((x)_u) \Rightarrow u + 1 < \ell h(x))$$

Cons(x, y, z): x and y are Gödel numbers of tape descriptions α and β, and z is the Gödel number of a Turing machine quadruple that transforms α into β:

$$\mathrm{TD}(x) \wedge \mathrm{TD}(y) \wedge \mathrm{Quad}(z) \wedge (\exists w)_{w<\ell h(x)\dot{-}1}[\mathrm{IS}((x)_w) \wedge$$
$$(x)_w = (z)_0 \wedge (x)_{w+1} = (z)_1 \wedge$$
$$\mathrm{I}\begin{cases}([\mathrm{Sym}((z)_2) \wedge (y)_{w+1} = (z)_2 \wedge (y)_w = (z)_3 \wedge \ell h(x) = \ell h(y) \\ \wedge (\forall u)_{u<\ell h(x)}(u \neq w \wedge u \neq w + 1 \Rightarrow (x)_u = (y)_u)] \vee\end{cases}$$
$$\mathrm{II}\begin{cases}[(z)_2 = 3 \wedge (y)_w = (x)_{w+1} \wedge (y)_{w+1} = (z)_3 \wedge \\ (\forall u)_{u<\ell h(x)}(u \neq w \wedge u \neq w + 1 \Rightarrow (y)_u = (x)_u) \wedge \\ ([w + 2 < \ell h(x) \wedge \ell h(y) = \ell h(x)] \vee [w + 2 = \ell h(x) \wedge \\ \ell h(y) = \ell h(x) + 1 \wedge (y)_{w+2} = 7])] \vee\end{cases}$$
$$\mathrm{III}\begin{cases}[(z)_2 = 5 \wedge \{[w \neq 0 \wedge (y)_{w\dot{-}1} = (z)_3 \wedge (y)_w = (x)_{w\dot{-}1} \\ \wedge \ell h(y) = \ell h(x) \wedge (\forall u)_{u<\ell h(x)}(u \neq w \dot{-} 1 \wedge u \neq w \Rightarrow \\ (y)_u = (x)_u)] \vee [w = 0 \wedge (y)_0 = (z)_3 \wedge (y)_1 = 7 \wedge \\ \ell h(y) = \ell h(x) + 1 \wedge (\forall u)_{0<u<\ell h(x)}(y)_{u+1} = (x)_u]\}])]\end{cases}$$

I corresponds to a quadruple $q_j a_i a_k q_r$, II to a quadruple $q_j a_i R q_r$, and III to a quadruple $q_j a_i L q_r$.

NTD(x): x is the Gödel number of a numerical tape description—that is, a tape description in which the tape has the form $E_1 \overline{k} E_2$, where each of E_1

and E_2 is empty or consists entirely of blanks, and the location of the reading head is arbitrary:

$$\mathrm{TD}(x) \wedge (\forall u)_{u<\ell h(x)}(\mathrm{Sym}((x)_u) \Rightarrow (x)_u = 7 \vee (x)_u = 11) \wedge$$
$$(\forall u)_{u<\ell h(x)}(\forall v)_{v<\ell h(x)}(\forall w)_{w<\ell h(x)}(u < v \wedge v < w \wedge (x)_u = 11 \wedge (x)_w = 11$$
$$\Rightarrow (x)_v \neq 7)$$

Stop(x, z): z is the Gödel number of a Turing machine $\mathcal{T}$ and x is the Gödel number of a tape description α such that $\mathcal{T}$ stops at α:

$$\mathrm{TM}(z) \wedge \mathrm{TD}(x) \wedge \neg(\exists u)_{u<\ell h(x)}[\mathrm{IS}((x)_u) \wedge (\exists v)_{v<\ell h(z)}(((z)_v)_0 = (x)_u \wedge$$
$$((z)_v)_1 = (x)_{u+1})]$$

Comp(y, z): z is the Gödel number of a Turing machine $\mathcal{T}$ and y is the Gödel number of a computation of $\mathcal{T}$:

$$y > 1 \wedge \mathrm{TM}(z) \wedge (\forall u)_{u<\ell h(y)}\mathrm{TD}((y)_u) \wedge \mathrm{Stop}((y)_{\ell h(y)\dot{-}1}, z) \wedge$$
$$(\forall u)_{u<\ell h(y)\dot{-}1}(\exists w)_{w<\ell h(z)}\mathrm{Cons}((y)_u, (y)_{u+1}, (z)_w) \wedge$$
$$(\forall v)_{v<\ell h((y)_0)}(\mathrm{IS}(((y)_0)_v) \Rightarrow ((y)_0)_v = 9)$$

Num(x): The Gödel number of the word $\bar{x}$—that is, of $|^{x+1}$:

$$\mathrm{Num}(x) = \prod_{u \leqslant x} p_u^{11}$$

TR$(x_1, \ldots, x_n)$: The Gödel number of the tape representation $\overline{(x_1, \ldots, x_n)}$ of the n-tuple $(x_1, \ldots, x_n)$:

$$\mathrm{TR}(x_1, \ldots, x_n) = \mathrm{Num}(x_1) * 2^7 * \mathrm{Num}(x_2) * 2^7 * \cdots * 2^7 * \mathrm{Num}(x_n)$$

$U(y)$: If y is the Gödel number of a computation that results in a numerical tape description, then $U(y)$ is the number represented on that final tape:†

$$U(y) = \left[\sum_{u<\ell h((y)_{\ell h(y)\dot{-}1})} \overline{sg}(|((y)_{\ell h(y)\dot{-}1})_u - 11|) \right] \dot{-} 1$$

[Let w be the number, represented by $|^{w+1}$, on the final tape. The calculation of $U(y)$ tallies a 1 for every stroke | that appears on the final tape. This yields a sum of $w + 1$, and then 1 is subtracted at the end.]

$T_n(z, x_1, \ldots, x_n, y)$: y is the Gödel number of a computation of a Turing machine with Gödel number z such that the computation begins on the tape $\overline{(x_1, x_2, \ldots, x_n)}$, with the reading head scanning the first | in $\bar{x}_1$, and ends with a numerical tape description:

$$\mathrm{Comp}(y, z) \wedge (y)_0 = 2^9 * \mathrm{TR}(x_1, \ldots, x_n) \wedge \mathrm{NTD}((y)_{\ell h(y)\dot{-}1})^{\ddagger}$$

[Observe that, if $T_n(z, x_1, \ldots, x_n, y_1)$ and $T_n(z, x_1, \ldots, x_n, y_2)$, then $y_1 = y_2$,

† If y is not the Gödel number of a computation that yields a numerical tape description, $U(y)$ is defined, but its value in such cases will be of no significance.

‡ When $n = 1$, replace TR$(x_1, \ldots, x_n)$ by Num(x_1).

since there is at most one computation of a Turing machine starting with a given initial tape.]

PROPOSITION 5.5 If $\mathcal{T}$ is a Turing machine that computes a number-theoretic function $f(x_1, \ldots, x_n)$ and e is a Gödel number of $\mathcal{T}$, then

$$f(x_1, \ldots, x_n) = U(\mu y\, T_n(e, x_1, \ldots, x_n, y))^*$$

Proof Let $k_1, \ldots, k_n$ be any natural numbers. Then $f(k_1, \ldots, k_n)$ is defined if and only if there is a computation of $\mathcal{T}$ beginning with $\overline{(k_1, \ldots, k_n)}$—that is, if and only if $(\exists y) T_n(e, k_1, \ldots, k_n, y)$. So, $f(k_1, \ldots, k_n)$ is defined if and only if $\mu y\, T_n(e, k_1, \ldots, k_n, y)$ is defined. Moreover, when $f(k_1, \ldots, k_n)$ is defined, $\mu y\, T_n(e, k_1, \ldots, k_n, y)$ is the Gödel number of a computation of $\mathcal{T}$ beginning with $\overline{(k_1, \ldots, k_n)}$ and $U(\mu y\, T_n(e, k_1, \ldots, k_n, y))$ is the value yielded by the computation—namely, $f(k_1, \ldots, k_n)$.

COROLLARY 5.6 Every Turing-computable function is partial recursive.

Proof Assume $f(x_1, \ldots, x_n)$ is Turing-computable by a Turing machine with Gödel number e. Then, $f(x_1, \ldots, x_n) = U(\mu y\, T_n(e, x_1, \ldots, x_n, y))$. Since T_n is primitive recursive, $\mu y\, T_n(e, x_1, \ldots, x_n, y)$ is partial recursive. Hence, $U(\mu y\, T_n(e, x_1, \ldots, x_n, y))$ is partial recursive.

COROLLARY 5.7 The Turing-computable functions are identical with the partial recursive functions.

Proof The proof is by Corollaries 5.6 and 5.3.

COROLLARY 5.8 Every total partial recursive function is recursive.

Proof Assume that the total partial recursive function $f(x_1, \ldots, x_n)$ is Turing-computable by the Turing machine with Gödel number e. Then, for all $x_1, \ldots, x_n$, $(\exists y) T_n(e, x_1, \ldots, x_n, y)$. Hence, $\mu y\, T_n(e, x_1, \ldots, x_n, y)$ is produced by an application of the restricted μ-operator and is, therefore, recursive. Hence, $U(\mu y\, T_n(e, x_1, \ldots, x_n, y))$ also is recursive. Now use Proposition 5.5.

COROLLARY 5.9 For any total number-theoretic function f, f is recursive if and only if f is Turing-computable.

Proof Use Corollaries 5.7–5.8.

Church's Thesis amounts to the assertion that the recursive functions are the same as the effectively computable total functions. By Corollary 5.9, this is

* Remember that an equality between two partial functions means that, whenever one of them is defined, the other is also and the two functions have the same value.

equivalent to the identity, for total functions, of effective computability and Turing-computability. This strengthens the case for Church's Thesis because of the plausibility of the identification of Turing-computability with effective computability. In fact, let us now widen Church's Thesis to assert that the effectively computable functions (partial or total) are the same as the Turing-computable functions. By Corollary 5.7, this also claims that a function is effectively computable if and only if it is partial recursive.

COROLLARY 5.10 Any number-theoretic function is Turing-computable if and only if it is standard Turing-computable.

Proof Use Proposition 5.1, Corollary 5.6, and Proposition 5.2.

COROLLARY 5.11 (KLEENE'S NORMAL FORM THEOREM) As z varies over all natural numbers, $U(\mu y\, T_n(z, x_1, \ldots, x_n, y))$ enumerates with repetitions all partial recursive functions of n variables.

Proof Use Corollary 5.3 and Proposition 5.5. The fact that every partial recursive function reappears for infinitely many z follows from Exercise 5.7 [Notice that, when z is not the Gödel number of a Turing machine, there is no y such that $T_n(z, x_1, \ldots, x_n, y)$, and, therefore, the corresponding partial recursive function is the empty function.*]

COROLLARY 5.12

(a) The partial recursive function $\mu y\, T_n(x_1, x_1, x_2, \ldots, x_n, y)$ is not recursive.
(b) The partial recursive function $\mu y\, T_n(z, x_1, x_2, \ldots, x_n, y)$ is not recursive.

Proof

(a) Assume $\mu y\, T_n(x_1, x_1, x_2, \ldots, x_n, y)$ is recursive. Then the function $f(x_1, x_2, \ldots, x_n) = U(\mu y\, T_n(x_1, x_1, x_2, \ldots, x_n, y)) + 1$ is recursive. By Proposition 5.5, there exists e such that $f(x_1, x_2, \ldots, x_n) = U(\mu y\, T_n(e, x_1, x_2, \ldots, x_n, y))$. Hence, $U(\mu y\, T_n(e, x_1, x_2, \ldots, x_n, y)) = U(\mu y\, T_n(x_1, x_1, x_2, \ldots, x_n, y)) + 1$. Let $x_1 = x_2 = \cdots = x_n = e$. Then, we obtain the contradiction

$$U(\mu y\, T_n(e, e, e, \ldots, e, y)) = U(\mu y\, T_n(e, e, e, \ldots, e, y)) + 1$$

(b) If $\mu y\, T_n(z, x_1, x_2, \ldots, x_n, y)$ were recursive, then, by substitution, $\mu y\, T_n(x_1, x_1, x_2, \ldots, x_n, y)$ would be recursive, contradicting part (a).†

COROLLARY 5.13 For any recursive relation $R(x_1, \ldots, x_n, y)$, there exist natural numbers z_0 and v_0 such that, for all natural numbers $x_1, \ldots, x_n$:

*The empty function is the empty set 0. It has the empty set as its domain.

†Easier proofs of parts (a) and (b) follow from the fact that recursive functions are total functions. If z_0 is not the Gödel number of a Turing machine, then $\mu y\, T_n(x_1, x_1, x_2, \ldots, x_n, y)$ is not defined when $x_1 = z_0$, and $\mu y\, T_n(z, x_1, x_2, \ldots, x_n, y)$ is not defined when $z = z_0$.

(a) $(\exists y)R(x_1, \ldots, x_n, y)$ if and only if $(\exists y)T_n(z_0, x_1, \ldots, x_n, y)$.
(b) $(\forall y)R(x_1, \ldots, x_n, y)$ if and only if $(\forall y)\neg T_n(v_0, x_1, \ldots, x_n, y)$.

Proof

(a) The function $f(x_1, \ldots, x_n) = \mu y\, R(x_1, \ldots, x_n, y)$ is partial recursive. Let z_0 be a Gödel number of a Turing machine that computes f. Hence, $f(x_1, \ldots, x_n)$ is defined if and only if $(\exists y)T_n(z_0, x_1, \ldots, x_n, y)$. But, $f(x_1, \ldots, x_n)$ is defined if and only if $(\exists y)R(x_1, \ldots, x_n, y)$.
(b) Applying part (a) to the recursive relation $\neg R(x_1, \ldots, x_n, y)$, we obtain a number v_0 such that:

$$(\exists y)\neg R(x_1, \ldots, x_n, y) \text{ if and only if } (\exists y)T_n(v_0, x_1, \ldots, x_n, y)$$

Now take the negations of both sides of this equivalence.

Exercise

5.19 Extend Corollary 5.13 to two or more quantifiers. For example, if $R(x_1, \ldots, x_n, y, z)$ is a recursive relation, show that there are natural numbers z_0 and v_0 such that, for all $x_1, \ldots, x_n$:
(a) $(\forall z)(\exists y)R(x_1, \ldots, x_n, y, z)$ if and only if $(\forall z)(\exists y)T_{n+1}(z_0, x_1, \ldots, x_n, y, z)$.
(b) $(\exists z)(\forall y)R(x_1, \ldots, x_n, y, z)$ if and only if $(\exists z)(\forall y)\neg T_{n+1}(v_0, x_1, \ldots, x_n, y, z)$.

COROLLARY 5.14
(a) $(\exists y)T_n(x_1, x_1, x_2, \ldots, x_n, y)$ is not recursive.
(b) $(\exists y)T_n(z, x_1, \ldots, x_n, y)$ is not recursive.

Proof

(a) Assume $(\exists y)T_n(x_1, x_1, x_2, \ldots, x_n, y)$ is recursive. Then the relation $\neg(\exists y)\, T_n(x_1, x_1, x_2, \ldots, x_n, y) \wedge z = z$ is recursive. So, by Corollary 5.13(a), there exists z_0 such that:

$$(\exists z)(\neg(\exists y)T_n(x_1, x_1, x_2, \ldots, x_n, y) \wedge z = z) \text{ if and only if}$$
$$(\exists z)T_n(z_0, x_1, x_2, \ldots, x_n, z)$$

Hence, since z obviously can be omitted on the left,

$$\neg(\exists y)T_n(x_1, x_1, x_2, \ldots, x_n, y) \text{ if and only if } (\exists z)T_n(z_0, x_1, x_2, \ldots, x_n, z)$$

Let $x_1 = x_2 = \cdots = x_n = z_0$. Then we obtain the contradiction

$$\neg(\exists y)T_n(z_0, z_0, z_0, \ldots, z_0, y) \text{ if and only if } (\exists z)T_n(z_0, z_0, z_0, \ldots, z_0, z)$$

(b) If $(\exists y)T_n(z, x_1, x_2, \ldots, x_n, y)$ were recursive, so would be, by substitution, $(\exists y)T_n(x_1, x_1, x_2, \ldots, x_n, y)$, contradicting part (a).

Exercises

5.20 Prove that there is a partial recursive function $g(z, x)$ such that, for any partial recursive function $f(x)$, there is a number z_0 for which $f(x) = g(z_0, x)$ holds for all x. Then show that there must exist a number v_0 such that $g(v_0, v_0)$ is not defined.

5.21 Let $h_1(x_1,\ldots,x_n)$ and $h_2(x_1,\ldots,x_n)$ be partial recursive functions and let $R(x_1,\ldots,x_n)$ be a recursive relation. Define

$$g(x_1,\ldots,x_n) = \begin{cases} h_1(x_1,\ldots,x_n) & \text{if } R(x_1,\ldots,x_n) \\ h_2(x_1,\ldots,x_n) & \text{if } \neg R(x_1,\ldots,x_n) \end{cases}$$

Prove that g is partial recursive.

5.22 A partial function $f(x)$ is said to be *recursively completable* if there is a recursive function $h(x)$ such that, for every x in the domain of f, $h(x) = f(x)$.

(a) Prove that $\mu y\, T_1(x, x, y)$ is not recursively completable.

(b) Prove that a partial recursive function $f(x)$ is recursively completable if the domain D of f is a recursive set—that is, if the property "$x \in D$" is recursive.

(c) Find a partial recursive function $f(x)$ that is recursively completable but whose domain is not recursive.

5.23 If $R(x, y)$ is a recursive relation, prove that there are natural numbers z_0 and v_0 such that:

(a) $\neg[(\exists y)R(z_0, y) \Leftrightarrow (\forall y)\neg T_1(z_0, z_0, y)]$

(b) $\neg[(\forall y)R(v_0, y) \Leftrightarrow (\exists y)T_1(v_0, v_0, y)]$

5.24 If $S(x)$ is a recursive property, show that there are natural numbers z_0 and v_0 such that:

(a) $\neg[S(z_0) \Leftrightarrow (\forall y)\neg T_1(z_0, z_0, y)]$

(b) $\neg[S(v_0) \Leftrightarrow (\exists y)T_1(v_0, v_0, y)]$

5.25 Show that there is no recursive function $B(z, x_1,\ldots,x_n)$ such that, if z is a Gödel number of a Turing machine $\mathcal{T}$ and $k_1, \ldots, k_n$ are natural numbers for which $f_{\mathcal{T},n}(k_1,\ldots,k_n)$ is defined, then the number of steps in the computation of $f_{\mathcal{T},n}(k_1,\ldots,k_n)$ is less than $B(z, k_1,\ldots,k_n)$.

Let $\mathcal{T}$ be a Turing machine. The *halting problem* for $\mathcal{T}$ is the problem of determining, for each tape description β, whether $\mathcal{T}$ is *applicable* to β—that is, whether there is a computation of $\mathcal{T}$ that begins with β.

We say that the halting problem for $\mathcal{T}$ is *algorithmically solvable* if there is an algorithm that, given a tape description β, determines whether $\mathcal{T}$ is applicable to β. Instead of a tape description β, we may assume that the algorithm is given the Gödel number of β. Then the desired algorithm will be an effectively computable function $H_{\mathcal{T}}$ such that:

$$H_{\mathcal{T}}(x) = \begin{cases} 0 & \text{if } x \text{ is the Gödel number of a tape description } \beta \\ & \text{to which } \mathcal{T} \text{ is applicable} \\ 1 & \text{otherwise} \end{cases}$$

If we accept Turing algorithms as exact counterparts of algorithms, then the halting problem for $\mathcal{T}$ is algorithmically solvable if and only if the function $H_{\mathcal{T}}$ is Turing-computable or, equivalently, by Corollary 5.9, recursive. When the function $H_{\mathcal{T}}$ is recursive, we say that the halting problem for $\mathcal{T}$ is *recursively solvable*. If $H_{\mathcal{T}}$ is not recursive, we say that the halting problem for $\mathcal{T}$ is *recursively unsolvable*.

PROPOSITION 5.15 There is a Turing machine with recursively unsolvable halting problem.

Proof By Proposition 5.2, let $\mathscr{T}$ be a Turing machine that ST-computes the partial recursive function $\mu y\, T_1(x, x, y)$. Remember that, by the definition of standard Turing-computability, if $\mathscr{T}$ is begun on the tape consisting of only $\overline{x}$ with its reading head scanning the leftmost |, then $\mathscr{T}$ stops if and only if $\mu y\, T_1(x, x, y)$ is defined. Assume that $\mathscr{T}$ has a recursively solvable halting problem—that is, that the function $H_{\mathscr{T}}$ is recursive. Recall that the Gödel number of the tape description $q_0\overline{x}$ is $2^9 * \text{Num}(x)$. Now,

$$\begin{aligned}(\exists y)T_1(x, x, y) \quad & \text{if and only if } \mu y\, T_1(x, x, y) \text{ is defined}\\ & \text{if and only if } \mathscr{T}\text{, begun on } q_0\overline{x}\text{, performs a computation}\\ & \text{if and only if } H_{\mathscr{T}}(2^9 * \text{Num}(x)) = 0\end{aligned}$$

Since $H_{\mathscr{T}}$, Num, and $*$ are recursive, $(\exists y)T_1(x, x, y)$ is recursive, contradicting Corollary 5.14(a) (when $n = 1$).

Exercises

5.26 Give an example of a Turing machine with a recursively solvable halting problem.

5.27 Show that the following *special halting problem* is recursively unsolvable: given a Gödel number of a Turing machine $\mathscr{T}$ and a natural number x, determine whether $\mathscr{T}$ is applicable to $q_0\overline{x}$.

5.28 Show that the following *self-halting problem* is recursively unsolvable: given a Gödel number z of a Turing machine $\mathscr{T}$, determine whether $\mathscr{T}$ is applicable to $q_0\overline{z}$.

5.29 The *printing problem* for a Turing machine $\mathscr{T}$ and a symbol a_k is the problem of determining, for any given tape description α, whether $\mathscr{T}$, if begun on α, ever prints the symbol a_k. Find a Turing machine $\mathscr{T}$ and a symbol a_k for which the printing problem is recursively unsolvable.

5.30 Show that the following decision problem is recursively unsolvable: given any Turing machine $\mathscr{T}$, if $\mathscr{T}$ is begun on an empty tape, determine whether $\mathscr{T}$ stops (that is, determine whether $\mathscr{T}$ is applicable to q_0B).

5.31^{D} Show that the problem of deciding, for any given Turing machine, whether it has a recursively unsolvable halting problem is itself recursively unsolvable.

To deal with more intricate decision problems and other aspects of the theory of computability, we need more powerful tools. First of all, let us introduce the notation

$$\phi_z^n(x_1, \ldots, x_n) = U(\mu y\, T_n(z, x_1, \ldots, x_n, y))$$

Thus, by Corollary 5.11, ϕ_0^n, ϕ_1^n, ϕ_2^n, ... is an enumeration of all partial recursive functions of n variables. The subscript j is called an *index* of the function ϕ_j^n. Each partial recursive function of n variables has infinitely many indices.

PROPOSITION 5.16 (ITERATION THEOREM OR *s-m-n* THEOREM) For any positive integers m and n, there is a primitive recursive function $s_n^m(z, y_1, \ldots, y_m)$ such that

$$\phi_z^{m+n}(x_1,\ldots,x_n,y_1,\ldots,y_m) = \phi_{s_n^m(z,y_1,\ldots,y_m)}(x_1,\ldots,x_n)$$

Thus, not only does assigning particular values to z, $y_1, \ldots, y_m$ in $\phi_z^{m+n}(x_1,\ldots,x_n,y_1,\ldots,y_m)$ yield a new partial recursive function of n variables, but also the index of the resulting function is a primitive recursive function of the old index z and of $y_1, \ldots, y_m$.

Proof If $\mathscr{T}$ is a Turing machine with Gödel number z, let $\mathscr{T}_{y_1,\ldots,y_m}$ be a Turing machine that, when begun on $\overline{(x_1,\ldots,x_n)}$, produces $\overline{(x_1,\ldots,x_n,y_1,\ldots,y_m)}$, moves back to the leftmost | of $\bar{x}_1$, and then behaves like $\mathscr{T}$. Such a machine is defined by the diagram

$$\mathscr{R}(\mathbf{ra_1})^{y_1+1}\mathbf{ra_0}(\mathbf{ra_1})^{y_2+1}\mathbf{ra_0}\ldots\mathbf{ra_0}(\mathbf{ra_1})^{y_m+1}\mathbf{r}\mathscr{L}\mathbf{r}\mathscr{T}$$

The index $s_n^m(z,y_1,\ldots,y_m)$ of this Turing machine can be effectively computed and, by Church's Thesis, would be partial recursive. In fact, after an exhausting analysis, s_n^m actually can be computed by a primitive recursive function $g(z,y_1,\ldots,y_m)$, defined in the following manner.

Let $t = y_1 + \cdots + y_m + m + 3$. Also, let $u(i) = 2^{9+4i}3^7 5^{11} 7^{9+4i}$ and $v(i) = 2^{9+4i}3^{11}5^3 7^{13+4i}$. Then take $g(z,y_1,\ldots,y_m)$ to be:

$$[2^{2^9 3^{11} 5^3 7^9} 3^{2^9 3^7 5^3 7^{13}} 5^{2^{13} 3^{11} 5^3 7^9} 7^{2^{13} 3^7 5^7 7^{17}}] * \left[\prod_{i=2}^{y_1+2} p_{|i-2|}^{u(i)} p_{|i-1|}^{v(i)}\right] *$$

$$\left[\prod_{i=y_1+3}^{y_1+y_2+3} p_{|i-(y_1+3)|}^{u(i)} p_{|i-(y_1+2)|}^{v(i)}\right] * \cdots *$$

$$\left[\prod_{i=y_1+\cdots+y_{m-1}+m+1}^{y_1+\cdots+y_m+m+2} p_{|i-(y_1+\cdots+y_m+m+1)|}^{u(i)} p_{|i-(y_1+\cdots+y_m+m)|}^{v(i)}\right] *$$

$$[2^{2^{9+4t}3^7 5^5 7^{13+4t}} 3^{2^{13+4t}3^{11}5^5 7^{13+4t}} 5^{2^{13+4t}3^7 5^5 7^{17+4t}} 7^{2^{17+4t}3^{11}5^5 7^{13+4t}}$$

$$11^{2^{17+4t}3^7 5^3 7^{21+4t}}] * \prod_{i=0}^{\ell h(z) \dot{-} 1} p_i^{2^{((z)_i)_0+4(t+3)} 3^{((z)_i)_1} 5^{((z)_i)_2} 7^{((z)_i)_3+4(t+3)}}$$

When z is not a Gödel number of a Turing machine, ϕ_z^{m+n} is the empty function, and, therefore, $s_n^m(z,y_1,\ldots,y_m)$ must be an index of the empty function and can be taken to be 0. Thus, we define:

$$s_n^m(z,y_1,\ldots,y_m) = \begin{cases} g(z,y_1,\ldots,y_m) & \text{if TM}(z) \\ 0 & \text{otherwise} \end{cases}$$

Hence, s_n^m is primitive recursive.

COROLLARY 5.17 For any partial recursive function $f(x_1,\ldots,x_n,y_1,\ldots,y_m)$, there is a recursive function $g(y_1,\ldots,y_m)$ such that

$$f(x_1,\ldots,x_n,y_1,\ldots,y_m) = \phi_{g(y_1,\ldots,y_m)}^n(x_1,\ldots,x_n)$$

Proof Let e be an index of f. By Proposition 5.16,

$$\phi_e^{m+n}(x_1, \ldots, x_n, y_1, \ldots, y_m) = \phi_{s_n^m(e, y_1, \ldots, y_m)}^n(x_1, \ldots, x_n)$$

Let $g(y_1, \ldots, y_m) = s_n^m(e, y_1, \ldots, y_m)$.

Examples

1. Let $G(x)$ be a fixed partial recursive function with nonempty domain. Consider the following decision problem: for any u, determine whether $\phi_u^1 = G$. Let us show that this problem is recursively unsolvable—that is, that the property $R(u)$, defined by $\phi_u^1 = G$, is not recursive. Assume, for the sake of contradiction, that R is recursive. Consider the function $f(x, u) = G(x) \cdot N(Z(\mu y\, T_1(u, u, y)))$. [Remember that $N(Z(t)) = 1$ for all t.] Applying Corollary 5.17 to $f(x, u)$, we obtain a recursive function $g(u)$ such that $f(x, u) = \phi_{g(u)}^1(x)$. For any fixed u, $\phi_{g(u)}^1 = G$ if and only if $(\exists y) T_1(u, u, y)$. (Here, we use the fact that G has nonempty domain.) Hence, $(\exists y) T_1(u, u, y)$ if and only if $R(g(u))$. Since $R(g(u))$ is recursive, $(\exists y) T_1(u, u, y)$ would be recursive, contradicting Corollary 5.14(a).

2. *A universal Turing machine.* Let the partial recursive function $U(\mu y\, T_1(z, x, y))$ be computed by a Turing machine $\mathscr{V}$ with Gödel number e. Thus, $U(\mu y\, T_1(z, x, y)) = U(\mu y\, T_2(e, z, x, y))$. $\mathscr{V}$ is *universal* in the following sense. First, it can compute *every* partial recursive function $f(x)$ of one variable. If z is a Gödel number of a Turing machine that computes f, then, if $\mathscr{V}$ begins on the tape $\overline{(z, x)}$, it will compute $U(\mu y\, T_1(z, x, y)) = f(x)$. Further, $\mathscr{V}$ can be used to compute any partial recursive function $h(x_1, \ldots, x_n)$. Let v_0 be a Gödel number of a Turing machine that computes h, and let $f(x) = h((x)_0, (x)_1, \ldots, (x)_{n-1})$. Then $h(x_1, \ldots, x_n) = f(p_0^{x_1} \cdots p_{n-1}^{x_n})$. By applying Corollary 5.17 to the partial recursive function $U(\mu y\, T_n(v, (x)_0, \ldots, (x)_{n-1}, y))$, we obtain a recursive function $g(v)$ such that $U(\mu y\, T_n(v, (x)_0, \ldots, (x)_{n-1}, y)) = \phi_{g(v)}^1(x)$. Hence, $f(x) = \phi_{g(v_0)}^1(x)$. So, $h(x_1, \ldots, x_n)$ is computed by applying $\mathscr{V}$ to the tape $\overline{(g(v_0), p_0^{x_1} \cdots p_{n-1}^{x_n})}$.

Exercises

5.32 Find a *superuniversal* Turing machine $\mathscr{V}^*$ such that, for any Turing machine $\mathscr{T}$, if z is a Gödel number of $\mathscr{T}$ and x is the Gödel number of an initial tape description α of $\mathscr{T}$, then $\mathscr{V}^*$ is applicable to $q_0\overline{(z, x)}$ if and only if $\mathscr{T}$ is applicable to α; moreover, if $\mathscr{T}$, when applied to α, ends with a tape description that has Gödel number w, then $\mathscr{V}^*$, when applied to $q_0\overline{(z, x)}$, produces $\overline{w}$.

5.33 Show that the following decision problem is recursively unsolvable: for any u and v, determine whether $\phi_u^1 = \phi_v^1$.

5.34 Show that the following decision problem is recursively unsolvable: for any u, determine whether ϕ_u^1 has empty domain. [Hence, the condition in Example 1 above, that $G(x)$ has nonempty domain, is unnecessary.]

5.35 (a) Prove that there is a recursive function $g(u, v)$ such that

$$\phi_{g(u,v)}^1(x) = \phi_u^1(x) \cdot \phi_v^1(x)$$

(b) Prove that there is a recursive function $C(u, v)$ such that

$$\phi_{C(u,v)}^1(x) = \phi_u^1(\phi_v^1(x))$$

4. THE KLEENE–MOSTOWSKI HIERARCHY. RECURSIVELY ENUMERABLE SETS

Consider the following array, where $R(x_1,\ldots,x_n,y_1,\ldots,y_m)$ expresses a recursive relation.

$$R(x_1,\ldots,x_n)$$

$$\begin{array}{ll}
(\exists y_1)R(x_1,\ldots,x_n,y_1) & (\forall y_1)R(x_1,\ldots,x_n,y_1) \\
(\exists y_1)(\forall y_2)R(x_1,\ldots,x_n,y_1,y_2) & (\forall y_1)(\exists y_2)R(x_1,\ldots,x_n,y_1,y_2) \\
(\exists y_1)(\forall y_2)(\exists y_3)R(x_1,\ldots,x_n,y_1,y_2,y_3) & (\forall y_1)(\exists y_2)(\forall y_3)R(x_1,\ldots,x_n,y_1,y_2,y_3) \\
\vdots & \vdots
\end{array}$$

Let $\Sigma_0^n = \Pi_0^n =$ the set of all n-place recursive relations. For $k > 0$, let Σ_k^n be the set of all n-place relations expressible in the prenex form $(\exists y_1)(\forall y_2)\ldots(Qy_k)R(x_1,\ldots,x_n,y_1,\ldots,y_k)$, consisting of k alternating quantifiers beginning with an existential quantifier and followed by a recursive relation R. [Here, "(Qy_k)" denotes an existential or universal quantifier, depending on whether k is odd or even.] Let Π_k^n be the set of all n-place relations expressible in the prenex form $(\forall y_1)(\exists y_2)\ldots(Qy_k)R(x_1,\ldots,x_n,y_1,\ldots,y_k)$, consisting of k alternating quantifiers beginning with a universal quantifier and followed by a recursive relation R. Then the array above can be written

$$\begin{array}{cc}
\multicolumn{2}{c}{\Sigma_0^n} \\
\Sigma_1^n & \Pi_1^n \\
\Sigma_2^n & \Pi_2^n \\
\Sigma_3^n & \Pi_3^n \\
\vdots & \vdots
\end{array}$$

This array of classes of relations is called the *Kleene-Mostowski hierarchy*, or the *arithmetical hierarchy*.

PROPOSITION 5.18

(a) Every relation expressible in any form listed above is expressible in all the forms in lower rows; that is, for all $j > k$,

$$\Sigma_k^n \subseteq \Sigma_j^n \cap \Pi_j^n \quad \text{and} \quad \Pi_k^n \subseteq \Sigma_j^n \cap \Pi_j^n$$

(b) There is a relation of each form, except Σ_0^n, that is not expressible in the other form in the same row and, hence, by part (a), not in any of the rows above; that is, for $k > 0$,

$$\Sigma_k^n - \Pi_k^n \neq 0 \quad \text{and} \quad \Pi_k^n - \Sigma_k^n \neq 0$$

(c) Every arithmetical relation is expressible in at least one of these forms.

(d) (Post) For any relation $Q(x_1,\ldots,x_n)$, Q is recursive if and only if both Q

and $\neg Q$ are expressible in the form $(\exists y_1)R(x_1, \ldots, x_n, y_1)$, where R is recursive; that is, $\Sigma_1^n \cap \Pi_1^n = \Sigma_0^n$.

(e) If $Q_1 \in \Sigma_k^n$ and $Q_2 \in \Sigma_k^n$, then $Q_1 \vee Q_2$ and $Q_1 \wedge Q_2$ are in Σ_k^n. If Q_1 and Q_2 are in Π_k^n, then $Q_1 \vee Q_2$ and $Q_1 \wedge Q_2$ are in Π_k^n.

(f) In contradistinction to part (d), if $k > 0$, then

$$(\Sigma_{k+1}^n \cap \Pi_{k+1}^n) - (\Sigma_k^n \cup \Pi_k^n) \neq 0$$

Proof

(a)
$$(\exists z_1)(\forall y_1) \ldots (\exists z_k)(\forall y_k)R(x_1, \ldots, x_n, z_1, y_1, \ldots, z_k, y_k) \Leftrightarrow$$
$$(\forall u)(\exists z_1)(\forall y_1) \ldots (\exists z_k)(\forall y_k)(R(x_1, \ldots, x_n, z_1, y_1, \ldots, z_k, y_k) \wedge u = u) \Leftrightarrow$$
$$(\exists z_1)(\forall y_1) \ldots (\exists z_k)(\forall y_k)(\exists u)(R(x_1, \ldots, x_n, z_1, y_1, \ldots, z_k, y_k) \wedge u = u)$$

Hence, any relation expressible in one of the forms in the array is expressible in both forms in any lower row.

(b) Let us take a typical case. Consider $(\exists v)(\forall z)(\exists y)T_{n+2}(x_1, x_1, x_2, \ldots, x_n, v, z, y)$. Assume that this is expressible in the form $(\forall v)(\exists z)(\forall y)R(x_1, \ldots, x_n, v, z, y)$, where R is recursive. By Exercise 5.19, this relation is equivalent to $(\forall v)(\exists z)(\forall y) \neg T_{n+2}(e, x_1, \ldots, x_n, v, z, y)$ for some e. When $x_1 = e$, this yields a contradiction.

(c) Every wf of the first-order theory S can be put into prenex normal form. Then, it suffices to note that $(\exists u)(\exists v)R(u, v)$ is equivalent to $(\exists z)R((z)_0, (z)_1)$, and $(\forall u)(\forall v)R(u, v)$ is equivalent to $(\forall z)R((z)_0, (z)_1)$. Hence, successive quantifiers of the same kind can be condensed into one such quantifier.

(d) If Q is recursive, so is $\neg Q$, and, if $P(x_1, \ldots, x_n)$ is recursive, then $P(x_1, \ldots, x_n) \Leftrightarrow (\exists y)(P(x_1, \ldots, x_n) \wedge y = y)$. Conversely, assume Q is expressible as $(\exists y)R_1(x_1, \ldots, x_n, y)$ and $\neg Q$ as $(\exists y)R_2(x_1, \ldots, x_n, y)$, where the relations R_1 and R_2 are recursive. Hence, $(\forall x_1) \ldots (\forall x_n)(\exists y)(R_1(x_1, \ldots, x_n, y) \vee R_2(x_1, \ldots, x_n, y))$. So, $\psi(x_1, \ldots, x_n) = \mu y(R_1(x_1, \ldots, x_n, y) \vee R_2(x_1, \ldots, x_n, y)$ is recursive. Then, $Q(x_1, \ldots, x_n) \Leftrightarrow R_1(x_1, \ldots, x_n, \psi(x_1, \ldots, x_n))$ and, therefore, Q is recursive.

(e) Use the following facts. If x is not free in $\mathscr{A}$,

$$\vdash (\exists x)(\mathscr{A} \vee \mathscr{B}) \Leftrightarrow (\mathscr{A} \vee (\exists x)\mathscr{B}), \quad \vdash (\exists x)(\mathscr{A} \wedge \mathscr{B}) \Leftrightarrow (\mathscr{A} \wedge (\exists x)\mathscr{B}),$$
$$\vdash (\forall x)(\mathscr{A} \vee \mathscr{B}) \Leftrightarrow (\mathscr{A} \vee (\forall x)\mathscr{B}), \quad \vdash (\forall x)(\mathscr{A} \wedge \mathscr{B}) \Leftrightarrow (\mathscr{A} \wedge (\forall x)\mathscr{B})$$

(f) We shall suggest a proof in the case $n = 1$; the other cases are then easy consequences. Let $Q(x) \in \Sigma_k^1 - \Pi_k^1$. Define $P(x)$ as $(\exists z)[(x = 2z \wedge Q(z)) \vee (x = 2z + 1 \wedge \neg Q(z))]$. It is easy to prove that $P \notin \Sigma_k^1 \cup \Pi_k^1$ and that $P \in \Sigma_{k+1}^1$. Observe that $P(x)$ holds if and only if

$$(\exists z)(x = 2z \wedge Q(z)) \vee ((\exists z)_{z<x}(x = 2z + 1) \wedge (\forall z)(x = 2z + 1 \Rightarrow \neg Q(z)))$$

Hence, $P \in \Pi_{k+1}^1$ (Rogers, 1959).

Exercises

5.36 For any relation W of n variables, prove that $W \in \Sigma_k^n$ if and only if $\overline{W} \in \Pi_k^n$, where $\overline{W}$ is the complement of W with respect to the set of all n-tuples of natural numbers.

5.37 For each $k > 0$, find a *universal* relation V_k in Σ_k^{n+1}; that is, for any relation W of n

variables: (a) if $W \in \Sigma_k^n$, then there exists z_0 such that, for all $x_1, \ldots, x_n$, $W(x_1, \ldots, x_n)$ if and only if $V_k(z_0, x_1, \ldots, x_n)$; and (b) if $W \in \Pi_k^n$, there exists v_0 such that, for all $x_1, \ldots, x_n$, $W(x_1, \ldots, x_n)$ if and only if $\neg V_k(v_0, x_1, \ldots, x_n)$. (*Hint:* Use Exercise 5.19.)

The *s-m-n* Theorem (Proposition 5.16) enables us to prove the following basic result of recursion theory.

PROPOSITION 5.19 (RECURSION THEOREM) If $n > 1$ and $f(x_1, \ldots, x_n)$ is a partial recursive function, then there exists a natural number e such that

$$f(x_1, \ldots, x_{n-1}, e) = \phi_e^{n-1}(x_1, \ldots, x_{n-1})$$

Proof Let d be an index of $f(x_1, \ldots, x_{n-1}, s_{n-1}^1(x_n, x_n))$. Then

$$f(x_1, \ldots, x_{n-1}, s_{n-1}^1(x_n, x_n)) = \phi_d^n(x_1, \ldots, x_{n-1}, x_n)$$

By the *s-m-n*-Theorem, $\phi_d^n(x_1, \ldots, x_n) = \phi_{s_{n-1}^1(d, x_n)}^{n-1}(x_1, \ldots, x_{n-1})$. Let $e = s_{n-1}^1(d, d)$. Then:

$$\begin{aligned} f(x_1, \ldots, x_{n-1}, e) &= f(x_1, \ldots, x_{n-1}, s_{n-1}^1(d, d)) = \phi_d^n(x_1, \ldots, x_{n-1}, d) \\ &= \phi_{s_{n-1}^1(d,d)}^{n-1}(x_1, \ldots, x_{n-1}) = \phi_e^{n-1}(x_1, \ldots, x_{n-1}) \end{aligned}$$

COROLLARY 5.20 (FIXED POINT THEOREM) If $h(x)$ is recursive, then there exists e such that $\phi_e^1 = \phi_{h(e)}^1$.

Proof Applying the Recursion Theorem to $f(x, u) = \phi_{h(u)}^1(x)$, we obtain a number e such that $f(x, e) = \phi_e^1(x)$. But, $f(x, e) = \phi_{h(e)}^1(x)$.

APPLICATION: RICE'S THEOREM (Rice, 1953) Let $\mathscr{F}$ be a set consisting of at least one, but not all, partial recursive function of one variable. Then the set $A = \{u \mid \phi_u^1 \in \mathscr{F}\}$ is not recursive.

Proof By hypothesis, there exist numbers u_1 and u_2 such that $u_1 \in A$ and $u_2 \notin A$. Now assume that A is recursive. Define

$$h(x) = \begin{cases} u_1 & \text{if } x \notin A \\ u_2 & \text{if } x \in A \end{cases}$$

Clearly, $h(x) \in A$ if and only if $x \notin A$. h is recursive, by Proposition 3.18. By the Fixed Point Theorem, there is a number e such that $\phi_e^1 = \phi_{h(e)}^1$. Then we obtain a contradiction as follows:

$$\begin{aligned} e \in A \quad & \text{if and only if } \phi_e^1 \in \mathscr{F} \\ & \text{if and only if } \phi_{h(e)}^1 \in \mathscr{F} \\ & \text{if and only if } h(e) \in A \\ & \text{if and only if } e \notin A \end{aligned}$$

Rice's Theorem can be used to show the recursive unsolvability of various decision problems.

Example Consider the following decision problem: for any u, determine whether ϕ_u^1 has an infinite domain. Let $\mathscr{F}$ be the set of all partial recursive functions of one variable that have infinite domain. By Rice's Theorem, $\{u|\phi_u^1 \in \mathscr{F}\}$ is not recursive. Hence, the problem is recursively unsolvable.

Notice that Example 1 on page 255 and Exercise 5.34 can be handled in the same way.

Exercises

5.38 Show that the following decision problems are recursively unsolvable.
(a) For any u, determine whether ϕ_u^1 has infinite range.
(b) For any u, determine whether ϕ_u^1 is a constant function.
(c) For any u, determine whether ϕ_u^1 is recursive.

5.39 (a) Show that there is a number e such that the domain of ϕ_e^1 is $\{e\}$.
(b) Show that there is a number e such that the domain of ϕ_e^1 is $\omega - \{e\}$.

5.40[D] Use the Recursion Theorem to show that the function $\psi(x, y)$ of Exercise 3.33 on page 148 is recursive. [*Hint:* Use Exercise 5.21 to show that there is a partial recursive function g such that $g(x, 0, u) = x + 1$, $g(0, y + 1, u) = \phi_u^2(1, y)$, $g(x + 1, y + 1, u) = \phi_u^2(\phi_u^2(x, y + 1), y)$. Then use the Recursion Theorem to find e such that $g(x, y, e) = \phi_e^2(x, y)$. By induction, show that $g(x, y, e) = \psi(x, y)$.]

A set of natural numbers is said to be *recursively enumerable* (r.e.) if and only if it is either empty or the range of a recursive function $g(x)$. If we accept Church's Thesis, a nonempty recursively enumerable set is a collection of natural numbers generated by some mechanical process or effective procedure.

PROPOSITION 5.21

(a) A set B is r.e. if and only if $x \in B$ is expressible in the form $(\exists y)R(x, y)$, where R is recursive. (We even can allow R here to be primitive recursive.)
(b) B is r.e. if and only if B is either empty or the range of a partial recursive function.*
(c) B is r.e. if and only if B is the domain of a partial recursive function.
(d) B is recursive if and only if both B and its complement $\bar{B}$ are r.e.†
(e) The set $K = \{x|(\exists y)T_1(x, x, y)\}$ is r.e. but not recursive.

Proof

(a) Assume B is r.e. If B is empty, then $x \in B \Leftrightarrow (\exists y)(x \neq x \wedge y \neq y)$. If B is nonempty, B is the range of a recursive function g. Then, $x \in B \Leftrightarrow (\exists y)(g(y) = x)$. Conversely, assume $x \in B \Leftrightarrow (\exists y)R(x, y)$, where R is recursive. If B is empty, B is r.e. If B is nonempty, let k be a fixed element of B. Define

$$\theta(z) = \begin{cases} k & \text{if } \neg R((z)_0, (z)_1) \\ (z)_0 & \text{if } R((z)_0, (z)_1) \end{cases}$$

* Since the empty function is partial recursive and has the empty set as its range, the condition that B is empty can be omitted.

† $\bar{B} = \omega - B$, where ω is the set of natural numbers.

θ is recursive, by Proposition 3.18. Clearly, B is the range of θ. [We can take R to be primitive recursive, since, if R is recursive, then, by Corollary 5.13(a), $(\exists y)R(x, y) \Leftrightarrow (\exists y)T_1(e, x, y)$ for some e, and $T_1(e, x, y)$ is primitive recursive.]

(b) Assume B is the range of a partial recursive function g. If B is empty, then B is r.e. If B is nonempty, let k be a fixed element of B. By Corollary 5.11, there is a number e such that $g(x) = U(\mu y\, T_1(e, x, y))$. Let

$$h(z) = \begin{cases} U((z)_1) & \text{if } T_1(e, (z)_0, (z)_1) \\ k & \text{if } \neg T_1(e, (z)_0, (z)_1) \end{cases}$$

By Proposition 3.18, h is primitive recursive. Clearly, B is the range of h. Hence, B is r.e.

(c) Assume B is r.e. If B is empty, B is the domain of the partial recursive function $\mu y(x + y + 1 = 0)$. If B is nonempty, B is the range of a recursive function g. Let G be the partial recursive function such that $G(y) = \mu x(g(x) = y)$. Then B is the domain of G. Conversely, assume B is the domain of a partial recursive function H. Then there is a number e such that $H(x) = U(\mu y\, T_1(e, x, y))$. Hence, $H(x) = z$ if and only if $(\exists y)(T_1(e, x, y) \wedge U(y) = z)$. But, $x \in B$ if and only if $(\exists z)(H(x) = z)$. So, $x \in B$ if and only if $(\exists z)(\exists y)(T_1(e, x, y) \wedge U(y) = z)$, and the latter is equivalent to $(\exists u)(T_1(e, x, (u)_1) \wedge U((u)_1) = (u)_0)$. Moreover, $T_1(e, x, (u)_1) \wedge U((u)_1) = (u)_0$ is recursive. Thus, by part (a), B is r.e.

(d) Use part (a) and Proposition 5.18(d). [The intuitive meaning of part (d) is the following: if there are mechanical procedures for generating B and $\bar{B}$, then to determine whether any number n is in B we need only wait until n is generated by one of the procedures and then observe which procedure produced it.]

(e) Use parts (a) and (d) and Corollary 5.14(a).

Remember that the functions $\phi_n^1(x) = U(\mu y\, T_1(n, x, y))$ form an enumeration of all partial recursive functions of one variable. If we designate the domain of ϕ_n^1 by W_n, then Proposition 5.21(c) tells us that $W_0, W_1, W_2, \ldots$ is an enumeration (with repetitions) of all r.e. sets. The number n is called an *index* of the set W_n.

Exercises

5.41 Prove that a set B is r.e. if and only if it is either empty or the range of a primitive recursive function. [*Hint:* See the proof of Proposition 5.21(b).]

5.42 (a) Prove that the inverse image of a r.e. set under a partial recursive function is r.e. [that is, if f is partial recursive and B is r.e., then $\{x | f(x) \in B\}$ is r.e.].

(b) Prove that the inverse image of a recursive set under a recursive function is recursive.

(c) Prove that the image of a r.e. set under a partial recursive function is r.e.

(d) Using Church's Thesis, give intuitive arguments for the results in parts (a)–(c).

(e) Show that the image of a recursive set under a recursive function need not be recursive.

5.43 Prove that an infinite set is recursive if and only if it is the range of a strictly increasing recursive function. [g is *strictly increasing* if $x < y$ implies $g(x) < g(y)$.]

5.44 Prove that an infinite set is r.e. if and only if it is the range of a one–one recursive function.

5.45 Prove that every infinite r.e. set contains an infinite recursive subset.

5.46 Assume that A and B are r.e. sets.

(a) Prove that $A \cup B$ is r.e. [In fact, show that there is a recursive function $g(u, v)$ such that $W_{g(u,v)} = W_u \cup W_v$.]

(b) Prove that $A \cap B$ is r.e. [In fact, show that there is a recursive function $h(u, v)$ such that $W_{h(u,v)} = W_u \cap W_v$.]

(c) Show that $\bar{A}$ need not be r.e.

(d) Prove that $\bigcup_{n \in A} W_n$ is r.e.

5.47 Show that the assertion

(∇) A set B is r.e. if and only if B is effectively enumerable (that is, there is a mechanical procedure for generating the numbers in B)

is equivalent to Church's Thesis.

5.48 Prove that the set $A = \{u \mid W_u = \omega\}$ is not r.e.

5.49 A set B is called *creative* if and only if B is r.e. and there is a partial recursive function h such that, for any n, if $W_n \subseteq \bar{B}$, then $h(n) \in \bar{B} - W_n$.

(a) Prove that $\{x \mid (\exists y) T_1(x, x, y)\}$ is creative.

(b) Show that every creative set is nonrecursive.

5.50^{D} A set B is called *simple* if B is r.e., $\bar{B}$ is infinite, and $\bar{B}$ contains no infinite r.e. set. Clearly, every simple set is nonrecursive. Show that a simple set exists.

5.51 A *recursive permutation* is a one–one recursive function from ω onto ω. Sets A and B are called *isomorphic* (written $A \simeq B$) if there is a recursive permutation that maps A onto B.

(a) Prove that the recursive permutations form a group under the operation of composition.

(b) Prove that $\simeq$ is an equivalence relation.

(c) Prove that, if A is recursive (r.e., creative, simple) and $A \simeq B$, then B is recursive (r.e., creative, simple).

Myhill (1955) proved that any two creative sets are isomorphic. (Also see Bernays, 1957.)

5.52 A is *many–one reducible* to B (written $A \mathrm{R}_{\mathrm{m}} B$) if there is a recursive function f such that $u \in A$ if and only if $f(u) \in B$. (Many–one reducibility of A to B implies that, if the decision problem for membership in B is recursively solvable, so is the decision problem for membership in A.) A and B are called *many–one equivalent* (written $A \equiv_{\mathrm{m}} B$) if $A \mathrm{R}_{\mathrm{m}} B$ and $B \mathrm{R}_{\mathrm{m}} A$. A is *one–one reducible* to B (written $A \mathrm{R}_1 B$) if there is a one–one recursive function f such that $u \in A$ if and only if $f(u) \in B$. A and B are called *one–one equivalent* (written $A \equiv_1 B$) if $A \mathrm{R}_1 B$ and $B \mathrm{R}_1 A$.

(a) Prove that $\equiv_{\mathrm{m}}$ and $\equiv_1$ are equivalence relations.

(b) Prove that, if A is creative, B is r.e., and $A \mathrm{R}_{\mathrm{m}} B$, then B is creative. [Myhill (1955) showed that, if A is creative and B is r.e., then $B \mathrm{R}_{\mathrm{m}} A$.]

(c) (Myhill, 1955) Prove that, if $A \mathrm{R}_1 B$ then $A \mathrm{R}_{\mathrm{m}} B$, and if $A \equiv_1 B$ then $A \equiv_{\mathrm{m}} B$. However, many–one reducibility does not imply one–one reducibility, and many–one equivalence does not imply one–one equivalence. [*Hint:* Let A be a simple set, C an infinite recursive subset of A, and $B = A - C$. Then $A \mathrm{R}_1 B$ and $B \mathrm{R}_m A$ but not$(B \mathrm{R}_1 A)$.] It can be shown that $A \equiv_1 B$ if and only if $A \simeq B$.

5.53 (Dekker, 1955) A is said to be *productive* if there is a partial recursive function f such that, if $W_n \subseteq A$, then $f(n) \in A - W_n$. Prove the following.

(a) If A is productive, then A is not r.e.; hence, both A and $\bar{A}$ are infinite.

(b)^{D} If A is productive, then A has an infinite r.e. subset. Hence, if A is productive, $\bar{A}$ is not simple.

(c) If A is r.e., then A is creative if and only if $\bar{A}$ is productive.

(d)^{D} There exist $2^{\aleph_0}$ productive sets.

5.54 (Dekker & Myhill, 1960) A is *recursively equivalent* to B (written $A \sim B$) if there is a one–one partial recursive function that maps A onto B.

(a) Prove that $\sim$ is an equivalence relation.

(b) A is said to be *immune* if A is infinite and A has no infinite r.e. subset. A is said to be *isolated* if A is not recursively equivalent to a proper subset of A. (The isolated sets may be considered the counterparts of the Dedekind-finite sets.) Prove that an infinite set is isolated if and only if it is immune.

(c)^D Prove that there exist $2^{\aleph_0}$ immune sets.

Recursively enumerable sets play an important role in logic because, if we assume Church's Thesis, the set T_K of Gödel numbers of the theorems of any axiomatizable first-order theory K is r.e. (The same holds true of arbitrary formal axiomatic systems.) In fact, the relation (see page 155)

$$\mathrm{Pf}_K(y, x)\text{: } y \text{ is the Gödel number of a proof in K of a wf with Gödel number } x$$

is recursive if the set of Gödel numbers of the axioms is recursive—that is, if there is a decision procedure for axiomhood and Church's Thesis holds. Now, $x \in T_K$ if and only if $(\exists y)\mathrm{Pf}_K(y, x)$ and, therefore, T_K is r.e. Thus, if we accept Church's Thesis, K is effectively decidable if and only if the r.e. set T_K is recursive. It was shown in Corollary 3.41 that every consistent extension K of the theory RR is recursively undecidable; that is, T_K is not recursive.

Much more general results along these lines can be proved (see Smullyan, 1961; Feferman, 1957; Putnam, 1957; Ehrenfeucht & Feferman, 1960; and Myhill, 1955). For example, (1) if every recursive set is expressible in K, then K is essentially recursively undecidable; that is, for every consistent extension K′ of K, $T_{K'}$ is not recursive (see Exercise 5.58); (2) for any consistent first-order theory with equality K in which every recursive function is representable and which satisfies conditions 4 and 5 on page 163, the set T_K is creative. (We assume that K has among its terms the numerals $\overline{0}, \overline{1}, \overline{2}, \ldots$.) For further study of r.e. sets, see Post (1944) and Rogers (1967); for the relationship between logic and recursion theory, see Yasuhara (1971) and Monk (1976, part III).

Exercises

5.55 Let K be a first-order theory with equality that contains all the symbols of formal number theory S. A number-theoretic relation $B(x_1, \ldots, x_n)$ is said to be *weakly expressible* in K if and only if there is a wf $\mathscr{B}(x_1, \ldots, x_n)$ of K such that, for any natural numbers $k_1, \ldots, k_n$, $B(k_1, \ldots, k_n)$ if and only if $\vdash_K \mathscr{B}(\overline{k}_1, \ldots, \overline{k}_n)$.

(a) Show that, if K is consistent, every relation expressible in K is weakly expressible in K.

(b) Prove that, if every recursive relation is expressible in K and K is ω-consistent, every r.e. set is weakly expressible in K. (Remember that, when we refer here to a r.e. set B, we mean the corresponding relation "$x \in B$".)

(c) If K is such that the relations (a)–(d) of Propositions 3.25 and 3.26 are recursive, prove that any set that is weakly expressible in K is r.e.

(d) If formal number theory S is ω-consistent, prove that a set B is r.e. if and only if B is weakly expressible in S.

5.56 (a) (Craig, 1953) Let K be a first-order theory such that the set T_K of Gödel numbers of theorems of K is r.e. Show that K is recursively axiomatizable.

(b) For any wf $\mathscr{A}$ of formal number theory S, let $\mathscr{A}\#$ represent its translation into axiomatic set theory NBG (see page 208). Prove that the set of wfs $\mathscr{A}$ such that $\vdash_{\text{NBG}} \mathscr{A}\#$ is a (proper) recursively axiomatizable extension of S. (However, no "natural" set of axioms for this theory is known.)

5.57 Given a set A of natural numbers, let $u \in A^*$ if and only if u is a Gödel number of a wf $\mathscr{A}(x_1)$ and the Gödel number of $\mathscr{A}(\bar{u})$ is in A. Prove that, if A is recursive, then A^* is recursive.

5.58 Let K be a consistent theory that has the same symbols as S.

(a) Prove that $(\bar{T}_{\text{K}})^*$ is not weakly expressible in K.

(b) If every recursive set is weakly expressible in K, show that K is recursively undecidable.

(c) If every recursive set is expressible in K, prove that K is essentially recursively undecidable.

5. OTHER NOTIONS OF EFFECTIVE COMPUTABILITY

Computability has been treated here in terms of Turing machines because Turing's definition probably is the one that makes clearest the equivalence between the precise mathematical concept and the intuitive notion. For further justification of this equivalence, see Turing (1936–37) and Kleene (1952, pp. 317–23, 376–81). We already have encountered other equivalent notions: standard Turing-computability and partial recursiveness. One of the strongest arguments for the rightness of Turing's definition is that all of the many definitions that have been proposed have turned out to be equivalent. We shall present a survey of some of these other definitions.

I. HERBRAND–GÖDEL COMPUTABILITY

The idea of defining computable functions in terms of fairly simple systems of equations was proposed by Herbrand, given a more precise form by Gödel (1934), and developed in detail by Kleene (1936a). The exposition given here is a version of the presentation in Kleene (1952, chap. XI).

We define first the *terms.*

1. All variables are terms.
2. 0 is a term.
3. If t is a term, then $(t)'$ is a term.
4. If $t_1, \ldots, t_n$ are terms and f_j^n is a function letter, $f_j^n(t_1, \ldots, t_n)$ is a term.

For every natural number n, we define the corresponding *numeral* $\bar{n}$ as follows: (1) $\bar{0}$ is 0 and (2) $\overline{n+1}$ is $(\bar{n})'$. Thus, every numeral is a term.

An *equation* is a formula $r = s$ where r and s are terms. A *system* E of equations is a finite sequence $r_1 = s_1, r_2 = s_2, \ldots, r_k = s_k$ of equations such that r_k is of the form $f_j^n(t_1, \ldots, t_n)$. The function letter f_j^n is called the *principal letter* of the system E. Those function letters (if any) that appear only on the right side of equations of E are called the *initial letters* of E; any function letter other than the principal letter that appears on the left side of some equations and also on the right side of some equations is called an *auxiliary letter* of E.

We have two rules of inference:

R_1: An equation e_2 is a consequence of an equation e_1 by R_1 if and only if e_2 arises from e_1 by substituting any numeral $\bar{n}$ for all occurrences of a variable.

R_2: An equation e is a consequence by R_2 of equations $f_h^m(\bar{n}_1, \ldots, \bar{n}_m) = \bar{p}$ and $r = s$ if and only if e arises from $r = s$ by replacing one or more occurrences of $f_h^m(\bar{n}_1, \ldots, \bar{n}_m)$ in s by $\bar{p}$, and $r = s$ contains no variables.

A *proof* of an equation e from a set B of equations is a sequence $e_0, \ldots, e_q$ of equations such that e_q is e and, if $0 \leqslant i \leqslant q$, then (1) e_i is an equation of B, or (2) e_i is a consequence by R_1 of a preceding equation e_j ($j < i$), or (3) e_i is a consequence by R_2 of two preceding equations e_j and e_m ($j < i, m < i$). We use the notation $B \vdash e$ to state that there is a proof from B of e (or, in other words, that e is *derivable* from B).

Example Let E be the system

$$f_1^1(x_1) = (x_1)'$$
$$f_1^2(x_1, x_2) = f_1^3(\bar{2}, x_2, f_1^1(x_1))$$

The principal letter of E is f_1^2, f_1^1 is an auxiliary letter, and f_1^3 is an initial letter. The sequence of equations

$$f_1^2(x_1, x_2) = f_1^3(\bar{2}, x_2, f_1^1(x_1))$$
$$f_1^2(\bar{2}, x_2) = f_1^3(\bar{2}, x_2, f_1^1(\bar{2}))$$
$$f_1^2(\bar{2}, \bar{1}) = f_1^3(\bar{2}, \bar{1}, f_1^1(\bar{2}))$$
$$f_1^1(x_1) = (x_1)'$$
$$f_1^1(\bar{2}) = (\bar{2})' \qquad [\text{i.e., } f_1^1(\bar{2}) = \bar{3}]$$
$$f_1^2(\bar{2}, \bar{1}) = f_1^3(\bar{2}, \bar{1}, \bar{3})$$

is a proof of $f_1^2(\bar{2}, \bar{1}) = f_1^3(\bar{2}, \bar{1}, \bar{3})$ from E.

A number-theoretic partial function $\varphi(x_1, \ldots, x_n)$ is said to be *computed by a system* E of equations if and only if the principal letter of E is a letter f_j^n and, for any natural numbers $k_1, \ldots, k_n, p$,

$$\text{E} \vdash f_j^n(\bar{k}_1, \ldots, \bar{k}_n) = \bar{p} \text{ if and only if } \varphi(k_1, \ldots, k_n) = p$$

The function φ is called *Herbrand-Gödel-computable* (for short, HG-computable) if and only if there is some system E of equations by which φ is computed.

Examples

1. Let E be the system $f_1^1(x_1) = 0$. Then E computes the zero function Z. Hence, Z is HG-computable.
2. Let E be the system $f_1^1(x_1) = (x_1)'$. Then E computes the successor function N. Hence, N is HG-computable.
3. Let E be the system $f_i^n(x_1, \ldots, x_n) = x_i$. Then E computes the projection function U_i^n. Hence, U_i^n is HG-computable.

4. Let E be the system

$$f_1^2(x_1, 0) = x_1$$
$$f_1^2(x_1, (x_2)') = (f_1^2(x_1, x_2))'$$

Then E computes the addition function.

5. Let E be the system

$$f_1^1(x_1) = 0$$
$$f_1^1(x_1) = x_1$$

The function $\varphi(x_1)$ computed by E is the partial function with domain $\{0\}$ such that $\varphi(0) = 0$. For every $k \neq 0$, $E \vdash f_1^1(\bar{k}) = \bar{0}$ and $E \vdash f_1^1(\bar{k}) = \bar{k}$. Hence, $\varphi(x_1)$ is not defined for $x_1 \neq 0$.

Exercises

5.59 (a) What functions are HG-computable by the following systems of equations?
   (i) $f_1^1(0) = 0, f_1^1((x_1)') = x_1$
   (ii) $f_1^2(x_1, 0) = x_1, f_1^2(0, x_2) = 0, f_1^2((x_1)', (x_2)') = f_1^2(x_1, x_2)$
   (iii) $f_1^1(x_1) = 0, f_1^1(x_1) = 0'$
   (iv) $f_1^2(x_1, 0) = x_1, f_1^2(x_1, (x_2)') = (f_1^2(x_1, x_2))', f_1^1(x_1) = f_1^2(x_1, x_1)$

(b) Show that the following functions are HG-computable.
   (i) $|x_1 - x_2|$
   (ii) $x_1 \cdot x_2$
   (iii) $\varphi(x) = \begin{cases} 0 & \text{when } x \text{ is even} \\ 1 & \text{when } x \text{ is odd} \end{cases}$

5.60 (a) Find a system E of equations that computes the n-place function that is nowhere defined.
(b) Let f be an n-place function defined on a finite domain. Find a system of equations that computes f.
(c) If $f(x)$ is an HG-computable total function and $g(x)$ is a partial function that coincides with $f(x)$ except on a finite set A, where g is undefined, find a system of equations that computes g.

PROPOSITION 5.22 Every partial recursive function is HG-computable.

Proof

(a) Examples 1–3 above show that the initial functions Z, N, and U_i^n are HG-computable.

(b) *Substitution Rule (IV)*. Let $\varphi(x_1, \ldots, x_n) = \eta(\psi_1(x_1, \ldots, x_n), \ldots, \psi_m(x_1, \ldots, x_n))$ where η, ψ_1, ..., ψ_m have been shown to be HG-computable. Let E_i be a system of equations computing ψ_i, with principal letter f_i^n, and let E_{m+1} be a system of equations computing η, with principal letter f_{m+1}^m. By changing indices, we may assume that no two of $E_1, \ldots, E_m, E_{m+1}$ have any function letters in common. Construct a system E for φ by listing E_1, ..., E_{m+1} and then adding the equation $f_{m+2}^n(x_1, \ldots, x_n) = f_{m+1}^m(f_1^n(x_1, \ldots, x_n), \ldots, f_m^n(x_1, \ldots, x_n))$. (We may assume that f_{m+2}^n does not occur in $E_1, \ldots, E_{m+1}$.) It is clear that, if $\varphi(k_1, \ldots, k_n) = p$, then $E \vdash$

$f^n_{m+2}(\overline{k_1},\ldots,\overline{k_n}) = \overline{p}$. Conversely, if $E \vdash f^n_{m+2}(\overline{k_1},\ldots,\overline{k_n}) = \overline{p}$, then $E \vdash f^n_1(\overline{k_1},\ldots,\overline{k_n}) = \overline{p}_1,\ldots, E \vdash f^n_m(\overline{k_1},\ldots,\overline{k_n}) = \overline{p}_m$ and $E \vdash f^m_{m+1}(\overline{p}_1,\ldots,\overline{p}_m) = \overline{p}$. Hence, it readily follows that $E_1 \vdash f^n_1(\overline{k_1},\ldots,\overline{k_n}) = \overline{p}_1$, ..., $E_m \vdash f^n_m(\overline{k_1},\ldots,\overline{k_n}) = \overline{p}_m$ and $E_{m+1} \vdash f^n_{m+1}(\overline{p}_1,\ldots,\overline{p}_m) = \overline{p}$. Consequently, $\psi_1(k_1,\ldots,k_n) = p_1,\ldots,\psi_m(k_1,\ldots,k_n) = p_m$ and $\eta(p_1,\ldots,p_m) = p$. So, $\varphi(k_1,\ldots,k_n) = p$. [The details of this proof are left as an exercise. Hints may be found in Kleene (1952, chap. XI, especially pp. 262–70).] Hence φ is HG-computable.

(c) *Recursion Rule (V).* Let

$$\varphi(x_1,\ldots,x_n,0) = \psi(x_1,\ldots,x_n)$$
$$\varphi(x_1,\ldots,x_n,x_{n+1}+1) = \theta(x_1,\ldots,x_{n+1},\varphi(x_1,\ldots,x_{n+1}))$$

where ψ and θ are HG-computable. Assume that E_1 is a system of equations computing ψ with principal letter f^n_1 and that E_2 is a system of equations computing θ with principal letter f^{n+2}_1. Then form a system for computing φ by adding to E_1 and E_2

$$f^{n+1}_1(x_1,\ldots,x_n,0) = f^n_1(x_1,\ldots,x_n)$$
$$f^{n+1}_1(x_1,\ldots,x_n,(x_{n+1})') = f^{n+2}_1(x_1,\ldots,x_{n+1},f^{n+1}_1(x_1,\ldots,x_{n+1}))$$

(We assume that E_1 and E_2 have no function letters in common.) Clearly, if $\varphi(k_1,\ldots,k_n,k) = p$, then $E \vdash f^{n+1}_1(\overline{k_1},\ldots,\overline{k_n},\overline{k}) = \overline{p}$. Conversely, one can prove easily by induction on k that, if $E \vdash f^{n+1}_1(\overline{k_1},\ldots,\overline{k_n},\overline{k}) = \overline{p}$, then $\varphi(k_1,\ldots,k_n,k) = p$. Therefore, φ is HG-computable. (The case when the recursion has no parameters is even easier to handle and is left as an exercise.)

(d) *μ-Operator Rule (VI).* Let $\varphi(x_1,\ldots,x_n) = \mu y(\psi(x_1,\ldots,x_n,y) = 0)$ and assume that ψ is HG-computable by a system E_1 of equations with principal letter f^{n+1}_1. By parts (a)–(c), we know that every primitive recursive function is HG-computable. In particular, multiplication is HG-computable; hence, there is a system E_2 of equations having no function letters in common with E_1 and with principal letter f^2_2 such that $E_2 \vdash f^2_2(\overline{k_1},\overline{k_2}) = \overline{p}$ if and only if $k_1 \cdot k_2 = p$. We form a system E_3 by adding to E_1 and E_2 the equations

$$f^{n+1}_3(x_1,\ldots,x_n,0) = 1$$
$$f^{n+1}_3(x_1,\ldots,x_n,(x_{n+1})') = f^2_2(f^{n+1}_3(x_1,\ldots,x_n,x_{n+1}),f^{n+1}_1(x_1,\ldots,x_n,x_{n+1}))$$

One can prove by induction that E_3 computes the function $\prod_{y<z}\psi(x_1,\ldots,x_n,y)$; that is, $E_3 \vdash f^{n+1}_3(\overline{k_1},\ldots,\overline{k_n},\overline{k}) = \overline{p}$ if and only if $\prod_{y<k}\psi(k_1,\ldots,k_n,y) = p$. Now construct the system E by adding to E_3 the equations

$$f^3_4((x_1)',0,x_3) = x_3$$
$$f^n_3(x_1,\ldots,x_n) = f^3_4(f^{n+1}_3(x_1,\ldots,x_n,x_{n+1}),f^{n+1}_3(x_1,\ldots,x_n,(x_{n+1})'),x_{n+1})$$

Then E computes the function $\varphi(x_1,\ldots,x_n) = \mu y(\psi(x_1,\ldots,x_n,y) = 0.)$ If $\mu y(\psi(k_1,\ldots,k_n,y) = 0) = q$, then $E_3 \vdash f^{n+1}_3(\overline{k_1},\ldots,\overline{k_n},\overline{q}) = \overline{p}'$, where $p+1 = \prod_{y<q}\psi(k_1,\ldots,k_n,y)$, and $E_3 \vdash f^{n+1}_3(\overline{k_1},\ldots,\overline{k_n},\overline{q}') = 0$. Hence $E \vdash f^n_3(\overline{k_1},\ldots,\overline{k_n}) = f^3_4(\overline{p}',0,\overline{q})$. But, $E \vdash f^3_4(\overline{p}',0,\overline{q}) = \overline{q}$, and so, $E \vdash f^n_3(\overline{k_1},\ldots,\overline{k_n}) = \overline{q}$. Conversely, if $E \vdash f^n_3(\overline{k_1},\ldots,\overline{k_n}) = \overline{q}$, then $E \vdash f^3_4(\overline{m}',0,\overline{q}) = \overline{q}$, where $E_3 \vdash f^{n+1}_3(\overline{k_1},\ldots,\overline{k_n},\overline{q}) =$

$(\overline{m})'$ and $E_3 \vdash f_3^{n+1}(\overline{k_1}, \ldots, \overline{k_n}, \overline{q}') = 0$. Hence $\prod_{y<q}\psi(k_1, \ldots, k_n, y) = m + 1 \neq 0$ and $\prod_{y<q+1}\psi(k_1, \ldots, k_n, y) = 0$. So, $\psi(k_1, \ldots, k_n, y) \neq 0$ for $y < q$, and $\psi(k_1, \ldots, k_n, q) = 0$. Thus, $\mu y(\psi(k_1, \ldots, k_n, y) = 0) = q$. Therefore, φ is HG-computable.

We now shall proceed to show that every Herbrand-Gödel-computable function is partial recursive by means of an arithmetization of the apparatus of Herbrand-Gödel computability. We shall use the same arithmetization that was used for first-order theories (see Chapter 3, Section 4). [We take the symbol $'$ to be an abbreviation for f_1^1. Remember that $r = s$ is an abbreviation for $A_1^2(r, s)$. The only individual constant is 0.] In particular, the following relations and functions are primitive recursive (see pages 150–152):

FL(x): x is the Gödel number of a function letter.

$$(\exists y)_{y<x}(\exists z)_{z<x}(x = 1 + 8(2^y \cdot 3^z) \wedge y > 0 \wedge z > 0)$$

EVbl(x): x is the Gödel number of an expression consisting of a variable.

EFL(x): x is the Gödel number of an expression consisting of a function letter.

Nu(x): x is the Gödel number of a numeral.

Trm(x): x is the Gödel number of a term.

Num(x) = the Gödel number of the numeral $\overline{x}$.

$\mathrm{Arg_T}(x)$ = the number of arguments of a function letter f, if x is the Gödel number of f.

$x * y$ = the Gödel number of an expression AB if x is the Gödel number of the expression A and y is the Gödel number of the expression B.

Subst(a, b, u, v): v is the Gödel number of a variable x_i, u is the Gödel number of a term t, b is the Gödel number of an expression $\mathscr{A}$, and a is the Gödel number of the result of substituting t for all occurrences of x_i in $\mathscr{A}$.

The following are also primitive recursive:

Eqt(x): x is the Gödel number of an equation:

$$\ell h(x) = 3 \wedge \mathrm{Trm}((x)_1) \wedge \mathrm{Trm}((x)_2) \wedge (x)_0 = 99$$

(Remember that = is A_1^2, whose Gödel number is 99.)

Syst(x): x is the Gödel number of a system of equations:

$$(\forall y)_{y<\ell h(x)}\mathrm{Eqt}((x)_y) \wedge \mathrm{FL}((((x)_{\ell h(x) \dot{-} 1})_1)_0)$$

Occ(u, v): u is the Gödel number of a term t or equation $\mathfrak{B}$ and v is the Gödel number of a term that occurs in t or $\mathfrak{B}$:

$$(\mathrm{Trm}(u) \vee \mathrm{Eqt}(u)) \wedge \mathrm{Trm}(v) \wedge (\exists x)_{x<u}(\exists y)_{y<u}(u = x * v * y \vee$$
$$u = x * v \vee u = v * y \vee u = v)$$

$\mathrm{Cons}_1(u, v)$: u is the Gödel number of an equation e_1, v is the Gödel number of an equation e_2, and e_2 is a consequence of e_1 by rule R_1:

$$\mathrm{Eqt}(u) \wedge \mathrm{Eqt}(v) \wedge (\exists x)_{x<u}(\exists y)_{y<v}(\mathrm{Nu}(y) \wedge \mathrm{Subst}(v, u, y, x) \wedge \mathrm{Occ}(u, x))$$

$\mathrm{Cons}_2(u, z, v)$: u, z, and v are Gödel numbers of equations e_1, e_2, and e_3, respectively, and e_3 is a consequence of e_1 and e_2 by rule R_2:

$$\mathrm{Eqt}(u) \wedge \mathrm{Eqt}(z) \wedge \mathrm{Eqt}(v) \wedge \neg(\exists x)_{x<z}(\mathrm{EVbl}(x) \wedge \mathrm{Occ}(z,x)) \wedge$$
$$\mathrm{FL}(((z)_1)_0) \wedge (\forall x)_{0<x<\ell h((z)_1)} \neg\mathrm{FL}(((z)_1)_x) \wedge$$
$$(\forall x)_{x<\ell h((z)_2)} \neg\mathrm{FL}(((z)_2)_x) \wedge \mathrm{Occ}((u)_2,(z)_1) \wedge$$
$$[(\exists y)_{y<u}(\exists w)_{w<u}((u)_2 = y * (z)_1 * w \wedge v = 2^{99}3^{(u)_1}5^{y*(z)_2*w}) \vee$$
$$((u)_2 = (z)_1 \wedge v = 2^{99}3^{(u)_1}5^{(z)_2})]$$

Ded(u, z): u is the Gödel number of a system of equations E and z is the Gödel number of a proof from E:

$$\mathrm{Syst}(u) \wedge (\forall x)_{x<\ell h(z)}((\exists w)_{w<\ell h(u)}(u)_w = (z)_x \vee$$
$$(\exists y)_{y<x}\mathrm{Cons}_1((z)_y,(z)_x) \vee (\exists y)_{y<x}(\exists v)_{v<x}\mathrm{Cons}_2((z)_y,(z)_v,(z)_x))$$

$\mathrm{S}_n(u, x_1, \ldots, x_n, z)$: u is the Gödel number of a system of equations E whose principal letter is of the form f_j^n, and z is the Gödel number of a proof from E of an equation of the form $f_j^n(\overline{x}_1, \ldots, \overline{x}_n) = \overline{p}$:

$$\mathrm{Ded}(u,z) \wedge \mathrm{Arg}_{\mathrm{T}}(((u)_{\ell h(u)\dot{-}1})_1)_0 = n \wedge (((z)_{\ell h(z)\dot{-}1})_1)_0 =$$
$$(((u)_{\ell h(u)\dot{-}1})_1)_0 \wedge (\forall y)_{0<y<\ell h(((z)_{\ell h(z)\dot{-}1})_1)} \neg\mathrm{FL}((((z)_{\ell h(z)\dot{-}1})_1)_y)$$
$$\wedge \mathrm{Nu}(((z)_{\ell h(z)\dot{-}1})_2) \wedge ((z)_{\ell h(z)\dot{-}1})_1 = 2^{(((u)_{\ell h(u)\dot{-}1})_1)_0} * 2^3 * 2^{\mathrm{Num}(x_1)} * 2^7 *$$
$$2^{\mathrm{Num}(x_2)} * 2^7 * \cdots * 2^7 * 2^{\mathrm{Num}(x_n)} * 2^5$$

Remember that $g(() = 3$, $g()) = 5$, and $g(,) = 7$.
$U(x) = \mu y_{y<x}(\mathrm{Num}(y) = ((x)_{\ell h(x)\dot{-}1})_2)$. [If x is the Gödel number of a proof of an equation $r = \overline{p}$, then $U(x) = p$.]

PROPOSITION 5.23 (Kleene, 1936a) If $\varphi(x_1, \ldots, x_n)$ is HG-computable by a system of equations E with Gödel number e, then

$$\varphi(x_1, \ldots, x_n) = U(\mu y(\mathrm{S}_n(e, x_1, \ldots, x_n, y)))$$

Hence, every HG-computable function φ is partial recursive, and, if φ is total, then φ is recursive.

Proof $\varphi(k_1, \ldots, k_n) = p$ if and only if $\mathrm{E} \vdash f_j^n(\overline{k_1}, \ldots, \overline{k_n}) = \overline{p}$, where f_j^n is the principal letter of E. $\varphi(k_1, \ldots, k_n)$ is defined if and only if $(\exists y)\mathrm{S}_n(e, k_1, \ldots, k_n, y)$. If $\varphi(k_1, \ldots, k_n)$ is defined, $\mu y(\mathrm{S}_n(e, k_1, \ldots, k_n, y))$ is the Gödel number of a proof from E of an equation $f_j^n(\overline{k_1}, \ldots, \overline{k_n}) = \overline{p}$. Hence, $U(\mu y(\mathrm{S}_n(e, k_1, \ldots, k_n, y))) = p = \varphi(k_1, \ldots, k_n)$. Also, since S_n is primitive recursive, $\mu y(\mathrm{S}_n(e, x_1, \ldots, x_n, y))$ is partial recursive. If φ is total, then $(\forall x_1)\ldots(\forall x_n)(\exists y)\mathrm{S}_n(e, x_1, \ldots, x_n, y)$; hence, $\mu y(\mathrm{S}_n(e, x_1, \ldots, x_n, y))$ is recursive, and then, so is $U(\mu y(\mathrm{S}_n(e, x_1, \ldots, x_n, y)))$.

Thus, the class of Herbrand-Gödel-computable functions is identical with the class of partial recursive functions. This is further evidence for Church's Thesis.

II. MARKOV ALGORITHMS

By an *algorithm in* an alphabet A, we mean an effectively computable function $\mathfrak{A}$ whose domain is a subset of the set of words of A and the values of which are also words in A. If P is a word in A, $\mathfrak{A}$ is said to be *applicable* to P if P is in the domain of $\mathfrak{A}$; if $\mathfrak{A}$ is applicable to P, we denote its value by $\mathfrak{A}(P)$. By an *algorithm over* an alphabet A, we mean an algorithm $\mathfrak{A}$ in an extension B of A.† Of course, the notion of algorithm is as hazy as that of effectively computable function.

Most familiar algorithms can be broken down into a few simple steps. Starting from this observation and following Markov (1954), we select a particularly simple operation, substitution of one word for another, as the basic unit from which algorithms are to be constructed. To this end, if P and Q are words of an alphabet A, then we call the expressions $P \to Q$ and $P \to \cdot Q$ *productions* in the alphabet A. We assume here that "$\to$" and "$\cdot$" are not symbols of A. Notice that P or Q can be the empty word. $P \to Q$ is called a *simple* production, whereas $P \to \cdot Q$ is a *terminal* production. Let us use $P \to (\cdot)Q$ to denote either $P \to Q$ or $P \to \cdot Q$. A finite list of productions in A

$$\begin{gathered} P_1 \to (\cdot)Q_1 \\ P_2 \to (\cdot)Q_2 \\ \vdots \\ P_r \to (\cdot)Q_r \end{gathered}$$

is called an *algorithm schema* and determines the following algorithm $\mathfrak{A}$ in A. As a preliminary definition, we say that a word T *occurs* in a word Q if there are words U, V (possibly empty) such that $Q = UTV$. Now, given a word P in A: (1) We write $\mathfrak{A} : P \sqsupset$ if none of the words $P_1, \ldots, P_r$ occurs in P. (2) Otherwise, if m is the least integer, with $1 \leqslant m \leqslant r$, such that P_m occurs in P, and if R is the word that results from replacing the leftmost occurrence of P_m in P by Q_m, then we write

$$\text{(a)} \quad \mathfrak{A} : P \vdash R$$

if $P_m \to (\cdot)Q_m$ is simple (and we say that $\mathfrak{A}$ simply transforms P into R);

$$\text{(b)} \quad \mathfrak{A} : P \vdash \cdot R$$

if $P_m \to (\cdot)Q_m$ is terminal (and we say that $\mathfrak{A}$ terminally transforms P into R). We then define $\mathfrak{A} : P \models R$ to mean that there is a sequence $R_0, R_1, \ldots, R_k$ such that $P = R_0$; $R = R_k$; if $0 \leqslant j \leqslant k-2$, $\mathfrak{A} : R_j \vdash R_{j+1}$; and either $\mathfrak{A} : R_{k-1} \vdash R_k$ or $\mathfrak{A} : R_{k-1} \vdash \cdot R_k$. (In the second case, we write $\mathfrak{A} : P \models \cdot R$.) We set $\mathfrak{A}(P) = R$ if and only if either $\mathfrak{A} : P \models \cdot R$, or $\mathfrak{A} : P \models R$ and $\mathfrak{A} : R \sqsupset$. The algorithm thus defined is called a *normal* algorithm (or Markov algorithm) in the alphabet A.

The action of $\mathfrak{A}$ can be described as follows: given a word P, we find the first production $P_m \to (\cdot)Q_m$ in the schema such that P_m occurs in P. We then substitute Q_m for the leftmost occurrence of P_m in P. Let R_1 be the new word obtained in this way. If $P_m \to (\cdot)Q_m$ is a terminal production, the process stops and the value of the algorithm is R_1. If $P_m \to (\cdot)Q_m$ is simple, then we apply the same process to R_1 as was just applied to P, and so on. If we ever obtain a word R_i such that $\mathfrak{A} : R_i \sqsupset$—that is, no P_m occurs in R_i for $1 \leqslant m \leqslant r$—then the process stops and the value of $\mathfrak{A}$ is R_i. It is possible that the process just described never stops. In that case, $\mathfrak{A}$ is not applicable to the given word P.

†An alphabet B is an extension of A if $A \subseteq B$.

Our exposition of the theory of normal algorithms will closely follow that of Markov (1954).

Examples

1. Let A be the alphabet $\{b, c\}$. Consider the schema

$$b \to \cdot \Lambda$$

$$c \to c$$

The normal algorithm $\mathfrak{A}$ defined by this schema transforms any word that contains at least one occurrence of b into the word obtained by erasing the leftmost occurrence of b. $\mathfrak{A}$ transforms the empty word Λ into itself. $\mathfrak{A}$ is not applicable to any nonempty word not containing b.

2. Let A be the alphabet $\{a_0, a_1, \ldots, a_n\}$. Consider the schema

$$a_0 \to \Lambda$$

$$a_1 \to \Lambda$$

$$\vdots$$

$$a_n \to \Lambda$$

We can abbreviate this schema as follows:

$$\xi \to \Lambda \qquad (\xi \text{ in A})$$

(Whenever we use such abbreviations, the productions intended may be listed in any order.) The corresponding normal algorithm transforms every word into the empty word. For example, $\mathfrak{A}: a_1a_2a_1a_3a_0 \vdash a_1a_2a_1a_3 \vdash a_2a_1a_3 \vdash a_2a_3 \vdash a_3 \vdash \Lambda$ and $\mathfrak{A}: \Lambda \sqsupset$. Hence $\mathfrak{A}(a_1a_2a_1a_3a_0) = \Lambda$.

3. Let A be an alphabet containing the symbol a_1, which we shall abbreviate $|$. For natural numbers n, we define $\bar{n}$ inductively as follows: $\bar{0} = |$ and $\overline{n+1} = \bar{n}|$. Thus, $\bar{1} = ||$, $\bar{2} = |||$, and so on. The words $\bar{n}$ are called *numerals*. Now consider the schema $\Lambda \to \cdot |$, defining a normal algorithm $\mathfrak{A}$. For any word P in A, $\mathfrak{A}(P) = |P$.†
In particular, for every natural number n, $\mathfrak{A}(\bar{n}) = \overline{n+1}$.

4. Let A be an arbitrary alphabet $\{a_0, a_1, \ldots, a_n\}$. Given a word $P = a_{j_0}a_{j_1}\ldots a_{j_k}$, let $\breve{P} = a_{j_k}\ldots a_{j_1}a_{j_0}$ be the *inverse* of P. We seek a normal algorithm $\mathfrak{A}$ such that $\mathfrak{A}(P) = \breve{P}$. Consider the following (abbreviated) algorithm schema in the alphabet $B = A \cup \{\alpha, \beta\}$.

(a)	$\alpha\alpha \to \beta$	
(b)	$\beta\xi \to \xi\beta$	(ξ in A)
(c)	$\beta\alpha \to \beta$	
(d)	$\beta \to \cdot \Lambda$	
(e)	$\alpha\eta\xi \to \xi\alpha\eta$	(ξ, η in A)
(f)	$\Lambda \to \alpha$	

† To see this, observe that Λ occurs at the beginning of any word P, since $P = \Lambda P$.

This determines a normal algorithm $\mathfrak{A}$ in B. Let $P = a_{j_0}a_{j_1}\ldots a_{j_k}$ be any word in A. Then, $\mathfrak{A} : P \vdash \alpha P$ by production (f); $\alpha P \vdash a_{j_1}\alpha a_{j_0}a_{j_2}\ldots a_{j_k} \vdash a_{j_1}a_{j_2}\alpha a_{j_0}a_{j_3}\ldots a_{j_k}\ldots \vdash a_{j_1}a_{j_2}\ldots a_{j_k}\alpha a_{j_0}$ by production (e). Thus, $\mathfrak{A} : P \models a_{j_1}a_{j_2}\ldots a_{j_k}\alpha a_{j_0}$. Then, by production (f), $\mathfrak{A} : P \models \alpha a_{j_1}a_{j_2}\ldots a_{j_k}\alpha a_{j_0}$. Applying, as before, production (e), $\mathfrak{A} : P \models a_{j_2}a_{j_3}\ldots a_{j_k}\alpha a_{j_1}\alpha a_{j_0}$. Iterating this process, we obtain $\mathfrak{A} : P \models \alpha a_{j_k}\alpha a_{j_{k-1}}\alpha\ldots\alpha a_{j_1}\alpha a_{j_0}$. Then, by production (f), $\mathfrak{A} : P \models \alpha\alpha a_{j_k}\alpha a_{j_{k-1}}\alpha\ldots\alpha a_{j_1}\alpha a_{j_0}$, and, by production (a), $\mathfrak{A} : P \models \beta a_{j_k}\alpha a_{j_{k-1}}\alpha\ldots\alpha a_{j_1}\alpha a_{j_n}$; applying productions (b) and (c) and finally, (d), we arrive at $\mathfrak{A} : P \models \cdot\breve{P}$. Thus, $\mathfrak{A}$ is a normal algorithm over A that inverts every word of A.†

Exercises

5.61 Let A be an alphabet. Describe the action of the normal algorithms given by the following schemas.

(a) Let Q be a fixed word in A and let the algorithm schema be: $\Lambda \to \cdot Q$.

(b) Let Q be a fixed word in A and let α be a symbol not in A. Let $B = A \cup \{\alpha\}$. Consider the schema

$$\alpha\xi \to \xi\alpha \qquad (\xi \text{ in A})$$
$$\alpha \to \cdot Q$$
$$\Lambda \to \alpha$$

(c) Let Q be a fixed word in A. Take the schema

$$\xi \to \Lambda \qquad (\xi \text{ in A})$$
$$\Lambda \to \cdot Q$$

(d) Let $B = A \cup \{|\}$. Consider the schema

$$\xi \to | \qquad (\xi \text{ in } A - \{|\})$$
$$\Lambda \to \cdot |$$

5.62 Let A be an alphabet not containing the symbols α, β, γ. Let $B = A \cup \{\alpha\}$ and $C = A \cup \{\alpha, \beta, \gamma\}$.

(a) Construct a normal algorithm $\mathfrak{A}$ in B such that $\mathfrak{A}(\Lambda) = \Lambda$ and $\mathfrak{A}(\xi P) = P$ for any symbol ξ in A and any word P in A. Thus, $\mathfrak{A}$ erases the first letter of any nonempty word in A.

(b) Construct a normal algorithm $\mathfrak{D}$ in B such that $\mathfrak{D}(\Lambda) = \Lambda$ and $\mathfrak{D}(P\xi) = P$ for any symbol ξ in A and word P in A. Thus, $\mathfrak{D}$ erases the last letter of any nonempty word in A.

(c) Construct a normal algorithm $\mathfrak{C}$ in B such that $\mathfrak{C}(P) = \Lambda$ if P contains exactly two occurrences of α and $\mathfrak{C}(P)$ is defined and $\neq \Lambda$ in all other cases.

(d) Construct a normal algorithm $\mathfrak{B}$ in C such that, for any word P of A, $\mathfrak{B}(P) = PP$.

5.63 Let A and B be alphabets and let α be a symbol in neither A nor B. For certain symbols $a_1, \ldots, a_k$ in A, let $Q_1, \ldots, Q_k$ be corresponding words in B. Consider the algorithm that

†The distinction between a normal algorithm in A and a normal algorithm over A is important. A normal algorithm in A uses only symbols of A, whereas a normal algorithm over A may use additional symbols not in A. Every normal algorithm in A is a normal algorithm over A, but there are algorithms in A that are determined by normal algorithms over A but that are not normal algorithms in A [for example, the algorithm of Exercise 5.62(d)].

associates with each word of A the word $\text{Sub}^{a_1,\ldots,a_k}_{Q_1,\ldots,Q_k}(P)$ obtained by simultaneous substitution of each Q_i for $a_i (i = 1, \ldots, k)$. Show that this is given by a normal algorithm in $A \cup B \cup \{\alpha\}$.

5.64 Let $H = \{|\}$ and $M = \{|, B\}$. Every natural number n is represented by its numeral $\overline{n}$, which is a word in H. We represent every k-tuple $(n_1, n_2, \ldots, n_k)$ of natural numbers by the word $\overline{n}_1 B \overline{n}_2 B \ldots B \overline{n}_k$ in M. We shall denote this word by $\overline{(n_1, \ldots, n_k)}$. For example, $\overline{(3, 1, 2)}$ is $||||B||B|||$.

(a) Show that the schema

$$B \to B$$

$$\alpha|| \to \alpha|$$

$$\alpha| \to \cdot|$$

$$\Lambda \to \alpha$$

defines a normal algorithm $\mathfrak{A}_Z$ over M such that $\mathfrak{A}_Z(\overline{n}) = \overline{0}$ for any n, and $\mathfrak{A}_Z$ is applicable only to numerals in M.

(b) Show that the schema

$$B \to B$$

$$\alpha| \to \cdot||$$

$$\Lambda \to \alpha$$

defines a normal algorithm $\mathfrak{A}_N$ over M such that $\mathfrak{A}_N(\overline{n}) = \overline{n+1}$ for all n, and $\mathfrak{A}_N$ is applicable only to numerals in M.

(c) Let $\alpha_1, \ldots, \alpha_{2k}$ be symbols not in M. Let $1 \leqslant j \leqslant k$. Let $\mathscr{S}_i$ be the list

$$\alpha_{2i-1}B \to \alpha_{2i-1}B$$

$$\alpha_{2i-1}| \to \alpha_{2i}|$$

$$\alpha_{2i}| \to \alpha_{2i}$$

$$\alpha_{2i}B \to \alpha_{2i+1}$$

If $1 < j < k$, consider the algorithm schema

$$\mathscr{S}_1$$
$$\vdots$$
$$\mathscr{S}_{j-1}$$
$$\alpha_{2j-1}B \to \alpha_{2j-1}B$$
$$\alpha_{2j-1}| \to \alpha_{2j}|$$
$$\alpha_{2j}| \to |\alpha_{2j}$$
$$\alpha_{2j}B \to \alpha_{2j+1}$$
$$\mathscr{S}_{j+1}$$
$$\vdots$$
$$\mathscr{S}_{k-1}$$
$$\alpha_{2k-1}B \to \alpha_{2k-1}B$$
$$\alpha_{2k-1}| \to \alpha_{2k}|$$
$$\alpha_{2k}| \to \alpha_{2k}$$
$$\alpha_{2k}B \to \alpha_{2k}B$$
$$\alpha_{2k} \to \cdot\Lambda$$
$$\Lambda \to \alpha_1$$

If $j = 1$, consider the schema

$$\alpha_1 B \to \alpha_1 B$$
$$\alpha_1| \to \alpha_2|$$
$$\alpha_2| \to |\alpha_2$$
$$\alpha_2 B \to \alpha_3$$
$$\mathscr{S}_2$$
$$\vdots$$
$$\mathscr{S}_{k-1}$$
$$\alpha_{2k-1}B \to \alpha_{2k-1}B$$
$$\alpha_{2k-1}| \to \alpha_{2k}|$$
$$\alpha_{2k}| \to \alpha_{2k}$$
$$\alpha_{2k}B \to \alpha_{2k}B$$
$$\alpha_{2k} \to \cdot\Lambda$$
$$\Lambda \to \alpha_1$$

If $j = k$, consider the schema

$$\mathscr{S}_1$$
$$\vdots$$
$$\mathscr{S}_{k-1}$$
$$\alpha_{2k-1}B \to \alpha_{2k-1}B$$
$$\alpha_{2k-1}| \to \alpha_{2k}|$$
$$\alpha_{2k}| \to |\alpha_{2k}$$
$$\alpha_{2k}B \to \alpha_{2k}B$$
$$\alpha_{2k} \to \cdot\Lambda$$
$$\Lambda \to \alpha_1$$

Show that the corresponding normal algorithm $\mathfrak{A}_j^k$ is such that $\mathfrak{A}_j^k(\overline{(n_1,\ldots,n_k)}) = \overline{n}_j$; and $\mathfrak{A}_j^k$ is applicable to only words of the form $\overline{(n_1,\ldots,n_k)}$.

(d) Construct a schema for a normal algorithm in M transforming $\overline{(n_1, n_2)}$ into $\overline{|n_1 - n_2|}$.

(e) Construct a normal algorithm in M for addition.

(f) Construct a normal algorithm over M for multiplication.

Given algorithms $\mathfrak{A}$ and $\mathfrak{B}$ and a word P, we write $\mathfrak{A}(P) \approx \mathfrak{B}(P)$ if and only if either $\mathfrak{A}$ and $\mathfrak{B}$ are both applicable to P and $\mathfrak{A}(P) = \mathfrak{B}(P)$ or neither $\mathfrak{A}$ nor $\mathfrak{B}$ is applicable to P. More generally, if C and D are expressions, then $C \approx D$ is to hold if and only if neither C nor D is defined or both C and D are defined and denote the same object. If $\mathfrak{A}$ and $\mathfrak{B}$ are algorithms over an alphabet A, then we say that $\mathfrak{A}$ and $\mathfrak{B}$ are *fully equivalent* relative to A if and only if $\mathfrak{A}(P) \approx \mathfrak{B}(P)$ for every word P in A; we say that $\mathfrak{A}$ and $\mathfrak{B}$ are *equivalent* relative to A if and only if, for any word P in A, whenever $\mathfrak{A}(P)$ or $\mathfrak{B}(P)$ exists and is in A, then $\mathfrak{A}(P) \approx \mathfrak{B}(P)$.

Let M be the alphabet $\{|, B\}$, as in Exercise 5.64; let ω be the set of natural numbers. Given a partial number-theoretic function φ of k arguments—that is, a function from a subset of ω^k into ω—we denote by $\mathfrak{B}_\varphi$ the corresponding function in M; that is, $\mathfrak{B}_\varphi(\overline{(n_1,\ldots,n_k)}) = \overline{\varphi(n_1,\ldots,n_k)}$ whenever either of the two sides of the equation is defined; $\mathfrak{B}_\varphi$ is assumed to be inapplicable to words not of the form $\overline{(n_1,\ldots,n_k)}$. The function φ is said to be *Markov-computable* if and only if there is a normal algorithm $\mathfrak{A}$ over M that is fully equivalent to $\mathfrak{B}_\varphi$ relative to M.†

A normal algorithm is said to be *closed* if and only if one of the productions in its schema has the form $\Lambda \to \cdot Q$. Such an algorithm can end only terminally—that is, by an application of a terminal production. Given an arbitrary normal algorithm $\mathfrak{A}$, add on at the end of the schema for $\mathfrak{A}$ the new production $\Lambda \to \cdot \Lambda$, and denote by $\mathfrak{A}\cdot$ the normal algorithm determined by this enlarged schema. $\mathfrak{A}\cdot$ is closed, and $\mathfrak{A}\cdot$ is fully equivalent to $\mathfrak{A}$ relative to the alphabet of $\mathfrak{A}$.

Let us show now that the composition of two normal algorithms is again a normal algorithm. Let $\mathfrak{A}$ and $\mathfrak{B}$ be normal algorithms in an alphabet A. For each symbol b in A, form a new symbol $\bar{b}$, called the correlate of b. Let $\bar{A}$ be the alphabet consisting of the correlates of the symbols of A. Let α and β be two symbols not in $A \cup \bar{A}$. Let $\mathscr{S}_{\mathfrak{A}}$ be the schema of $\mathfrak{A}\cdot$ except that the terminal dot in terminal productions is replaced by α. Let $\mathscr{S}_{\mathfrak{B}}$ be the schema of $\mathfrak{B}\cdot$ except that every symbol is replaced by its correlate, every terminal dot is replaced by β, productions of the form $\Lambda \to Q$ are replaced by $\alpha \to \alpha Q$, and productions $\Lambda \to \cdot Q$ are replaced by $\alpha \to \alpha\beta Q$. Consider the abbreviated schema

$$
\begin{array}{ll}
a\alpha \to \alpha a & (\text{a in A}) \\
\alpha a \to \alpha \bar{a} & (\text{a in A}) \\
\bar{a} b \to \bar{a}\bar{b} & (\text{a, b in A}) \\
\bar{a}\beta \to \beta \bar{a} & (\text{a in A})
\end{array}
$$

† In this and in all other definitions in this chapter, the existential quantifier "there is" is meant in the ordinary, "classical" sense. When we assert that there exists an object of a certain kind, we do not necessarily imply that any human being has found or ever will find such an object. Thus, a function φ may be Markov-computable without our ever knowing it to be so.

$$\beta\bar{a} \rightarrow \beta a \qquad (a \text{ in } A)$$

$$a\bar{b} \rightarrow ab \qquad (a, b \text{ in } A)$$

$$\alpha\beta \rightarrow \cdot\Lambda$$

$$\mathscr{S}_{\mathfrak{B}}$$

$$\mathscr{S}_{\mathfrak{A}}$$

This schema determines a normal algorithm $\mathfrak{C}$ over A such that $\mathfrak{C}(P) \approx \mathfrak{B}(\mathfrak{A}(P))$ for any word P in A. $\mathfrak{C}$ is called the *composition* of $\mathfrak{A}$ and $\mathfrak{B}$ and is denoted $\mathfrak{B} \circ \mathfrak{A}$. In general, by $\mathfrak{A}_n \circ \cdots \circ \mathfrak{A}_1$, we mean $\mathfrak{A}_n \circ (\cdots \circ (\mathfrak{A}_3 \circ (\mathfrak{A}_2 \circ \mathfrak{A}_1))\cdots)$.

Let $\mathfrak{D}$ be a normal algorithm in an alphabet A and let B be an extension of A. If we take a schema for $\mathfrak{D}$ and prefix to it the productions $b \rightarrow b$ for each symbol b in B − A, then the new schema determines a normal algorithm $\mathfrak{D}_B$ in B such that $\mathfrak{D}_B(P) \approx \mathfrak{D}(P)$ for every word P in A, and $\mathfrak{D}_B$ is not applicable to any word in B that contains any symbol of B − A. $\mathfrak{D}_B$ is fully equivalent to $\mathfrak{D}$ relative to A and is called the *propagation* of $\mathfrak{D}$ onto B.

Assume that $\mathfrak{A}$ is a normal algorithm in an alphabet A_1 and $\mathfrak{B}$ is a normal algorithm in an alphabet A_2. Let $A = A_1 \cup A_2$. Let $\mathfrak{A}_A$ and $\mathfrak{B}_A$ be the propagations of $\mathfrak{A}$ and $\mathfrak{B}$, respectively, onto A. Then the composition $\mathfrak{C}$ of $\mathfrak{A}_A$ and $\mathfrak{B}_A$ is called the *normal composition* of $\mathfrak{A}$ and $\mathfrak{B}$ and is denoted by $\mathfrak{B} \circ \mathfrak{A}$. (When $A_1 = A_2$, the normal composition of $\mathfrak{A}$ and $\mathfrak{B}$ is identical with the composition of $\mathfrak{A}$ and $\mathfrak{B}$; hence, the notation $\mathfrak{B} \circ \mathfrak{A}$ is unambiguous.) $\mathfrak{C}$ is a normal algorithm over A such that $\mathfrak{C}(P) \approx \mathfrak{B}(\mathfrak{A}(P))$ for any word P in A_1, and $\mathfrak{C}$ is applicable to only those words P of A such that P is a word of A_1, $\mathfrak{A}$ is applicable to P, and $\mathfrak{B}$ is applicable to $\mathfrak{A}(P)$.

PROPOSITION 5.24 Let $\mathscr{T}$ be a Turing machine with alphabet A. Then there is a normal algorithm $\mathfrak{A}$ over A that is fully equivalent to the Turing algorithm $Alg_{\mathscr{T}}$ relative to A.

Proof Let $D = A \cup \{q_{k_0}, \ldots, q_{k_m}\}$, where $q_{k_0}, \ldots, q_{k_m}$ are the internal states of $\mathscr{T}$ and $q_{k_0} = q_0$. Write the algorithm schema for $\mathfrak{A}$ as follows: first, for all quadruples $q_j a_i a_k q_r$ of $\mathscr{T}$, take the production $q_j a_i \rightarrow q_r a_k$. Second, for each quadruple $q_j a_i L q_r$, take the productions $a_l q_j a_i \rightarrow q_r a_l a_i$ for all symbols a_l of A; then take the production $q_j a_i \rightarrow q_r a_0 a_i$. Third, for each quadruple $q_j a_i R q_r$, take the productions $q_j a_i a_l \rightarrow a_i q_r a_l$ for all symbols a_l of A; then take the production $q_j a_i \rightarrow a_i q_r a_0$. Fourth, write the productions $q_{k_i} \rightarrow \cdot\Lambda$ for each internal state q_{k_i} of $\mathscr{T}$, and finally, take $\Lambda \rightarrow q_0$. This schema defines a normal algorithm $\mathfrak{A}$ over A, and it is easy to see that, for any word P of A, $Alg_{\mathscr{T}}(P) \approx \mathfrak{A}(P)$.

COROLLARY 5.25 Every Turing-computable function is Markov-computable.

Proof Let $f(x_1, \ldots, x_n)$ be standard Turing-computable by a Turing machine $\mathscr{T}$ with alphabet $A \supseteq \{|, B\}$. (Remember that B is a_0 and | is a_1.) We know that, for any natural numbers $k_1, \ldots, k_n$, if $f(k_1, \ldots, k_n)$ is not defined, then $Alg_{\mathscr{T}}$ is not applicable to $\overline{(k_1, \ldots, k_n)}$, whereas, if $f(k_1, \ldots, k_n)$ is defined, then $Alg_{\mathscr{T}}(\overline{(k_1, \ldots, k_n)})) \approx$

$R_1\overline{(k_1,\ldots,k_n)}B\overline{f(k_1,\ldots,k_n)}R_2$, where R_1 and R_2 are (possibly empty) sequences of B's. Let $\mathfrak{B}$ be a normal algorithm over A that is fully equivalent to $Alg_{\mathcal{T}}$ relative to A. Let $\mathfrak{C}$ be the normal algorithm over $\{1, B\}$ determined by the schema:

$$\alpha B \to \alpha$$
$$\alpha| \to \beta|$$
$$\beta| \to |\beta$$
$$\beta B \to B\gamma$$
$$\gamma| \to \beta|$$
$$\gamma B \to \gamma$$
$$B\gamma \to \cdot\Lambda$$
$$\beta \to \cdot\Lambda$$
$$\Lambda \to \alpha$$

If R_1 and R_2 are possibly empty sequences of B's, then $\mathfrak{C}$, when applied to $R_1\overline{(k_1,\ldots,k_n)}B\overline{f(k_1,\ldots,k_n)}R_2$, will erase R_1 and R_2. Finally, let $\mathfrak{A}_{n+1}^{n+1}$ be the normal algorithm defined in Exercise 5.64(c). Then the normal composition $\mathfrak{A}_{n+1}^{n+1} \circ \mathfrak{C} \circ \mathfrak{B}$ is a normal algorithm that computes $f(x_1,\ldots,x_n)$.

Let $\mathfrak{A}$ be any algorithm over an alphabet $A = \{a_{j_0},\ldots,a_{j_m}\}$. We can associate with $\mathfrak{A}$ a partial number-theoretic function $\psi_{\mathfrak{A}}$ such that $\psi_{\mathfrak{A}}(n) = m$ if and only if either n is not the Gödel number* of a word of A and $m = 0$, or n and m are Gödel numbers of words P and Q of A such that $\mathfrak{A}(P) = Q$.

PROPOSITION 5.26 If $\mathfrak{A}$ is a normal algorithm over $A = \{a_{j_0},\ldots,a_{j_m}\}$, then $\psi_{\mathfrak{A}}$ is partial recursive.

Proof We may assume that the symbols of the alphabet of $\mathfrak{A}$ are of the form a_i. Given a simple production $P \to Q$, we call $2^1 3^{g(P)} 5^{g(Q)}$ its *index*; given a terminal production $P \to \cdot Q$, we let $2^2 3^{g(P)} 5^{g(Q)}$ be its *index*. If $P_0 \to (\cdot)Q_0, \ldots, P_r \to (\cdot)Q_r$ is an algorithm schema, we let its *index* be $2^{k_0}3^{k_1}\ldots p_r^{k_r}$, where k_i is the index of $P_i \to (\cdot)Q_i$. Let Word(u) be the recursive predicate that holds if and only if u is the Gödel number of a finite sequence of symbols of the form a_i:

$$u \neq 0 \wedge [u = 1 \vee (\forall z)(z < \ell h(u) \Rightarrow (\exists y)(y < u \wedge (u)_z = 7 + 4y))]$$

Let SI(u) be the recursive predicate that holds when u is the index of a simple production: $\ell h(u) = 3 \wedge (u)_0 = 1 \wedge \text{Word}((u)_1) \wedge \text{Word}((u)_2)$. Similarly, TI($u$) is the recursive predicate that holds when u is the index of a terminal production: $\ell h(u) = 3 \wedge (u)_0 = 2 \wedge \text{Word}((u)_1) \wedge \text{Word}((u)_2)$. Let Ind($u$) be the recursive predicate that

*Here and below, we use the Gödel numbering of the language of Turing-computability given on page 246. Thus, the Gödel number $g(a_i)$ of a_i is $7 + 4i$. In particular, $g(B) = g(a_0) = 7$ and $g(|) = g(a_1) = 11$.

holds when u is the index of an algorithm schema: $u > 1 \wedge (\forall z)(z < \ell h(u) \Rightarrow \mathrm{SI}((u)_z) \vee \mathrm{TI}((u)_z))$. Let $\mathrm{Lsub}(x, y, e)$ be the recursive predicate that holds if and only if e is the index of a production $\mathrm{P} \to (\cdot)\mathrm{Q}$ and x and y are Gödel numbers of words U and V such that P occurs in U, and V is the result of substituting Q for the leftmost occurrence of P in U:

$$\mathrm{Word}(x) \wedge \mathrm{Word}(y) \wedge (\mathrm{SI}(e) \vee \mathrm{TI}(e)) \wedge (\exists u)_{u \leqslant x}(\exists v)_{v \leqslant x}(x = u * (e)_1 * v$$

$$\wedge\ y = u * (e)_2 * v \wedge \neg(\exists w)_{w \leqslant x}(\exists z)_{z \leqslant x}(x = w * (e)_1 * z \wedge w < u))$$

Let $\mathrm{Occ}(x, y)$ be the recursive predicate that holds when x and y are Gödel numbers of words U and V such that V occurs in U: $\mathrm{Word}(x) \wedge \mathrm{Word}(y) \wedge (\exists v)_{v \leqslant x}(\exists z)_{z \leqslant x}(x = v * y * z)$. Let $\mathrm{End}(e, z)$ be the recursive predicate that holds when and only when z is the Gödel number of a word P, and e is the index of an algorithm schema defining an algorithm $\mathfrak{A}$ that cannot be applied to P (i.e., $\mathfrak{A} : \mathrm{P} \sqsupset$): $\mathrm{Ind}(e) \wedge \mathrm{Word}(z) \wedge (\forall w)_{w < \ell h(e)} \neg \mathrm{Occ}(z, ((e)_w)_1)$. Let $\mathrm{SCons}(e, y, x)$ be the recursive predicate that holds if and only if e is the index of an algorithm schema and y and x are Gödel numbers of words V and U such that V arises from U by a simple production of the schema:

$$\mathrm{Ind}(e) \wedge \mathrm{Word}(x) \wedge \mathrm{Word}(y) \wedge (\exists v)_{v < \ell h(e)}[\mathrm{SI}((e)_v) \wedge \mathrm{Lsub}(x, y, (e)_v)$$

$$\wedge (\forall z)_{z < v} \neg \mathrm{Occ}(x, ((e)_z)_1)]$$

Similarly, one defines the recursive predicate $\mathrm{TCons}(e, y, x)$, which differs from $\mathrm{SCons}(e, y, x)$ only in that the production in question is terminal. Let $\mathrm{Der}(e, x, y)$ be the recursive predicate that is true when and only when e is the index of an algorithm schema that determines an algorithm $\mathfrak{A}$, x is the Gödel number of a word U_0, y is the Gödel number of a sequence of words $\mathrm{U}_0, \ldots, \mathrm{U}_k (k \geqslant 0)$ such that, for $0 \leqslant i < k \dot{-} 1$, U_{i+1} arises from U_i by a production of the schema, and, either $\mathfrak{A} : \mathrm{U}_{k \dot{-} 1} \vdash \cdot \mathrm{U}_k$ or $\mathfrak{A} : \mathrm{U}_{k \dot{-} 1} \vdash \mathrm{U}_k$ and $\mathfrak{A} : \mathrm{U}_k \sqsupset$ (or, if $k = 0$, just $\mathfrak{A} : \mathrm{U}_k \sqsupset$):

$$\mathrm{Ind}(e) \wedge \mathrm{Word}(x) \wedge (\forall z)_{z < \ell h(y)} \mathrm{Word}((y)_z) \wedge (y)_0 = x$$

$$\wedge (\forall z)_{z < \ell h(y) \dot{-} 2} \mathrm{Scons}(e, (y)_{z+1}, (y)_z) \wedge [(\ell h(y) = 1 \wedge \mathrm{End}(e, (y)_0))$$

$$\vee (\ell h(y) > 1 \wedge \{\mathrm{TCons}(e, (y)_{\ell h(y) \dot{-} 1}, (y)_{\ell h(y) \dot{-} 2}) \vee (\mathrm{SCons}(e, (y)_{\ell h(y) \dot{-} 1}, (y)_{\ell h(y) \dot{-} 2})$$

$$\wedge \mathrm{End}(e, (y)_{\ell h(y) \dot{-} 1}))\})]$$

Let $W_{\mathrm{A}}(u)$ be the recursive predicate that holds if and only if u is the Gödel number of a word of A:

$$u \neq 0 \wedge (u = 1 \vee (\forall z)_{z < \ell h(u)}((u)_z = 7 + 4j_0 \vee \cdots \vee (u)_z = 7 + 4j_m)$$

Let e be the index of an algorithm schema for $\mathfrak{A}$. Now define the partial recursive function $\phi(x) = \mu y((W_{\mathrm{A}}(x) \wedge \mathrm{Der}(e, x, y)) \vee \neg W_{\mathrm{A}}(x))$. But $\psi_{\mathfrak{A}}(x) = (\phi(x))_{\ell h(\phi(x)) \dot{-} 1}$. Therefore, $\psi_{\mathfrak{A}}$ is partial recursive.

COROLLARY 5.27 Every Markov-computable function ϕ is partial recursive.

Proof Let $\mathfrak{A}$ be a normal algorithm over $\{1, \mathrm{B}\}$ such that $\phi(k_1, \ldots, k_n) = l$ if and only if $\mathfrak{A}(\overline{(k_1, \ldots, k_n)}) = \bar{l}$. By Proposition 5.26, the function $\psi_{\mathfrak{A}}$ is partial recursive. Define the recursive function $\gamma(x) = \ell h(x) \dot{-} 1$. If $x = \prod_{i=0}^{n} p_i^{11}$, then $n = \gamma(x)$. [Remember that a stroke |, which is an abbreviation for a_1, has Gödel number 11. So, if x is the Gödel number of the numeral $\bar{n}$, then $\gamma(x) = n$.] Let $\xi(k_1, \ldots, k_n)$ be the Gödel number of $\overline{(k_1, \ldots, k_n)}$:

$$\xi(k_1, \ldots, k_n) = g(\overline{(k_1, \ldots, k_n)}) = g(|^{k_1+1}\mathrm{B}|^{k_2+1}\mathrm{B} \ldots \mathrm{B}|^{k_n+1})$$

$$= \left(\prod_{i=0}^{k+1} p_i^{11}\right) \cdot (p_{k_1+2})^7 \cdot \left(\prod_{i=0}^{k_2+1} (p_{i+k_1+3})^{11}\right) \cdot (p_{k_1+k_2+5})^7 \cdots$$

$$\cdot (p_{k_1+\cdots+k_n+2n\dot{-}3})^7 \cdot \left(\prod_{i=0}^{k_n+1} (p_{i+k_1+\cdots+k_n+2n\dot{-}2})^{11}\right)$$

ξ is clearly recursive. Then $\phi = \gamma \circ \psi_{\mathfrak{A}} \circ \xi$ is partial recursive.

The equivalence of Markov-computability and Turing-computability follows from Corollaries 5.25 and 5.27 and the known equivalence of Turing-computability and partial recursiveness.

III. REGISTER MACHINES

To bring the theory of computability closer to the viewpoint of those who work with modern digital computers, let us consider certain computing devices called *register machines.*

A register machine $\mathscr{M}$ contains a finite number of registers or compartments, $\mathrm{R}_1, \mathrm{R}_2, \ldots, \mathrm{R}_m$. At any given time, each register contains a word in a given alphabet $\mathrm{A} = \{\mathrm{b}_1, \ldots, b_r\}$. There is also a program, a list of instructions, $\mathrm{L}_0, \mathrm{L}_1, \ldots, \mathrm{L}_k$. The instructions may be of the following types:

1. *Print instruction.* Print symbol b_l to the left of the word in register R_j. Then go to instruction L_u. (Abbreviation: $\mathrm{P}j, \mathrm{b}_l, u$)
2. *Decision instruction.* If register R_j is empty, go to instruction L_v. Otherwise, if the first symbol of the word in R_j is b_t, erase that symbol and proceed to instruction $\mathrm{L}_{u_t}(t = 1, \ldots, r)$. (Abbreviation: $\mathrm{D}j, v, u_1, \ldots, u_r$)
3. *Stop.*

We assume that the first instruction to be read by the machine is always L_0. We also fix the last register R_m as the *output register*. A register machine determines the following algorithm. Given any initial setting of the contents of the registers, the machine begins to change the contents of the registers according to the instructions in its program. If the machine ever reaches the instruction *Stop*, it actually does stop, and the content of the output register is considered the result of the algorithm. If the instruction *Stop* is never reached, the machine never stops and the algorithm produces no result.

Although register machines resemble digital computers, they differ from actual computers in having unbounded *memory capacity*. In fact, each register is allowed to hold arbitrarily long words, whereas real computers have a fixed upper limit on the lengths of such words. Another unrealistic assumption is that we permit register machines to have any number of instructions.

Register machines allow us to compute number-theoretic functions. A partial number-theoretic function $f(x_1, \ldots, x_n)$ is said to be *register-computable* if there is a register machine $\mathcal{M}$ that satisfies the following conditions:

1. $\mathcal{M}$ has at least n registers.
2. The alphabet of $\mathcal{M}$ contains at least the symbol a_1 (abbreviated, as usual, by the stroke |).
3. For any natural numbers $k_1, \ldots, k_n$, if we start $\mathcal{M}$ off with $\overline{k_1}$ in register R_1, $\overline{k_2}$ in register $R_2, \ldots, \overline{k_n}$, in register R_n, and any other register is empty, then $\mathcal{M}$ stops if and only if $f(k_1, \ldots, k_n)$ is defined; moreover, if $\mathcal{M}$ stops, the content of the output register is $\overline{f(k_1, \ldots, k_n)}$.

The essential ideas of register-computability seem to have been put forth by Shepherdson and Sturgis (1963) and independently by Melzak (1961) and Lambek (1961).

Examples

1. The zero function $Z(x) = 0$ is register-computable by the register machine with alphabet {|}, registers R_1 and R_2, and instructions:

   L_0. P2, |, 1

   L_1. Stop.

   L_0 prints | in register R_2 and transfers to L_1, the Stop instruction. We are left with $\overline{0}$ in the output register R_2.
2. The successor function $N(x) = x + 1$ is register-computable by the register machine with alphabet {|}, register R_1, and instructions:

   L_0. P1, |, 1

   L_1. Stop.

3. The projection function $U_j^n(x_1, \ldots, x_n) = x_j$ is register-computable by the register machine with alphabet {|}, registers $R_1, \ldots, R_{n+1}$, and instructions:

   L_0. Dj, 2, 1

   L_1. P(n + 1), |, 0

   L_2. Stop.

   L_0 and L_1 move the content of R_j to R_{n+1}.

Exercises

5.65 Show that the following functions are register-computable.
(a) $2x_1$ (b) $x_1 + x_2$ (c) $x_1 \dot{-} x_2$ (d) $x_1 x_2$

5.66 Find instructions that will copy the content of one register R_u into an empty register

R_v without ultimately changing the content of R_u. Assume that the alphabet is $\{b_1, \ldots, b_r\}$. (*Hint:* It will be necessary to use an additional register.)

PROPOSITION 5.28 Every Turing-computable function is register-computable.

Proof Assume that $f(x_1, \ldots, x_n)$ is standard Turing-computable by a Turing machine $\mathcal{T}$ in the alphabet $\{a_0, a_1, \ldots, a_r\}$. We shall construct a register machine $\mathcal{M}$, with the same alphabet and $n + 3$ registers, that will compute f. We are given an n-tuple $k_1, \ldots, k_n$ of natural numbers. The registers $R_1, \ldots, R_n$ must start off with the contents $\overline{k_1}, \ldots, \overline{k_n}$, respectively. Let us call the three other registers R_α, R_β, and R_γ. First, we specify instructions that will print $\overline{k_1}a_0\overline{k_1}a_0 \ldots a_0\overline{k_n}$ in register R_α (so that R_α will contain the initial tape of the Turing machine $\mathcal{T}$).

L_0.	Dn, $r+2, 1, 2, \ldots, r+1$	L_{r+3}.	D$(n-1)$, $2r+5, r+4, \ldots, 2r+4$ $\ldots$
L_1.	Pα, a_0, 0	L_{r+4}.	Pα, a_0, $r+3$
L_2.	Pα, a_1, 0	L_{r+5}.	Pα, a_1, $r+3$
$\vdots$	$\vdots$	$\vdots$	$\vdots$
L_{r+1}.	Pα, a_r, 0	L_{2r+4}.	Pα, a_r, $r+3$
L_{r+2}.	Pα, a_0, $r+3$	L_{2r+5}.	Pα, a_0, $2r+6$ $\ldots$

The first group of instructions on the left prints $a_0\overline{k_n}$ in R_α. The next vertical grouping moves the content of R_{n-1} and one a_0 into R_α; thus, R_α now contains $a_0\overline{k_{n-1}}a_0\overline{k_n}$. We add more instructions in a similar way so as to finally obtain $\overline{k_1}a_0\overline{k_2}a_0 \ldots a_0\overline{k_n}$ in R_α. [The last vertical grouping will start with an instruction of the form D1, $n(r+3)-1, \ldots, n(r+3)-2$, so that the first instruction label not yet used will be $L_{n(r+3)-1}$.] Now that we have the initial tape of the Turing machine $\mathcal{T}$ inscribed in register R_α, we wish to mimic the behavior of $\mathcal{T}$ inside the register machine $\mathcal{M}$. The basic idea is that the following conditions must hold: (1) register R_α will contain everything that is on the Turing machine tape to the right of the scanned symbol (including that symbol); and (2) register R_β will contain, in reverse order, everything that is on the tape to the left of the scanned symbol. In particular, the symbol (if any) immediately to the left of the scanned symbol on the tape will be the first symbol in R_β. At the point we have reached, these conditions hold, since R_α contains the initial tape of $\mathcal{T}$ and R_β is empty. As the Turing machine continues to operate, we must add appropriate instructions to ensure that conditions 1 and 2 remain true.

We shall associate with every quadruple of $\mathcal{T}$ certain instructions of $\mathcal{M}$. To this end, for every pair of natural numbers j, i such that q_j is an internal state and a_i is in the alphabet of $\mathcal{T}$, we choose a natural number $\zeta(j, i)$ that will serve as a subscript of an instruction $L_{\zeta(j,i)}$. In what follows, it is assumed that these and all other new subscripts of instructions are chosen so that they are different from all other previously chosen subscripts.

If $q_j a_i R q_u$ is a quadruple of $\mathscr{T}$, we put the following instructions in $\mathscr{M}$:

$$\begin{array}{ll} L_{\zeta(j,i)}. & P\beta, a_i, \mu \\ L_{\mu}. & D\alpha, 0, \tau, \tau, \ldots, \tau \\ L_{\tau}. & D\alpha, \sigma, s_0, s_1, \ldots, s_r \\ L_{s_0}. & P\alpha, a_0, \zeta(u,0) \\ L_{s_1}. & P\alpha, a_1, \zeta(u,1) \\ \vdots & \vdots \\ L_{s_r}. & P\alpha, a_r, \zeta(u,r) \\ L_{\sigma}. & P\alpha, a_0, \zeta(u,0) \end{array}$$

These instructions insert a_i as the new first symbol in R_β, delete the first symbol a_i of R_α, and, if a_l is the symbol on the tape to the right of a_i, transfer to the instruction $L_{\zeta(u,l)}$ corresponding to the new internal state q_u and the new scanned symbol a_l. If there was no symbol on the tape to the right of a_i, a_0 is printed in R_α and we move to instruction $\zeta(u,0)$. Thus, conditions 1 and 2 remain true.

If $q_j a_i L q_u$ is a quadruple of $\mathscr{T}$, put the following instructions in $\mathscr{M}$.

$$\begin{array}{ll} L_{\zeta(j,i)}. & D\beta, t, t_0, t_1, \ldots, t_r \\ L_{t_0}. & P\alpha, a_0, \zeta(u,0) \\ L_{t_1}. & P\alpha, a_1, \zeta(u,1) \\ \vdots & \vdots \\ L_{t_r}. & P\alpha, a_r, \zeta(u,r) \\ L_{t}. & P\alpha, a_0, \zeta(u,0) \end{array}$$

These instructions move the first symbol of R_β into position as the first symbol of R_α and then transfer to the instruction corresponding to the new internal state q_u and the new scanned symbol. If R_β is empty, a_0 is inserted as the new first symbol in R_α. So, conditions 1 and 2 still hold.

If $q_j a_i a_l q_u$ is a quadruple of $\mathscr{T}$, add the following instructions:

$$\begin{array}{ll} L_{\zeta(j,i)}. & D\alpha, 0, \xi, \xi, \ldots, \xi \\ L_{\xi}. & P\alpha, a_l, \zeta(u,l) \end{array}$$

These instructions change the first symbol in R_α from a_i to a_l and then move to the instruction corresponding to the new internal state q_u and the new scanned symbol a_l. Conditions 1 and 2 are met.

If no quadruple of $\mathscr{T}$ begins with $q_j a_i$, add the following instructions:

$$\begin{array}{ll} L_{\zeta(j,i)}. & D\alpha, \chi, \chi, \nu, \nu, \ldots, \nu \\ L_{\nu}. & P\gamma, |, \zeta(j,i) \\ L_{\chi}. & \text{Stop.} \end{array}$$

If $f(k_1, \ldots, k_n)$ is defined, $\mathcal{T}$ will stop with its reading head scanning the first | of $\overline{f(k_1, \ldots, k_n)}$, possibly followed by several blanks. Since $\mathcal{T}$ has stopped, if its internal state at that moment is q_j, no quadruple of $\mathcal{T}$ will begin with $q_j a_1$. Hence, the three instructions above, corresponding to $\zeta(j, 1)$, will print $\overline{f(k_1, \ldots, k_n)}$ in the output register R_γ and then stop. If $f(k_1, \ldots, k_n)$ is not defined, $\mathcal{T}$ will never stop and neither will $\mathcal{M}$. Thus, $\mathcal{M}$ will compute the function $f(x_1, \ldots, x_n)$. [We have to specify that $\zeta(0, 1)$ is $n(r + 3) - 1$. $\mathcal{T}$ starts off in state q_0, scanning the first stroke of $\overline{k_1}$. So, the instruction associated with q_0, a_1—namely $L_{\zeta(0,1)}$—must be the first instruction available after we put the initial tape into R_α.]

Now we wish to show that every register-computable function is Turing-computable. As a preliminary, we shall need a Turing machine **T*** that performs a right translation: $\mathrm{BW\underline{B}} \sim \;\rightarrow \mathrm{B\underline{B}W} \sim$, where W is a nonempty sequence of non-blanks. **T*** is given by the diagram in Figure 5.21. In this diagram, it is assumed that the alphabet is $\{a_0, a_1, \ldots, a_k\}$.

PROPOSITION 5.29 Every register-computable function is Turing-computable.

Proof Assume that $f(x_1, \ldots, x_n)$ is register-computable by a register machine $\mathcal{M}$ with alphabet $\{a_1, \ldots, a_r\}$ and registers $R_1, \ldots, R_m$. We shall show that f can be computed by a Turing machine $\mathcal{T}$ in the alphabet $\{a_0, a_1, \ldots, a_r, \alpha\}$. We shall construct $\mathcal{T}$ by means of a diagram. With the instructions L_i of the program of $\mathcal{M}$ we shall associate Turing machines $\mathcal{T}_i$ that will be used in the construction of the diagram. The computation starts with $\overline{k_1} a_0 \overline{k_2} a_0 \ldots a_0 \overline{k_n}$ on the tape and the machine scanning the first | of $\overline{k_1}$. First, we prefix α to each $\overline{k_i}$ and also add α's to the right, separated by a_o's, to indicate the remaining $m - n$ registers of $\mathcal{M}$; that is, we must construct

$$(\#) \quad \alpha\overline{k_1} a_0 \alpha \overline{k_2} a_0 \ldots a_0 \alpha \overline{k_n} a_0 \alpha a_0 \alpha a_0 \ldots a_0 \alpha$$

in which there are, all together, m α's. Our aim is to imitate the operation of $\mathcal{M}$ on such a tape containing m nonempty words separated by a_0's; these nonempty words (excluding the "punctuation symbol" α) are to be the contents of the m registers of $\mathcal{M}$. To construct (#), begin the diagram with $\mathcal{R}\mathbf{r}(\mathbf{T^*}\alpha)^n\mathcal{R}(\mathbf{r}\alpha\mathbf{a_0})^{m-n}\mathbf{l}$. Draw an arrow from **l** to the Turing machine $\mathcal{T}_0$ that will be constructed below to correspond to

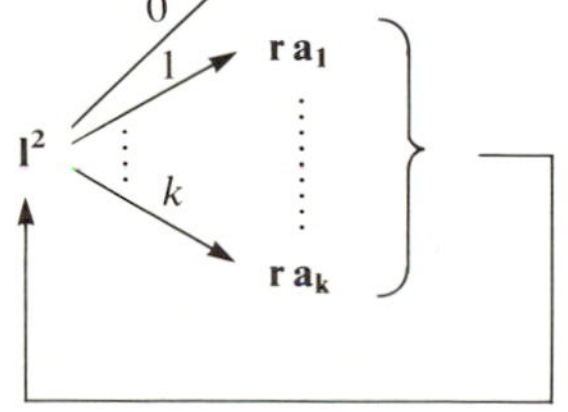

Figure 5.21

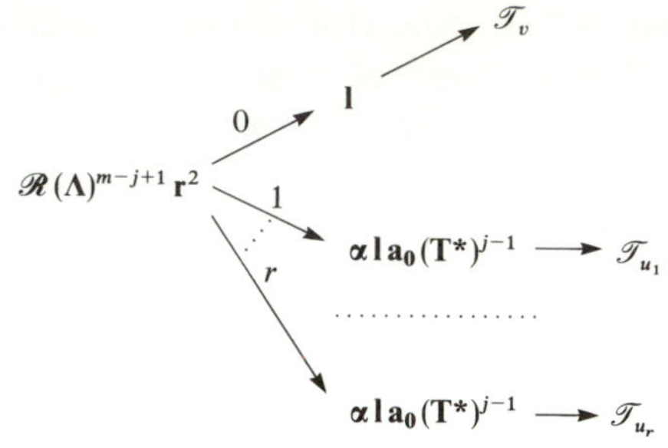

Figure 5.22

the initial instruction L_0 of $\mathscr{M}$. Now we shall describe the construction of the $\mathscr{T}_i$'s and their placement in the diagram.

If L_i is Pj, i, u, we construct a Turing machine $\mathscr{T}_i$ that moves to the right end, then moves to the beginning of the $(m - j + 1)$th word from the right, makes one square available after the α, and prints a_i in that square. As $\mathscr{T}_i$ we can take $\mathscr{R}\mathbf{r}(\mathbf{T}^*)^{m-j+1}\alpha\mathbf{r}\mathbf{a_i}$. Finally, we draw an arrow from $\mathscr{T}_i$ to $\mathscr{T}_u$.

If L_i is Dj, v, $u_1, \ldots, u_r$, construct a Turing machine $\mathscr{T}_i$ that moves to the right end, then moves $m - j + 1$ words to the left, and looks at the square after the α in that word. If it contains a_0, draw an arrow to $\mathscr{T}_v$. If it contains $a_i (1 \leqslant i \leqslant r)$, erase a_i, close up the resulting gap, and draw an arrow to $\mathscr{T}_{u_i}$. As $\mathscr{T}_i$ we can take the machine given by the diagram in Figure 5.22. (Of course, $\mathscr{T}_v, \mathscr{T}_{u_1}, \ldots, \mathscr{T}_{u_r}$ and the arrows leading to them are not part of $\mathscr{T}_i$. They just show how $\mathscr{T}_i$ is connected to the rest of the diagram.)

If L_i is Stop, then $\mathscr{M}$ has finished its computation and the value $\overline{f(k_1, \ldots, k_n)}$ is in output register R_m. So, our Turing machine tape will have the form $\alpha W_1 a_0 \alpha W_2 a_0 \ldots a_0 \alpha \overline{f(k_1, \ldots, k_n)}$. Hence, $\mathscr{T}_i$ should move to the right end, use the clean-up machine **C** to erase everything except $\alpha\overline{f(k_1, \ldots, k_n)}$, erase α, and move right into the first square of $\overline{f(k_1, \ldots, k_n)}$. Thus, our Turing machine will have computed $f(k_1, \ldots, k_n)$. As $\mathscr{T}_i$ take the machine $\mathscr{R}\mathbf{C}\Lambda\mathbf{r}\mathbf{a}_0\mathbf{r}$.

Propositions 5.28 and 5.29 show that the register-computable functions are identical with the Turing-computable functions.

The following notions of computability will be treated in less detail than the ones discussed above.

IV. RANDOM ACCESS MACHINES (RAMs)

Another idealized form of digital computer is a *random access machine* (RAM). As in the case of a register machine, a RAM uses an alphabet $\{b_1, \ldots, b_r\}$, a finite set of registers $\{R_1, \ldots, R_m\}$, and a finite list of instructions (its *program*). However, the instructions of a RAM are carried out in the order in which they are listed, except in certain indicated cases. The instructions are labeled L_0, L_1, ..., but the list of instructions is allowed to repeat the same instruction (possibly with the same label) any number of times. There are five types of instructions:

1. **Add$_j$R$_k$.** Print the symbol b_j to the right of the word in the register R_k.
2. **Del R$_k$.** Delete the first letter (if any) of the word in the register R_k.
3. **R$_u$ jmp$_j$ L$_v$ a.** If the first letter of the word in R_u is b_j, proceed to the first instruction labeled L_v that lies above this instruction. (The letter "a" stands for "above".)
4. **R$_u$ jmp$_j$ L$_v$ b.** This is the same as instruction 3 except that the new instruction is the first instruction labeled L_v that lies *below* this instruction.
5. **Continue.** This is a Stop instruction when it is the last Continue instruction in the program. Otherwise, it is ignored.

Any RAM program is required to contain the instructions indicated in instructions of forms 3 and 4. For example, in case 3, there must be an instruction labeled L_v lying above the given instruction.

To define the algorithm determined by a RAM, we place arbitrary words in the registers and start the RAM working according to its program. The first instruction carried out is the one at the top of the program. The RAM stops if it reaches the last Continue instruction.* If the RAM stops, the word in register R_1 is the result of the algorithm.

A RAM *computes* a partial number-theoretic function $f(x_1, \ldots, x_n)$ if, for any natural numbers $k_1, \ldots, k_n$, when $\overline{k_1}, \ldots, \overline{k_n}$ are placed in registers $R_1, \ldots, R_n$, respectively, and the other registers are empty, then: (1) the RAM stops if and only if $f(k_1, \ldots, k_n)$ is defined, and (2) if the RAM stops, the content of R_1 is $\overline{f(k_1, \ldots, k_n)}$. A function computed by a RAM is said to be *RAM-computable.*

Exercises

5.67 Show that the following instructions can be added to the types allowed in RAMs without extending the class of algorithms that can be carried out by RAMs.

6. **Clear R$_j$.** Erase the word in register R_j.
7. **R$_k$ ← R$_j$.** Copy the word in R_j into R_k. The previous word in R_k is erased. R_j contains the same word that it had before.
8. **Jump L$_v$ a.** Proceed to the first instruction labeled L_v lying above this instruction.
9. **Jump L$_v$ b.** Proceed to the first instruction labeled L_v lying below this instruction.

5.68 Show that the initial functions $Z(x)$, $N(x)$, and $U_j^n(x_1, \ldots, x_n)$ are RAM-computable.

The RAM-computable functions turn out to be the same as the partial recursive functions. Readers should try to work this out on their own, but a proof may be found in Machtey and Young (1978, pp. 28–38). RAMs were first introduced by Cook (1972). They are more useful than Turing machines or register machines when studying *computational complexity*, especially when we are interested in the speed with which computations can be carried out.

Many other definitions of effective computability have been given, all of them turning out to be equivalent to Turing-computability. One of the earliest definitions, λ-computability, was developed by Church and Kleene as part of the theory of λ-conversion (see Church, 1941). Its equivalence with effective computability is not

* We allow several Continue operations in a program to make it easier to combine programs to form larger programs.

immediately plausible and gained credence only when λ-computability was shown to be equivalent to partial recursiveness and Turing-computability (see Kleene, 1936b, and Turing, 1937).

All reasonable variations of Turing-computability, register-computability, and so on seem to yield equivalent notions. Accounts of such variations, in the case of Turing machines, may be found in W. Oberschelp (1958) and Fischer (1965).

6. DECISION PROBLEMS

A class of problems is said to be *unsolvable* if there is no effective procedure for solving each problem in the class. For example, given any polynomial $f(x)$ with integral coefficients (for example, $3x^5 - 4x^4 + 7x^2 - 13x + 12$), is there an integer k such that $f(k) = 0$? We can certainly answer this question for certain special polynomials, but is there a single general procedure that will solve the problem for *every* polynomial $f(x)$? (The answer is given on page 285, (4).)

If we can arithmetize the formulation of a class of problems and assign to each problem a natural number, then this class is unsolvable if and only if there is no effectively computable function h such that, if n is the number of a given problem, then $h(n)$ yields the solution of the problem. If Church's Thesis is assumed, the function h has to be partial recursive, and we then have a precise mathematical question.

Davis (1977b) gives an excellent survey of research on unsolvable problems. We shall discuss a few decision problems, some of which we already have solved.

1. Is a statement form of the propositional calculus a tautology? Truth tables provide an easy, effective procedure for answering any such question.

2. *Decidable and undecidable theories.* Is there a procedure for determining whether an arbitrary wf of a formal system $\mathscr{S}$ is a theorem of $\mathscr{S}$? If so, $\mathscr{S}$ is called *decidable*; otherwise, it is *undecidable*.

- **(a)** The system L of Chapter 1 is decidable. The theorems of L are the tautologies, and we can apply the truth table method.
- **(b)** The pure predicate calculus PP and the full predicate calculus PF were both shown to be recursively undecidable (Proposition 3.52).
- **(c)** The theory RR and all its consistent extensions (including Peano arithmetic S) have been shown to be recursively undecidable (Corollary 3.44).
- **(d)** The axiomatic set theory NBG and all its consistent extensions are recursively undecidable (pages 208–209).
- **(e)** Various theories concerning order structures or algebraic structures have been shown to be decidable (often by the method of quantifier elimination). Examples are the theory of unbounded densely ordered sets (see page 90 and Langford, 1927), the theory of abelian groups (Szmielew, 1955), and the theory of real closed fields (Tarski, 1951). For further information, consult Kreisel and Krivine (1967, chap. 4); Chang and Keisler (1973, chap. 1.5); Monk (1976, chap. 13); Ershov et al. (1965); Rabin (1977); and Baudisch

et al. (1985). The undecidability of many algebraic theories can be derived from the results in Chapter 3 (see Tarski, Mostowski & Robinson, 1953, II.6, III; and Monk, 1976, chap. 16).

3. *Logical validity.* Is a given wf of quantification theory logically valid? By Gödel's Completeness Theorem (Corollary 2.19), a wf is logically valid if and only if it is provable in the full predicate calculus PF. Since PF is recursively undecidable (Proposition 3.52), the problem of logical validity is recursively unsolvable.

However, there is a decision procedure for the logical validity of wfs of the pure monadic predicate calculus (Exercise 3.61, page 173).

There have been extensive investigations of decision procedures for various important subclasses of wfs of the pure predicate calculus; for example, the class $(\forall\exists\forall)$ of all closed wfs of the form $(\forall x)(\exists y)(\forall z)\mathscr{A}(x, y, z)$, where $\mathscr{A}(x, y, z)$ contains no quantifiers. See Ackermann (1954), Dreben and Goldfarb (1980), and Lewis (1979).

4. *Hilbert's Tenth Problem.* If $f(x_1, \ldots, x_n)$ is a polynomial with integral coefficients, are there integers $k_1, \ldots, k_n$ such that $f(k_1, \ldots, k_n) = 0$? This difficult decision problem is known as *Hilbert's Tenth Problem.*

For one variable, the solution is easy. When a_0, a_1, ..., a_n are integers, any integer x such that $a_n x^n + \cdots + a_1 x + a_0 = 0$ must be a divisor of a_0. Hence, when $a_0 \neq 0$, we can test each of the finite number of divisors of a_0. If $a_0 = 0$, $x = 0$ is a solution. However, there is no analogous procedure when the polynomial has more than one variable. It was finally shown by Matiyasevich (1970) that there is no decision procedure for determining whether a polynomial with integral coefficients has a solution consisting of integers. His proof was based in part on some earlier work of Davis, Putnam, and Robinson (1961). Of course, the proof ultimately relies on basic facts of recursion theory, particularly the existence of a nonrecursive r.e. set [Proposition 5.21(e)]. Clear expositions may be found in Davis (1973) and Matiyasevich (1972, 1973).

5. *Word problems*

(a) *Semi-Thue systems.* Let $\mathrm{B} = \{\mathrm{b}_1, \ldots, \mathrm{b}_n\}$ be a finite alphabet. Remember that a word of B is a finite sequence of elements of B. Moreover, the empty sequence Λ is considered a word of B. By a *production* of B we mean an ordered pair $\langle \mathrm{u}, \mathrm{v} \rangle$, where u and v are words of B. If $\mathrm{p} = \langle \mathrm{u}, \mathrm{v} \rangle$ is a production of B, and if w and w′ are words of B, we write $\mathrm{w} \Rightarrow_{\mathrm{p}} \mathrm{w}'$ if w′ arises from w by replacing a part u of w by v. [Recall that u is a part of w if there exist (possibly empty) words w_1 and w_2 such that $\mathrm{w} = \mathrm{w}_1 \mathrm{u} \mathrm{w}_2$.]

By a *semi-Thue system* on B we mean a finite set $\mathscr{S}$ of productions of B. For words w and w′ of B, we write $\mathrm{w} \Rightarrow_{\mathscr{S}} \mathrm{w}'$ if there is a finite sequence w_0, w_1, ..., w_k ($k \geqslant 0$) of words of B such that $\mathrm{w} = \mathrm{w}_0$, $\mathrm{w}' = \mathrm{w}_k$, and, for $0 \leqslant i < k$, there is a production p of $\mathscr{S}$ such that $\mathrm{w}_i \Rightarrow_{\mathrm{p}} \mathrm{w}_{i+1}$. Observe that $\mathrm{w} \Rightarrow_{\mathscr{S}} \mathrm{w}$ for any word w. Moreover, if $\mathrm{w}_1 \Rightarrow_{\mathscr{S}} \mathrm{w}_2$ and $\mathrm{w}_2 \Rightarrow_{\mathscr{S}} \mathrm{w}_3$, then $\mathrm{w}_1 \Rightarrow_{\mathscr{S}} \mathrm{w}_3$. In addition, if $\mathrm{w}_1 \Rightarrow_{\mathscr{S}} \mathrm{w}_2$ and $\mathrm{w}_3 \Rightarrow_{\mathscr{S}} \mathrm{w}_4$, then $\mathrm{w}_1 \mathrm{w}_3 \Rightarrow_{\mathscr{S}} \mathrm{w}_2 \mathrm{w}_4$. Notice that there is no fixed order in which the productions have to be applied and that many different productions of $\mathscr{S}$ might be applicable to the same word.

By a *Thue system* we mean a semi-Thue system such that, for every production

$\langle u, v\rangle$, the *inverse* $\langle v, u\rangle$ is also a production. Clearly, if $\mathcal{S}$ is a Thue system and $w \Rightarrow_{\mathcal{S}} w'$, then $w' \Rightarrow_{\mathcal{S}} w$. Hence, $\Rightarrow_{\mathcal{S}}$ is an equivalence relation on the set of words of the alphabet of $\mathcal{S}$.

Example Let $\mathcal{S}^{\#}$ be the Thue system that has alphabet $\{b\}$ and productions $\langle b^3, \Lambda\rangle$ and $\langle \Lambda, b^3\rangle$. It is easy to see that every word is transformable into b^2, b, or Λ.

By a *semigroup* we mean a nonempty set G together with a binary operation on G (denoted by the juxtaposition uv of elements u and v) that satisfies the associative law: $x(yz) = (xy)z$. An element y such that $xy = yx = x$ for all x in G is called an *identity element*. If an identity element exists, it is unique and is denoted 1.

A Thue system $\mathcal{S}$ on an alphabet $B = \{b_1, \ldots, b_m\}$ determines a semigroup G with an identity element. In fact, for each word w of B, let [w] be the set of all words w′ such that $w \Rightarrow_{\mathcal{S}} w'$. [w] is just the equivalence class of w with respect to $\Rightarrow_{\mathcal{S}}$. Let G consist of the sets [w] for all words w of B. If U and V are elements of G, choose a word u in U and a word v in V. Let UV stand for the set [uv]. This defines an operation on G, since, if u′ is any word in U and v′ is any word in V, $[uv] = [u'v']$.

Exercises

5.69 For the set G determined by the Thue system $\mathcal{S}$, prove the following.
(a) $(UV)W = U(VW)$ for all members U, V, W of G.
(b) The equivalence class $[\Lambda]$ of the empty word Λ acts as an identity element of G.

5.70 (a) Show that a semigroup contains at most one identity element.
(b) Give an example of a semigroup without an identity element.

A Thue system $\mathcal{S}$ provides what is called a *finite presentation* of the corresponding semigroup G. The elements $b_1, \ldots, b_m$ of the alphabet are called *generators*, and the productions $\langle u, v\rangle$ of $\mathcal{S}$ are written in the form of equations $u = v$. These equations are called the *relations* of the presentation. Thus, in the example above, b is the only generator and $b^3 = 1$ can be taken as the only relation. The corresponding semigroup is a cyclic group of order 3.

If $\mathcal{S}$ is a semi-Thue or Thue system, the *word problem* for $\mathcal{S}$ is the problem of determining, for any words w and w′, whether $w \Rightarrow_{\mathcal{S}} w'$.

Exercises

5.71 Show that, for the Thue system $\mathcal{S}^{\#}$ in the example, the word problem is solvable.

5.72 Consider the following Thue system $\mathcal{S}$. The alphabet is $\{a, b, c, d\}$ and the productions are $\langle ac, \Lambda\rangle$, $\langle ca, \Lambda\rangle$, $\langle bd, \Lambda\rangle$, $\langle db, \Lambda\rangle$, $\langle a^3, \Lambda\rangle$, $\langle b^2, \Lambda\rangle$, $\langle ab, ba\rangle$, and their inverses.
(a) Show that $c \Rightarrow_{\mathcal{S}} a^2$ and $d \Rightarrow_{\mathcal{S}} b$.
(b) Show that every word of $\mathcal{S}$ can be transformed into one of the words a, a^2, b, ab, a^2b, and Λ.
(c) Show that the word problem for $\mathcal{S}$ is solvable. [*Hint:* To show that the six words of part (b) cannot be transformed into one another, use the cyclic group of order 6 generated by an element g, with $a = g^2$ and $b = g^3$.]

PROPOSITION 5.30 (Post, 1947) There exists a Thue system with a recursively unsolvable word problem.

Proof Let $\mathscr{T}$ be a Turing machine with alphabet $\{a_0, a_1, \ldots, a_n\}$ and internal states $\{q_0, q_1, .., q_m\}$. Remember that a tape description is a sequence of symbols describing the condition of $\mathscr{T}$ at any given moment; it consists of symbols of the alphabet of $\mathscr{T}$ plus one internal state q_j, and q_j is not the last symbol of the description. $\mathscr{T}$ is in state q_j, scanning the symbol following q_j, and the alphabet symbols, read from left to right, constitute the entire tape at the given moment. We shall construct a semi-Thue system $\mathscr{S}$ that will reflect the operation of $\mathscr{T}$: each action induced by quadruples of $\mathscr{T}$ will be copied by productions of $\mathscr{S}$. The alphabet of $\mathscr{S}$ consists of $\{a_0, a_1, \ldots, a_n, q_0, q_1, \ldots, q_m, \beta, \delta, \xi\}$. The symbol β will be placed at the beginning and end of a tape description in order to "alert" the semi-Thue system when it is necessary to add an extra blank square on the left or right end of the tape. We wish to ensure that, if $W \underset{\mathscr{T}}{\mapsto} W'$, then $\beta W \beta \Rightarrow_{\mathscr{S}} \beta W' \beta$. The productions of $\mathscr{S}$ are constructed from the quadruples of $\mathscr{T}$ in the following manner.

(a) If $q_j a_i a_k q_r$ is a quadruple of $\mathscr{T}$, let $\langle q_j a_i, q_r a_k \rangle$ be a production of $\mathscr{S}$.
(b) If $q_j a_i R q_r$ is a quadruple of $\mathscr{T}$, let $\langle q_j a_i a_l, a_i q_r a_l \rangle$ be a production of $\mathscr{S}$ for every a_l. In addition, let $\langle q_j a_i \beta, a_i q_r a_0 \beta \rangle$ be a production of $\mathscr{S}$. (This last production adds a blank square when $\mathscr{T}$ reaches the right end of the tape and is ordered to move right.)
(c) If $q_j a_i L q_r$ is a quadruple of $\mathscr{T}$, let $\langle a_l q_j a_i, q_r a_l a_i \rangle$ be a production of $\mathscr{S}$ for each a_l. In addition, let $\langle \beta q_j a_i, \beta q_r a_0 a_i \rangle$ be a production of $\mathscr{S}$. (This last production adds a blank square to the left of the tape when this is required.)
(d) If there is no quadruple of $\mathscr{T}$ beginning with $q_j a_i$, let $\mathscr{S}$ contain the following productions: $\langle q_j a_i, \delta \rangle$, $\langle \delta a_l, \delta \rangle$ for all a_l; $\langle \delta \beta, \xi \rangle$, $\langle a_l \xi, \xi \rangle$ for all a_l; and $\langle \beta \xi, \xi \rangle$.

$\mathscr{T}$ stops when it is in a state q_j, scanning a symbol a_i, such that $q_j a_i$ does not begin a quadruple of $\mathscr{T}$. In such a case, $\mathscr{S}$ would replace $q_j a_i$ in the final tape description of $\mathscr{T}$ by δ. Then δ proceeds to annihilate all the other symbols to its right, including the rightmost β, whereupon it changes to ξ. ξ then annihilates all symbols to its left, including the remaining β. The final result is ξ alone. Hence:

($\Box$) For any initial tape description α, $\mathscr{T}$ halts when and only when $\beta\alpha\beta \Rightarrow_{\mathscr{S}} \xi$

Now, enlarge $\mathscr{S}$ to a Thue system $\mathscr{S}'$ by adding to $\mathscr{S}$ the inverses of all the productions of $\mathscr{S}$. Let us show that:

(∇) For any initial tape description α of $\mathscr{T}$, $\beta\alpha\beta \Rightarrow_{\mathscr{S}'} \xi$ if and only if $\beta\alpha\beta \Rightarrow_{\mathscr{S}} \xi$

Clearly, if $\beta\alpha\beta \Rightarrow_{\mathscr{S}} \xi$, then $\beta\alpha\beta \Rightarrow_{\mathscr{S}'} \xi$. Conversely, assume for the sake of contradiction that $\beta\alpha\beta \Rightarrow_{\mathscr{S}'} \xi$, but it is not the case that $\beta\alpha\beta \Rightarrow_{\mathscr{S}} \xi$. Consider a sequence of words leading from $\beta\alpha\beta$ to ξ in $\mathscr{S}'$:

$$\beta\alpha\beta = w_0 \Rightarrow_{\mathscr{S}'} w_1 \Rightarrow_{\mathscr{S}'} \cdots \Rightarrow_{\mathscr{S}'} w_{t-1} \Rightarrow_{\mathscr{S}'} w_t = \xi$$

Here, each arrow is intended to indicate a single application of a production. It is clear from the definition of $\mathscr{S}$ that no production of $\mathscr{S}$ applies to ξ alone. Hence, the last step in the sequence $w_{t-1} \Rightarrow_{\mathscr{S}'} \xi$ must be the result of a production of $\mathscr{S}$. So, $w_{t-1} \Rightarrow_{\mathscr{S}} \xi$. Working backward, let us find the least p such that $w_p \Rightarrow_{\mathscr{S}} \xi$. Since we have assumed that it is not true that $\beta\alpha\beta \Rightarrow_{\mathscr{S}} \xi$, we must have $p > 0$. By the minimality of p, it is not true that $w_{p-1} \Rightarrow_{\mathscr{S}} w_p$. Therefore, $w_p \Rightarrow_{\mathscr{S}} w_{p-1}$. Examination of the productions of $\mathscr{S}$ shows that each of the words $w_0, w_1, \ldots, w_t$ must contain exactly one of the symbols $q_0, q_1, \ldots, q_m, \delta$, or ξ, and that, to such a word, at most one production of $\mathscr{S}$ is applicable. But, w_p is transformed into both w_{p+1} and w_{p-1} by productions of $\mathscr{S}$. Hence, $w_{p-1} = w_{p+1}$. But, $w_{p+1} \Rightarrow_{\mathscr{S}} \xi$. Hence, $w_{p-1} \Rightarrow_{\mathscr{S}} \xi$, contradicting the definition of p. This establishes (∇).

Now, let $\mathscr{T}$ be a Turing machine with a recursively unsolvable halting problem (Proposition 5.15). Construct the corresponding Thue system $\mathscr{S}'$ as above. Then, by ($\square$) and (∇), for any tape description α, $\mathscr{T}$ halts if and only if $\beta\alpha\beta \Rightarrow_{\mathscr{S}'} \xi$. So, if the word problem for $\mathscr{S}'$ were recursively solvable, the halting problem for $\mathscr{T}$ would be recursively solvable. (The function that assigns to the Gödel number of α the Gödel number of $\langle \beta\alpha\beta, \xi \rangle$ is clearly recursive under a suitable arithmetization of the symbolism of Turing machines and Thue systems.) Thus, $\mathscr{S}'$ has a recursively unsolvable word problem.

That the word problem is unsolvable even for certain Thue systems on a *two-element* alphabet (semigroups with two generators) was proved by Hall (1949).

(b) *Finitely presented groups.* A *finite presentation* of a group consists of a finite set of generators $g_1, \ldots, g_r$ and a finite set of equations $W_1 = W'_1, \ldots, W_t = W'_t$ between words of the alphabet $B = \{g_1, \ldots, g_r, g_1^{-1}, \ldots, g_r^{-1}\}$. What is really involved here is a Thue system $\mathscr{S}$ with alphabet B, productions $\langle W_1, W'_1 \rangle, \ldots, \langle W_t, W'_t \rangle$ and their inverses, and all the productions $\langle g_i g_i^{-1}, \Lambda \rangle, \langle g_i^{-1} g_i, \Lambda \rangle$ and their inverses. The corresponding semigroup G is actually a group and is called a *finitely presented group*. The word problem for G (or, rather, for the finite presentation of G) is the word problem for the Thue system $\mathscr{S}$.

Problems that concern word problems for finitely presented groups are generally much more difficult than corresponding problems for finitely presented semigroups (Thue systems). The existence of a finitely presented group with a recursively unsolvable word problem was proved, independently, by Novikov (1955) and Boone (1959). Other proofs have been given by Higman (1961), Britton (1963), and McKenzie and Thompson (1973). (See also Rotman, 1973.) Results on other decision problems connected with groups may be found in Rabin (1958). For corresponding problems in general algebraic systems, consult Evans (1951).

Bibliography

Listed here are not only books and papers mentioned in the text but also other material that will be helpful in a further study of mathematical logic. Additional references may be found in the reviews in the *Journal of Symbolic Logic* and *Mathematical Reviews*. We shall use the following abbreviations:

AML for *Annals of Mathematical Logic*
AMS for American Mathematical Society
Arch for *Archiv für mathematische Logik und Grundlagenforschung*
FM for Fundamenta Mathematicae
HML for *Handbook of Mathematical Logic*, Springer Verlag (Editor: J. Barwise)
HPL for *Handbook of Philosophical Logic*, Reidel (Editors: D. Gabbay, F. Guenthner)
JSL for *Journal of Symbolic Logic*
MTL for *Model-Theoretic Logics*, Springer (Editors: J. Barwise, S. Feferman)
NDJFL for *Notre Dame Journal of Formal Logic*
N-H for North-Holland Publishing Company
ZML for *Zeitschrift für mathematische Logik und Grundlagen der Mathematik*

Aberth, O. 1980. *Computable Analysis*. McGraw-Hill.

Ackermann, W. 1928. Zum Hilbertsche Aufbau der reelen Zahlen, *Math. Annalen*, 99, 118–133. 1940. Zur Widerspruchfreiheit der Zahlentheorie, *Math. Annalen*, 117, 162–194. 1954. *Solvable Cases of the Decision Problem*. N-H. Also see Hilbert, D.

Aczel, P. 1977. An introduction to inductive definitions, *HML*, 739–782.

Andrews, P. 1965. *Transfinite Type Theory with Type Variables*. N-H.

Asser, G. 1955. Das Repräsentantenproblem im Predikätenkalkül der ersten Stufe mit Identität, *ZML*, 1, 252–263. 1959. Turing-Maschinen und Markowsche Algorithmen, *ZML*, 5, 346–365.

Bachmann, H. 1955. *Transfinite Zahlen*. Springer. (Second Edition, 1967)

Barendregt, H. P. 1977. *The Lambda Calculus: Its Syntax and Semantics*. N-H.

Bar-Hillel, Y. See Fraenkel, A.

Barwise, J. 1975. *Admissible Sets and Structures*. Springer. 1977. An introduction to first-order logic, *HML*, 5–46.

Baudisch, A., D. Seese, P. Tuschik, and M. Weese. 1985. Decidability and quantifier elimination, *MTL*, 235–268.

Beeson, M. J. 1985. *Foundations of Constructive Mathematics*. Springer.

Bell, J. L. 1977. *Boolean-Valued Models and Independence Proofs in Set Theory*. Oxford.

Bell, J. L., and A. B. Slomson. 1969. *Models and Ultraproducts*. N-H.

Benacerraf, P. 1963. God, the devil, and Gödel, *The Monist*, 51, 9–32.

Bernardi, C. 1975. The fixed-point theorem for diagonalizable algebras, *Studia Logica*, 34, 239–251. 1976. The uniqueness of the fixed-point in every diagonalizable algebra, *Studia Logica*, 35, 335–343.

Bernays, P. 1937–1954. A system of axiomatic set theory, *JSL*, I. Vol. 2 (1937), 65–77; II. Vol. 6 (1941), 1–17; III. Vol. 7 (1942), 65–89; IV. Vol. 7 (1942), 133–145; V. Vol. 8 (1943), 89–104; VI. Vol. 13 (1948), 65–79; VII. Vol. 19 (1954), 81–96. 1957. Review of Myhill (1954), *JSL*, 22, 73–76. 1958. *Axiomatic Set Theory*. N-H. 1961. Zur Frage der Unendlichkeitsschemata in der axiomatischen Mengenlehre, *Essays on the Foundations of Mathematics*, Jerusalem, 3–49. 1976. On the problem of schemata of infinity in axiomatic set theory, *Sets and Classes*, N-H, 121–172. Also see Hilbert, D.

Bernstein, A. R. 1973. Non-standard analysis, *Studies in Model Theory*, Math. Assoc. of America, 35–58.

Beth, E. 1951. A topological proof of the theorem of Löwenheim-Skolem-Gödel, *Indag. Math.*, 13, 436–444. 1953. Some consequences of the theorem of Löwenheim-Skolem-Gödel-Malcev, *Indag. Math.*, 15, 66–71. 1959. *The Foundations of Mathematics*. N-H.

Bezboruah, A., and J. C. Shepherdson. 1976. Gödel's second incompleteness theorem for Q, *JSL*, 41, 503–512.

Birkhoff, G. 1948. *Lattice Theory*. AMS Colloquium Publications. (Revised, 1979)

Bishop, E. 1967. *Foundations of Constructive Analysis*. McGraw-Hill. 1970. Mathematics as a numerical language, *Intuitionism and Proof Theory*. N-H, 53–71.

Blum, M. 1967. A machine-independent theory of the complexity of recursive functions, *J. of the Assoc. for Computing Mach.*, 14, 322–336.

Boffa, M. 1969. Sur la théorie des ensembles sans axiome de fondement, *Bull. Soc. Math. Belgique*, 21, 16–56. 1977. The consistency problem for NF, *JSL*, 42, 215–220.

Boolos, G. 1971. The iterative conception of set, *J. of Philosophy*, 68, 215–231. 1979. *The Unprovability of Consistency*. Cambridge.

Boolos. G., and R. C. Jeffrey. 1981. *Computability and Logic*. Cambridge. (Second Edition)

Boone, W. 1959. The word problem, *Annals of Math.*, 70, 207–265.

Bourbaki, N. 1947. *Algebre*, Livre II, Chap. II. Herrmann, Paris.

Bradford, R. 1971. Cardinal addition and the axiom of choice, *AML*, 3, 111–196.

Bridge, J. 1977. *Beginning Model Theory*. Oxford.

Britton, J. L. 1958. The word problem for groups, *Proc. London Math. Soc.*, 8, 493–506. 1963. The word problem, *Annals of Math*, 77, 16–32.

Brouwer, L. E. J. 1976. *Collected Works, Vol. 1, Philosophy and Foundations of Mathematics*. N-H.

Bruijn, N. G. de, and P. Erdös. 1951. A colour problem for infinite graphs and a problem in the theory of relations, *Indag. Math.*, 13, 369–373.

Büchi, J. R. 1962. Turing machines and the Entscheidungsproblem, *Math. Annalen*, 148, 201–213.

Bunder, M. W. 1984. A simpler preparation for the proof of Gödel's incompleteness theorems, Preprint 17–84, Math. Dept., Univ. of Wollongong.

Burgess, J. P. 1977. Forcing, *HML*, 404–452.

Carnap, R. 1934. *Die Logische Syntax der Sprache*. Springer. (English translation, *The*

Logical Syntax of Language, Routledge & Kegan Paul, 1937. Text edition, Humanities, 1964)

Chang, C. C., and H. J. Keisler. 1966. *Continuous Model Theory*. Princeton U. Press. 1973. *Model Theory*. N-H. (Second Edition, 1977)

Chang, C. L., and R. C. Lee. 1973. *Symbolic Logic and Mechanical Theorem Proving*. Academic.

Cherlin, G. 1976. *Model Theoretic Algebra, Selected Topics*. Springer.

Chuquai, R. 1972. Forcing for the impredicative theory of classes, *JSL*, 37, 1–18. 1981. *Axiomatic Set Theory. Impredicative Theories of Classes*. N-H.

Church, A. 1936a. A note on the Entscheidungsproblem, *JSL*, 1, 40–41; Correction, ibid., 101–102. (Reprinted in Davis, 1965) 1936b. An unsolvable problem of elementary number theory, *Am. J. Math.*, 58, 345–363. (Reprinted in Davis, 1965) 1940. A formulation of the simple theory of types, *JSL*, 5, 56–68. 1941. *The Calculi of Lambda Conversion*. Princeton. (Second printing, 1951) 1956. *Introduction to Mathematical Logic, I*. Princeton.

Chwistek, L. 1924–25. The theory of constructive types, *Annales de la Soc. Polonaise de Math.*, 2, 9–48; 3, 92–141.

Cohen, P. J. 1963. A minimal model for set theory, *Bull. AMS*, 69, 537–540. 1963–64. The independence of the continuum hypothesis, *Proc. Natl. Acad. Sci. USA*, 50, 1143–1148; 51, 105–110. 1966. *Set Theory and the Continuum Hypothesis*. Benjamin. 1969. Decision procedures for real and p-adic fields, *Comm. Pure and Applied Math.*, 22, 131–151.

Cohn, P. M. 1965. *Universal Algebra*. Harper & Row. (Revised Edition, 1982, Reidel)

Collins, G. E. 1955. The modeling of Zermelo set theories in New Foundations, Ph.D. Thesis, Cornell.

Collins, G. E., and J. D. Halpern. 1970. On the interpretability of arithmetic in set theory, *NDJFL*, 11, 477–483.

Comfort, W. W., and S. Negrepontis. 1974. *The Theory of Ultrafilters*. Springer.

Cook, S. A. 1971. The complexity of theorem proving procedures, *Proc. 3rd Annual ACM Symp. on Theory of Computing*, 151–158. 1972. Linear time simulation of two-way pushdown automata, *Proceedings of IFIP Congress 71, Foundations of Information Processing*. N-H, 75–80.

Corcoran, J. 1980. Categoricity, *History and Philosophy of Logic*, 1, 187–207.

Cowen, R. 1973. Some combinatorial theorems equivalent to the prime ideal theorem, *Proc. AMS*, 41, 268–273.

Cowles, J. 1979. The relative expressive power of some logics extending first-order logics, *JSL*, 44, 129–146.

Crabbe, M. 1982. On the consistency of an impredicative subsystem of Quine's NF, *JSL*, 47, 131–136.

Craig, W. 1953. On axiomatizability within a system, *JSL*, 18, 30–32.

Cresswell, M. J. See Hughes, G. E.

Crossley, J. N. 1969. *Constructive Order Types*. N-H.

Curry, H. B. 1963. *Foundations of Mathematical Logic*. McGraw-Hill. (Second Edition, 1977, Dover)

Curry, H. B., and R. Feys. 1958. *Combinatory Logic, I*. N-H.

Curry, H. B., J. R. Hindley, and J. Seldin. 1972. *Combinatory Logic, II*. N-H.

Cutland, N. 1980. *Computability*. Cambridge.

Da Costa, N. C. A. 1965. On two systems of set theory, *Proc. Koninkl. Nederl. Akad. Wetensch.*, A, 68, 95–99.

Davis, M. 1958. *Computability and Unsolvability*. McGraw-Hill. (Dover, 1983) 1965. (Editor) *The Undecidable: Basic Papers on Undecidable Propositions, Unsolvable Problems, and Computable Functions*. Raven. 1973. Hilbert's tenth problem is unsolvable, *Amer. Math. Monthly*, 80, 233–269. 1977a. *Applied Nonstandard Analysis*. Wiley. 1977b. Unsolvable problems, *HML*, 567–594.

Davis, M., Yu. Matiyasevich, and J. Robinson. 1976. Hilbert's tenth problem, Diophantine equations: positive aspects of a negative solution, *Proc. Symp. on the Hilbert Problems*, AMS, 323–378.

Davis, M., H. Putnam, and J. Robinson. 1961. The decision problem for exponential diophantine equations, *Annals of Math.*, 74, 425–436.

Dedekind, R. 1901. *Essays on the Theory of Numbers*. Open Court. (Dover, 1963)

Dekker, J. C. E. 1953. Two notes on recursively enumerable sets, *Proc. AMS*, 4, 495–501. 1955. Productive sets, *Trans. AMS*, 78, 129–149. 1966. *Les Fonctions Combinatoires et les Isols*. Gauthier-Villars.

Dekker, J. C. E., and J. Myhill. 1960. Recursive equivalence types, *Univ. Calif. Publ. Math.*, 3, 67–213.

Detlovs, V. K. 1958. Equivalence of normal algorithms and recursive functions, *Tr. Mat. Inst. Steklov*, LII, 75–139. (In Russian)

Devlin, K. J. 1973. *Aspects of Constructibility*. Springer. 1977. *The Axiom of Constructibility*. Springer. 1984. *Constructibility*. Springer.

Devlin, K. J., and H. Johnsbraten. 1974. *The Souslin Problem*. Springer.

Dickmann, M. 1976. *Large Infinitary Languages*. N-H.

Di Paola, R. 1967. Some theorems on extensions of arithmetic, *JSL*, 32, 180–189.

Doets, K. See Van Benthem, J.

Drake, F. 1974. *Set Theory: An Introduction to Large Cardinals*. N-H.

Dreben, B. 1952. On the completeness of quantification theory, *Proc. Natl. Acad. Sci. USA*, 38, 1047–1052.

Dreben, B., and W. D. Goldfarb. 1980. *Decision Problems. Solvable Classes of Quantificational Formulas*. Addison-Wesley.

Dummett, M. 1977. *Elements of Intuitionism*. Oxford.

Easton, W. B. 1970. Powers of regular cardinals, *AML*, 1, 139–178.

Ehrenfeucht, A. 1957. On theories categorical in power, *FM*, 44, 241–248. 1958. Theories having at least continuum non-isomorphic models in each infinite power (abstract), *Notices AMS*, 5, 680. 1961. An application of games to the completeness problem for formalized theories, *Arch.*, 49, 129–141.

Ehrenfeucht, A., and S. Feferman. 1960. Representability of recursively enumerable sets in formal theories, *Arch.*, 5, 37–41.

Ehrenfeucht, A., and D. Jensen. See Jensen, D.

Ehrenfeucht, A., and G. Kreisel. 1966. Strong models of arithmetic, *Bull. Acad. Pol. des Sci.*, 14, 107–110.

Ehrenfeucht, A., and A. Mostowski. 1957. Models of axiomatic theories admitting automorphisms, *FM*, 43, 50–68.

Eilenberg, S., and C. C., Elgot. 1970. *Recursiveness*. Academic.

Eisenberg, M. 1971. *Axiomatic Theory of Sets and Classes*. Holt, Rinehart & Winston.

Eklof, P. C. 1977. Ultraproducts for algebraists, *HML*, 105–136.

Eklof, P. C., and E. R. Fischer. 1972. The elementary theory of abelian groups, *AML*, 4, 115–171.

Elgot, C. C. See Eilenberg, S.

Engeler, E. 1968. *Formal Languages: Automata and Structures*. Markham. 1973. *Introduction to the Theory of Computability*. Academic. 1975. On the solvability of algorithmic problems, *Logic Colloquium '73*, Springer, 231–251.

Erdös, P. See Bruijn, N. G. de.

Erdös, P., and A. Tarski. 1961. On some problems involving inaccessible cardinals, *Essays on the Foundations of Mathematics*, Magnes, Jerusalem, 50–82.

Ershov, Yu., Lavrov, I., Taimanov, A., and Taitslin, M. 1965. Elementary theories, *Russian Mathematical Surveys*, 20, 35–105.

Evans, T. 1951. The word problem for abstract algebras, *J. London Math. Soc.*, 26, 64–71.

Feferman, S. 1957. Degrees of unsolvability associated with classes of formalized theories, *JSL*, 22, 165–175. 1960. Arithmetization of metamathematics in a general setting, *FM*, 49, 35–92. 1962. Transfinite recursive progressions of axiomatic theories, *JSL*, 27, 259–316. 1965. Some applications of the notion of forcing and generic sets, *FM*, 56, 325–345. 1975. A language and axioms for explicit mathematics, *Algebra and Logic*, Springer, 87–139. 1982. Inductively presented systems and the formalisation of metamathematics, *Logic Colloquium '80*, N-H, 95–128.

Feferman, S., and A. Ehrenfeucht. See Ehrenfeucht, A.

Felgner, U. 1971a. *Models of ZF-Set Theory*. Springer. 1971b. Comparison of the axioms of local and universal choice, *FM*, 71, 43–62. 1976. Choice functions on sets and classes, *Sets and Classes*, N-H, 217–255.

Feys, R. See Curry, H. B.

Fischer, E. R. See Eklof, P. C.

Fischer, P. C. 1965. On formalisms for Turing machines, *J. Assoc. Comp. Mach.*, 12, 570–580.

Fisher, A. 1982. *Formal Number Theory and Computability. A Workbook*. Oxford.

Fitting, M. 1982. *Fundamentals of Generalized Recursion Theory*. N-H. 1983. *Proof Methods for Modal and Intuitionistic Logics*. Reidel.

Flagg, R. C. 1985. Church's Thesis is consistent with epistemic arithmetic, *Intensional Mathematics*, N-H, 121–172.

Forster, T. E. 1983. *Quine's New Foundations (An Introduction)*. Cahiers du Centre de Logique 5, Univ. Catholique de Louvain.

Fraenkel, A. A. 1922. Zu den Grundlagen der Cantor-Zermeloschen Mengenlehre, *Math. Annalen*, 86, 230–237. 1976. *Abstract Set Theory*. N-H. (Fourth Edition)

Fraenkel, A., Y. Bar-Hillel, and A. Lévy. 1973. *Foundations of Set Theory*. N-H. (Second Revised Edition)

Frayne, T., A. Morel, and D. Scott. 1956. Reduced direct products, *FM*, 51, 195–228.

Frege, G. 1893, 1903. *Grundgesetze der Arithmetik, Begriffsschriftlich Abgeleitet*, Vols. 1–2. Jena. (Partial English translation in *The Basic Laws of Arithmetic: Exposition of the System*, Univ. of California Press, 1964)

Friedberg, R. 1957. Two recursively enumerable sets of incomparable degrees of unsolvability, *Proc. Natl. Acad. Sci. USA*, 43, 236–238.

Friedman, H. 1971. Higher set theory and mathematical practice, *AML*, 2, 325–357. 1973. Countable models of set theory, *Lecture Notes in Math.*, Springer, 337, 539–573. 1975. One hundred and two problems in mathematical logic, *JSL*, 40, 113–129.

Friedman, J. 1969. Proper classes as members of extended sets, *Math. Ann.*, 183, 232–246.

Gabriel, P. 1962. Des categories abeliennes, *Bull. Soc. Math. France*, 90, 323–448.

Gaifman, H. 1972. A note on models and submodels of arithmetic, *Conf. in Math. Logic—London 1970*, Springer, 128–144. 1976. Models and types of Peano arithmetic, *AML*, 9, 223–306.

Galler, B. 1957. Cylindric and polyadic algebras, *Proc. AMS*, 8, 176–183.

Gandy, R. 1956, 1959. On the axiom of extensionality, Part I, *JSL*, 21, 36–48; Part II, *JSL*, 24, 287–300.

Garey, M. R., and D. S. Johnson. 1978. *Computers and Intractability: A Guide to the Theory of NP-Completeness*. Freeman.

Geiser, J. R. 1970. Nonstandard analysis, *ZML*, 16, 297–318.

Gentzen, G. 1936. Die Widerspruchsfreiheit der reinen Zahlentheorie, *Math. Ann.*, 112, 493–565. 1938. Neue Fassung des Widerspruchsfreiheitsbeweises für die reine Zahlentheorie, *Forschungen zur Logik*, 4, 5–18. 1969. *Collected Papers* (Edited by M. E. Szabo). N-H.

Gödel, K. 1930. Die Vollständigkeit der Axiome des logischen Funktionenkalküls, *Monatsh. Math. Phys.*, 37, 349–360. 1931. Ueber formal unentscheidbare Sätze der Principia Mathematica und verwandter Systeme, I, ibid., 38, 173–198. (English translation in Davis, 1965) 1933. Zum intuitionistischen Aussagenkalkül; Zur intuitionistischen Arithmetik und Zahlentheorie, *Ergeb. Math. Koll.*, 4, 34–38 and 40. (Translation in Davis, 1965) 1934. On undecidable propositions of formal mathematical systems, *Lecture Notes, Institute for Advanced Study*, Princeton. (Reprinted in Davis, 1965, 39–73) 1936. Über die Länge der Beweise, ibid., 7, 23–24. (Translation in Davis, 1965) 1938. The consistency of the axiom of choice and the generalized continuum hypothesis, *Proc. Natl. Acad. Sci. USA*, 24, 556–557. 1939. Consistency proof for the generalized continuum hypothesis, ibid., 25, 220–226. 1940. *The Consistency of the Axiom of Choice and of the Generalized Continuum Hypothesis with the Axioms of Set Theory*. Princeton. 1944. Russell's mathematical logic, *The Philosophy of Bertrand Russell*, Tudor, 125–153. 1947. What is Cantor's continuum problem? *Amer. Math. Monthly*, 54, 515–525. 1958. Über eine bisher noch nicht benutzte Erweiterung des finiten Standpunkts, *Dialectica*, 12, 280–287.

Goldblatt, R. 1980. *Topoi: The Categorial Analysis of Logic*. N-H.

Goldfarb, W. D. 1979. Logic in the twenties: nature of the quantifier, *JSL*, 44, 351–368. Also see Dreben, B.

Goodman, N. D. 1985. A genuinely intensional set theory, *Intensional Mathematics*, N-H, 63–79.

Grandy, R. 1977. *Advanced Logic for Applications*. Reidel.

Grätzer, G. 1968. *Universal Algebra*. Van Nostrand. (Second Edition, Springer, 1979)

Grzegorczyk, A., A. Mostowski, and C. Ryll-Nardzewski. 1958. The classical and the ω-complete arithmetic, *JSL*, 23, 188–206.

Guaspari, D. 1979. Partially conservative extensions of arithmetic, *Trans. AMS*, 254, 47–68. 1983. Sentences implying their own provability, *JSL*, 48, 777–789.

Guaspari, D., and R. Solovay. 1979. Rosser sentences, *AML*, 81–99.

Hailperin, T. 1944. A set of axioms for logic, *JSL*, 9, 1–19.

Hajek, P. See Vopenka, P.

Hall, M., Jr. 1949. The word problem for semigroups with two generators, *JSL*, 14, 115–118.

Halmos, P. 1960. *Naive Set Theory*. Van Nostrand. (Springer, 1974) 1962. *Algebraic Logic*. Chelsea. 1963. *Lectures on Boolean Algebra*. Van Nostrand. (Springer, 1977)

Halmos, P., and H. Vaughn. 1950. The marriage problem, *Amer. J. Math.*, 72, 214–215.

Halpern, J. D. 1964. The independence of the axiom of choice from the Boolean prime ideal theorem, *FM*, 55, 57–66.

Halpern, J. D., and G. E. Collins. See Collins, G. E.

Halpern, J. D., and P. E. Howard. 1970. Cardinals m such that 2m = m, *Proc. AMS*, 26, 487–490.

Halpern, J. D., and A. Lévy. 1971. The Boolean prime ideal theorem does not imply the axiom of choice, *Proc. Symp. in Pure Math.*, 13, AMS, 83–134.

Hamlet, R. G. 1974. *Introduction to Computation Theory*. Intext.

Hanf, W. 1965. Model-theoretic methods in the study of elementary logic, *The Theory of Models*, N-H, 132–145.

Harrington, L. See Paris, J.

Hartogs, F. 1915. Über das Problem der Wohlordnung, *Math. Annalen*, 76, 438–443.

Hasenjaeger, G. 1952. Über ω-Unvollständigkeit in der Peano-Arithmetik, *JSL*, 17, 81–97. 1953. Eine Bemerkung zu Henkin's Beweis für die Vollständigkeit des Prädikatenkalküls der ersten Stufe, *JSL*, 18, 42–48.

Hasenjaeger, G., and H. Scholz. 1961. *Grundzüge der mathematischen Logik*. Springer.

Hatcher, W. 1982. *The Logical Foundations of Mathematics*. Pergamon.

Hazen, A. 1983. Predicative logics, *HPL*, I, 331–407.

Hechler, S. 1973. Powers of singular cardinals and a strong form of the negation of the generalized continuum hypothesis, *ZML*, 19, 83–84.

Heijenoort, J. van. 1967. (Editor) *From Frege to Gödel (A Source Book in Mathematical Logic, 1879–1931)*. Harvard.

Hellman, M. 1961. A short proof of an equivalent form of the Schröder-Bernstein Theorem, *Amer. Math. Monthly*, 68, 770.

Henkin, L. 1949. The completeness of the first-order functional calculus, *JSL*, 14, 159–166. 1950. Completeness in the theory of types, ibid., 15, 81–91. 1953. Some interconnections between modern algebra and mathematical logic, *Trans. Amer. Math. Soc.*, 74, 410–427. 1954. Boolean representation through propositional calculus, *FM*, 41, 89–96. 1955a. The representation theorem for cylindrical algebras, *Mathematical Interpretations of Formal Systems*, N-H, 85–97. 1955b. On a theorem of Vaught, *JSL*, 20, 92–93.

Henkin, L., J. D. Monk, and A. Tarski. 1971. *Cylindric Algebras, I*. N-H.

Henson, C. W. 1973. Permutation methods applied to NF, *JSL*, 38, 69–76.

Herbrand, J. 1930. Recherches sur la theorie de la demonstration, *Travaux de la Soc. des Sci. et des Lettres de Varsovie, III*, 33, 33–160. 1971. *Logical Writings*. Harvard & Reidel.

Hermes, H. 1965. *Enumerability, Decidability, Computability*. Springer. (Second Edition, 1969)

Heyting, A. 1956. *Intuitionism*. N-H.

Higman, G. 1961. Subgroups of finitely presented groups, *Proc. Royal Soc.*, Ser. A, 262, 455–475.

Hilbert, D., and W. Ackermann. 1950. *Principles of Mathematical Logic*. Chelsea.

Hilbert, D., and P. Bernays. 1934, 1939. *Grundlager der Mathematik*, Vol. I (1934), Vol. II (1939), Springer. (Second Edition, 1968, 1970)

Hindley, J. R., B. Lercher, and J. Seldin. 1972. *Introduction to Combinatory Logic*. Cambridge. Also see Curry, H. B.

Hinman, P. G. 1977. *Recursion-Theoretic Hierarchies*. Springer.

Hintikka, K. J. 1954. An application of logic to algebra, *Math. Scand.*, 2, 243–246. 1955a. Form and content in quantification theory, *Acta Phil. Fennica*, 11–55. 1955b. Notes on the quantification theory, *Comment Phys.-Math., Soc. Sci. Fennica*, 17, 1–13.

Hirschfeld, J., and W. H. Wheeler. 1975. *Forcing, Arithmetic, Division Rings*. Springer.

Hodges, W. 1983. Elementary predicate logic, *HPL*, I, 1–132.

Howard, P. E., and J. D. Halpern. See Halpern, J. D.

Howard, P. E., A. L. Rubin, and J. E. Rubin. 1978. Independence results for class forms of the axiom of choice, *JSL*, 43, 673–683.

Hrbacek, K., and T. Jech. 1978. *Introduction to Set Theory*. Academic. (Second Edition, Dekker, 1984)

Hughes, G. E., and M. J. Cresswell. 1968. *An Introduction to Modal Logic*. Methuen.

Isbell, J. 1966. Structure of categories, *Bull. AMS*, 72, 619–655.

Jaskowski, S. 1936. Recherches sur le systeme de la logique intuitioniste, *Act. Sci. I Ind.*, 393, Paris, 58–61.

Jech, T. 1973. *The Axiom of Choice*. N-H. 1978. *Set Theory*. Academic.

Jech, T., and K. Hrbacek. See Hrbacek, K.

Jeffrey, R. See Boolos, G.

Jensen, D., and A. Ehrenfeucht. 1976. Some problems in elementary arithmetic, *FM*, 92, 223–245.

Jensen, R. B. 1967. *Modelle der Mengenlehre*. Springer. 1968–69. On the consistency of a slight (?) modification of Quine's New Foundations, *Synthèse*, 19, 250–263.

Jeroslow, R. G. 1971. Consistency statements in formal theories, *FM*, 72, 17–40. 1972. On the encodings used in the arithmetization of mathematics, unpublished manuscript. 1973. Redundancies in the Hilbert-Bernays derivability conditions for Gödel's second incompleteness theorem, *JSL*, 38, 359–367.

Johnsbraten, H. See Devlin, K. J.

Johnson, D. S. See Garey, M. R.

Jonsson, B. 1962. Algebraic extensions of relational systems, *Math. Scandinavica*, 11, 179–205.

Kalmár, L. 1935. Über die Axiomatisierbarkeit des Aussagenkalküls, *Acta Sci. Math.*, 7, 222–243. 1936. Zuruckführung des Entscheidungsproblems auf den Fall von Formeln mit einer einziger binären Funktionsvariablen, *Comp. Math.*, 4, 137–144.

Kamke, E. 1950. *Theory of Sets*. Dover.

Kamo, S. 1981. Nonstandard natural number systems and nonstandard models, *JSL*, 46, 365–376.

Kanamori, A., and M. Magidor. 1978. The evolution of large cardinal axioms in set theory, *Higher Set Theory*, Springer, 99–275.

Karp, C. 1964. *Languages with Expressions of Infinite Length*. N-H. 1967. A proof of the relative consistency of the continuum hypothesis, *Sets, Models, and Recursion Theory*, N-H, 1–32.

Kaufmann, M. 1985. The quantifier "there exist uncountably many" and some of its relatives, *MTL*, 123–176.

Keisler, H. J. 1971. *Model Theory for Infinitary Logic*. N-H. 1976. *Elementary Calculus: An Approach Using Infinitesimals*. Prindle, Weber & Schmidt. 1977. Fundamentals of model theory, *HML*, 47–103.

Keisler, H. J., and C. C. Chang. See Chang, C. C.

Keisler, H. J., and M. Morley. 1968. Elementary extensions of models of set theory, *Israel J. of Math.*, 6, 49–65.

Keisler, H. J., and A. Tarski. 1964. From accessible to inaccessible cardinals, *FM*, 53, 225–308.

Kelley, J. 1955. *General Topology*. Van Nostrand. (Springer, 1975)

Kemeny, J. 1948. Models of logical systems, *JSL*, 13, 16–30. 1949. Type theory vs. set theory, Ph.D. Thesis, Princeton. 1958. Undecidable problems of elementary number theory, *Math. Ann.*, 135, 160–169.

Kent, C. F. 1973. The relation of A to Prov $\ulcorner A \urcorner$ in the Lindenbaum sentence algebra, *JSL*, 38, 295–298.

Kirby, L., and J. Paris. 1977. Initial segments of models of Peano's axioms, *Proc. Bierutowice Conf. 1976, Lecture Notes in Math., 619*, Springer, 211–226.

Kleene, S. C. 1936a. General recursive functions of natural numbers, *Math. Ann.*, 112, 727–742. (Reprinted in Davis, 1965) 1936b. λ-definability and recursiveness, *Duke Math. J.*, 2, 340–353. 1938. On notation for ordinal numbers, *JSL*, 3, 150–155. 1943. Recursive predicates and quantifiers, *Trans. AMS*, 53, 41–73. (Reprinted in Davis, 1965) 1945. On the interpretation of intuitionistic number theory, *JSL*, 10, 109–124. 1952. *Introduction to Metamathematics*. Van Nostrand. 1955a. Hierarchies of number-theoretic predicates, *Bull. AMS*, 61, 193–213. 1955b. Arithmetical predicates and function quantifiers, *Trans. AMS*, 79, 312–340. 1976. The work of Kurt Gödel, *JSL*, 41, 761–778. (Addendum, *JSL*, 43, 1978, 613)

Kleene, S. C., and E. L. Post. 1954. The upper semi-lattice of degrees of recursive unsolvability, *Ann. of Math.*, 59, 379–407.

Kleene, S. C., and R. E. Vesley. 1965. *The Foundations of Intuitionistic Mathematics*. N-H.

Kneale, W., and M. Kneale. 1962. *The Development of Logic*. Clarendon, Oxford.

Knight, J. F. 1975. Types omitted in uncountable models of arithmetic, *JSL*, 40, 307–320.

Kochen, S. 1961. Ultraproducts in the theory of models, *Ann. of Math.*, 74, 221–261.

Kopperman, R. 1972. *Model Theory and Its Applications*. Allyn & Bacon.

Kreider, D. L., and H. Rogers, Jr. 1961. Constructive versions of ordinal number classes, *Trans. AMS*, 100, 325–369.

Kreisel, G. 1950. Note on arithmetic models for consistent formulae of the predicate calculus, *FM*, 37, 265–285. 1951–52. On the interpretation of non-finitist proofs, *JSL*, 16, 241–267; 17, 43–58. 1952a. On the concepts of completeness and interpretations of formal systems, *FM*, 39, 103–127. 1952b. Some concepts concerning formal systems of number theory, *Math. Zeitschr.*, *57*, 1–12. 1953. On a problem of Henkin's, *Indag. Math.*, 15, 405–406. 1955. Models, translations, and interpretations, *Mathematical Interpretations of Formal Systems*, N-H, 26–50. 1958a. The metamathematical significance of consistency proofs, *JSL*, 23, 155–182. 1958b. Hilbert's programme, *Dialectica*, 12, 346–372. 1965. Mathematical logic, *Lectures on Modern Mathematics, Vol. 3*, Wiley, 95–195. 1967. Informal rigor and consistency proofs, *Problems in the Philosophy of Mathematics*, N-H, 138–171. 1968. A survey of proof theory I, *JSL*, 33, 321–388. 1969. Two notes on the foundations of set theory, *Dialectica*, 23, 93–114. 1971. A survey of proof theory, II, *Proc. Second Scandinavian Logic Symp.*, N-H, 109–170.

Kreisel, G., and J. L. Krivine. 1967. *Elements of Mathematical Logic*. N-H.

Kreisel, G., and A. Lévy. 1968. Reflection principles and their use for establishing the complexity of axiomatic systems, *ZML*, 14, 97–191.

Kreisel, G., and H. Wang. 1955. Some applications of formalized consistency proofs, *FM*, 42, 101–110.

Kripke, S. 1962. "Flexible" predicates of formal number theory, *Proc. AMS*, 13, 647–650. 1975. Outline of a theory of truth, *J. of Philosophy*, 690–716.

Krivine, J.-L. 1971. *Introduction to Axiomatic Set Theory*. Reidel. Also see Kreisel, G.

Kruse, A. H. 1966. Grothendieck universes and the super-complete models of Shepherdson, *Comp. Math.*, 17, 86–101.

Kunen, K. 1970. Some applications of iterated ultrapowers in set theory, *AML*, I, 179–227. 1977. Combinatorics, *HML*, 371–401. 1980. *Set Theory. An Introduction to Independence Proofs*. N-H.

Kunen, K., and J. Vaughan. (Editors) 1984. *Handbook of Set-Theoretic Topology*. N-H.

Lambek, J. 1961. How to program an infinite abacus, *Canadian Math. Bull.*, 4, 295–302; 5, 297.

Langford, C. H. 1927. Some theorems on deducibility, *Ann. of Math.*, I, 28, 16–40; II, 28, 459–471.

Langford, C. H., and C. I. Lewis. See Lewis, C. I.

Laüchli, H. 1962. Auswahlaxiom in der Algebra, *Comment. Math. Helvetica*, 37, 1–18.

Lavrov, I. See Ershov, Yu.

Lawvere, F. W. 1964. An elementary theory of the category of sets, *Proc. Natl. Acad. Sci. USA*, 52, 1506–1511.

Leblanc, H. 1983. Alternatives to standard first-order semantics, *HPL*, I, 189–274.

Lee, R. C. See Chang, C. L.

Leivant, D. 1981. On the proof theory of the modal logic for arithmetic provability, *JSL*, 46, 531–538.

Lercher, B. See Hindley, R.

Lerman, M. 1983. *Degrees of Unsolvability Local and Global Theory*. Springer.

Lévy, A. 1960. Axiom schemata of strong infinity, *Pacific J. Math.*, 10, 223–238. 1964. The independence of certain consequences of the axiom of choice, *FM*, 54, 135–157. 1965. A hierarchy of formulas in set theory, *Memoirs AMS*, No. 57. 1969. The definability of cardinal numbers, *Foundations of Mathematics, Gödel-Festschrift*, Springer, 15–38. 1978. *Basic Set Theory*. Springer. Also see Fraenkel, A. A.; Halpern, J. D.; Kreisel, G.

Lewis, C. I., and C. H. Langford. 1960. *Symbolic Logic*. Dover. (Reprint of 1932 edition)

Lewis, H. R. 1979. *Unsolvable Classes of Quantificational Formulas*. Addison-Wesley.

Lewis, H. R., and C. H. Papadimitriou. 1981. *Elements of the Theory of Computation*. Prentice-Hall.

Lipshitz, L., and M. Nadel. 1978. The additive structure of models of arithmetic, *Proc. AMS*, 68, 331–336.

Löb, M. H. 1955. Solution of a problem of Leon Henkin, *JSL*, 20, 115–118.

Lorenzen, P. 1955. *Einführung in die Operative Logik und Mathematik*. Springer.

Loś, J. 1954a. Sur la théorème de Gödel pour les théories indénombrables, *Bull. de l'Acad. Polon. des Sci.*, III, 2, 319–320. 1954b. On the existence of a linear order in a group, ibid., 21–23. 1954c. On the categoricity in power of elementary deductive systems and some related problems, *Coll. Math.*, 3, 58–62. 1955a. The algebraic treatment of the methodology of elementary deductive systems, *Studia Logica*, 2, 151–212. 1955b. Quelques remarques, théorèmes et problèmes sur les classes definissables d'algèbres, *Math. Interpretations of Formal Systems*, N-H, 98–113.

Loś, J., and C. Ryll-Nardzewski. 1954. Effectiveness of the representation theory for Boolean algebras, *FM*, 41, 49–56.

Loveland, D. W. 1978. *Automated Theorem Proving. A Logical Basis*. N-H.

Löwenheim, L. 1915. Ueber Möglichkeiten im Relativkalkül, *Math. Ann.*, 76, 447–470.

Luxemburg, W. A. J. 1962. *Non-Standard Analysis*. Caltech Bookstore, Pasadena. 1969. *Applications of Model Theory to Algebra, Analysis, and Probability*. Holt, Rinehart & Winston. 1973. What is Non-Standard Analysis? *Papers in the Foundations of Mathematics, Amer. Math. Monthly*, 80, No. 6, Part II, 38–67. Also see Stroyan, K. D.

Lyndon, R. 1959. Properties preserved under algebraic constructions, *Bull. AMS*, 65, 287–299.

Macdowell, R., and E. Specker. 1961. Modelle der Arithmetik, *Infinitistic Methods*, Pergamon, 257–263.

Machtey, M., and P. Young. 1978. *An Introduction to the General Theory of Algorithms*. N-H.

MacIntyre, A. 1977. Model completeness, *HML*, 139–176.

MacIntyre, A., and H. Simmons. 1973. Gödel's diagonalization technique and related properties of theories, *Colloq. Math.*, 28, 165–180.

Maclane, S. 1971. Categorical algebra and set-theoretic foundations, *Proc. Symp. Pure Math.*, AMS, XIII, Part I, 231–240.

Maclaughlin, T. 1961. A muted variation on a theme of Mendelson, *ZML*, 17, 57–60.

Maddy, P. 1983. Proper classes, *JSL*, 48, 113–139.

Magari, R. 1975. The diagonalizable algebras, *Boll. Unione Mat. Italiana* (4), 12, 117–125.

Magidor, M. See Kanamori, A.

Makkai, M., and G. Reyes. 1977. *First Order Categorical Logic*. Springer.

Malcev, A. 1936. Untersuchungen aus dem Gebiet der mathematischen Logik, *Mat. Sbornik*, 2, 323–336.

Marek, W. 1973. On the metamathematics of impredicative set theory, *Dissertationes Math.*, 98.

Margaris, A. 1967. *First-Order Mathematical Logic*. Blaisdell.

Markov, A. A. 1954. *The Theory of Algorithms*, Tr. Mat. Inst. Steklov, XLII. (Translation: Office of Technical Services, U.S. Department of Commerce, 1962)

Markwald, S. 1954. Zur Theorie der konstruktiven Wohlordnungen, *Math. Ann.*, 127, 135–149.

Martin, D. A. 1977. Descriptive set theory: projective sets, *HML*, 783–815.

Martin, D. A., and R. M. Solovay. 1971. Internal Cohen extensions, *AML*, 2, 143–178.

Martin-Löf, P. 1970. *Notes on Constructive Mathematics*. Almqvist & Wiksell, Stockholm.

Matiyasevich, Yu. 1970. Enumerable sets are Diophantine, *Doklady Akad. Nauk SSSR*, 191, 279–282. (English translation: *Soviet Math. Doklady*, 1970, 354–357) 1971. Diophantine representation of recursively enumerable predicates, *Izv. Akad. Nauk SSSR*, Ser. Mat. 35, 3–30. 1972. Diophantine sets, *Uspekhi Mat. Nauk*, 27, 185–222. (English translation: *Russian Math. Surveys*, 27, 124–164) 1973. Hilbert's tenth problem, *Proc. IV Int. Cong. on Logic, Methodology and Phil. of Sci., Bucharest 1971*, N-H, 89–110.

Matiyasevich, Yu., M. Davis, and J. Robinson. See Davis, M.

McAloon, K. 1971. Consistency results about ordinal definability, *AML*, 2, 449–467. 1978. Completeness theorems, incompleteness theorems and models of arithmetic, *Trans. AMS*, 239, 253–277. 1980. Les rapports entre la methode des indicatrices et la methode de Gödel pour obtenir résultats d'indépendence, *Modèles de l'Arithmetique, Astérisque*, 73.

McKenzie, R., and R. J. Thompson. 1973. An elementary construction of unsolvable word problems in group theory, *Word Problems* (editors: W. W. Boone, F. B. Cannonito, R. C. Lyndon), N-H.

McKinsey, J. C. C., and A. Tarski. 1948. Some theorems about the sentential calculi of Lewis and Heyting, *JSL*, 13, 1–15.

McNaughton, R. See Wang, H.

Melzak, Z. A. 1961. An informal arithmetical approach to computability and computation, *Canadian Math. Bull.*, 4, 279–293.

Mendelson, E. 1956a. Some proofs of independence in axiomatic set theory, *JSL*, 21, 291–303. 1956b. The independence of a weak axiom of choice, ibid., 350–366. 1958. The axiom of Fundierung and the axiom of choice, *Arch.*, 4, 65–70. 1961. On non-standard models of number theory, *Essays on the Foundations of Mathematics*, Jerusalem, Magnes, 259–268. 1970. *Introduction to Boolean Algebra and Switching Circuits*. Schaum, McGraw-Hill. 1973. *Number Systems and the Foundations of Analysis*. Academic. (Reprint, 1985, Krieger)

Meredith, C. A. 1953. Single axioms for the systems (C, N), (C, O) and (A, N) of the two-valued propositional calculus, *J. Comp. Syst.*, 3, 155–164.

Miller, C. F., III. 1971. *On Group-Theoretic Decision Problems and Their Classification.* Princeton.

Minsky, M. L. 1967. *Computation: Finite and Infinite Machines.* Prentice-Hall.

Monk, J. D. 1976. *Mathematical Logic.* Springer. 1980. *Introduction to Set Theory.* Krieger. Also see Henkin, L.

Montague, R. 1961a. Semantic closure and non-finite axiomatizability, *Infinitistic Methods*, Pergamon, 45–69. 1961b. Fraenkel's addition to the axioms of Zermelo, *Essays on the Foundations of Mathematics*, Magnes, Jerusalem, 91–114.

Montague, R., and R. L. Vaught. 1959. Natural models of set theories, *FM*, 47, 219–242.

Moore, G. H. 1982. *Zermelo's Axiom of Choice: Its Origin, Development, and Influence.* Springer.

Morel, A. See Frayne, T.

Morley, M. 1965. Categoricity in power, *Trans. AMS*, 114, 514–538.

Morley M. See Keisler, H. J.

Morse, A. 1965. *A Theory of Sets.* Academic.

Moschovakis, Y. N. 1974. *Elementary Induction on Abstract Structures.* N-H. 1980. *Descriptive Set Theory.* N-H.

Mostowski, A. 1939. Ueber die Unabhängigkeit des Wohlordnungsatzes vom Ordnungsprinzip, *FM*, 32, 201–252. 1947. On definable sets of positive integers, *FM*, 34, 81–112. 1947a. On absolute properties of relations, *JSL*, 12, 33–42. 1948. On the principle of dependent choices, *FM*, 35, 127–130. 1949. An undecidable arithmetic statement, *FM*, 36, 143–164. 1951. Some impredicative definitions in the axiomatic set theory, *FM*, 37, 111–124 (also, 38, 1942, 238) 1952. On models of axiomatic systems, *FM*, 39, 133–158. 1952a. On direct powers of theories, *JSL*, 17, 1–31. 1956. *Thirty Years of Foundational Studies.* Blackwell, Oxford. 1957. On a generalization of quantifiers, *FM*, 44, 12–36. 1969. *Constructible Sets with Applications.* N-H. 1979. *Foundational Studies. Selected Works*, Vols. I–II. N-H. Also see Ehrenfeucht, A.; Grzegorczyk, A.; Tarski, A.

Mundici, D. 1985. Other quantifiers: an overview, *MTL*, 211–234.

Myhill, J. 1955. Creative sets, *ZML*, 1, 97–108. 1985. Intensional set theory, *Intensional Mathematics*, N-H, 47–61.

Myhill, J., and J. Dekker. See Dekker, J. C. E.

Myhill, J., and D. Scott. 1971. Ordinal definability, *Proc. Symp. in Pure Math.*, AMS, 13, 271–278.

Nadel, M. 1980. On a problem of Macdowell and Specker, *JSL*, 45, 612–622.

Nadel, M., and L. Lipshitz. See Lipshitz, L.

Nagel, E., and J. R. Newman. 1958. *Gödel's Proof.* NYU Press.

Nagornyi, N. 1953. Stronger reduction theorems for the theory of normal algorithms, *Dokl. Akad. Nauk SSSR*, 90, 341–342.

Negrepontis, S. See Comfort, W. W.

Nelson, E. 1977. Internal set theory: a new approach to nonstandard analysis, *Bull. AMS*, 83, 1165–1198.

Nerode, A. 1961. Extensions to isols, *Ann. of Math.*, 73, 362–403.

Neumann, J. von. 1925. Eine Axiomatisierung der Mengenlehre, *J. für Math.*, 154, 219–240 (also, 155, 128). 1928. Die Axiomatisierung der Mengenlehre, *Math. Zeitschr.*, 27, 669–752.

Newman, J. R. See Nagel, E.

Nicod, J. G. 1917. A reduction in the number of primitive propositions of logic, *Proc. Camb. Phil. Soc.*, 19, 32–41.

Novak, I. L. (Gal, L. N.) 1951. A construction for models of consistent systems, *FM*, 37, 87–110.

Novikov, P. S. 1955. On the algorithmic unsolvability of the word problem for group theory, *Tr. Mat. Inst. Steklov*, 44 (*Amer. Math. Soc. Translations*, Series 2, 9, 1–124).

Oberschelp, A. 1964. Eigentliche Klassen als Urelemente in der Mengenlehre, *Math. Ann.*, 157, 234–260.

Oberschelp, W. 1958. Varianten von Turingmaschinen, *Arch.*, 4, 53–62.

Orey, S. 1956. On ω-consistency and related properties, *JSL*, 21, 246–252. 1956a. On the relative consistency of set theory, ibid., 280–290. 1961. Relative interpretations, *ZML*, 7, 146–153.

Papadimitriou, C. H. See Lewis, H. R.

Parikh, R. 1971. Existence and feasibility in arithmetic, *JSL*, 36, 494–508.

Paris, J. B. 1972. On models of arithmetic, *Conference in Math. Logic—London, 1970*, Springer, 251–280. 1978. Some independence results for Peano arithmetic, *JSL*, 43, 725–731.

Paris, J., and L. Harrington. 1977. A mathematical incompleteness in Peano arithmetic, *HML*, 1133–1142.

Parsons, C. 1972. On n-quantifier induction, *JSL*, 37, 466–482.

Peano, G. 1891. Sul concetto di numero, *Rivista di Mat*, 1, 87–102.

Péter, R. 1935. Konstruktion nichtrekursiver Funktionen, *Math. Ann.*, 111, 42–60. 1967. *Recursive Functions*. Academic.

Phillips, R. G. 1971. On the structure of nonstandard models of arithmetic, *Proc. AMS*, 27, 359–363.

Pincus, D., and R. M. Solovay. 1977. Definability of measures and ultrafilters, *JSL*, 42, 179–190.

Post, E. L. 1921. Introduction to a general theory of elementary propositions, *Amer. J. Math.*, 43, 163–185. 1936. Finite combinatory process-formulation 1, *JSL*, 1, 103–105. (Reprinted in Davis, 1965) 1941. *The Two-Valued Iterative Systems of Mathematical Logic*. Princeton. 1943. Formal reductions of the general combinatorial decision problem, *Amer. J. Math.*, 65, 197–215. 1944. Recursively enumerable sets of positive integers and their decision problems, *Bull. AMS*, 50, 284–316. (Reprinted in Davis, 1965) 1947. Recursive unsolvability of a problem of Thue, *JSL*, 12, 1–11. (Reprinted in Davis, 1965)

Post, E. L., and S. C. Kleene. See Kleene, S. C.

Prawitz, D. 1971. Ideas and results of proof theory, *Proc. Second Scandinavian Logic Symp*, N-H, 235–307.

Presburger, M. 1929. Ueber die Vollständigkeit eines gewissen Systems der Arithmetik ganzer Zahlen in welchem die Addition als einziger Operation hervortritt, *Comptes Rendus, 1 Congrès des Math. des Pays Slaves*, Warsaw, 192–201, 395.

Putnam, H. 1957. Decidability and essential undecidability, *JSL*, 22, 39–54. Also see Davis, M.

Quine, W. V. 1937. New foundations for mathematical logic, *Amer. Math. Monthly*, 44, 70–80. 1950. *Methods of Logic*. Holt. (4th Edition, 1982, Harvard) 1951. *Mathemat-*

ical Logic. Harvard. 1953. *From a Logical Point of View*. Harvard. 1963. *Set Theory and Its Logic* Harvard. (Revised Edition, 1969) 1965. *Selected Logical Papers*. Random House.

Rabin, M. 1958. On recursively enumerable and arithmetic models of set theory, *JSL*, 23, 408–416. 1958a. Recursive unsolvability of group-theoretic problems. *Ann. of Math.*, 67, 172–194. 1959. Arithmetical extensions with prescribed cardinality, *Indag. Math.*, 21, 439–446. 1960. Computable algebra, general theory and theory of computable fields, *Trans. AMS*, 95, 341–360. 1961. Nonstandard models and independence of the induction axiom, *Essays in the Foundations of Mathematics*, Magnes, Jerusalem, 287–299. 1962. Diophantine equations and nonstandard models of arithmetic, *Logic, Methodology, and Philosophy of Science, Proc. Int. Cong. 1960*, Stanford, 151–158. 1977. Decidable theories, *HML*, 595–629.

Ramsey, F. P. 1925. New foundations of mathematics, *Proc. London Math. Soc.*, 25, 338–384.

Rasiowa, H. 1956. On the ε-theorems, *FM*, 43, 156–165. 1974. *An Algebraic Approach to Non-Classical Logics*. N-H.

Rasiowa, H., and R. Sikorski. 1951. A proof of the completeness theorem of Gödel, *FM*, 37, 193–200. 1952. A proof of the Skolem-Löwenheim theorem, *FM*, 38, 230–232. 1953. Algebraic treatment of the notion of satisfiability, *FM*, 40, 62–95. 1963. *The Mathematics of Metamathematics*. Warsaw.

Reinhardt, W. N. 1974. Set existence principles of Shoenfield, Ackermann, and Powell, *FM*, 84, 5–34.

Rescher, N. 1969. *Many-Valued Logics*. McGraw-Hill.

Ressayre, J. P. 1969. Sur les théories du premier ordre categorique en un cardinal, *Trans. AMS*, 142, 481–505.

Reyes, G. See Makkai, M.

Rice, H. G. 1953. Classes of recursively enumerable sets and their decision problems, *Trans. AMS*, 74, 358–366.

Richman, F. 1983. Church's Thesis without tears, *JSL*, 48, 797–803.

Ritchie, R. W., and P. R. Young. 1968. Strong representability of partial functions in arithmetic theories, *Information Science*, 1, 189–204.

Ritter, W. E. 1967. Representability of partial recursive functions in formal theories, *Proc. AMS*, 18, 647–651.

Robinson, A. 1951. *On the Metamathematics of Algebra*. N-H. 1952. On the application of symbolic logic to algebra, *Int. Cong. Math., Cambridge*, 1, 686–694. 1955. On ordered fields and definite functions, *Math. Ann.*, 130, 257–271. 1956. *Complete Theories*. N-H. (Second Edition, 1977) 1961. Model theory and nonstandard arithmetic, *Infinitistic Methods*, Warsaw, 266–302. 1965. *Introduction to Model Theory and to the Metamathematics of Algebra*. N-H. 1966. *Non-Standard Analysis*. N-H. 1979. *Selected Papers*. Vol. 1, *Model Theory and Algebra*, N-H.; Vol. 2, *Nonstandard Analysis and Philosophy*, Yale.

Robinson, J. 1949. Definability and decision problems in arithmetic, *JSL*, 14, 98–114. 1950. General recursive functions, *Proc. AMS*, 1, 703–718. 1952. Existential definability in arithmetic, *Trans. AMS*, 72, 437–449. 1969. Diophantine decision problems. *Studies in Number Theory*, MAA Studies in Math., 6, 76–116. Also see Davis, M.

Robinson, J. A. 1965. A machine-oriented logic based on the resolution principle, *J. Assoc. for Computing Mach.*, 12, 23–41.

Robinson, R. M. 1937. The theory of classes. A modification of von Neumann's system, *JSL*, 2, 69–72. 1947. Primitive recursive functions, *Bull. AMS*, 53, 925–942. 1948.

Recursion and double recursion, ibid., 54, 987–993. 1950. An essentially undecidable axiom system, *Proc. Int. Cong. Math., Cambridge*, 1, 729–730. 1956. Arithmetical representations of recursively enumerable sets, *JSL*, 21, 162–186. Also see Tarski, A.

Rogers, H., Jr. 1958. Gödel numberings of partial recursive functions, *JSL*, 23, 331–341. 1959. Computing degrees of unsolvability, *Math. Ann.*, 138, 125–140. 1967. *Theory of Recursive Functions and Effective Computability*. McGraw-Hill.

Rogers, H., Jr., and D. L. Kreider. See Kreider, D. L.

Rosenbloom, P. 1950. *Elements of Mathematical Logic*. Dover.

Rosser, J. B. 1936a. Constructibility as a criterion for existence, *JSL*, 1, 36–39. 1936b. Extensions of some theorems of Gödel and Church, *JSL*, 87–91. (Reprinted in Davis, 1965) 1937. Gödel theorems for non-constructive logics, *JSL*, 2, 129–137. 1939a. On the consistency of Quine's "New foundations for mathematical logic", *JSL*, 4, 15–24. 1939b. An informal exposition of proofs of Gödel's Theorem and Church's Theorem, *JSL*, 53–60. (Reprinted in Davis, 1965) 1942. The Burali-Forti paradox, *JSL*, 7, 1–17. 1953. *Logic for Mathematicians*. McGraw-Hill. (Second Edition, 1978, Chelsea) 1954. The relative strength of Zermelo's set theory and Quine's New Foundations, *Proc. Int. Cong. Math., Amsterdam*, III, 289–294. 1955. *Deux Esquisses de Logique*. Gauthiers-Villars. 1969. *Simplified Independence Proofs*. Academic.

Rosser, J. B., and A. Turquette. 1952. *Many-Valued Logics*. N-H. (Second Edition, 1977, Greenwood)

Rosser, J. B., and H. Wang. 1950. Non-standard models for formal logics, *JSL*, 15, 113–129.

Rotman, J. J. 1973. *The Theory of Groups*. Allyn & Bacon. (Second Edition)

Rubin, H., and J. Rubin. 1963. *Equivalents of the Axiom of Choice*. N-H.

Rubin, J. 1967. *Set Theory for the Mathematician*. Holden-Day. Also see Howard, P. E.

Russell, B. 1908. Mathematical logic as based on the theory of types, *Amer. J. Math.*, 30, 222–262.

Russell, B., and A. N. Whitehead. See Whitehead, A. N.

Ryll-Nardzewski, C. 1953. The role of the axiom of induction in elementary arithmetic, *FM*, 39, 239–263. Also see Grzegorczyk, A.; Loś, J.

Sacks, G. E. 1963. *Degrees of Unsolvability*. Princeton. 1972. *Saturated Model Theory*. Benjamin.

Sambin, G. 1976. An effective fixed point theorem in intuitionistic diagonalizable algebras, *Studia Logica*, 35, 345–361.

Šanin, N. A. 1958. On the constructive interpretation of mathematical judgments, *Tr. Mat. Inst. Steklov*, 52, 226–311. (English translation: *AMS Translations* (2), 23, 1963, 109–189)

Scarpellini, B. 1969. On the metamathematics of rings and integral domains, *Trans. AMS*, 138, 71–96.

Schütte, K. 1951. Beweistheoretische Erfassung der unendlichen Induktion in der Zahlentheorie, *Math. Ann.*, 122, 368–389. 1960. *Beweistheorie*. Springer. (English translation: *Proof Theory*, 1977)

Scott, D. 1961. On constructing models for arithmetic, *Infinitistic Methods*, Warsaw, 235–255. 1962a. Algebras of sets binumerable in complete extensions of arithmetic, *Proc. Symp. Pure Math.*, 5, AMS, 117–121. 1962b. Quine's individuals, *Logic, Methodology, and Philosophy of Science*, Stanford, 111–115. 1967. A proof of the independence of the continuum hypothesis, *Math. Systems Theory*, 1, 89–111. 1974. Axiomatizing set theory, *Proc. Symp. Pure Math.*, 13 AMS, 11, 207–214. Also see Frayne, T.; Myhill, J.

Seese, D. See Baudisch, A.

Seidenberg, A. 1954. A new decision method for elementary algebra, *Ann. of Math.*, 60, 365–374.

Seldin, J. See Curry, H. B.; Hindley, J. R.

Shannon, C. 1938. A symbolic analysis of relay and switching circuits, *Trans. Amer. Inst. Elect. Eng.*, 57, 713–723.

Shapiro, S. 1980. On the notion of effectiveness, *History and Philosophy of Logic*, 1, 209–230. 1985. Epistemic and intuitionistic arithmetic, *Intensional Mathematics*, N-H, 11–46.

Shelah, S. 1971. Every two elementarily equivalent models have isomorphic ultrapowers, *Israel J. Math.*, 10, 224–233.

Shepherdson, J. C. 1951–53. Inner models for set theory, *JSL*, I, 16, 161–190; II, 17, 225–237; III, 18, 145–167. 1961. Representability of recursively enumerable sets in formal theories, *Arch.*, 5, 119–127. Also see Bezboruah, A.

Shepherdson, J. C., and H. E. Sturgis. 1963. Computability of recursive functions, *J. Assoc. Comp. Mach.*, 10, 217–255.

Shoenfield, J. 1954. A relative consistency proof, *JSL*, 19, 21–28. 1958. Degrees of formal systems, *JSL*, 23, 389–392. 1959. On a restricted ω-rule, *Bull. Acad. Polon. Sci., Ser. Sci. Math. Astr. Phys.*, 7, 405–407. 1961a. Undecidable and creative theories, *FM*, 49, 171–179. 1961b. The problem of predicativity, *Essays on the Foundations of Mathematics*, Magnes, Jerusalem, 132–139. 1967. *Mathematical Logic*. Addison-Wesley. 1971a. *Degrees of Unsolvability*. N-H. 1971b. Unramified forcing, *Proc. Symp. Pure Math.*, 13, AMS, 357–381. 1977. Axioms of set theory, *HML*, 321–344.

Sierpinski, W. 1947. L'hypothèse généralisée du continu et l'axiome du choix, *FM*, 34, 1–5. 1958. *Cardinal and Ordinal Numbers*. Warsaw.

Sigler, L. E. 1977. *Exercises in Set Theory*. Springer.

Sikorski, R. 1960. *Boolean Algebras*. Springer. (Third Edition, 1969) Also see Rasiowa, H.

Silver, J. 1971. Some applications of model theory in set theory, *AML*, 3, 45–110.

Simmons, H. 1974–75. Topological aspects of suitable theories, *Proc. Edinburgh Math. Soc.*, 19, 383–391. Also see MacIntyre, A.

Simpson, S. 1977. Degrees of unsolvability: a survey of results, *HML*, 631–652.

Skolem, T. 1919. Logisch-kombinatorische Untersuchungen über die Erfüllbarkeit oder Beweisbarkeit mathematischer Sätze, *Skrifter-Vidensk*, Kristiana, I, 1–36. 1923. Einige Bemerkungen zur axiomatischen Begründung der Mengenlehre, *Wiss. Vorträge gehalten auf dem 5. Kongress der skandinav. Mathematiker in Helsingförs*, 1922, 217–232. 1934. Ueber die Nicht-Charakterisierbarkeit der Zahlenreihe mittels endlich oder abzählbar unendlich vieler Aussagen mit ausschliesslich Zahlenvariablen, *FM*, 23, 150–161. 1955. Peano's axioms and models of arithmetic, *Mathematical Interpretations of Formal Systems*, N-H, 1–14. 1970. *Selected Works in Logic*. Universitetsforlaget, Oslo.

Slomson, A. B. See Bell, J. L.

Smorynski, C. 1977. The incompleteness theorems, *HML*, 821–866. 1978. Avoiding self-referential statements, *Proc. AMS*, 70, 181–184. 1979. Calculating self-referential statements, I: explicit calculations, *Studia Logica*, 38, 7–36. 1980. Calculating self-referential statements: non-explicit calculations, *FM*, 109, 189–210. 1981a. Calculating self-referential statements: Guaspari sentences of the first kind, *JSL*, 46, 329–344. 1981b. Fifty years of self-reference, *NDJFL*, 22, 357–374. 1982. Fixed point algebras, *Bull. AMS*, 6, 317–356.

Smullyan, R. 1957. Languages in which self-reference is possible, *JSL*, 22, 55–67. 1961. *Theory of Formal Systems*. Princeton. 1968. *First-Order Logic*. Springer. 1978.

What Is the Name of This Book. Prentice-Hall. 1985. *To Mock a Mockingbird*. Knopf. 1985a. Modality and self-reference, *Intensional Mathematics*, N-H, 191–211. 1985b. Some principles related to Löb's Theorem, *Intensional Mathematics*, N-H, 213–230.

Solovay, R. M. 1970. A model of set theory in which every set of reals is Lebesgue measurable, *Ann. of Math.*, 92, 1–56. 1976. Provability interpretations of modal logic, *Israel J. Math.*, 25, 287–304. 1985. Explicit Henkin sentences, *JSL*, 50, 91–93. 1985a. Infinite fixed point algebras, *Proc. Symp. in Pure Math.*, 42, AMS, 473–486. Also see Guaspari, D.; Pincus, D.

Sonner, J. 1962. On the formal definition of categories, *Math. Z.*, 80, 163–176.

Specker, E. 1949. Nicht-konstruktiv beweisbare Sätze der Analysis, *JSL*, 14, 145–148. 1953. The axiom of choice in Quine's "New foundations for mathematical logic", *Proc. Acad. Sci. USA*, 39, 972–975. 1954. Verallgemeinerte Kontinuumhypothese und Auswahlaxiom, *Archiv der Math.*, 5, 332–337. 1957. Zur Axiomatik der Mengenlehre (Fundierungs und Auswahlaxiom), *ZML*, 3, 173–210. 1958. Dualität, *Dialectica*, 12, 451–465. 1962. Typical ambiguity, *Logic, Methodology and Philosophy of Science, Proc. Int. Cong., 1960*, Stanford, 116–124. Also see Macdowell, R.

Spector, C. 1955. Recursive well-orderings, *JSL*, 20, 151–163.

Stone, M. 1936. The representation theorem for Boolean algebras, *Trans. AMS*, 40, 37–111.

Stroyan, K. D., and W. A. J. Luxemburg. 1976. *Introduction to the Theory of Infinitesimals*. Academic.

Sturgis, H. E. See Shepherdson, J. C.

Sundholm, G. 1983. Systems of deduction, *HPL*, I, 133–188.

Suppes, P. 1960. *Axiomatic Set Theory*. Van Nostrand. (Reprint, 1972, Dover)

Szmielew, W. 1955. Elementary properties of abelian groups, *FM*, 41, 203–271.

Taimanov, A. See Ershov, Yu.

Taitslin, M. See Ershov, Yu.

Takeuti, G. 1975. *Proof Theory*. N-H. 1978. *Two Applications of Logic to Mathematics*. Princeton.

Takeuti, G., and W. M. Zaring. 1971. *Introduction to Axiomatic Set Theory*. Springer. 1973. *Axiomatic Set Theory*. Springer.

Tarski, A. 1923. Sur quelques théorèmes qui equivalent a l'axiome de choix, *FM*, 5, 147–154. 1925. Sur les ensembles finis, *FM*, 6, 45–95. 1933. Einige Betrachtungen über die Begriffe der ω-Widerspruchsfreiheit und der ω-Vollständigkeit, *Monats. Math. Phys.*, 40, 97–112. 1936. Der Wahrheitsbegriff in den formalisierten Sprachen, *Studia Philos.*, 1, 261–405. (English translation in Tarski, 1956) 1938. Ueber unerreichbare Kardinalzahlen, *FM*, 30, 68–89. 1944. The semantic conception of truth and the foundations of semantics, *Philos. and Phenom. Res.*, 4, 341–376. 1951. *A. Decision Method for Elementary Algebra and Geometry*. Berkeley. 1952. Some notions and methods on the borderline of algebra and metamathematics, *Int. Cong. Math.*, Cambridge, 705–720. 1954–55. Contributions to the theory of models, *Indag. Math.*, 16, 572–588; 17, 56–64. 1956. *Logic, Semantics, Metamathematics*. Oxford. (Second Edition, 1983, J. Corcoran (Editor), Hackett) Also see Erdös, P.; Henkin, L.; Keisler, H. J.; McKinsey, J. C. C.

Tarski, A., A. Mostowski, and R. Robinson. 1953. *Undecidable Theories*. N-H.

Tarski, A., and R. Vaught. 1957. Arithmetical extensions of relational systems, *Comp. Math.*, 18, 81–102.

Tennenbaum, S. 1968. Souslin's problem, *Proc. Natl. Acad. Sci. USA*, 59, 60–63.

Tharp, L. 1967. On a set theory of Bernays, *JSL*, 32, 319–321.

Thompson, R. J. See McKenzie, R.

Troelstra, A. S. 1969. *Principles of Intuitionism*. Springer.

Truss, J. K. 1973. The well-ordered and well-orderable subsets of a set, *ZML*, 19, 211–214.

Turing, A. 1936–37. On computable numbers, with an application to the Entscheidungsproblem, *Proc. London Math. Soc.*, 42, 230–265; 43, 544–546. 1937. Computability and λ-definability, *JSL*, 2, 153–163. 1939. Systems of logic based on ordinals, *Proc. London Math. Soc.*, 45, 161–228. 1948. Practical forms of type theory, *JSL*, 13, 80–94. 1950a. The word problem in semigroups with cancellation, *Ann. of Math.*, 52, 491–505. (Review by W. W. Boone, *JSL*, 1952, 74–76) 1950b. Computing machinery and intelligence, *Mind*, 59, 433–460.

Turquette, A. See Rosser, J. B.

Tuschik, P. See Baudisch, A.

Ulam, S. 1930. Zur Masstheorie in der allgemeinen Mengenlehre, *FM*, 16, 140–150. 1974. *Sets, Numbers, and Universes. Selected Works*. MIT Press.

Van Benthem, J., and K. Doets. 1983. Higher-order logic, *HPL*, I, 275–330.

Van Dalen, D. 1983. Algorithms and decision problems. A crash course in recursion theory, *HPL*, I, 409–478.

Vaughan, J. See Kunen, K.

Vaughn, H. See Halmos, P.

Vaught, R. 1954. Applications of the Löwenheim-Skolem-Tarski theorem to problems of completeness and decidability, *Indag. Math.*, 16, 467–472. 1963. Models of complete theories, *Bull. AMS*, 69, 467–472. Also see Montague, R.; Tarski, A.

Vesley, R. E. See Kleene, S. C.

Vopenka, P., and P. Hajek. 1972. *The Theory of Semisets*. N-H.

Waerden, B. L. van der. 1949. *Modern Algebra*. Ungar.

Wajsberg, M. 1933. Untersuchungen über den Funktionenkalkül für endliche Individuenbereiche, *Math Annalen*, 108, 218–228.

Wang, H. 1949. On Zermelo's and von Neumann's axioms for set theory, *Proc. Natl. Acad. Sci. USA*, 35, 150–155. 1950. A formal system of logic, *JSL*, 15, 25–32. 1955. Undecidable sentences generated by semantical paradoxes, *JSL*, 20, 31–43. 1957. The axiomatization of arithmetic, *JSL*, 22, 145–158. 1970. *Logic, Computers, and Sets*. Chelsea. 1973. *From Mathematics to Philosophy*. Humanities. Also see Kreisel, G.; Rosser, J. B.

Wang, H., and R. McNaughton. 1953. *Les Systèmes Axiomatiques de la Theorie des Ensembles*. Gauthiers-Villars.

Webb, J. C. 1980. *Mechanism, Mentalism, and Metamathematics*. Reidel.

Weese, M. See Baudisch, A.

Whitehead, A. N., and B. Russell. 1910–13. *Principia Mathematica*, Vols. I–III. Cambridge.

Williams, N. H. 1977. *Combinatorial Set Theory*. N-H.

Yasuhara, A. 1971. *Recursive Function Theory and Logic*. Academic.

Yasuhara, M. 1984. Extensionality in Bernays set theory, *NDJFL*, 25, 357–363.

Young, P. 1985. Gödel's theorem, exponential difficulty, and the undecidability of arithmetic theories: an exposition, *Recursion Theory, Proc. Symp. Pure Math.*, 42, AMS, 503–522. Also see Machtey, M.; Ritchie, R. W.

Zakon, E. 1969. Remark on the nonstandard real axis, *Applications of Model Theory to Algebra, Analysis, and Probability*, Holt, Rinehart & Winston, 195–227.

Zaring, W. M. See Takeuti, G.

Zeeman, E. C. 1955. On direct sums of free cycles, *J. London Math. Soc.*, 30, 195–212.

Zermelo, E. 1908. Untersuchungen über die Grundlagen der Mengenlehre, *Math. Ann.*, 65, 261–281.

Zuckerman, M. 1974. *Sets and Transfinite Numbers*. Macmillan.

Answers to Selected Exercises

Chapter 1

1.1

A	B	
T	T	F
F	T	T
T	F	T
F	F	F

1.2

A	B	$\neg A$	$A \Rightarrow B$	$(A \Rightarrow B) \vee \neg A$
T	T	F	T	T
F	T	T	T	T
T	F	F	F	F
F	F	T	T	T

1.3 $((A \Rightarrow B) \wedge A)$

A	$\Rightarrow$	B	$\wedge$	A
T	T	T	T	T
F	T	T	F	F
T	F	F	F	T
F	T	F	F	F

1.4 (a) $((A \Rightarrow \neg B) \wedge (\neg A \Rightarrow \neg B))$
(d) $A \Rightarrow B$, A: Fiorello goes to the movies.
(e) $A \Rightarrow B$, A: x is prime.
(f) $A \Rightarrow B$, A: s converges.
(h) $\neg A \Rightarrow B$, A: Kasparov wins today.

1.5 (a) No (c) No (e) Yes (g) Yes

1.11 All except (i)

1.13 Only (c) and (e)

1.15 (a) $(B \Rightarrow \neg A) \wedge C$ (e) $A \Leftrightarrow B \Leftrightarrow \neg(C \vee D)$
(c) Drop all parentheses. (g) $\neg(\neg\neg(B \vee C) \Leftrightarrow (B \Leftrightarrow C))$

1.16 (a) $(C \vee ((\neg A) \wedge B))$ (c) $((C \Rightarrow ((\neg(A \vee C)) \wedge A)) \Leftrightarrow B)$

1.17 (a) $(((\neg(\neg A)) \Leftrightarrow A) \Leftrightarrow (B \vee C))$
(d) No, extra right parenthesis at the end

1.18 (a) $\vee \Rightarrow C \neg AB$ and $\vee C \Rightarrow \wedge B \neg DC$
(c) (a) $\wedge \Rightarrow B \neg AC$ (b) $\vee A \vee BC$
(d) (i) No (ii) $(A \Rightarrow B) \Rightarrow ((B \Rightarrow C) \Rightarrow (\neg A \Rightarrow C))$
(iii) $((\neg A \vee \neg B) \wedge C) \vee ((A \vee C) \wedge (\neg C \vee \neg A))$

1.19 (a), (d), (f), (g), and (h) are tautologies; (e) is a contradiction.

1.21 (a) and (b) are true; (c) can be either T or F.

1.29 (c) (i) One possible answer is $A \wedge (B \Leftrightarrow C)$. (ii) $A \wedge B \wedge \neg C$

1.30 (a) If $\mathscr{A}$ is a tautology, replace all statement letters by their negations, and then move all the negation signs outward by using Exercise 1.27 (a, b). The result is $\neg \mathscr{A}'$.

Conversely, if $\neg\mathscr{A}'$ is a tautology, let $\mathscr{B}$ be $\neg\mathscr{A}'$. By the first part, $\neg\mathscr{B}'$ is a tautology. But, $\neg\mathscr{B}'$ is $\neg\neg\mathscr{A}$.

1.32 (a) For Figure 1.4:

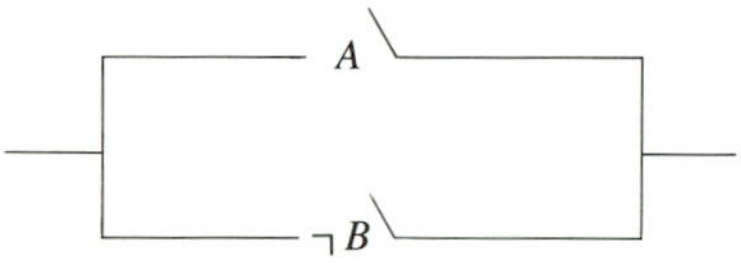

1.33 (a), (d), and (i) are not correct.

1.34 (a) Consistent; let A, B, and C be F, and let D be T.

1.36 $(A \wedge B \wedge C) \vee (\neg A \wedge B \wedge C) \vee (A \wedge \neg B \wedge C) \vee (\neg A \wedge \neg B \wedge \neg C)$ is the statement form for f.

1.37 For $\Rightarrow$ and $\vee$, notice that any statement form built up using $\Rightarrow$ and $\vee$ will always take the value T when the statement letters in it are T. In the case of $\neg$ and $\Leftrightarrow$, using only the statement letters A and B, find all the truth functions of two variables that can be generated by applying $\neg$ and $\Leftrightarrow$ any number of times.

1.39 (a) $2^4 = 16$ (b) 2^{2^n}

1.40 $h(C, C, C) = \neg C$, $h(\neg C, \neg C, C) = T$, $h(A, B, \neg T) = \neg(A \vee B)$

1.41 (b) For $\neg(A \Rightarrow B) \vee (\neg A \wedge C)$, a disjunctive normal form is $(A \wedge \neg B) \vee (\neg A \wedge C)$, and a conjunctive normal form is $(A \vee C) \wedge (\neg B \vee \neg A) \wedge (\neg B \vee C)$.

(c) (i) For $(A \wedge B) \vee \neg A$, a full dnf is $(A \wedge B) \vee (\neg A \wedge B) \vee (\neg A \wedge \neg B)$, and a full cnf is $B \vee \neg A$.

1.42 (b) (i) $(A \vee \neg B) \wedge B \wedge A$

1.45 (b)

1.	$\mathscr{B} \Rightarrow \mathscr{C}$	Hypothesis
2.	$\mathscr{A} \Rightarrow \mathscr{B}$	Hypothesis
3.	$\mathscr{A} \Rightarrow (\mathscr{B} \Rightarrow \mathscr{C}) \Rightarrow ((\mathscr{A} \Rightarrow \mathscr{B}) \Rightarrow (\mathscr{A} \Rightarrow \mathscr{C}))$	Axiom (A2)
4.	$(\mathscr{B} \Rightarrow \mathscr{C}) \Rightarrow (\mathscr{A} \Rightarrow (\mathscr{B} \Rightarrow \mathscr{C}))$	Axiom (A1)
5.	$\mathscr{A} \Rightarrow (\mathscr{B} \Rightarrow \mathscr{C})$	1, 4, MP
6.	$(\mathscr{A} \Rightarrow \mathscr{B}) \Rightarrow (\mathscr{A} \Rightarrow \mathscr{C})$	3, 5, MP
7.	$\mathscr{A} \Rightarrow \mathscr{C}$	1, 6, MP

1.46 (a)

1.	$\mathscr{A} \Rightarrow \neg\neg\mathscr{A}$	Lemma 1.10(b)
2.	$\neg\neg\mathscr{A} \Rightarrow (\neg\mathscr{A} \Rightarrow \mathscr{B})$	Lemma 1.10(c)
3.	$\mathscr{A} \Rightarrow (\neg\mathscr{A} \Rightarrow \mathscr{B})$	1, 2, Corollary 1.9(a)
4.	$\mathscr{A} \Rightarrow (\mathscr{A} \vee \mathscr{B})$	3, Abbreviation

(c)

1.	$\neg\mathscr{B} \Rightarrow \mathscr{A}$	Hypothesis
2.	$(\neg\mathscr{B} \Rightarrow \mathscr{A}) \Rightarrow (\neg\mathscr{A} \Rightarrow \neg\neg\mathscr{B})$	Lemma 1.10(e)
3.	$\neg\mathscr{A} \Rightarrow \neg\neg\mathscr{B}$	1, 2, MP
4.	$\neg\neg\mathscr{B} \Rightarrow \mathscr{B}$	Lemma 1.10(a)
5.	$\neg\mathscr{A} \Rightarrow \mathscr{B}$	3, 4, Corollary 1.9(a)
6.	$\neg\mathscr{B} \Rightarrow \mathscr{A} \vdash \neg\mathscr{A} \Rightarrow \mathscr{B}$	1–5
7.	$\vdash (\neg\mathscr{B} \Rightarrow \mathscr{A}) \Rightarrow (\neg\mathscr{A} \Rightarrow \mathscr{B})$	1–6, Deduction Theorem
8.	$\vdash \mathscr{B} \vee \mathscr{A} \Rightarrow \mathscr{A} \vee \mathscr{B}$	7, Abbreviation

1.49 Take any assignment of truth values to the statement letters of $\mathscr{A}$ that makes $\mathscr{A}$ false. Replace in $\mathscr{A}$ each letter having the value T by $A_1 \vee A_1$, and each letter having the value F by $A_1 \wedge \neg A_1$. Call the resulting statement form $\mathscr{B}$. Observe that $\mathscr{B}$ always has the value F for any truth assignment.

1.50 (Deborah Moll) Use two truth values. Let $\Rightarrow$ have its usual table and let $\neg$ be interpreted as the constant function F. When B is F, $(\neg B \Rightarrow \neg A) \Rightarrow ((\neg B \Rightarrow A) \Rightarrow B)$ is F.

1.51 The theorems of P are the same as the axioms. Assume that P is suitable for some n-valued logic. Then, for all values k, $k * k$ will be a designated value. Consider the sequence of formulas $\mathscr{A}_0 = A$, $\mathscr{A}_{j+1} = A * \mathscr{A}_j$. Since there are n^n possible truth functions of one variable, among $\mathscr{A}_0, \ldots, \mathscr{A}_{n^n}$ there must be two different formulas $\mathscr{A}_j$ and $\mathscr{A}_k$ that determine the same truth function. Hence, $\mathscr{A}_j * \mathscr{A}_k$ will be an exceptional formula that is not a theorem.

1.52 Take as axioms all exceptional formulas and the identity function as the only rule of inference.

Chapter 2

2.1 (a) $((\forall x_1)(A_1^1(x_1) \wedge (\neg A_1^1(x_2))))$ (b) $(((\forall x_2)A_1^1(x_2)) \Leftrightarrow A_1^1(x_2))$
(d) $(((\forall x_1)((\forall x_3)((\forall x_4)A_1^1(x_1)))) \Rightarrow (A_1^1(x_2) \wedge (\neg A_1^1(x_1))))$

2.2 (a) $((\forall x_1)(A_1^1(x_1) \Rightarrow A_1^1(x_1))) \vee (\exists x_1)A_1^1(x_1)$

2.3 (a) The only free occurrence of a variable is that of x_2.
(b) The first occurrence of x_3 is free, as is the last occurrence of x_2.

2.6 Yes, in parts (a), (c), and (e)

2.8 (a) $(\forall x)(P(x) \Rightarrow L(x))$
(b) $(\forall x)(P(x) \Rightarrow \neg H(x))$ or $\neg(\exists x)(P(x) \wedge H(x))$ (c) $\neg(\forall x)(B(x) \Rightarrow F(x))$
(g) $(\forall x)(\neg H(x,x) \Rightarrow H(j,x))$ or $(\forall x)(P(x) \wedge \neg H(x,x) \Rightarrow H(j,x))$ [In the second wf, we have specified that John hates those *persons* who do not hate themselves, where $P(x)$ means x *is a person*.]

2.9 (a) All bachelors are unhappy. (c) There is no greatest integer.

2.10 (a) (i) is satisfied by all pairs $\langle x_1, x_2 \rangle$ of positive integers such that $x_1 \cdot x_2 \geqslant 2$.
(ii) is satisfied by all pairs $\langle x_1, x_2 \rangle$ of positive integers such that either $x_1 < x_2$ (when the antecedent is false) or $x_1 = x_2$ (when the antecedent is true).
(iii) is true.

2.11 (a) Between any two real numbers there is a rational number.

2.12 (I) A sequence s satisfies $\neg\mathscr{A}$ if and only if s does not satisfy $\mathscr{A}$. Hence, all sequences satisfy $\neg\mathscr{A}$ if and only if no sequence satisfies $\mathscr{A}$; that is, $\neg\mathscr{A}$ is true if and only if $\mathscr{A}$ is false.
(II) There is at least one sequence s in $\sum$. If s satisfies $\mathscr{A}$, $\mathscr{A}$ cannot be false for M. If s does not satisfy $\mathscr{A}$, $\mathscr{A}$ cannot be true for M.
(III) If a sequence s satisfies both $\mathscr{A}$ and $\mathscr{A} \Rightarrow \mathscr{B}$, then s satisfies $\mathscr{B}$ by condition 3 of the definition.
(V) (a) s satisfies $\mathscr{A} \wedge \mathscr{B}$ if and only if s satisfies $\neg(\mathscr{A} \Rightarrow \neg\mathscr{B})$
if and only if s does not satisfy $\mathscr{A} \Rightarrow \neg\mathscr{B}$
if and only if s satisfies $\mathscr{A}$ but not $\neg\mathscr{B}$
if and only if s satisfies $\mathscr{A}$ and s satisfies $\mathscr{B}$
(VI) (a) Assume $\models_{\mathrm{M}} \mathscr{A}$. Then every sequence satisfies $\mathscr{A}$. In particular, every sequence that differs from a sequence s in at most the ith place satisfies $\mathscr{A}$. So, every sequence satisfies $(\forall x_i)\mathscr{A}$; that is, $\models_{\mathrm{M}} (\forall x_i)\mathscr{A}$.
(b) Assume $\models_{\mathrm{M}} (\forall x_i)\mathscr{A}$. If s is a sequence, then any sequence that differs from s in at most the ith place satisfies $\mathscr{A}$, and, in particular, s satisfies $\mathscr{A}$. Then every sequence satisfies $\mathscr{A}$; that is, $\models_{\mathrm{M}} \mathscr{A}$.
(VIII) *Lemma.* If all the variables in a term t occur in the list $x_{i_1}, \ldots, x_{i_k}$ ($k \geqslant 0$; when $k = 0$, t has no variables), and if the sequences s and s' have the same components in the i_1th, …, i_kth places, then $s^*(t) = (s')^*(t)$.
Proof. Induction on the number m of function letter in t. Assume the result holds for all integers $<m$.
Case 1. t is an individual constant a_p. Then $s^*(a_p) = (a_p)^{\mathrm{M}} = (s')^*(a_p)$.
Case 2. t is a variable x_{i_j}. Then $s^*(x_{i_j}) = s_{i_j} = s'_{i_j} = (s')^*(x_{i_j})$.
Case 3. t is of the form $f_j^n(t_1, \ldots, t_n)$. For $q \leqslant n$, each t_q has fewer than m function letters and all its variables occur among $x_{i_1}, \ldots, x_{i_k}$. By inductive hypothesis, $s^*(t_q) = (s')^*(t_q)$. Then $s^*(f_j^n(t_1, \ldots, t_n)) = (f_j^n)^{\mathrm{M}}(s^*(t_1), \ldots, s^*(t_n)) = (f_j^n)^{\mathrm{M}}((s')^*(t_1), \ldots, (s')^*(t_n)) = (s')^*(f_j^n(t_1, \ldots, t_n))$.
Proof of (VIII). Induction on the number r of connectives and quantifiers in $\mathscr{A}$. Assume the result holds for all $q < r$.
Case 1. $\mathscr{A}$ is of the form $A_j^n(t_1, \ldots, t_n)$; that is, $r = 0$. All the variables of each t_i occur among $x_{i_1}, \ldots, x_{i_k}$. Hence, by the lemma, $s^*(t_i) = (s')^*(t_i)$. But s satisfies

$A_j^n(t_1,\ldots,t_n)$ if and only if $\langle s^*(t_1),\ldots,s^*(t_n)\rangle$ is in $(A_j^n)^M$—that is, if and only if $\langle (s')^*(t_1),\ldots,(s')^*(t_n)\rangle$ is in $(A_j^n)^M$, which is equivalent to s' satisfying $A_j^n(t_1,\ldots,t_n)$.

Case 2. $\mathscr{A}$ is of the form $\neg\mathscr{B}$.

Case 3. $\mathscr{A}$ is of the form $\mathscr{B} \Rightarrow \mathscr{C}$. Both cases 2 and 3 are easy.

Case 4. $\mathscr{A}$ is of the form $(\forall x_j)\mathscr{B}$. The free variables of $\mathscr{B}$ occur among $x_{i_1}, \ldots, x_{i_k}$ and x_j. Assume s satisfies $\mathscr{A}$. Then every sequence that differs from s in at most the jth place satisfies $\mathscr{B}$. Let $s^\#$ be any sequence that differs from s' in at most the jth place. Let s^b be a sequence that has the same components as s in all but the jth place, where it has the same component as $s^\#$. Hence, s^b satisfies $\mathscr{B}$. Since s^b and $s^\#$ agree in the i_1th, ..., i_kth and jth places, it follows by inductive hypothesis that s^b satisfies $\mathscr{B}$ if and only if $s^\#$ satisfies $\mathscr{B}$. Hence, $s^\#$ satisfies $\mathscr{B}$. Thus, s' satisfies $\mathscr{A}$. By symmetry, the converse also holds.

(IX) Assume $\mathscr{A}$ is closed. By (VIII), for any sequences s and s', s satisfies $\mathscr{A}$ if and only if s' satisfies $\mathscr{A}$. If $\neg\mathscr{A}$ is not true for M, some sequence s' does not satisfy $\neg\mathscr{A}$; that is, s' satisfies $\mathscr{A}$. Hence, every sequence s satisfies $\mathscr{A}$; that is, $\models_M \mathscr{A}$.

(X) *Proof of Lemma 1:* induction on the number m of function letters in t.

Case 1. t is a_j. Then t' is a_j. Hence,

$$s^*(t') = s^*(a_j) = (a_j)^M = (s')^*(a_j) = (s')^*(t)$$

Case 2. t is x_j, where $j \neq i$. Then t' is x_j. By the lemma of (VIII), $s^*(t') = (s')^*(t)$, since s and s' have the same component in the jth place.

Case 3. t is x_i. Then t' is u. Hence, $s^*(t') = s^*(u)$, while $(s')^*(t) = (s')^*(x_i) = s'_i = s^*(u)$.

Case 4. t is of the form $f_j^n(t_1,\ldots,t_n)$. For $1 \leqslant q \leqslant n$, let t'_q result from t_q by the substitution of u for x_i. By inductive hypothesis, $s^*(t'_q) = (s')^*(t_q)$. But

$$\begin{aligned} s^*(t') &= s^*(f_j^n(t'_1,\ldots,t'_n)) = (f_j^n)^M(s^*(t'_1),\ldots,s^*(t'_n)) \\ &= (f_j^n)^M((s')^*(t_1),\ldots,(s')^*(t_n)) = (s')^*(f_j^n(t_1,\ldots,t_n)) = (s')^*(t) \end{aligned}$$

Proof of Lemma 2(a): induction on the number m of connectives and quantifiers in $\mathscr{A}(x_i)$.

Case 1. $m = 0$. Then $\mathscr{A}(x_i)$ is $A_j^n(t_1,\ldots,t_n)$. Let t'_q be the result of substituting t for all occurrences of x_i in t_q. Thus, $\mathscr{A}(t)$ is $A_j^n(t'_1,\ldots,t'_n)$. By Lemma 1, $s^*(t'_q) = (s')^*(t_q)$. Now, s satisfies $\mathscr{A}(t)$ if and only if $\langle s^*(t'_1),\ldots,s^*(t'_n)\rangle$ belongs to $(A_j^n)^M$, which is equivalent to $\langle (s')^*(t_1),\ldots,(s')^*(t_n)\rangle$ belonging to $(A_j^n)^M$—that is, to s' satisfying $\mathscr{A}(x_i)$.

Case 2. $\mathscr{A}(x_i)$ is $\neg\mathscr{B}(x_i)$; this is straightforward.

Case 3. $\mathscr{A}(x_i)$ is $\mathscr{B}(x_i) \Rightarrow \mathscr{C}(x_i)$; this is straightforward.

Case 4. $\mathscr{A}(x_i)$ is $(\forall x_j)\mathscr{B}(x_i)$.

Case 4a. x_j is x_i. Then x_i is not free in $\mathscr{A}(x_i)$, and $\mathscr{A}(t)$ is $\mathscr{A}(x_i)$. Since x_i is not free in $\mathscr{A}(x_i)$, it follows by (VIII) that s satisfies $\mathscr{A}(t)$ if and only if s' satisfies $\mathscr{A}(x_i)$.

Case 4b. x_j is different from x_i. Since t is free for x_i in $\mathscr{A}(x_i)$, t is also free for x_i in $\mathscr{B}(x_i)$.

Assume s satisfies $(\forall x_j)\mathscr{B}(t)$. We must show that s' satisfies $(\forall x_j)\mathscr{B}(x_i)$. Let $s^\#$ differ from s' in at most the jth place. It suffices to show that $s^\#$ satisfies $\mathscr{B}(x_i)$. Let s^b be the same as $s^\#$ except that it has the same ith component as s. Hence, s^b is the same as s except in its jth component. Since s satisfies $(\forall x_j)\mathscr{B}(t)$, s^b satisfies $\mathscr{B}(t)$. Now, since t is free for x_i in $(\forall x_j)\mathscr{B}(x_i)$, t does not contain x_j. [The other possibility, that x_i is not free in $\mathscr{B}(x_i)$, is handled as in Case 4a.] Hence, by the lemma of (VIII), $(s^b)^*(t) = s^*(t)$. Hence, by the inductive hypothesis and the fact that $s^\#$ is obtained from s^b by substituting $(s^b)^*(t)$ for the ith component of s^b, it follows that $s^\#$ satisfies $\mathscr{B}(x_i)$ if and only if s^b satisfies $\mathscr{B}(t)$. Since s^b satisfies $\mathscr{B}(t)$, $s^\#$ satisfies $\mathscr{B}(x_i)$.

Conversely, assume s' satisfies $(\forall x_j)\mathscr{B}(x_i)$. Let s^b differ from s in at most the jth place. Let $s^\#$ be the same as s' except in the jth place, where it is the same as s^b. Then $s^\#$ satisfies $\mathscr{B}(x_i)$. As above, $s^*(t) = (s^b)^*(t)$. Hence, by the inductive hypothesis, s^b satisfies $\mathscr{B}(t)$ if and only if $s^\#$ satisfies $\mathscr{B}(x_i)$. Since $s^\#$ satisfies $\mathscr{B}(x_i)$, s^b satisfies $\mathscr{B}(t)$. Therefore, s satisfies $(\forall x_j)\mathscr{B}(t)$.

Proof of Lemma 2(b). Assume s satisfies $(\forall x_i)\mathscr{A}(x_i)$. We must show that s satisfies $\mathscr{A}(t)$. Let s' arise from s by substituting $s^*(t)$ for the ith component of s. Since s satisfies $(\forall x_i)\mathscr{A}(x_i)$ and s' differs from s in at most the ith place, s' satisfies $\mathscr{A}(x_i)$. By Lemma 2(a), s satisfies $\mathscr{A}(t)$.

2.13 Assume $\mathscr{A}$ is satisfied by a sequence s. Let s' be any sequence. By (VIII), s' also satisfies $\mathscr{A}$. Hence, $\mathscr{A}$ is satisfied by all sequences; that is, $\models_M \mathscr{A}$.

2.14 (a) x is a common divisor of y and z.
(d) x_1 is a bachelor.

2.15 (a) (i) Every nonnegative integer is even or odd. True.
(ii) If the product of two nonnegative integers is zero, at least one of them is zero. True.
(iii) 1 is even. False.

2.16 (a) Consider an interpretation with the set of integers as its domain. Let $A_1^1(x)$ mean that x is even, and let $A_2^1(x)$ mean that x is odd. Then $(\forall x_1)A_1^1(x_1)$ is false, and so $(\forall x_1)A_1^1(x_1) \Rightarrow (\forall x_1)A_2^1(x_1)$ is true. However, $(\forall x_1)(A_1^1(x_1) \Rightarrow A_2^1(x_1))$ is false, since it asserts that all even integers are odd.

2.17 (a) $[(\forall x_i)\neg\mathscr{A}(x_i) \Rightarrow \neg\mathscr{A}(t)] \Rightarrow [\mathscr{A}(t) \Rightarrow \neg(\forall x_i)\neg\mathscr{A}(x_i)]$ is logically valid because it is an instance of the tautology $(A \Rightarrow \neg B) \Rightarrow (B \Rightarrow \neg A)$. By (X), $(\forall x_i)\neg\mathscr{A}(x_i) \Rightarrow \neg\mathscr{A}(t)$ is logically valid. Hence, by (III), $\mathscr{A}(t) \Rightarrow \neg(\forall x_i)\neg\mathscr{A}(x_i)$ is logically valid.
(b) Intuitive proof: If $\mathscr{A}$ is true for all x_i, then $\mathscr{A}$ is true for some x_i. Rigorous proof: Assume $(\forall x_i)\mathscr{A} \Rightarrow (\exists x_i)\mathscr{A}$ is not logically valid. Then there is an interpretation M for which it is not true. Hence, there is a sequence s in Σ such that s satisfies $(\forall x_i)\mathscr{A}$ and s does not satisfy $\neg(\forall x_i)\neg\mathscr{A}$. From the latter, s satisfies $(\forall x_i)\neg\mathscr{A}$. Since s satisfies $(\forall x_i)\mathscr{A}$, s satisfies $\mathscr{A}$, and, since s satisfies $(\forall x_i)\neg\mathscr{A}$, s satisfies $\neg\mathscr{A}$. But then s satisfies both $\mathscr{A}$ and $\neg\mathscr{A}$, which is impossible.

2.19 (a) Let the domain be the set of integers and let $A_1^2(x, y)$ mean that $x < y$.
(b) Same interpretation as in part (a).

2.21 (a) The premises are (i) $(\forall x)(S(x) \Rightarrow N(x))$ and (ii) $(\forall x)(V(x) \Rightarrow \neg N(x))$, and the conclusion is $(\forall x)(V(x) \Rightarrow \neg S(x))$. Intuitive proof: Assume $V(x)$. By (ii), $\neg N(x)$. By (i), $\neg S(x)$. Thus, $\neg S(x)$ follows from $V(x)$, and the conclusion holds. A more rigorous proof can be given along the lines of (I)–(XI), but a better proof will become available after the study of predicate calculi.

2.25 (a) $(\exists x)(\exists y)(A_1^1(x) \wedge \neg A_1^1(y))$

2.26 (a)

1.	$(\forall x)(\mathscr{A} \Rightarrow \mathscr{B})$	Hyp
2.	$(\forall x)\mathscr{A}$	Hyp
3.	$(\forall x)(\mathscr{A} \Rightarrow \mathscr{B}) \Rightarrow (\mathscr{A} \Rightarrow \mathscr{B})$	Axiom (4)
4.	$\mathscr{A} \Rightarrow \mathscr{B}$	1, 3, MP
5.	$(\forall x)\mathscr{A} \Rightarrow \mathscr{A}$	Axiom (4)
6.	$\mathscr{A}$	2, 5, MP
7.	$\mathscr{B}$	4, 6, MP
8.	$(\forall x)\mathscr{B}$	7, Gen
9.	$(\forall x)(\mathscr{A} \Rightarrow \mathscr{B}), (\forall x)\mathscr{A} \vdash (\forall x)\mathscr{B}$	1–8
10.	$(\forall x)(\mathscr{A} \Rightarrow \mathscr{B}) \vdash (\forall x)\mathscr{A} \Rightarrow (\forall x)\mathscr{B}$	1–9, Corollary 2.5
11.	$\vdash (\forall x)(\mathscr{A} \Rightarrow \mathscr{B}) \Rightarrow ((\forall x)\mathscr{A} \Rightarrow (\forall x)\mathscr{B})$	1–10, Corollary 2.5

2.27 *Hint:* Assume $\vdash_K \mathscr{A}$. By induction on the number of steps in a proof of $\mathscr{A}$ in K, prove that, for any variables $y_1, \ldots, y_n$ $(n \geqslant 0)$, $\vdash_{K^\#} (\forall y_1)\ldots(\forall y_n)\mathscr{A}$.

2.29 (a)

1.	$(\forall x)(\forall y)A_1^2(x, y)$	Hyp
2.	$(\forall y)A_1^2(x, y)$	1, Rule A4
3.	$A_1^2(x, x)$	2, Rule A4
4.	$(\forall x)A_1^2(x, x)$	3, Gen
5.	$(\forall x)(\forall y)A_1^2(x, y) \vdash (\forall x)A_1^2(x, x)$	1–4
6.	$\vdash (\forall x)(\forall y)A_1^2(x, y) \Rightarrow (\forall x)A_1^2(x, x)$	1–5, Corollary 2.5

2.34 (b) $(\exists\varepsilon)(\varepsilon > 0 \wedge (\forall\delta)(\delta > 0 \Rightarrow (\exists x)(|x - c| < \delta \wedge \neg|f(x) - f(c)| < \varepsilon)))$

2.35 (a) (i) Assume $\vdash \mathscr{A}$. By moving the negation step-by-step inward to the atomic wfs, show that $\vdash \neg\mathscr{A}^* \Leftrightarrow \mathscr{B}$, where $\mathscr{B}$ is obtained from $\mathscr{A}$ by replacing all atomic wfs by their negations. But, from $\vdash \mathscr{A}$ it can be shown that $\vdash \mathscr{B}$. Hence, $\vdash \neg\mathscr{A}^*$. The converse follows by noting that $(\mathscr{A}^*)^*$ is $\mathscr{A}$.

(ii) Apply (i) to $\neg\mathscr{A} \vee \mathscr{B}$.

2.37

1.	$(\exists y)(\forall x)(A_1^2(x, y) \Leftrightarrow \neg A_1^2(x, x))$	Hyp
2.	$(\forall x)(A_1^2(x, b) \Leftrightarrow \neg A_1^2(x, x))$	1, Rule C
3.	$A_1^2(b, b) \Leftrightarrow \neg A_1^2(b, b)$	2, Rule A4
4.	$\mathscr{C} \wedge \neg\mathscr{C}$	3, Tautology

($\mathscr{C}$ is any wf not containing b.) Use Proposition 2.23 and Proof by Contradiction.

2.44 (a) In step 4, b is not a *new* individual constant. It was already used in step 2.

2.47 Assume K is complete and let $\mathscr{A}$ and $\mathscr{B}$ be closed wfs of K such that $\vdash_K \mathscr{A} \vee \mathscr{B}$. Assume not-$\vdash_K \mathscr{A}$. Then, by completeness, $\vdash_K \neg\mathscr{A}$. Hence, by the tautology $\neg A \Rightarrow ((A \vee B) \Rightarrow B)$, $\vdash_K \mathscr{B}$. Assume K is not complete. Then there is a sentence $\mathscr{A}$ of K such that not-$\vdash_K \mathscr{A}$ and not-$\vdash_K \neg\mathscr{A}$. However, $\vdash_K \mathscr{A} \vee \neg\mathscr{A}$.

2.48 See Tarski, Mostowski, and Robinson (1953, pp. 15–16).

2.53 Assume $\mathscr{A}$ is not logically valid. Then the closure $\mathscr{B}$ of $\mathscr{A}$ is not logically valid. Hence, the theory K with $\neg\mathscr{B}$ as its only proper axiom has a model. By the Skolem-Löwenheim Theorem, K has a denumerable model, and, by the lemma in the proof of Corollary 2.22, K has a model of cardinality α. Hence, $\mathscr{B}$ is false in this model, and, therefore, $\mathscr{A}$ is not true in some model of cardinality α.

2.55 (b) It suffices to assume $\mathscr{A}$ is a closed wf. (Otherwise, look at the closure of $\mathscr{A}$.) We can effectively write all the interpretations on a finite domain $\{b_1, \ldots, b_k\}$. (We need only specify the interpretations of the symbols that occur in $\mathscr{A}$.) For every such interpretation, replace every wf $(\forall x)\mathscr{B}(x)$, where $\mathscr{B}(x)$ has no quantifiers, by $\mathscr{B}(b_1) \wedge \cdots \wedge \mathscr{B}(b_k)$, and continue until no quantifiers are left. One can then evaluate the truth of the resulting wf for the given interpretation.

2.59 Assume K is not finitely axiomatizable. Let the axioms of K_1 be $\mathscr{A}_1, \mathscr{A}_2, \ldots,$ and let the axioms of K_2 be $\mathscr{B}_1, \mathscr{B}_2, \ldots$. Then $\{\mathscr{A}_1, \mathscr{B}_1, \mathscr{A}_2, \mathscr{B}_2, \ldots\}$ is consistent. (If not, some finite subset $\{\mathscr{A}_1, \mathscr{A}_2, \ldots, \mathscr{A}_k, \mathscr{B}_1, \ldots, \mathscr{B}_m\}$ is inconsistent. Since K_1 is not finitely axiomatizable, there is a theorem $\mathscr{A}$ of K_1 such that $\mathscr{A}_1, \mathscr{A}_2, \ldots, \mathscr{A}_k \vdash \mathscr{A}$ does not hold. Hence, the theory with axioms $\{\mathscr{A}_1, \mathscr{A}_2, \ldots, \mathscr{A}_k, \neg\mathscr{A}\}$ has a model M. Since $\vdash_K \mathscr{A}$, M must be a model of K_2, and, therefore, M is a model of $\{\mathscr{A}_1, \mathscr{A}_2, \ldots, \mathscr{A}_k, \mathscr{B}_1, \ldots, \mathscr{B}_m\}$, contradicting the inconsistency of this set of wfs.) Since $\{\mathscr{A}_1, \mathscr{B}_1, \mathscr{A}_2, \mathscr{B}_2, \ldots\}$ is consistent, it has a model, which must be a model of both K_1 and K_2.

2.60 *Hint:* Let the closures of the axioms of K be $\mathscr{A}_1, \mathscr{A}_2, \ldots$. Choose a subsequence $\mathscr{A}_{j_1}$, $\mathscr{A}_{j_2}, \ldots$ such that $\mathscr{A}_{j_{n+1}}$ is the first sentence (if any) after $\mathscr{A}_{j_n}$ that is not deducible from $\mathscr{A}_{j_1} \wedge \cdots \wedge \mathscr{A}_{j_n}$. Let $\mathscr{B}_k$ be $\mathscr{A}_{j_1} \wedge \mathscr{A}_{j_2} \wedge \cdots \wedge \mathscr{A}_{j_k}$. Then the $\mathscr{B}_k$'s form an axiom set for the theorems of K such that $\vdash \mathscr{B}_{k+1} \Rightarrow \mathscr{B}_k$ but not-$\vdash \mathscr{B}_k \Rightarrow \mathscr{B}_{k+1}$. Then $\{\mathscr{B}_1, \mathscr{B}_1 \Rightarrow \mathscr{B}_2, \mathscr{B}_2 \Rightarrow \mathscr{B}_3, \ldots\}$ is an independent axiomatization of K.

2.62 (c)

1.	$x = x$	Proposition 2.23(a)
2.	$(\exists y)(x = y)$	1, Rule E4
3.	$(\forall x)(\exists y(x = y))$	2, Gen

2.65 (a) The problem obviously reduces to the case of substitution for a single variable at a time: $\vdash x_1 = y_1 \Rightarrow t(x_1) = t(y_1)$. From (A7), $\vdash x_1 = y_1 \Rightarrow (t(x_1) = t(x_1) \Rightarrow t(x_1) = t(y_1))$. By Proposition 2.23(a), $\vdash t(x_1) = t(x_1)$. Hence, $\vdash x_1 = y_1 \Rightarrow t(x_1) = t(y_1)$.

2.67 (a) By Exercise 2.62(c), $\vdash (\exists y)(x = y)$. By Proposition 2.23(b, c), $\vdash (\forall y)(\forall z)(x = y \wedge x = z \Rightarrow y = z)$. Hence, $\vdash (\exists_1 y)(x = y)$. By Gen, $\vdash (\forall x)(\exists_1 y)(x = y)$.

2.68 (b) (i) Let $\bigwedge_{1 \leqslant i < j \leqslant n} x_i \neq x_j$ stand for the conjunction of all wfs of the form $x_i \neq x_j$, where $1 \leqslant i < j \leqslant n$. Let $\mathscr{B}_n$ be $(\exists x_1)\ldots(\exists x_n)\bigwedge_{1 \leqslant i < j \leqslant n} x_i \neq x_j$.

(ii) Assume there is a theory with axioms $\mathscr{A}_1, \ldots, \mathscr{A}_n$ that has the same theorems as K. Each of $\mathscr{A}_1, \ldots, \mathscr{A}_n$ is provable from K_1 plus a finite number of the wfs $\mathscr{B}_1, \mathscr{B}_2, \ldots$. Hence, K_1 plus a finite number of wfs $\mathscr{B}_{j_1}, \ldots, \mathscr{B}_{j_n}$ suffices to prove all theorems of K. We may assume $j_1 < \cdots < j_n$. Then an interpretation whose domain consists of j_n objects would be a model of K, contradicting the fact that $\mathscr{B}_{j_n+1}$ is an axiom of K.

2.71 For the independence of axioms (1)–(3), replace all $t = s$ by the statement form $A \Rightarrow A$; then erase all quantifiers, terms, and associated commas and parentheses; axioms (4)–(6) go over into statement forms of the form $P \Rightarrow P$, and axiom (7) into $(P \Rightarrow P) \Rightarrow (Q \Rightarrow Q)$. For the independence of axiom (1), the following four-valued logic, due to Dr. D. K. Roy, works, where 0 is the only designated value.

A	B	$A \Rightarrow B$	A	B	$A \Rightarrow B$	A	B	$A \Rightarrow B$	A	B	$A \Rightarrow B$	A	$\neg A$
0	0	0	1	0	0	2	0	0	3	0	0	0	1
0	1	1	1	1	0	2	1	0	3	1	1	1	0
0	2	1	1	2	0	2	2	0	3	2	1	2	0
0	3	1	1	3	0	2	3	0	3	3	0	3	0

When A and B take the values 3 and 0, respectively, axiom (A1) takes the value 1. For the independence of axiom (2), Dr. Roy devised the following four-valued logic, where 0 is the only designated value.

A	B	$A \Rightarrow B$	A	B	$A \Rightarrow B$	A	B	$A \Rightarrow B$	A	B	$A \Rightarrow B$	A	$\neg A$
0	0	0	1	0	0	2	0	0	3	0	0	0	1
0	1	1	1	1	0	2	1	0	3	1	0	1	0
0	2	1	1	2	0	2	2	0	3	2	1	2	0
0	3	1	1	3	0	2	3	0	3	3	0	3	0

If A, B, and C take the values 3, 0, and 2, respectively, then axiom (A2) is 1. For the independence of axiom (3), the proof on page 36 works. For axiom (4), replace all universal quantifiers by existential quantifiers. For axiom (5), change all terms t to x_1 and replace all universal quantifiers by $(\forall x_1)$. For axiom (6), replace all wfs $t = s$ by the negation of some fixed theorem. For axiom (7), consider an interpretation in which the interpretation of $=$ is a reflexive nonsymmetric relation.

2.74 (a) $(\forall x)(\exists y)((\exists z)(\mathscr{A}(z, x, y, \ldots, y) \wedge A_1^3(x, y, z)) \Rightarrow (\exists z)(\mathscr{A}(z, y, x, \ldots, x) \wedge z = x))$

2.75 (a) $(\exists z)(\forall w)(\exists x)([A_1^1(x) \Rightarrow A_1^2(x, y)] \Rightarrow [A_1^1(w) \Rightarrow A_1^1(y, z)])$

2.78 $\mathscr{B}$ has the form $(\exists x)(\exists y)(\forall z)([A_1^2(x, y) \Rightarrow A_1^1(x)] \Rightarrow A_1^1(z))$. Let the domain D be $\{1, 2\}$, let A_1^2 be $<$, and let $A_1^1(u)$ stand for $u = 2$. Then $\mathscr{B}$ is true, but $(\forall x)(\exists y)A_1^2(x, y)$ is false.

2.79 Let g be a one–one correspondence between D' and D. Define: $(a_j)^{M'} = g((a_j)^M)$; $(f_j^n)^{M'}(b_1, \ldots, b_n) = g^{-1}[(f_j^n)^M(g(b_1), \ldots, g(b_n))]$; $\models_{M'} A_j^n[b_1, \ldots, b_n]$ if and only if $\models_M A_j^n[g(b_1), \ldots, g(b_n)]$.

2.86 *Hint:* Extend K by adding axioms $\mathscr{B}_n$, where $\mathscr{B}_n$ asserts that there are at least n elements. The new theory has no finite models.

2.87 (a) *Hint:* Consider the wfs $\mathscr{B}_n$, where $\mathscr{B}_n$ asserts that there are at least n elements. Use elimination of quantifiers, treating the $\mathscr{B}_n$'s as if they were atomic wfs.

2.92 Let W be any set. For each b in W, let a_b be an individual constant. Let the theory K have as its proper axioms: $a_b \neq a_c$ for all b, c in W such that $b \neq c$, plus the axioms for a total order. K is consistent, since any finite subset of its axioms has a model. (Any such finite subset contains only a finite number of individual constants. One can define a total order on any finite set B by using the one–one correspondence between B and a set $\{1, 2, 3, \ldots, n\}$ and carrying over to B the total order $<$ on $\{1, 2, 3, \ldots, n\}$.) Since K is consistent, K has a model M by the Generalized Completeness Theorem. The domain D of M is totally ordered by the relation $<^M$; hence, the subset D_w of D consisting of the objects $(a_b)^M$ is totally ordered by $<^M$. This total ordering of D_w can then be carried over to a total ordering of W: $b <_w c$ if and only if $a_b <^M a_c$.

2.95 Assume M_1 is finite and $M_1 \equiv M_2$. Let the domain D_1 of M_1 have n elements. Then, since the assertion that a model has exactly n elements can be written as a sentence, the domain D_2 of M_2 must also have n elements. Let $D_1 = \{b_1, \ldots, b_n\}$ and $D_2 = \{c_1, \ldots, c_n\}$.

Assume M_1 and M_2 are not isomorphic. Let φ be any one of the $n!$ one–one correspondences between D_1 and D_2. Since φ is not an isomorphism, either: (1) there is an individual constant a and an element b_j of D_1 such that either (i) $b_j = a^{M_1} \wedge \varphi(b_j) \neq a^{M_2}$ or (ii) $b_j \neq a^{M_1} \wedge \varphi(b_j) = a^{M_2}$; or (2) there is a function letter f_k^m and $b_l, b_{j_1}, \ldots, b_{j_m}$ in D_1 such that

$$b_l = (f_k^m)^{M_1}(b_{j_1},\ldots,b_{j_m}) \quad \text{and} \quad \varphi(b_l) \neq (f_k^m)^{M_2}(\varphi(b_{j_1}),\ldots,\varphi(b_{j_m}))$$

or (3) there is a predicate letter A_k^m and $b_{j_1}, \ldots, b_{j_m}$ in D_1 such that either
(i) $\models_{M_1} A_k^m[b_{j_1},\ldots,b_{j_m}]$ and $\models_{M_2} \neg A_k^m[\varphi(b_{j_1}),\ldots,\varphi(b_{j_m})]$ or
(ii) $\models_{M_1} \neg A_k^m[b_{j_1},\ldots,b_{j_m}]$ and $\models_{M_2} A_k^m[\varphi(b_{j_1}),\ldots,\varphi(b_{j_m})]$. Construct a wf $\mathscr{B}_\varphi$ as follows:

$$\mathscr{B}_\varphi \text{ is } \begin{cases} x_j = a & \text{if (1) (i) holds} \\ x_j \neq a & \text{if (1) (ii) holds} \\ x_l = f_k^m(x_{j_1},\ldots,x_{j_m}) & \text{if (2) holds} \\ A_k^m(x_{j_1},\ldots,x_{j_m}) & \text{if (3) (i) holds} \\ \neg A_k^m(x_{j_1},\ldots,x_{j_m}) & \text{if (3) (ii) holds} \end{cases}$$

Let $\varphi_1, \ldots, \varphi_{n!}$ be the one–one correspondences between D_1 and D_2. Let $\mathscr{A}$ be the wf

$$(\exists x_1)\ldots(\exists x_n)\left(\bigwedge_{1 \leqslant i < j \leqslant n} x_i \neq x_j \wedge \mathscr{B}_{\varphi_1} \wedge \mathscr{B}_{\varphi_2} \wedge \cdots \wedge \mathscr{B}_{\varphi_{n!}}\right)$$

Then $\mathscr{A}$ is true for M_1 but not for M_2.

2.96 (a) There are $\aleph_\alpha$ sentences of K. Hence, there are $2^{\aleph_\alpha}$ sets of sentences. If $M_1 \equiv M_2$ does not hold, then the set of sentences true for M_1 is different from the set of sentences true for M_2.

2.97 Let K′ be the theory with $\aleph_\gamma$ new symbols b_τ and, as axioms, all sentences true for M and all $b_\tau \neq b_\rho$ for $\tau \neq \rho$. Prove K' is consistent and apply Corollary 2.34.

2.100 (a) Let M be the field of rational numbers, and $x = \{-1\}$.

2.102 Consider the wf $(\exists x_2)(x_2 < x_1)$.

2.103 (a) (ii) Introduce a new individual constant b and form a new theory by adding to the complete diagram of M all the sentences $b \neq t$ for all closed terms of K.

2.104 If $0 \notin \mathscr{F}$, $\mathscr{F} \neq \mathscr{P}(A)$. Conversely, if $0 \in \mathscr{F}$, then, by clause (3) of the definition of *filter*, $\mathscr{F} = \mathscr{P}(A)$.

2.105 If $\mathscr{F} = \mathscr{F}_B$, then $\bigcap_{C \in \mathscr{F}} C = B \in \mathscr{F}$. Conversely, if $B = \bigcap_{C \in \mathscr{F}} C \in \mathscr{F}$, then $\mathscr{F} = \mathscr{F}_B$.

2.106 Use Exercise 2.105.

2.107 (a) $A \in \mathscr{F}$, since $A = A - 0$.

(b) If $B = A - W_1 \in \mathscr{F}$ and $C = A - W_2 \in \mathscr{F}$, where W_1 and W_2 are finite, then $B \cap C = A - (W_1 \cup W_2) \in \mathscr{F}$, since $W_1 \cup W_2$ is finite.

(c) If $B = A - W \in \mathscr{F}$, where W is finite, and if $B \subseteq C$, then $C = A - (W - C) \in \mathscr{F}$, since $W - C$ is finite.

(d) Let $B \in \mathscr{F}$. So, $B = A - W$, where W is finite. Let $b \in B$. Then $W \cup \{b\}$ is finite. Hence, $C = A - (W \cup \{b\}) \in \mathscr{F}$. But $B \nsubseteq C$, since $b \notin C$. Therefore, $\mathscr{F} \neq \mathscr{F}_B$.

2.110 Let $\mathscr{F}' = \{D \mid D \subseteq A \wedge (\exists C)(C \in \mathscr{F} \wedge B \cap C \subseteq D)\}$.

2.111 Assume that, for every $B \subseteq A$, either $B \in \mathscr{F}$ or $A - B \in \mathscr{F}$. Let $\mathscr{G}$ be a filter such that $\mathscr{F} \subset \mathscr{G}$. Let $B \in \mathscr{G} - \mathscr{F}$. Then $A - B \in \mathscr{F}$. Hence, $A - B \in \mathscr{G}$. So, $0 = B \cap (A - B) \in \mathscr{G}$ and $\mathscr{G}$ is improper. The converse follows from Exercise 2.110.

2.112 Assume $\mathscr{F}$ is an ultrafilter and $B \notin \mathscr{F}$, $C \notin \mathscr{F}$. By Exercise 2.111, $A - B \in \mathscr{F}$ and $A - C \in \mathscr{F}$. Hence, $A - (B \cup C) = (A - B) \cap (A - C) \in \mathscr{F}$. Since $\mathscr{F}$ is proper, $B \cup C \notin \mathscr{F}$. Conversely, assume $B \notin \mathscr{F} \wedge C \notin \mathscr{F} \Rightarrow B \cup C \notin \mathscr{F}$. Since $B \cup (A - B) = A \in \mathscr{F}$, this implies that, if $B \notin \mathscr{F}$, then $A - B \in \mathscr{F}$. Use Exercise 2.111.

2.113 (a) Assume $\mathscr{F}_C$ is a principal ultrafilter. Let $a \in C$ and assume $C \neq \{a\}$. Then $\{a\} \notin \mathscr{F}_C$ and $C - \{a\} \notin \mathscr{F}_C$. By Exercise 2.112, $C = \{a\} \cup (C - \{a\}) \notin \mathscr{F}_C$, which yields a contradiction.

(b) Assume a nonprincipal ultrafilter $\mathscr{F}$ contains a finite set, and let B be a finite set in $\mathscr{F}$ of least cardinality. Since $\mathscr{F}$ is nonprincipal, the cardinality of B is >1. Let

$b \in B$. Then $B - \{b\} \neq 0$. Both $\{b\}$ and $B - \{b\}$ are finite sets of lower cardinality than B. Hence, $\{b\} \notin \mathscr{F}$ and $B - \{b\} \notin \mathscr{F}$. By Exercise 2.112, $B = \{b\} \cup (B - \{b\}) \notin \mathscr{F}$, which contradicts the definition of B.

2.116 Let J be the set of all finite subsets of Γ. For each Δ in J, choose a model M_Δ of Δ. For Δ in J, let $\Delta^* = \{\Delta' | \Delta' \in J \wedge \Delta \subseteq \Delta'\}$. The collection $\mathscr{G}$ of all Δ^*'s has the finite-intersection property. By Exercise 2.109, there is a proper filter $\mathscr{F} \supseteq \mathscr{G}$. By the Ultrafilter Theorem, there is an ultrafilter $\mathscr{F}' \supseteq \mathscr{F} \supseteq \mathscr{G}$. Consider $\prod_{\Delta \in J} M_\Delta/\mathscr{F}'$. Let $\mathscr{A} \in \Gamma$. Then $\{\mathscr{A}\}^* \in \mathscr{G} \subseteq \mathscr{F}'$. Therefore, $\{\mathscr{A}\}^* \subseteq \{\Delta | \Delta \in J \wedge \models_{M_\Delta} \mathscr{A}\} \in \mathscr{F}'$. By Loś's Theorem, $\mathscr{A}$ is true in $\prod_{\Delta \in J} M_\Delta/\mathscr{F}'$.

2.117 (a) Assume $\mathscr{W}$ is closed under elementary equivalence and ultraproducts. Let Δ be the set of all sentences of K that are true in every model in $\mathscr{W}$. Let M be any model of Δ. We must show that M is in $\mathscr{W}$. Let Γ be the set of all sentences true for M. Let J be the set of finite subsets of Γ. For $\Gamma' = \{\mathscr{A}_1, \ldots, \mathscr{A}_n\} \in J$, choose a model $N_{\Gamma'} \in \mathscr{W}$ such that $\mathscr{A}_1 \wedge \cdots \wedge \mathscr{A}_n$ is true in $N_{\Gamma'}$. (If there were no such model, $\neg(\mathscr{A}_1 \wedge \cdots \wedge \mathscr{A}_n)$, though false in M, would be in Δ.] As in Exercise 2.116, there is an ultrafilter $\mathscr{F}'$ such that $N^* = \prod_{\Gamma' \in J} N_{\Gamma'}/\mathscr{F}'$ is a model of Γ. Now, $N^* \in \mathscr{W}$. Moreover, $M \equiv N^*$. Hence, $M \in \mathscr{W}$.

(b) Use part (a) and Exercise 2.59.

(c) Let $\mathscr{W}$ be the class of all fields of characteristic 0. Let J be a nonprincipal ultrafilter on the set P of primes, and consider $M = \prod_{p \in P} Z_p/\mathscr{F}$, where Z_p is the field of integers modulo p. Apply part (b).

2.118 $R^\# \subseteq R^*$. Hence, the cardinality of R^* is $\geqslant 2^{\aleph_0}$. On the other hand, R^ω is equinumerous with 2^ω and, therefore, has cardinality $2^{\aleph_0}$. But the cardinality of R^* is at most that of R^ω.

2.119 Assume x and y are infinitesimals. Let ε be any positive real. Then $|x| < \varepsilon/2$ and $|y| < \varepsilon/2$. So, $|x + y| \leqslant |x| + |y| < \varepsilon/2 + \varepsilon/2 = \varepsilon$; $|xy| = |x||y| < 1 \cdot \varepsilon = \varepsilon$; $|x - y| \leqslant |x| + |-y| < \varepsilon/2 + \varepsilon/2 = \varepsilon$.

2.120 Assume $|x| < r_1$ and $|y| < \varepsilon$ for all positive real ε. Let ε be a positive real. Then ε/r_1 is a positive real. Hence $|y| < \varepsilon/r_1$, and so, $|xy| = |x||y| < r_1(\varepsilon/r_1) = \varepsilon$.

2.122 Assume $x - r_1$ and $x - r_2$ are infinitesimals, with r_1 and r_2 real. Then $(x - r_1) - (x - r_2) = r_2 - r_1$ is infinitesimal and real. Hence, $r_2 - r_1 = 0$.

2.123 (a) $x - \mathrm{st}(x)$ and $y - \mathrm{st}(y)$ are infinitesimals. Hence, their sum $(x + y) - (\mathrm{st}(x) + \mathrm{st}(y))$ is an infinitesimal. Since $\mathrm{st}(x) + \mathrm{st}(y)$ is real, $\mathrm{st}(x) + \mathrm{st}(y) = \mathrm{st}(x + y)$ by Exercise 2.122.

2.124 (a) By Proposition 2.45, $s^*(n) \approx c_1$ and $u^*(n) \approx c_2$ for all $n \in \omega^* - \omega$. Hence, $s^*(n) + u^*(n) \approx c_1 + c_2$ for all $n \in \omega^* - \omega$. But $s^*(n) + u^*(n) = (s + u)^*(n)$. Apply Proposition 2.45.

2.125 Assume f continuous at c. Take any positive real ε. Then there is a positive real δ such that $(\forall x)(x \in B \wedge |x - c| < \delta \Rightarrow |f(x) - f(c)| < \varepsilon)$ holds in $\mathscr{R}$. Therefore, $(\forall x)(x \in B^* \wedge |x - c| < \delta \Rightarrow |f^*(x) - f(c)| < \varepsilon)$ holds in $\mathscr{R}^*$. So, if $x \in B^*$ and $x \approx c$, then $|x - c| < \delta$ and, therefore, $|f^*(x) - f(c)| < \varepsilon$. Since ε was arbitrary, $f^*(x) \approx f(c)$. Conversely, assume $x \in B^* \wedge x \approx c \Rightarrow f^*(x) \approx f(c)$. Take any positive real ε. Let δ_0 be a positive infinitesimal. Then $(\forall x)(x \in B^* \wedge |x - c| < \delta_0 \Rightarrow |f^*(x) - f(c)| < \varepsilon)$ holds for $\mathscr{R}^*$. Hence, $(\exists \delta)(\delta > 0 \wedge (\forall x)(x \in B^* \wedge |x - c| < \delta \Rightarrow |f'(x) - f(c)| < \varepsilon))$ holds for $\mathscr{R}^*$, and so, $(\exists \delta)(\delta > 0 \wedge (\forall x)(x \in B \wedge |x - c| < \delta \Rightarrow |f(x) - f(c)| < \varepsilon))$ holds in $\mathscr{R}$.

2.126 (a) Since $x \in B^* \wedge x \approx c \Rightarrow (f^*(x) \approx f(c) \wedge g^*(x) \approx g(c))$ by Proposition 2.46, we can conclude $x \in B^* \wedge x \approx c \Rightarrow (f + g)^*(x) \approx (f + g)(c)$, and so, by Proposition 2.46, $f + g$ is continuous at c.

2.131 Consider $s_{\mathscr{F}} \in R^*$. Since s is bounded by b, $|s_{\mathscr{F}}| \leqslant b$. So, $s_{\mathscr{F}} \in R_1$. Let $r = \mathrm{st}(s_{\mathscr{F}})$. Let ε be any positive real. Then $|r - s_{\mathscr{F}}| < \varepsilon$, since $r - s$ is an infinitesimal. Hence, $\{j | |s_j - r| \leqslant \varepsilon\} \in \mathscr{F}$ [remembering that r stands for $(r^\#)_{\mathscr{F}}$]. Since the empty set does not belong to $\mathscr{F}$, there exists j such that $|s_j - r| < \varepsilon$.

2.132 (a) (i) $\neg[(\forall x)(A_1^1(x) \vee A_2^1(x)) \Rightarrow ((\forall x)A_1^1(x)) \vee (\forall x)A_2^1(x)]$

(ii)	$(\forall x)(A_1^1(x) \vee A_2^1(x))$	(i)
(iii)	$\neg[((\forall x)A_1^1(x)) \vee (\forall x)A_2^1(x)]$	(i)

(iv)	$\neg(\forall x)A_1^1(x)$	(iii)
(v)	$\neg(\forall x)A_2^1(x)$	(iii)
(vi)	$(\exists x)\neg A_1^1(x)$	(iv)
(vii)	$(\exists x)\neg A_2^1(x)$	(v)
(viii)	$\neg A_1^1(b)$	(vi)
(ix)	$\neg A_2^1(c)$	(vii)
(x)	$A_1^1(b) \vee A_2^1(b)$	(ii)
(xi)	$A_1^1(b) \swarrow \quad \searrow A_2^1(b)$	(x)
(xii)	$\times \qquad A_1^1(c) \vee A_2^1(c)$	(ii)
(xiii)	$A_1^1(c) \swarrow \quad \searrow A_2^1(c)$	(xii)
	$\times$	

No further rules are applicable and there is an unclosed branch. Let the model M have domain $\{b, c\}$, let $(A_1^1)^M$ hold only for c, and let $(A_2^1)^M$ hold for only b. Then, $(\forall x)(A_1^1(x) \vee A_2^1(x))$ is true for M, but $(\forall x)A_1^1(x)$ and $(\forall x)A_2^1(x)$ are both false for M. Hence, $(\forall x)(A_1^1(x) \vee A_2^1(x)) \Rightarrow ((\forall x)A_1^1(x)) \vee (\forall x)A_2^1(x)$ is not logically valid.

Chapter 3

3.4 Consider the interpretation that has as its domain the set of polynomials with integral coefficients such that the leading coefficient is nonnegative. The usual operations of addition and multiplication are the interpretations of + and ·. Verify that (S1)–(S8) hold but that Proposition 3.11 is false (substituting the polynomial x for x and 2 for y).

3.5 (a) Form a new theory S′ by adding to S a new individual constant b and the axioms $b \neq 0, b \neq \bar{1}, b \neq \bar{2}, \ldots, b \neq \bar{n}, \ldots$. Show that S′ is consistent, and apply Proposition 2.26 and Corollary 2.34(c).

(b) By a *cortège* let us mean any denumerable sequence of 0's and 1's. There are $2^{\aleph_0}$ cortèges. An element c of a denumerable model M of S determines a cortège $(s_0, s_1, s_2, \ldots)$ as follows: $s_i = 0$ if $\models_M p_i | c$, and $s_i = 1$ if $\models_M \neg(p_i | c)$. Consider now any cortège s. Add a new constant b to S, together with the axioms $\mathscr{B}_i(b)$, where $\mathscr{B}_i(b)$ is $p_i | b$ if $s_i = 0$ and $\mathscr{B}_i(b)$ is $\neg(p_i | b)$ if $s_i = 1$. This theory is consistent and, therefore, has a denumerable model M_s, in which the interpretation of b determines the cortège s. Thus, each of the $2^{\aleph_0}$ cortèges is determined by an element of some denumerable model. Every denumerable model determines denumerably many cortèges. Therefore, if a maximal collection of mutually nonisomorphic denumerable models had cardinality $\mathfrak{m} < 2^{\aleph_0}$, then the total number of cortèges represented in all denumerable models would be $\leqslant \mathfrak{m} \times \aleph_0 < 2^{\aleph_0}$. (We use the fact that the elements of a denumerable model determine the same cortèges as the elements of an isomorphic model.)

3.6 Let $(D, 0, ')$ be one model of Peano's Postulates, with $0 \in D$ and $'$ the successor operation, and let $(D\#, 0\#, *)$ be another such model. For each x in D, by an *x-mapping* we mean a function f from $S_x = \{u | u \in D \wedge u \leqslant x\}$ into $D\#$ such that $f(0) = 0\#$ and $f(u') = (f(u))^*$ for all $u < x$. Show by induction that, for every x in D, there is a unique x-mapping (which will be denoted f_x). It is easy to see that, if $x_1 < x_2$, then the restriction of f_{x_2} to S_{x_1} must be f_{x_1}. Define $F(x) = f_x(x)$ for all x in D. Then F is a function from D into $D\#$ such that $F(0) = 0\#$ and $F(x') = (F(x))^*$ for all x in D. It is easy to prove that F is one–one. [If not, a contradiction results when we consider the least x in D for which there is some y in D such that $x \neq y$ and $F(x) = F(y)$.] To see that F is an isomorphism, it only remains to show that the range of F is $D\#$. If not, let z be the least element of $D\#$ not in the range of F. Clearly $z \neq 0\#$. Hence, $z = w^*$ for some w. Then w is in the range of F, and so $w = F(u)$ for some u in D. Therefore, $F(u') = (F(u))^* = w^* = z$, contradicting the fact that z is not in the range of F.

The reason that this proof does not work for models of first-order number theory S is that the proof uses mathematical induction and the least number principle several times, and these uses involve properties that cannot be formulated within the language of S. Since the validity of mathematical induction and the least number principle in models of S is guaranteed to hold, by virtue of axiom (S9), only for wfs of S, the categoricity proof is not applicable. For example, in a nonstandard model for S, the property of being the interpretation of one of the standard integers $\overline{0}, \overline{1}, \overline{2}, \overline{3}, \ldots$ is not expressible by a wf of S. If it were, then, by axiom (S9), one could prove that $\{\overline{0}, \overline{1}, \overline{2}, \overline{3}, \ldots\}$ constitutes the whole model.

3.7 Use a reduction procedure similar to that given for the theory K_2 on pages 90–91. For any number k, define $k \cdot t$ by induction: $0 \cdot t$ is 0 and $(k+1) \cdot t$ is $(k \cdot t) + t$; thus, $k \cdot t$ is the sum of t taken k times. Also, for any given k, let $t \equiv s \pmod{k}$ stand for $(\exists x)(t = s + k \cdot x \vee s = t + k \cdot x)$. In the reduction procedure, consider all such wfs $t \equiv s \pmod{k}$, as well as the wfs $t < s$, as atomic wfs, although they actually are not. Given any wfs of S_+, we may assume by Proposition 2.30 that it is in prenex normal form. Describe a method that, given a wf $(\exists y)\mathscr{C}$, where $\mathscr{C}$ contains no quantifiers [remembering the convention that $t \equiv s \pmod{k}$ and $t < s$ are considered atomic], finds an equivalent wf without quantifiers (again remembering our convention). For help on details, see Hilbert and Bernays (1934, I, pp. 359–66).

3.8 (a) *Hint:* Show that, for any term r not containing variables, there is a natural number m such that $\vdash_S r = \overline{m}$.

(b) Use part (a) and Lemma 1.12.

3.13 Assume $f(x_1, \ldots, x_n) = x_{n+1}$ is expressible in S by $\mathscr{B}(x_1, \ldots, x_{n+1})$. Let $\mathscr{C}(x_1, \ldots, x_{n+1})$ be $\mathscr{B}(x_1, \ldots, x_{n+1}) \wedge (\forall z)(z < x_{n+1} \Rightarrow \neg\mathscr{B}(x_1, \ldots, x_{n+1}))$. Show that $\mathscr{C}$ represents $f(x_1, \ldots, x_n)$ in S. [Use Proposition 3.8(b).] Assume, conversely, that $f(x_1, \ldots, x_n)$ is representable in S by $\mathscr{A}(x_1, \ldots, x_{n+1})$. Show that the same wf expresses $f(x_1, \ldots, x_n) = x_{n+1}$ in S.

3.16 $(\exists y)_{u<y<v} R(x_1, \ldots, x_n, y)$ is equivalent to $(\exists z)_{z<v \dot{-} (u+1)} R(x_1, \ldots, x_n, z + u + 1)$, and similarly for the other cases.

3.18 If the relation $R(x_1, \ldots, x_n, y)$: $f(x_1, \ldots, x_n) = y$ is recursive, then C_R is recursive and, therefore, so is $f(x_1, \ldots, x_n) = \mu y(C_R(x_1, \ldots, x_n, y) = 0)$. Conversely, if $f(x_1, \ldots, x_n)$ is recursive, $C_R(x_1, \ldots, x_n, y) = \mathrm{sg}\,|f(x_1, \ldots, x_n) - y|$ is recursive.

3.19

$$[\sqrt{n}] = \delta(\mu y_{y \leqslant n+1}(y^2 > n))$$

$$\Pi(n) = \sum_{y \leqslant n} \overline{\mathrm{sg}}(C_{Pr}(y))$$

3.20 $[ne] = \left[n\left(1 + 1 + \frac{1}{2!} + \frac{1}{3!} + \cdots + \frac{1}{n!}\right)\right]$, since $n\left(\frac{1}{(n+1)!} + \frac{1}{(n+2)!} + \cdots\right) < \frac{1}{n!}$. Let $1 + 1 + \frac{1}{2!} + \cdots + \frac{1}{n!} = \frac{g(n)}{n!}$. Then $g(0) = 1$ and $g(n+1) = (n+1)g(n) + 1$. Hence, g is primitive recursive. Therefore, so is $[ne] = \left[\frac{ng(n)}{n!}\right] = \mathrm{qt}(n!, ng(n))$.

3.21 $RP(y, z)$ stands for $(\forall x)_{x \leqslant y+z}(x|y \wedge x|z \Rightarrow x = 1)$.

$$\varphi(n) = \sum_{y \leqslant n} \overline{\mathrm{sg}}(C_{RP}(y, n))$$

3.22 $Z(0) = 0$, $Z(y+1) = U_2^2(y, Z(y))$.

3.23 Let $v = (p_0 p_1 \cdots p_k) + 1$. Some prime q is a divisor of v. Hence, $q \leqslant v$. But q is different from $p_0, p_1, \ldots, p_k$. If $q = p_j$, then $p_j | v$ and $p_j | p_0 p_1 \cdots p_k$ would imply that $p_j | 1$ and, therefore, $p_j = 1$. Thus, $p_{k+1} \leqslant q \leqslant (p_0 p_1 \cdots p_k) + 1$.

3.26 If Fermat's Last Theorem is true, h is the constant function 2. If Fermat's Last Theorem is false, h is the constant function 1. In either case, h is primitive recursive.

3.28 List the recursive functions step-by-step in the following way. In the first step, start with the finite list consisting of $Z(x)$, $N(x)$, and $U_1^1(x)$. At the $(n+1)$st step, make one application of substitution, recursion, and the μ-operator to all appropriate sequences of functions already in the list after the nth step, and then add the $n+1$ functions $U_j^{n+1}(x_1, \ldots, x_{n+1})$ to the list. Every recursive function eventually appears in the list.

3.29 Assume $f_x(y)$ is primitive recursive (or recursive). Then so is $f_x(x) + 1$. Hence, $f_x(x) + 1$

is equal to $f_k(x)$ for some k. Therefore, $f_k(x) = f_x(x) + 1$ for all x and, in particular, $f_k(k) = f_k(k) + 1$.

3.30 (a) Let d be the least positive integer in the set Y of integers of the form $au + bv$, where u and v are arbitrary integers—say, $d = au_0 + bv_0$. Then $d|a$ and $d|b$. [To see this for a, let $a = qd + r$, where $0 \leqslant r < d$. Then $r = a - qd = a - q(au_0 + bv_0) = (1 - qu_0)a + (-qv_0)b \in Y$. Since d is the least positive integer in Y and $r < d$, r must be 0. Hence $d|a$.] If a and b are relatively prime, then $d = 1$. Hence, $1 = au_0 + bv_0$. Therefore, $au_0 \equiv 1 \pmod b$.

3.32 Assume that a function $f(x_1, \ldots, x_n)$ is representable in S by the wf $\mathscr{A}(x_1, \ldots, x_n, y)$. Then the wf $\mathscr{B}(x_1, \ldots, x_n, y)$:

$$[((\exists_1 y)\mathscr{A}(x_1, \ldots, x_n, y)) \wedge \mathscr{A}(x_1, \ldots, x_n, y)] \vee$$
$$[(\neg(\exists_1 y)\mathscr{A}(x_1, \ldots, x_n, y)) \wedge y = 0]$$

strongly represents $f(x_1, \ldots, x_n)$.

3.34 (a) $1944 = 2^3 3^5$. Hence, 1944 is the Gödel number of the expression ( ).

(b) $49 = 1 + 8(2^1 3^1)$. Hence, 49 is the Gödel number of the function letter f_1^1.

3.36 (a) $g(f_1^1) = 49$ and $g(a_1) = 15$. So, $g(f_1^1(a_1)) = 2^{49}3^3 5^{15} 7^5$.

3.39 Take as a normal model for RR, but not for S, the set of polynomials with integral coefficients such that the leading coefficient is nonnegative. Note that $(\forall x)(\exists y)(x = y + y \vee x = y + y + 1)$ is false in this model but is provable in S.

3.40 Let ∞ be an object that is not a natural number. Let $\infty' = \infty$, $\infty + x = x + \infty = \infty$ for all natural numbers x, $\infty \cdot 0 = 0 \cdot \infty = 0$, and $\infty \cdot x = x \cdot \infty = \infty$ for all $x \neq 0$.

3.44 Assume S is consistent. By Proposition 3.35(a), $\mathscr{G}$ is not provable in S. Hence, by Lemma 2.12, the theory S_g is consistent. Now, $\neg\mathscr{G}$ is equivalent to $(\exists x_2)\mathscr{P}\!f(x_2, \ulcorner\mathscr{G}\urcorner)$. Since there is no proof of $\mathscr{G}$ in S, Pf(k, q) is false for all natural numbers k, where $q = \ulcorner\mathscr{G}\urcorner$. Hence, $\vdash_S \neg\mathscr{P}\!f(\bar{k}, \bar{q})$ for all natural numbers k. Therefore, $\vdash_{S_g} \neg\mathscr{P}\!f(\bar{k}, \bar{q})$. But, $\vdash_{S_g} (\exists x_2)\mathscr{P}\!f(x_2, \bar{q})$. Thus, S_g is ω-inconsistent.

3.47 (a) Assume the "function" form of Church's Thesis and let A be an effectively decidable set of natural numbers. Then the characteristic function C_A is effectively computable and, therefore, recursive. Hence, by definition, A is a recursive set.

(b) Assume the "set" form of Church's Thesis and let $f(x_1, \ldots, x_n)$ be any effectively computable function. Then the relation $f(x_1, \ldots, x_n) = y$ is effectively decidable. Using the functions σ^k, σ_i^k of pages 140–141 let A be the set of all z such that $f(\sigma_1^{n+1}(z), \ldots, \sigma_n^{n+1}(z)) = \sigma_{n+1}^{n+1}(z)$. Then A is an effectively decidable set and, therefore, recursive. Hence, $f(x_1, \ldots, x_n) = \sigma_{n+1}^{n+1}(\mu z(C_A(z) = 0))$ is recursive.

3.49 Let K be the extension of S that has as proper axioms all wfs that are true in the standard model. If Tr were recursive, then, by Proposition 3.36, K would have an undecidable sentence, which is impossible.

3.50 Use Corollary 3.37.

3.51 Let $f(x_1, \ldots, x_n)$ be a recursive function. So, $f(x_1, \ldots, x_n) = y$ is a recursive relation, expressible in K by a wf $\mathscr{A}(x_1, \ldots, x_n, y)$. Then f is representable by $\mathscr{A}(x_1, \ldots, x_n, y) \wedge (\forall z)(z < y \Rightarrow \neg\mathscr{A}(x_1, \ldots, x_n, z))$, where $z < y$ stands for $z \leqslant y \wedge z \neq y$.

3.55 (a) $\vdash 0 = \bar{1} \Rightarrow \mathscr{G}$. Hence, $\vdash \mathscr{B}ew(\ulcorner 0 = \bar{1}\urcorner) \Rightarrow \mathscr{B}ew(\ulcorner\mathscr{G}\urcorner)$ and, therefore, $\vdash \neg\mathscr{B}ew(\ulcorner\mathscr{G}\urcorner) \Rightarrow \neg\mathscr{B}ew(\ulcorner 0 = \bar{1}\urcorner)$. Thus, $\vdash \mathscr{G} \Rightarrow \neg\mathscr{B}ew(\ulcorner 0 = \bar{1}\urcorner)$.

(b) $\vdash \mathscr{B}ew(\ulcorner\mathscr{G}\urcorner) \Rightarrow \mathscr{B}ew(\ulcorner\mathscr{B}ew(\ulcorner\mathscr{G}\urcorner)\urcorner)$. Also, $\vdash \neg\mathscr{G} \Leftrightarrow \mathscr{B}ew(\ulcorner\mathscr{G}\urcorner)$, and so, $\vdash \mathscr{B}ew(\ulcorner\neg\mathscr{G}\urcorner) \Leftrightarrow \mathscr{B}ew(\ulcorner\mathscr{B}ew(\ulcorner\mathscr{G}\urcorner)\urcorner)$. Hence, $\vdash \mathscr{B}ew(\ulcorner\mathscr{G}\urcorner) \Rightarrow \mathscr{B}ew(\ulcorner\neg\mathscr{G}\urcorner)$. By a tautology, $\vdash \mathscr{G} \Rightarrow (\neg\mathscr{G} \Rightarrow (\mathscr{G} \wedge \neg\mathscr{G}))$; hence, $\vdash \mathscr{B}ew(\ulcorner\mathscr{G}\urcorner) \Rightarrow \mathscr{B}ew(\ulcorner\neg\mathscr{G} \Rightarrow (\mathscr{G} \wedge \neg\mathscr{G})\urcorner)$. Therefore, $\vdash \mathscr{B}ew(\ulcorner\mathscr{G}\urcorner) \Rightarrow (\mathscr{B}ew(\ulcorner\neg\mathscr{G}\urcorner) \Rightarrow \mathscr{B}ew(\ulcorner(\mathscr{G} \wedge \neg\mathscr{G})\urcorner))$. It follows that $\vdash \mathscr{B}ew(\ulcorner\mathscr{G}\urcorner) \Rightarrow \mathscr{B}ew(\ulcorner(\mathscr{G} \wedge \neg\mathscr{G})\urcorner)$. But, $\vdash \mathscr{G} \wedge \neg\mathscr{G} \Rightarrow 0 = \bar{1}$; so, $\vdash \mathscr{B}ew(\ulcorner(\mathscr{G} \wedge \neg\mathscr{G})\urcorner) \Rightarrow \mathscr{B}ew(\ulcorner 0 = \bar{1}\urcorner)$. Thus, $\vdash \mathscr{B}ew(\ulcorner\mathscr{G}\urcorner) \Rightarrow \mathscr{B}ew(\ulcorner 0 = \bar{1}\urcorner)$, and $\vdash \neg\mathscr{B}ew(\ulcorner 0 = \bar{1}\urcorner) \Rightarrow \neg\mathscr{B}ew(\ulcorner\mathscr{G}\urcorner)$. Hence, $\vdash \neg\mathscr{B}ew(\ulcorner 0 = \bar{1}\urcorner) \Rightarrow \mathscr{G}$.

3.58 If a theory K is recursively decidable, the set of Gödel numbers of theorems of K is recursive. Taking the theorems of K as axioms, we obtain a recursive axiomatization.

3.60 Assume there is a recursive set C such that $T_K \subseteq C$ and $\mathrm{Ref}_K \subseteq \bar{C}$. Let C be expressible in K by $\mathscr{A}(x)$. Let $\mathscr{F}$, with Gödel number k, be a fixed point for $\neg\mathscr{A}(x)$. Then, $\vdash_K \mathscr{F} \Leftrightarrow \neg\mathscr{A}(\bar{k})$. Since $\mathscr{A}(x)$ expresses C in K, $\vdash_K \mathscr{A}(\bar{k})$ or $\vdash_K \neg\mathscr{A}(\bar{k})$.

(a) If $\vdash_K \mathscr{A}(\bar{k})$, then $\vdash_K \neg\mathscr{F}$. Therefore, $k \in \mathrm{Ref}_K \subseteq \bar{C}$. Hence, $\vdash_K \neg\mathscr{A}(\bar{k})$, contradicting the consistency of K.

(b) If $\vdash_K \neg\mathscr{A}(\bar{k})$, then $\vdash_K \mathscr{F}$. So, $k \in T_K \subseteq C$ and, therefore, $\vdash_K \mathscr{A}(\bar{k})$, contradicting the consistency of K.

3.62 Let K_2 be the theory whose axioms are those wfs of K_1 that are provable in K*. The theorems of K_2 are the axioms of K_2. Hence, $x \in T_{K_2}$ if and only if $\mathrm{Fml}_{K_1}(x) \wedge x \in T_{K^*}$. So, if K* were recursively decidable—that is, if T_{K^*} were recursive—T_{K_2} would be recursive. Since K_2 is a consistent extension of K_1, this would contradict the essential recursive undecidability of K_1.

3.63 (a) Compare the proof of Proposition 2.28.

(b) By part (a), K* is consistent. Hence, by Exercise 3.62, K* is essentially recursively undecidable. So, by part (a), K is recursively undecidable.

3.64 (b) Take $(\forall x)(A_j^1(x) \Leftrightarrow x = x)$ as a possible definition of A_j^1.

3.65 Use Exercises 3.63(b) and 3.64.

3.66 Use Corollary 3.44, Exercise 3.65, and Proposition 3.45.

Chapter 4

4.16 Let $X = \{\langle y_1, y_2\rangle | y_1 = y_2 \wedge y_1 \in Y\}$; that is, X is the class of all ordered pairs $\langle u, u\rangle$ with $u \in Y$. Clearly $\mathrm{Un}(X)$ and, for any set x, $(\exists v)(\langle v, u\rangle \in X \wedge v \in x) \Leftrightarrow u \in Y \cap x$, so, by Axiom R, $\mathrm{M}(Y \cap x)$.

4.17 $\mathscr{D}(x) \subseteq \bigcup(\bigcup(x))$ and $\mathscr{R}(x) \subseteq \bigcup(\bigcup(x))$. Apply Proposition 4.6.

4.18 (a) Assume $u \in x \times y$. Then $u = \langle v, w\rangle = \{\{v\}, \{v, w\}\}$ for some $v \in x$, $w \in y$. Then $v \in x \cup y$ and $w \in x \cup y$. So, $\{v\} \in \mathscr{P}(x \cup y)$ and $\{v, w\} \in \mathscr{P}(x \cup y)$. Hence, $\{\{v\}, \{v, w\}\} \in \mathscr{P}(\mathscr{P}(x \cup y))$.

(b) Use part (a), Exercise 4.12, Axiom W, and Proposition 4.6.

4.19 (a) If $\mathrm{Rel}(Y)$, then $Y \subseteq \mathscr{D}(Y) \times \mathscr{R}(Y)$. Use Exercise 4.18(b) and Proposition 4.6.

(b) Assume $\mathrm{Fnc}(Y)$. Then $\mathrm{Fnc}(x \upharpoonright Y)$ and $\mathscr{D}(x \upharpoonright Y) \subseteq x$. By Axiom R, $\mathrm{M}(Y``x)$.

4.20 (a) Let 0 be the class $\{u | u \neq u\}$. Assume $\mathrm{M}(X)$. Then $0 \subseteq X$. So, $0 = 0 \cap X$. By Axiom S, $\mathrm{M}(0)$.

4.21 Assume $\mathrm{M}(V)$. Let $Y = \{x | x \notin x\}$. It was proved above that $\neg\mathrm{M}(Y)$. But $Y \subseteq V$. Hence, by Proposition 4.6, $\neg\mathrm{M}(V)$.

4.33 Let u be the least $\in$-element of $X - Z$.

4.36 By Proposition 4.10(a), $\mathrm{Trans}(\omega)$. By Proposition 4.10(b) and Proposition 4.7(j), $\omega \in On$. If $\omega \in K_1$, then $\omega \in \omega$, contradicting Proposition 4.7(a). Hence, $\omega \notin K_1$.

4.39 Let $X_1 = X \times \{0\}$ and $Y_1 = Y \times \{1\}$.

4.40 For any $u \subseteq x$, let the characteristic function C_u be the function with domain x such that $C_u`y = 0$ if $y \in u$ and $C_u`y = 1$ if $y \in x - u$. Let F be the function with domain $\mathscr{P}(x)$, taking u into C_u. Then $\mathscr{P}(x) \underset{F}{\cong} 2^x$.

4.41 For any set u, $\mathscr{D}(u)$ is a set, by Exercise 4.17.

4.42 If $u \in x^y$, then $u \subseteq y \times x$. So, $x^y \subseteq \mathscr{P}(y \times x)$.

4.43 (a) 0 is the only function with domain 0.

(b) Define a function F with domain X such that, for any x_0 in X, $F(x_0)$ is the function g in $X^{\{u\}}$ such that $g`u = x_0$. Then $X \underset{F}{\cong} X^{\{u\}}$.

4.44 If $\mathscr{D}(u) \neq 0$, then $\mathscr{R}(u) \neq 0$.

4.45 Assume $X \underset{F}{\cong} Y$ and $Z \underset{G}{\cong} Z_1$. If $\neg\mathrm{M}(Z_1)$, then $\neg\mathrm{M}(Z)$ and $X^Z = Y^{Z_1} = 0$, by Exercise 4.41. Hence, we may assume $\mathrm{M}(Z_1)$ and $\mathrm{M}(Z)$. Define a function Φ on X^Z: if $f \in X^Z$, let $\Phi`f = F \circ f \circ G^{-1}$. Then $X^Z \underset{\Phi}{\cong} Y^{Z_1}$.

4.46 If X or Y is not a set, then $Z^{X \cup Y}$ and $Z^X \times Z^Y$ are both 0. We may assume then that X and Y are sets. Define a function Ψ with domain $Z^{X \cup Y}$ as follows: if $f \in Z^{X \cup Y}$, let $\Psi`f = \langle X \upharpoonright f, Y \upharpoonright f\rangle$. Then $Z^{X \cup Y} \underset{\Psi}{\cong} Z^X \times Z^Y$.

4.47 There are five cases:
When $Y = 0 \wedge \neg M(Z)$, $(X^Y)^Z = 0$ and $X^{Y \times Z} = X^0 = \{0\}$.
When $Y \neq 0 \wedge \neg M(Z)$, then $\neg M(Y \times Z)$ and $(X^Y)^Z = 0 = X^{Y \times Z}$.
When $\neg M(Y) \wedge Z = 0$, $(X^Y)^Z = 1 = X^0 = X^{Y \times Z}$.
When $\neg M(Y) \wedge Z \neq 0$, $\neg M(Y \times Z)$ and $(X^Y)^Z = 0^Z = 0 = X^{Y \times Z}$.
Finally, when $M(Y) \wedge M(Z)$, define a function Θ with domain $(X^Y)^Z$ as follows: for any $f \in (X^Y)^Z$, $\Theta`f$ is the function in $X^{Y \times Z}$ such that $(\Theta`f)`\langle y, z\rangle = (f`z)`y$ for all $\langle y, z\rangle \in Y \times Z$. Then $(X^Y)^Z \underset{\Theta}{\cong} X^{Y \times Z}$.

4.48 If $\neg M(Z)$, $(X \times Y)^Z = 0 = 0 \times 0 = X^Z \times Y^Z$. Assume then that $M(Z)$. Define a function $F : X^Z \times Y^Z \to (X \times Y)^Z$ as follows: for any $f \in X^Z$, $g \in Y^Z$, $(F`\langle f, g\rangle)`z = \langle f`z, g`z\rangle$ for all z in Z. Then $X^Z \times Y^Z \underset{F}{\cong} (X \times Y)^Z$.

4.49 This is a direct consequence of Proposition 4.17.

4.54 Use the Schröder–Bernstein Theorem [Proposition 4.21(d)].

4.55 Use Proposition 4.21(c and d).

4.56 Define a function F from V into 2_c as follows: $F`u = \{u, 0\}$ if $u \neq 0$; $F`0 = \{1, 2\}$. Since F is one–one, $V \preccurlyeq 2_c$. Hence, by Exercises 4.21 and 4.50, $\neg M(2_c)$.

4.57 (h) Use Exercise 4.46.
(i) $2^x \preccurlyeq 2^x +_c x \preccurlyeq 2^x +_c 2^x = 2^x \times 2 \cong 2^x \times 2^1 \cong 2^{x +_c 1} \cong 2^x$.
Hence, by the Schröder–Bernstein Theorem, $2^x +_c x \cong 2^x$.

4.58 Under the assumption of the Axiom of Infinity, ω is a set such that $(\exists u)(u \in \omega) \wedge (\forall y)(y \in \omega \Rightarrow (\exists z)(z \in \omega \wedge y \subset z))$. Conversely, assume (*) and let b be a set such that (i) $(\exists u)(u \in b)$ and (ii) $(\forall y)(y \in b \Rightarrow (\exists z)(z \in b \wedge y \subset z))$. Let $d = \{u | (\exists z)(z \in b \wedge u \subseteq z)\}$. Since $d \subseteq \mathscr{P}(\bigcup(b))$, d is a set. Define a relation $R = \{\langle n, v\rangle | n \in \omega \wedge v = \{u | u \in d \wedge u \cong n\}\}$. Thus, $\langle n, v\rangle \in R$ if and only if $n \in \omega$ and v consists of all elements of d that are equinumerous with n. R is a one–one function with domain ω and range a subset of $\mathscr{P}(d)$. Hence, by the Replacement Axiom applied to R^{-1}, ω is a set and, therefore, Axiom I holds.

4.59 Induction on α in $(\forall x)(x \cong \alpha \wedge \alpha \in \omega \Rightarrow \text{Fin}(\mathscr{P}(x)))$.

4.60 Induction on α in $(\forall x)(x \cong \alpha \wedge \alpha \in \omega \wedge (\forall y)(y \in x \Rightarrow \text{Fin}(y)) \Rightarrow \text{Fin}(\bigcup(x)))$.

4.61 Use Proposition 4.25(a).

4.62 $x \subseteq \mathscr{P}(\bigcup(x))$ and $y \in x \Rightarrow y \subseteq \bigcup(x)$.

4.63 Induction on α in $(\forall x)(x \cong \alpha \wedge \alpha \in \omega \Rightarrow (x \preccurlyeq y \vee y \preccurlyeq x))$.

4.65 Induction on α in $(\forall x)(x \cong \alpha \wedge \alpha \in \omega \wedge \text{Inf}(Y) \Rightarrow x \preccurlyeq Y)$.

4.66 Use Proposition 4.24(c).

4.67 Use Exercise 4.17(b).

4.68 $x^y \subseteq \mathscr{P}(y \times x)$

4.70 Let Z be a set such that every nonempty set of subsets of Z has a minimal element. Assume Inf(Z). Let Y be the set of all infinite subsets of Z. Then Y is a nonempty set of subsets of Z without a minimal element. Conversely, prove by induction that, for all α in ω, any nonempty subset of $\mathscr{P}(\alpha)$ has a minimal element. The result then carries over to nonempty subsets of $\mathscr{P}(z)$, where z is any finite set.

4.71 (a) Induction on α in $(\forall x)(x \cong \alpha \wedge \alpha \in \omega \wedge \text{Den}(y) \Rightarrow \text{Den}(x \cup y))$.
(b) Induction on α in $(\forall x)(x \cong \alpha \wedge x \neq 0 \wedge \text{Den}(y) \Rightarrow \text{Den}(x \times y))$.
(c) Assume $z \subseteq x$ and Den(z). Let $z \underset{f}{\cong} \omega$. Define a function g on x: $g`u = u$ if $u \in x - z$; $g`u = (\check{f})`((f`u)')$ if $u \in z$. Assume x is Dedekind-infinite. Assume $z \subset x$ and $x \underset{f}{\cong} z$. Let $v \in x - z$. Define a function h on ω such that $h`0 = v$ and $h`(\alpha') = f`(h`\alpha)$ if $\alpha \in \omega$. Then h is one–one; so, Den($h``\omega$) and $h``\omega \subseteq x$.
(d) Assume $y \notin x$. (i) Assume $x \cup \{y\} \cong x$. Define by induction a function g on ω such that $g`0 = y$ and $g`(n + 1) = f`(g`n)$. g is a one–one function from ω into x. Hence, x contains a denumerable subset and, by part (c), x is Dedekind-infinite. (ii) Assume x is Dedekind-infinite. Then, by part (c), there is a denumerable subset z of x. Assume $z \underset{f}{\cong} \omega$. Let $c_0 = (f^{-1})`0$. Define a function $F$ as follows: $F`u = u$ for $u \in x - z$; $F`c_0 = y$; $F`u = (f^{-1})`((f`u) - 1)$ for $u \in z - \{c_0\}$. Then $x \underset{F}{\cong} x \cup \{y\}$. If z is $\{c_0, c_1, c_2, \ldots\}$, F takes c_{i+1} into c_i and moves c_0 into y.

(e) Assume $\omega \preccurlyeq x$. By part (c), x is Dedekind-infinite. Choose $y \notin x$. By part (d), $x \cong x \cup \{y\}$. Hence, $x +_c 1 = (x \times \{0\}) \cup \{\langle 0, 1 \rangle\} \cong x \cup \{y\} \cong x$.

4.72 Assume M is a model of NBG with denumerable domain D. Let z be the element of D satisfying the wf $x = 2^{\omega}$. Hence, z satisfies the wf $\neg(x \cong \omega)$. This means that there is no object in D that satisfies the condition of being a one–one correspondence between z and ω. Since D is denumerable, there is a one–one correspondence between the set of "elements" of z (that is, the set of objects v in D such that $\models_{\mathrm{M}} v \in z$) and the set of natural numbers. However, no such one–one correspondence exists within M.

4.73 NBG is finitely axiomatizable and has only the binary predicate letter A_2^2. The argument on page 209 shows that NBG is recursively undecidable. Hence, by Proposition 3.47, the predicate calculus with A_2^2 as its only predicate letter is recursively undecidable.

4.74 (a) Assume $x \preccurlyeq \omega_\alpha$. If $2 \preccurlyeq x$, then, by Propositions 4.32(b) and 4.35, $\omega_\alpha \preccurlyeq x \cup \omega_\alpha \preccurlyeq x \times \omega_\alpha \preccurlyeq \omega_\alpha \times \omega_\alpha \cong \omega_\alpha$. If x contains one element, use Exercise 4.71 (c and d).

(b) Use Corollary 4.36.

4.77 (a) $\mathscr{P}(\omega_\alpha) \times \mathscr{P}(\omega_\alpha) \cong 2^{\omega_\alpha} \times 2^{\omega_\alpha} \cong 2^{\omega_\alpha +_c \omega_\alpha} \cong 2^{\omega_\alpha} \cong \mathscr{P}(\omega_\alpha)$

(b) $(\mathscr{P}(\omega_\alpha))^x \cong (2^{\omega_\alpha})^x \cong 2^{\omega_\alpha \times x} \cong 2^{\omega_\alpha} \cong \mathscr{P}(\omega_\alpha)$

4.78 (a) If y were nonempty and finite, $y \cong y +_c y$ would contradict Exercise 4.66.

(b) By part (c), let $y = u \cup v$, $u \cap v = 0$, $u \cong y$, $v \cong y$. Let $y \underset{f}{\cong} v$. Define a function g on $\mathscr{P}(y)$ as follows: for $x \subseteq y$, let $g`x = u \cup (f``x)$. Then $g`x \subseteq y$ and $y \cong u \preccurlyeq g`x \preccurlyeq y$. Hence, $g`x \cong y$. So, g is a one–one function from $\mathscr{P}(y)$ into $A = \{z | z \subseteq y \wedge z \cong y\}$. Thus, $\mathscr{P}(y) \preccurlyeq A$. Since $A \subseteq \mathscr{P}(y)$, $\mathscr{A} \preccurlyeq \mathscr{P}(y)$.

(e) Use part (d): $\{z | z \subseteq y \wedge z \cong y\} \subseteq \{z | z \subseteq y \wedge \mathrm{Inf}(z)\}$.

(f) By part (c), let $y = u \cup v$, $u \cap v = 0$, $u \cong y$, $v \cong y$. Let $u \underset{h}{\cong} v$. Define f on y as follows: $f`x = h`x$ if $x \in u$ and $f`x = (h^{-1})`x$ if $x \in v$.

4.79 (a) Use Proposition 4.32(b).

(b) (i) $\mathrm{Perm}\,(y) \subseteq y^y \preccurlyeq (2^y)^y \cong 2^{y \times y} \cong 2^y \cong \mathscr{P}(y)$.

(ii) By part (a), we may use Exercise 4.78(c). Let $y = u \cup v$, $u \cap v = 0$, $u \cong y$, $v \cong y$. Let $u \underset{H}{\cong} v$ and $y \underset{G}{\cong} u$. Define a function $F\colon \mathscr{P}(y) \to \mathrm{Perm}(y)$ in the following way: assume $z \in \mathscr{P}(y)$. Let $\psi_z\colon y \to y$ be defined as follows: $\psi_z`x = H`x$ if $x \in G``z$; $\psi_z`x = (H^{-1})`x$ if $(H^{-1})`x \in G``z$; $\psi_z`x = x$ otherwise. Then $\psi_z \in \mathrm{Perm}(y)$. Let $F`z = \psi_z$. F is one–one. Hence, $\mathscr{P}(y) \preccurlyeq \mathrm{Perm}(y)$.

4.80 (a) Use WO and Proposition 4.17.

(b) The proof of $\vdash$ Zorn $\Rightarrow$ (WO) in Proposition 4.37 uses only this special case of Zorn's Lemma.

(c) To prove the Hausdorff Maximal Principal (HMP) from Zorn, consider some $\subset$-chain C_0 in x. Let y be the set of all $\subset$-chains C in x such that $C_0 \subseteq C$ and apply part (b) to y. Conversely, assume HMP. To prove part (b), assume that the union of each nonempty $\subset$-chain in a given nonempty set x is also in x. By HMP applied to the $\subset$-chain 0, there is some maximal $\subset$-chain C in x. Then $\bigcup(C)$ is an $\subset$-maximal element of x.

(d) Assume the Teichmüller-Tukey Lemma (TT). To prove part (b), assume that the union of each nonempty $\subset$-chain in a given nonempty set x is also in x. Let y be the set of all $\subset$-chains in x. y is easily seen to be a set of finite character. Therefore, y contains a $\subset$-maximal element C. Then $\bigcup(C)$ is a $\subset$-maximal element of x. Conversely, let x be any set of finite character. In order to prove TT by means of part (b), we must show that, if C is a $\subset$-chain in x, then $\bigcup(C) \in x$. By the finite character of x, it suffices to show that every finite subset z of $\bigcup(C)$ is in x. Now, since z is finite, z is a subset of the union of a finite subset W of C. Since C is a $\subset$-chain, W has a $\subset$-greatest element $w \in x$, and z is a subset of w. Since x is of finite character, $z \in x$.

(e) Assume Rel(x). Let $u = \{z | (\exists v)(v \in \mathscr{D}(x) \wedge z = \{v\} \upharpoonright x\}$; that is, $z \in u$ if z is the set of all ordered pairs $\langle v, w \rangle$ in x, for some fixed v. Apply the Multiplicative Axiom to u. The resulting choice set $y \subseteq x$ is a function with domain $\mathscr{D}(x)$. Conversely, the given property easily yields the Multiplicative Axiom. If x is a set of disjoint nonempty sets, let r be the set of all ordered pairs $\langle u, v\rangle$ such that $u \in x$ and $v \in u$. By part (e), there is a function $f \subseteq r$ such that $\mathscr{D}(f) = \mathscr{D}(r) = x$. The range $\mathscr{R}(f)$ is the required choice set for x.

(f) By Trichotomy, either $x \prec y$ or $y \prec x$. If $x \prec y$, there is a function with domain y and range x. [Assume $x \underset{f}{\cong} y_1 \subseteq y$. Take $c \in x$. Define $g`u = c$ if $u \in y - y_1$, and $g`u = (f^{-1})`u$ if $u \in y_1$.] Similarly, if $y \prec x$, there is a function with domain $x$ and range $y$. Conversely, to prove WO, apply the assumption (f) to $x$ and $\mathscr{H}`(\mathscr{P}(x))$. Note that, if $(\exists f)(f\colon u \to v \wedge \mathscr{R}(f) = v)$, then $\mathscr{P}(v) \prec \mathscr{P}(u)$. Therefore, if there were a function f from x onto $\mathscr{H}`(\mathscr{P}(x))$, we would have $\mathscr{H}`(\mathscr{P}(x)) \prec \mathscr{P}(\mathscr{H}`(\mathscr{P}(x))) \prec \mathscr{P}(x)$, contradicting the definition of $\mathscr{H}`(\mathscr{P}(x))$. Hence, there is a function from $\mathscr{H}`(\mathscr{P}(x))$ onto $x$. Since $\mathscr{H}`(\mathscr{P}(x))$ is an ordinal, one can define a one–one function from x into $\mathscr{H}`(\mathscr{P}(x))$. Thus $x \prec \mathscr{H}`(\mathscr{P}(x))$ and, therefore, x can be well-ordered.

4.83 If $<$ is a partial ordering of x, use Zorn's Lemma to obtain a maximal partial ordering $<^*$ of x with $< \subseteq <^*$. But a maximal partial ordering must be a total ordering. [If u, v were distinct elements of x unrelated by $<^*$, we could add to $<^*$ all pairs $\langle u_1, v_1\rangle$ such that $u_1 \leqslant^* u$ and $v \leqslant^* v_1$. The new relation would be a partial ordering properly containing $<^*$.]

4.86 (b) Since $x \times y \cong x +_c y$, $x \times y = a \cup b$ with $a \cap b = 0$, $a \cong x$, $b \cong y$. Let r be a well-ordering of y. (i) Assume there exists u in x such that $\langle u, v\rangle \in a$ for all v in y. Then $y \preccurlyeq a$. Since $a \cong x$, $y \preccurlyeq x$, contradicting $\neg(y \preccurlyeq x)$. Hence, (ii) for any u in x, there exists v in y such that $\langle u, v\rangle \in b$. Define $f\colon x \to b$ such that $f`u = \langle u, v\rangle$, where v is the r-least element of y such that $\langle u, v\rangle \in b$. Since f is one–one, $x \preccurlyeq b \cong y$.

(c) Clearly Inf(z) and Inf($x +_c z$). Then

$$x +_c z \cong (x +_c z)^2 \cong x^2 +_c 2 \times (x \times z) +_c z^2 \cong x +_c 2 \times (x \times z) +_c z$$

Therefore, $x \times z \preccurlyeq 2 \times (x \times z) \preccurlyeq x +_c 2 \times (x \times z) +_c z \cong x +_c z$. Conversely, $x +_c z \preccurlyeq x \times z$ by Proposition 4.32(b).

(d) If AC holds $(\forall y)(\mathrm{Inf}(y) \Rightarrow y \cong y \times y)$ follows from Proposition 4.35 and Exercise 4.80(a). Conversely, if we assume $y \cong y \times y$ for all infinite y, then, by parts (c) and (b), it follows that $x \preccurlyeq \mathscr{H}`x$ for any infinite set $x$. Since $\mathscr{H}`x$ is an ordinal, x can be well-ordered. Thus, WO holds.

4.88 (a) Let $<$ be a well-ordering of the range of r. Let $f`0$ be the $<$-least element of $\mathscr{R}(r)$, and let $f`(n')$ be the $<$-least element of those v in $\mathscr{R}(r)$ such that $\langle f`n, v\rangle \in r$.

(b) Assume Den(x) $\wedge$ $(\forall u)(u \in x \Rightarrow u \neq 0)$. Let $\omega \underset{g}{\cong} x$. Let r be the set of all pairs $\langle a, b\rangle$ such that a and b are finite sequences $\langle v_0, v_1, \ldots, v_n\rangle$ and $\langle v_0, v_1, \ldots, v_{n+1}\rangle$ such that, for $0 \leqslant i \leqslant n + 1$, $v_i \in g`i$. Since $\mathscr{R}(r) \subseteq \mathscr{D}(r)$, PDC produces a function $h\colon \omega \to \mathscr{D}(r)$ such that $\langle h`n, h`(n')\rangle \in r$ for all $n$ in $\omega$. Define the choice function $f$ by taking, for each $u$ in $x$, $f`u$ to be the $(g`u)$th component of the sequence $h`(g`u)$.

(c) Assume PDC and Inf(x). Let r consist of all ordered pairs $\langle u, u \cup \{a\}\rangle$, where $u \cup \{a\} \subseteq x$, Fin($u \cup \{a\}$), and $a \notin u$. By PDC, there is a function $f\colon \omega \to \mathscr{D}(r)$ such that $\langle f`n, f`(n')\rangle \in r$ for all n in ω. Define $g\colon \omega \to x$ by setting $g`n$ equal to the unique element of $f`(n') - f`n$. Then g is one–one, and so, $\omega \preccurlyeq x$.

(d) In the proof of Proposition 4.39(b), instead of using the choice function h, apply PDC to obtain the function f. As the relation r, use the set of all pairs $\langle u, v\rangle$ such that $u \in c$, $v \in c$, $v \in u \cap X$.

4.89 Use transfinite induction.

4.92 Use induction on β.

4.93, 4.94 Use transfinite induction and Exercise 4.89.

4.96 Assume $u \subseteq H$. Let v be the set of ranks $\rho`x$ of elements $x$ in $u$. Let $\beta = \bigcup(v)$. Then $u \subseteq \Psi`\beta$. Hence, $u \in \mathscr{P}(\Psi`\beta) = \Psi`(\beta') \subseteq H$.

4.97 Assume $X \neq 0 \wedge \neg(\exists y)(y \in X \wedge y \cap X = 0)$. Choose $u \in X$. Define a function g: $g`0 = u \cap X$, $g`(n') = (\bigcup(g`n)) \cap X$. Let $x = \bigcup(\mathscr{R}(g))$. Then $x \neq 0$ and $(\forall y)(y \in x \Rightarrow y \cap x \neq 0)$.

4.102 *Hint:* Assume that the other axioms of NBG are consistent and that the Axiom of Infinity is provable from them. Show that H_ω is a model for the other axioms but not for the Axiom of Infinity.

4.103 Use $H_{\omega +_0 \omega}$.
4.109 (a) Let $C = \{x | \neg(\exists y)(x \in y \wedge y \in x)\}$.

Chapter 5

5.1 $q_0 | B q_0$
$q_0 B R q_1$
$q_1 | | q_0$
$q_1 B R q_2$

5.2 (a) U_2^3 (b) $\delta(x)$

5.7 Let a Turing machine $\mathscr{T}$ compute the function f. Replace all occurrences of q_0 in the quadruples of $\mathscr{T}$ by a new internal state q_r. Then add the quadruples $q_0 a_i a_i q_r$ for all symbols a_i of the alphabet of $\mathscr{T}$. The Turing machine defined by the enlarged set of quadruples also computes the function f.

5.10 (a) $N(x) = x + 1$ (c) $2x$

5.12 (a)

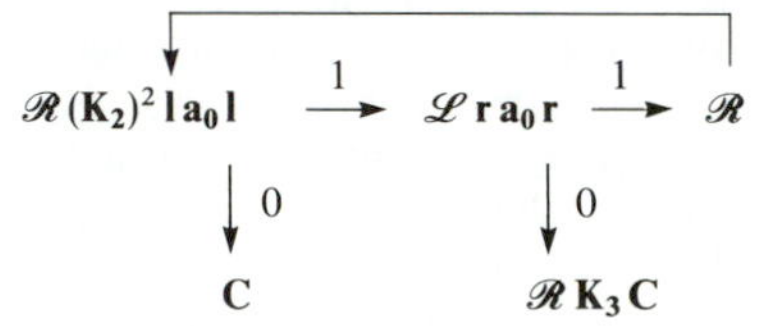

5.14 (a) The empty function (b) $N(x) = x + 1$ (c) $Z(x)$

5.16 If $f(a_1) = b_1, \ldots, f(a_n) = b_n$, then

$$f(x) = \mu y[(x = a_1 \wedge y = b_1) \vee \cdots \vee (x = a_n \wedge y = b_n)]$$

5.20 Let $g(z, x) = U(\mu y T_1(z, x, y))$ and use Corollary 5.11. Let v_0 be a number such that $g(x, x) + 1 = g(v_0, x)$. Then, if $g(v_0, v_0)$ is defined, $g(v_0, v_0) + 1 = g(v_0, v_0)$, which is impossible.

5.21 $g(x_1, \ldots, x_n) = h_1(x_1, \ldots, x_n) \cdot \overline{\mathrm{sg}}(C_R(x_1, \ldots, x_n)) + h_2(x_1, \ldots, x_n) \cdot C_R(x_1, \ldots, x_n)$

5.22 (a) Assume that $h(x)$ is a recursive function such that $h(x) = \mu y T_1(x, x, y)$ for every x in the domain of $\mu y T_1(x, x, y)$. Then $(\exists y) T_1(x, x, y)$ if and only if $T_1(x, x, h(x))$. Since $T_1(x, x, h(x))$ is a recursive relation, this contradicts Corollary 5.14(a).
(b) Use Exercise 5.21.
(c) $Z(\mu y T_1(x, x, y))$ is recursively completable, but its domain is $\{x | (\exists y) T_1(x, x, y)\}$, which, by Corollary 5.14(a), is not recursive.

5.29 Let $\mathscr{T}$ be a Turing machine with a recursively unsolvable halting problem. Let a_k be a symbol not in the alphabet of $\mathscr{T}$. Let q_r be an internal state symbol that does not occur in the quadruples of $\mathscr{T}$. For each q_i of $\mathscr{T}$ and a_j of $\mathscr{T}$, if no quadruple of $\mathscr{T}$ begins with $q_i a_j$, then add the quadruple $q_i a_j a_k q_r$. Call the new Turing machine $\mathscr{T}^*$. Then, for any initial tape description α of $\mathscr{T}$, $\mathscr{T}^*$, begun on α, prints a_k if and only if $\mathscr{T}$ is applicable to α. Hence, if the printing problem for $\mathscr{T}^*$ and a_k were recursively solvable, then the halting problem for $\mathscr{T}$ would be recursively solvable.

5.31 Let $\mathscr{T}$ be a Turing machine with a recursively unsolvable halting problem. For any initial tape description α for $\mathscr{T}$, construct a Turing machine $\mathscr{T}_\alpha$ that does the following: for any initial tape description β, start $\mathscr{T}$ on β; if $\mathscr{T}$ stops, erase the result and then start $\mathscr{T}$ on α. It is easy to check that $\mathscr{T}$ is applicable to α if and only if $\mathscr{T}_\alpha$ has a recursively unsolvable halting problem. It is very tedious to show how to construct $\mathscr{T}_\alpha$ and to prove that the Gödel number of $\mathscr{T}_\alpha$ is a recursive function of the Gödel number of α.

5.33 Let v_0 be the index of a partial recursive function $G(x)$ with nonempty domain. If the given decision problem were recursively solvable, so would be the decision problem of Example 1 on page 255.

5.34 By Corollary 5.17, there is a recursive function $g(u)$ such that $\phi^1_{g(u)}(x) = x \cdot \mu y T_1(u, u, y)$. Then $\phi^1_{g(u)}$ has an empty domain if and only if $\neg(\exists y)T_1(u, u, y)$. But, $\neg(\exists y)T_1(u, u, y)$ is not recursive by Corollary 5.14(a).

5.39 (a) By Corollary 5.17, there is a recursive function $g(u)$ such that $\phi^1_{g(u)}(x) = \mu y(x = u \wedge y = x)$. The domain of $\phi^1_{g(u)}$ is $\{u\}$. Apply the Fixed Point Theorem to g.

(b) There is a recursive function $g(u)$ such that $\phi^1_{g(u)}(x) = \mu y(x \neq u \wedge y = 0)$. Apply the Fixed Point Theorem to g.

5.42 (a) Let $A = \{x \mid f(x) \in B\}$. By Proposition 5.21(c), B is the domain of a partial recursive function g. Then A is the domain of the composition $g \circ f$. Since $g \circ f$ is partial recursive by substitution, A is r.e. by Proposition 5.21(c).

(b) Let B be a recursive set and let D be the inverse image of B under a recursive function f. Then $x \in D$ if and only if $C_B(f(x)) = 0$ and $C_B(f(x)) = 0$ is a recursive relation.

(c) Let B be an r.e. set and let A be the image $\{f(x) \mid x \in B\}$ under a partial recursive function f. If B is empty, so is A. If B is nonempty, then B is the range of a recursive function g. Then A is the range of the partial recursive function $f(g(x))$ and, by Proposition 5.21(b), A is r.e.

(d) Consider part (b). Given any natural number x, compute the value $f(x)$ and determine whether $f(x)$ is in B. This is an effective procedure for determining membership in the inverse image of B. Hence, by Church's Thesis, B is recursive.

(e) Any nonempty r.e. set that is not recursive [such as that of Proposition 5.21(e)] is the range of a recursive function g and is, therefore, the image of the recursive set ω of all natural numbers under the function g.

5.43 The proof has two parts:

1. Let A be an infinite recursive set. Then A is the range of a recursive function f, by Proposition 5.21(d). Since A is infinite, $h(u) = \mu y(f(y) > u)$ is recursive. Let a_0 be the least element of A. Define $g(0) = a_0$, $g(n + 1) = f(h(g(n)))$. Then g is a strictly increasing function with range A.
2. Let A be the range of a strictly increasing recursive function g. Then $g(x) \geqslant x$ for all x (by the special case of Proposition 4.14). Hence, $x \in A$ if and only if $(\exists u)_{u \leqslant x} g(u) = x$. So, A is recursive by Proposition 3.17.

5.44 Assume A is an infinite r.e. set. Let A be the range of the recursive function $g(x)$. Define the function f by the following course-of-values recursion:

$$f(n) = g(\mu y((\forall z)_{z<n} g(y) \neq f(z))) = g(\mu y((\forall z)_{z<n} g(y) \neq (f\#(n))_z))$$

Then A is the range of h, h is one–one, and h is recursive by Propositions 3.17 and 3.19. Intuitively, $f(0) = g(0)$ and, for $n > 0$, $f(n) = g(y)$, where y is the least number for which $g(y)$ is different from $f(0), f(1), \ldots, f(n - 1)$.

5.45 Let A be an infinite r.e. set, and let A be the range of the recursive function g. Since A is infinite, $F(u) = \mu y(g(y) > u)$ is a recursive function. Define $G(0) = g(0)$, $G(n + 1) = g(\mu y(g(y) > G(n))) = g(F(G(n)))$. G is a strictly increasing recursive function whose range is infinite and included in A. By Exercise 5.43, the range of G is an infinite recursive subset of A.

5.46 (a) By Corollary 5.17, there is a recursive function $g(u, v)$ such that $\phi^1_{g(u,v)}(x) = \mu y(T_1(u, x, y) \vee T_1(v, x, y))$.

5.47 Assume (∇). Let $f(x_1, \ldots, x_n)$ be effectively computable. Then the set $B = \{u \mid f((u)_1, \ldots, (u)_n) = (u)_{n+1}\}$ is effectively enumerable and, therefore, by (∇), r.e. Hence, $u \in B \Leftrightarrow (\exists y)R(u, y)$ for some recursive relation R. Then

$$f(x_1, \ldots, x_n) = ([\mu v(((v)_0)_1 = x_1 \wedge \cdots \wedge ((v)_0)_n = x_n \wedge R((v)_0, (v)_1))]_0)_{n+1}$$

So, f is recursive. Conversely, assume Church's Thesis and let W be an effectively enumerable set. If W is empty, then W is r.e. If W is nonempty, let W be the range of the effectively computable function g. By Church's Thesis, g is recursive. But, $x \in W \Leftrightarrow (\exists u)(g(u) = x)$. Hence, W is r.e. by Proposition 5.21(a).

5.48 Assume A is r.e. Since $A \neq 0$, A is the range of a recursive function $g(z)$. So, for each z, $U(\mu y T_1(g(z), x, y))$ is total and, therefore, recursive. Hence, $U(\mu y T_1(g(x), x, y)) + 1$ is recursive. Then there must be a number z_0 such that $U(\mu y T_1(g(x), x, y)) + 1 = U(\mu y T_1(g(z_0), x, y))$. A contradiction results when $x = z_0$.

5.49 (a) Let $\phi(n) = n$ for all n.

5.50 Let $\phi(z) = \sigma_1^2(\mu y[T_1(z, \sigma_1^2(y), \sigma_2^2(y)) \wedge \sigma_1^2(y) > 2z])$, and let B be the range of ϕ.

5.55 (b) Let A be r.e. Then $x \in A \Leftrightarrow (\exists y)R(x, y)$, where R is recursive. Let $\mathscr{R}(x, y)$ express $R(x, y)$ in K. Then $k \in A \Leftrightarrow \vdash_K (\exists y)\mathscr{R}(\bar{k}, y)$.

(c) Assume $k \in A \Leftrightarrow \vdash_K \mathscr{A}(\bar{k})$ for all natural numbers k. Then $k \in A \Leftrightarrow (\exists y)B_{\mathscr{A}}(k, y)$ and $B_{\mathscr{A}}$ is recursive (see the proof of Proposition 3.28 on page 156).

5.56 (a) Clearly T_K is infinite. Let $f(x)$ be a recursive function with range T_K. Let $\mathscr{B}_0, \mathscr{B}_1, \dots$ be the theorems of K, where $\mathscr{B}_j$ is the wf of K with Gödel number $f(j)$. Let $g(x, y)$ be the recursive function such that, if x is the Gödel number of a wf $\mathscr{C}$, then $g(x, j)$ is the Gödel number of the conjunction $\mathscr{C} \wedge \mathscr{C} \wedge \cdots \wedge \mathscr{C}$ consisting of j conjuncts; and, otherwise, $g(x, j) = 0$. Then $g(f(j), j)$ is the Gödel number of the j-fold conjunction $\mathscr{B}_j \wedge \mathscr{B}_j \wedge \cdots \wedge \mathscr{B}_j$. Let K′ be the theory whose axioms are all these j-fold conjunctions, for $j = 0, 1, 2, \dots$. Then K′ and K have the same theorems. Moreover, the set of axioms of K′ is recursive. In fact, x is the Gödel number of an axiom of K′ if and only if $x \neq 0 \wedge (\exists y)_{y \leqslant x}(g(f(y), y) = x)$. From an intuitive standpoint using Church's Thesis, we observe that, given any wf $\mathscr{A}$, one can decide whether $\mathscr{A}$ is a conjunction $\mathscr{C} \wedge \mathscr{C} \wedge \cdots \wedge \mathscr{C}$; if it is such a conjunction, one can determine the number j of conjuncts and check whether $\mathscr{C}$ is $\mathscr{B}_j$.

(b) Part (b) follows from part (a).

5.58 (a) Assume $\mathscr{B}(x_1)$ weakly expresses $(\bar{T}_K)^*$ in K. Then, for any n, $\vdash_K \mathscr{B}(\bar{n})$ if and only if $n \in (\bar{T}_K)^*$. Let p be the Gödel number of $\mathscr{B}(x_1)$. Then $\vdash_K \mathscr{B}(\bar{p})$ if and only if $p \in (\bar{T}_K)^*$. Hence, $\vdash_K \mathscr{B}(\bar{p})$ if and only if the Gödel number of $\mathscr{B}(\bar{p})$ is in $\bar{T}_K$; that is, $\vdash_K \mathscr{B}(\bar{p})$ if and only if not-$\vdash_K \mathscr{B}(\bar{p})$.

(b) If K is recursively decidable, T_K is recursive. Hence, $\bar{T}_K$ is recursive and, by Exercise 5.57, $(\bar{T}_K)^*$ is recursive. So, $(\bar{T}_K)^*$ is weakly expressible in K, contradicting part (a).

(c) Use part (b); every recursive set is expressible, and, therefore, weakly expressible, in every consistent extension of K.

5.59 (a) (i) $\delta(x)$.

(ii) $x_1 \dot{-} x_2$

(iii) The function with empty domain.

(iv) The doubling function.

(b) (i)

$$f_1^2(x_1, 0) = x_1$$
$$f_1^2(0, x_2) = x_2$$
$$f_1^2((x_1)', (x_2)') = f_1^2(x_1, x_2)$$

(ii)

$$f_1^2(x_1, 0) = x_1$$
$$f_1^2(x_1, (x_2)') = (f_1^2(x_1, x_2))'$$
$$f_2^2(x_1, 0) = 0$$
$$f_2^2(x_1, (x_2)') = f_1^2(f_2^2(x_1, x_2), x_1)$$

(iii)

$$f_1^1(0) = \bar{1}$$
$$f_1^1((x_1)') = 0$$
$$f_2^1(0) = 0$$
$$f_2^1((x_1)') = f_1^1(f_2^1(x_1))$$

5.61 (a) Any word P is transformed into QP.

(b) Any word P in A is transformed into PQ.

(c) Any word P in A is transformed into Q.

(d) Any word P in A is transformed into $\bar{n}$, where n is the number of symbols in P.

5.62 (a)

$$\alpha\xi \to \cdot\Lambda \qquad (\xi \text{ in A})$$
$$\alpha \to \cdot\Lambda$$
$$\Lambda \to \alpha$$

(b)

$$\alpha\xi \to \xi\alpha \qquad (\xi \text{ in A})$$
$$\xi\alpha \to \cdot\Lambda \qquad (\xi \text{ in A})$$
$$\alpha \to \cdot\Lambda$$
$$\Lambda \to \alpha$$

(c)

$$\xi \to \Lambda \qquad (\xi \text{ in A})$$
$$\alpha\alpha \to \cdot\Lambda$$
$$\Lambda \to \cdot\alpha$$

(d)

$$\xi\eta\beta \to \eta\beta\xi \qquad (\xi, \eta \text{ in A})$$
$$\alpha\xi \to \xi\beta\xi\alpha \qquad (\xi \text{ in A})$$
$$\beta \to \gamma$$
$$\gamma \to \Lambda$$
$$\alpha \to \cdot\Lambda$$
$$\Lambda \to \alpha$$

5.63

$$\alpha a_i \to Q_i\alpha \qquad (i = 1, \dots, k)$$
$$\alpha\xi \to \xi\alpha \qquad (\xi \text{ in A} - \{a_1, \dots, a_k\})$$
$$\alpha \to \cdot\Lambda$$
$$\Lambda \to \alpha$$

5.64 (d) $|B| \to B$

$B \to |$

(e) $|B| \to |$

(f) Let α, β, and δ be new symbols.

$$\beta| \to |\beta$$
$$\alpha| \to |\beta\alpha$$
$$\alpha \to \Lambda$$
$$\|\delta \to |\delta\alpha$$
$$|\delta \to |$$
$$\delta\| \to \delta|$$
$$\delta| \to |$$
$$\delta \to |$$
$$\beta \to |$$
$$|B| \to \delta$$

5.65 (b)

L_0. D1, 2, 1

L_1. P2, |, 0

L_2. D2, 3, 3

L_3. Stop.

5.68 To compute $Z(x)$, the alphabet is $\{|\}$ and the only register is R_1. The instructions are:

L_0. Del R_1

L_1. R_1 jmp_1 L_0 a

L_2. Add_1 R_1

L_3. Continue

Notation

Index